CODATA Secretariat

Executive Director: K. Cass
51, Boulevard de Montmorency
75016 Paris, France

E-mail: codata@dial.oleane.com
http://www.codata.org

M.Z. Zgurovsky · N.D. Pankratova

System Analysis: Theory and Applications

With 53 Figures and 26 Tables

Springer

I
CODATA
S
U

CODATA Secretariat
K. Cass, Exec. Director
51, Boulevard de Montmorency
F-75016 Paris

Authors:

M.Z. Zgurovsky
N.D. Pankratova

Institute for Applied System Analysis
National Technical University of Ukraine "KPI"
Av. Pobedy 37, Kiev 03056, Ukraine

E-mail: zgur@zgurov.kiev.ua
ndp@mmsa.ntu-kpi.kiev.ua

The image on the front cover comes from an animation which shows worldwide Internet traffic. The color and height of the arcs between the countries encode the data-packet counts and destinations, while the "skyscraper" glyphs (or icons) encode total traffic volume at any site. This image was generated by Stephen G. Eick at the AT&T Bell Laboratories.

Library of Congress Control Number: 2006936347

ISBN 978-3-540-48879-8 Springer Berlin Heidelberg New York

Springer is a part of Springer Science+Business Media
springer.com

Typesetting and production: LE-TEX Jelonek, Schmidt & Vöckler GbR, Leipzig
Cover design: eStudio Calamar, Spain

Printed on acid-free paper 52/3100/YL - 5 4 3 2 1 0

Introduction to the Series

CODATA is an interdisciplinary Scientific Committee of the International Council for Science (ICSU). The mission of CODATA is to strengthen international science for the benefit of society by promoting improved scientific and technical data management and use. It works to improve the quality, reliability, management and accessibility of data of importance to all fields of science and technology. CODATA is a resource that provides scientists and engineers with access to international data activities for increased awareness, direct cooperation and new knowledge. CODATA, established over 40 years ago by ICSU, promotes and encourages on a world-wide basis the compilation, evaluation and dissemination of reliable numerical data of importance to science and technology. This includes data initiatives and modeling of an interdisciplinary nature such as that encountered in far-reaching projects e.g. Global Change, various Genome projects, environmental and biodiversity issues, etc. Today 23 Countries are members, 15 International Scientific Unions have assigned liaison delegates, there are 4 co-opted delegates and 20 supporting organizations from industry, government and academia.

CODATA is concerned with all types of data resulting from experimental measurements, observations and calculations in every field of science and technology, including the physical sciences, biology, geology, astronomy, engineering, environmental science, ecology and others. Particular emphasis is given to data management problems common to different disciplines and to data used outside the field in which they were generated.

CODATA's primary purpose in launching the series "Data and Knowledge in a changing World" is to collect data and the wealth of information pertaining to the intelligent exploitation of data in the conduct of science and technology and to make these data and information available to a multidisciplinary community. This series in support of that goal provides a forum made up of many contributions which can be theoretical treatments, compilations or applied outlines. This includes computer related handling and visualization of data to the major scientific and technical fields.

To this end, the series on Data and Knowledge is open to contributions of various kinds, in particular:

- Fostering the improvement of the accessibility and quality of quantitative and qualitative data;

- Treating classical and ground breaking methods by which numeric and symbolic data are acquired, analyzed and managed;
- Presenting new data and knowledge interfaces designed to optimize interoperability and thereby increase the potential for sharing data among databases and networks;
- Promoting international cooperation in communication and data sharing. This includes works dealing with standardization, data quality agreements and conceptual data descriptions (metadata, syntactic and semantic approaches) along with papers dealing with the evolution of internet based facilities, other forms of worldwide communications and electronic publishing;
- Providing new insights into, or interpretations of, processes leading to creative design in the field of concurrent and/or cooperative engineering, including cognitive aspects critical to data based decision making.

In the evolving information world we live in, where the traditional ways of transferring information as an essential resource are rapidly changing, this Series aims to identify emerging and innovative concepts for data compilation, handling, management, and evaluation. Its ambition is to be a catalyst for change while simultaneously nurturing a thought-provoking forum.

Preface

The practical necessity of this book is dictated by an increasing need to solve various complex interdisciplinary problems emerging under the conditions of world economic globalization, rapid development of machinery and technology, and a growing threat of natural and human-caused catastrophes and cataclysms.

This determines the urgency and practical needs of up-to-date foresight, forecasting, and objective analysis of emerging socioeconomic, scientific-technical, and ecological conditions and on this basis working out efficient strategies aimed, on the one hand, at their improvement and, on the other hand, at an adequate response to and prevention of possible human-caused and natural disasters.

Thus, greater importance is attached to the creation and application of a methodology of system analysis for the formation and multicriterion estimation of efficient alternatives of complex solutions and strategies of control in different areas of human practical activity. Such an approach is of great importance for solving the problems of medium- and long-term quantitative and qualitative forecasting of socioeconomic and scientific-technical processes. We should emphasize problems of technology foresight whose solution basically presents two alternatives: either making a timely and qualitative leap in technological development and scientific and technological progress or hopelessly lagging behind global socioeconomic and scientific-technical processes. The successful solution to these problems depends mostly on analysts' and decision makers' skills to rationally use current possibilities of system methodology, heuristic methods and expert procedures, achievements in theoretical and applied disciplines, computers, and application software.

At the same time, the theoretical validation and practical application of the system methodology are fraught with considerable difficulties. Thus, in various publications on system studies, essential discrepancies are found not only in the interpretation of the fundamental notions of systematicity, but also in determining its place in modern science and its relationship with other disciplines, approaches, and methods. Only a few generally accepted theoretical statements and methodological recommendations have been worked out; existing methods and decision-supporting systems do not take into account a number of important features of applied system problems, in particular fuzziness and uncertainty of initial information, the presence of multiple risk factors, interdisciplinarity, and others.

These shortcomings are aggravated by the absence of a systematized description of the system methodology from the point of view of an integral multifactor, multicriterion understanding of properties and features of applied system problems in different areas of practical activity.

This book makes an attempt to eliminate the above mentioned flaws and present a new methodology for system analysis. This methodology will allow researchers to examine modern scientific, technical, and socioeconomic problems based on how they are presented using a set of interconnected problems of a various nature under conditions of limited time, material, and other resources given conceptual uncertainty and various risk factors.

The book is comprised of nine chapters.

In Chapter 1 the objective necessity of establishing and developing system analysis is presented as a methodology for solving increasingly complex interdisciplinary problems in different areas of life. The notions and properties of systematicity that determine the efficiency and effectiveness of the subjective interaction between a person and the objective environment in the process of his/her practical activities are considered. On the basis of K. Godel's principle of complementarity, the system analyst's role is defined and his or her place in solving general problems of system analysis for real systems with various goals is shown. The role and place of system analysis in science and in practice and its connection with other fundamental disciplines are examined. The general development tendencies of the modern science and system analysis are considered. The principal differences between system analysis as a scientific methodology and axiomatic disciplines are established.

In Chapter 2 the basic notions, axioms, and definitions of system analysis are stated. The brief characteristics of the main system analysis's objects are given: material and abstract, natural and artificial, active and passive. Basic statements are introduced for typical objects of system analysis: a system, a complex system, a grand system, a supersystem, and a global system. Examples of different systems are considered. A conceptual statement of the general problem of system analysis is given. The fundamental principles to be followed in the formation of a system methodology and its practical implementation are presented. The conceptual functional spaces for the conditions and properties of a system analysis objects based on specific features of complex, task-oriented system analysis procedures are introduced. The notion of complexity as a fundamental property of system analysis problems is given. Different complexity types are considered. Principles and techniques are proposed for solving problems with transcomputational complexity within the framework of general system analysis problems.

In Chapter 3 the problems associated with disclosing the uncertainties of the goals, situations, conflicts are stated. Methods for disclosing the uncertainties of goals are described: linear convolution, use of engineering constraints, reduction to a system of nonlinear equations, and reduction to the Chebyshev approximation problem. The notion of the Pareto rational compromise is given. Methods and techniques for Pareto set determination are presented. Problems and methods for disclosing natural and situational uncertainties are considered: disclosing uncertainties with known characteristics of random factors and disclosing uncertainties with incomplete information on unexpected factors. Problems and methods related to the disclosure of uncertainties in conflict situations are considered: the problem of disclosing uncertainty with respect to the active interaction of partners and the problem of opponents' counteraction, the problem of multipurpose interaction of partners under situational uncertainty conditions, and the problem of multipurpose active counteraction of opponents under situational uncertainty conditions. A problem statement and method of system disclosing uncertainty are given. The notion of rational compromise under conditions of conceptual uncertainty is introduced.

An approach to functional dependency recovery in conceptual uncertainty problems disclosing is proposed. This problem differs from a typical problem of functional dependency recovery by its principal complexity, which is determined not only by the heterogeneity of the initial information but also by the heterogeneity of factors with considered characteristic features. To overcome transcomputational complexity, we suggest that approximating functions in the form of a hierarchical multilevel system of models should be formed. An approach to system concordance of contradictory purposes in the problems associated with searching for rational compromises is suggested. Methods and techniques of the Pareto set formation under conceptual uncertainty conditions are presented. Practical problems based on the methodology of functional relationships and the Pareto set formation under conceptual uncertainty conditions are solved.

Chapter 4 is devoted to disclosing the uncertainties in problems of coalition interaction and counteraction. The main notions, methods, and principles of the system analysis of active interaction of partners and counteractions of opponents are described. A mathematical problem statement is given for the active interaction of partners and the counteraction of opponents in coalitions as well as for problems related to uncertainties of system coalition interaction and counteraction. Procedures for formalizing the objectives and strategy of interactions among partners, the counteraction of opponents, and coalition interactions and counteractions, taking into account risk factors, are described. Examples of solving problems related to coalition interactions and counteractions, taking into account risk factors, are presented.

Chapter 5 is devoted to solving problems of informational provision of system analysis practical problems. It is shown that for this class of problems a mere quantitative description of information that is used in information theory and related fields is insufficient. It is demonstrated that under real conditions of uncertainties and risks there is a practical necessity to describe and estimate the qualitative characteristics of information in the formalization of the practical problems of system analysis. The notions and definitions of qualitative characteristics of decision-makers' informedness: completeness, reliability and timeliness are introduced. Formalization problems are considered for qualitative characteristics of informedness. A method is suggested for solving a problem of classification and recognition of situations using an integral and partial attributes of informedness. Problems of situation recognition under conditions of information fuzziness and varying qualitative characteristics are studied. The problem of recognizing situations of incomplete and fuzzy information is stated, and methods are presented for solving it.

In Chapter 6 a conceptual statement of analyzing problems related to the structure and functions of complex multilevel hierarchical systems is suggested. Principles and techniques to structuring the formalized description of properties, structures, and functions of such a class of systems are considered. A mathematical statement of the problem of system analysis related to a complex multilevel hierarchical system is given and a general strategy is proposed for solving it. A mathematical statement is given for the problem of rational selection of the structure and functional elements of a complex multilevel hierarchical system, and a method is suggested and techniques and procedures analyzed for solving this problem. Practical examples for solving problems of the above mentioned type are given.

In Chapter 7 the basic notions, techniques, and principles of system analysis of multifactor risks under conditions of uncertainty are described. A general statement of the system analysis problem for multifactor risks is given. Based on the decomposition principle this problem is presented in the form of a sequence of tasks, systemically concordant with respect to their purpose, fixed times, and expected results. An approach to forming the fundamental methodological concepts of system analysis and risk forecasting for problems of safety control of complex technical systems is considered. We propose a mathematical apparatus for the systemically consistent solution to problems of detection, identification, forecasting, and minimization of risks, for abnormal, critical situations, accidents, and catastrophes. Example solutions to such problems are presented in a real time scale under conditions of incompleteness, fuzziness, and inconsistency of the initial information. Such examples assume the presence of unavoidable threshold time constraints on the cycle of formation and on the implementation of solutions for eliminating the consequences of risk factors.

Chapter 8 deals with the problems of system control of a complex hierarchical system under conditions of multifactor risks. A wide range of theoretical and practical problems of system control of a complex multilevel hierarchical system under conditions of uncertainty and risk is considered. The practical importance and specific features of the given class of problems are substantiated. The basic functional properties of system control are analyzed: controllability, adaptability, stability, coordinability, survivability, and efficiency. The general problem of a complex multilevel hierarchical system control under multifactor risk conditions is formulated and a strategy for solving it is suggested. The technical and economic analysis of a complex multilevel hierarchical system functioning under conditions of multifactor risks are given; in the framework of such a system various aspects of analysis and estimation of technical and economic efficiency of various complex systems are considered.

In Chapter 9 the most efficient practical techniques and methods of solving technology foresight problems are systematized and unified. Methodological and mathematical principles of foresight and methods and techniques of their strategic development and implementation are proposed; this will open up new possibilities for increasing the efficiency of innovation activities. To construct future scenarios, tools are developed in the form of a human-machine information platform of scenario analysis, which is a set of mathematical, software, logical, and organizational facilities for determining the sequence of application of individual methods and their interconnections and creating the foresight process as a whole. The developed tools allow one to increase the efficacy of making and implementing strategically important decisions in controlling innovation in enterprises and industries. Practical examples in applying the technology foresight methodology to the creation and comprehensive analysis of scenarios of future developments in various areas of human endeavor are given.

The book's material was formed during the conducting research and was refined by more than 10 years of lecturing for a course called the Foundation of System Analysis in the Department of Mathematical Methods of System Analysis at the Institute for Applied System Analysis of the National Technical University of Ukraine "Kiev Polytechnical Institute".

The book is addressed to scientists and specialists, as well as postgraduates and students specializing in the field of system research and in the analysis of various interdisciplinary problems.

Kiev, January 2007

M.Z. Zgurovsky
N.D. Pankratova

Contents

List of Figures

List of Tables

1 The Subject Domain of System Analysis

The emergence of the aerospace industry and space exploration; the development and wide introduction of computers, information technologies, and the Internet; the fast development and wide application of aircraft; nuclear research and the development of nuclear engineering; the very fast development of advanced technologies and equipment in various industries and services including health care, pharmacology, genetic engineering; and a number of other outstanding achievements in science and technology have entered the history of civilization as a triumphal result of the coordinated activities of science, education, and industry. These achievements are the result of the efforts of many scientists and researchers working in various fields during different historic periods. Some sciences originated in the ancient world; others appeared only in the 20th century. System analysis belongs to the second category along with other new directions of science appearing in the 20th century, such as cybernetics, the general theory of systems, system engineering, the theory of decision making, and the theory of optimization.

Therefore, it is important to analyze the reasons and factors that have allowed the formation of system analysis and the trends in its development as a universal scientific methodology, so as to define its role and place among other sciences and various fields of human endeavor.

1.1 Formation and Development of System Analysis

General characteristics of the problem. For a correct understanding of the role and an objective assessment of system research and system analysis in different areas of practical activities it is necessary to have a precise and substantiated answer to the question of whether the ideas and principles of systemness are the result of a "temporary fashion" in science or whether the process of formation and development of system analysis is an objective necessity. In the second half of the 20th century one of the founders of ideas of systemness, L. von Bertalanffy, drew attention to the following formulation of the question [7]: "If we want to correctly present and estimate the current system approach, then the idea of systemness should be considered not as a result of "a temporary fashion", but as a phenomenon the development of which is intertwined with the history of human thought". However, this statement leaves two very important questions open. First, what circumstances suggest "a temporary fashion"? Second, what connects "the idea of systemness" and "the history of human thought"?

The answer to the **first question** is simple enough. Discussions on systemness were prompted by the situation in science in the second half of the 20th century. On the one hand, in the foreground of scientific research were works united by the common name "system analysis" [104]. On the other hand, an interesting feature was observed in daily life that can be described as the following [6]: "Everyone who would like to analyse the most common current notions and catch-phrases would find the word "system" at the beginning of the list. This notion spread in all fields of science and penetrated into everyday thinking, in colloquial speech and in mass media". Such a process has resulted not only in an infinite number of publications, conferences, and symposiums but also a great variety of interpretations of such basic notions as system, systemness, system approach, system analysis, and system thinking, which involuntarily invokes an analogy with fashion [191]. The same situation has often been observed in the early 21st century. The essential differences and disagreements are still found in interpretations of many practically important notions and definitions, which is characteristic even of those authors who directly participate in system research or are supporters of system methodologies. The formation of system analysis as a scientific discipline is not yet complete; its directions have not been definitively shaped. This is evident from the absence of a common understanding of many key positions and problems of system analysis, namely, the essence and specificity of system research, its orientation and place in modern science, the potentialities and unavoidable limitations of different approaches and methods of system analysis, and their various applications in comparison with the methodologies of other disciplines.

The **subject domain** of system analysis is formed as a complex multilevel set of various types and classes of systems with diverse properties and relations among them; it is so extensive that it defies any strict and univocal classification and ordering.

The **subject** of study and the apparatus of research, as a set of knowledge, descriptions, explanations, and predictions of properties, possibilities, processes, and phenomena taking place in modern systems with various objectives have not yet been precisely formalized or uniquely defined.

The **methods** of system research as a means or way of obtaining practical or theoretical knowledge of phenomena and regularities of functioning and development of complex systems vary considerably and cannot be outlined by any framework.

At present we can speak only about the formation of **system analysis**, in a wide practical sense, **as a scientific methodology** or a set of methods and ways of studying the structure, logic organization, properties, and characteristics of the behavior and development of complex systems. Unfortunately, there is no generally accepted understanding of many key notions of problems and methodology of system analysis, including notions that reflect the specificity and toolkit of system analysis, features and properties of objects of system research, conditions of formalization, and solving of system problems. Meanwhile, it is really amazing how widely and quickly system research is penetrating different fields of science and areas of practical activities. It confirms high quantitative and qualitative indicators of development rates of system research during more than 50 years [4, 6, 8–10, 17, 22, 24–27, 30, 68–71, 80, 82, 94, 104, 105, 149, 158–160, 191]. The same tendencies of increasing the number of the above-indicated problems remain in the 21st century. Ideas and possibilities of the system methodology have attracted the attention of experts from weakly structured subject domains, including health care, ecology, sociology, the financial sphere, and other important fields of knowledge. There are numerous examples of system thinking, system methodology, and system analysis having solved complicated and practically important system problems and ensured qualitatively new results.

An answer to the **second question** is not so obvious and some explanations will be given here. First, it is necessary to point out that the idea of systemness in the quoted phrase is interpreted not as an ordinary event in social life but as a phenomenon that is "intertwined in the history of a human thought". Therefore, it is necessary to reveal the reasons and factors determining the great importance of the idea of systemness. Then, we will define the essence and novelty of system thinking and its practical value as an original contribution to the development of human thought. The estimation of the role of system thinking, as the antithesis to classical rationalism and mechanistic thinking is known, and the achievements in the development of human thought have been studied [161]. However, to solve real system problems successfully, another, **pragmatic**, role of system thinking is considered. Thus, it seems practically important to further study the process of system analysis formation and its development from a **pragmatic point of view**.

An examination of scientific development shows that the following groups of properties determine the effectiveness of any branch of science:

- Urgency: the scientific and practical importance of problems under consideration.
- The scientific level and practical possibilities of a theoretical and methodological toolkit for conducting research.
- The researcher's ability to master the corresponding theory and methodology, his/her skills in rationally using and improving the capacities of the research toolkit.

In the course of the formation and development of any branch of science, the components of the above groups are continuously changing, being specified and improved on the basis of mutual stimulation and rational coordination of requirements of the practice and capacities of the science. Development goes on in the form of a kind of tandem: new practical problems stimulate the development of the science, and new achievements in the science create the possibility for solving new practical problems, expanding the area of research and developing new products and technologies. Hence, the theoretical importance and practical utility of the scientific discipline is determined both by the level of capabilities of each group of properties and the level of mutual coordination of each group's capabilities. Experience shows that the realization of new ideas and technical solutions depends on the capabilities of a given branch of science and a researcher's ability to formalize and solve corresponding design tasks, as well as on the potential of industrial technologies to realize a project and to create a product with the required quality attributes. It is necessary to note the special importance of a procedure for problem formalization, for example, during the implementation of innovative ideas and technical solutions, while designing new technology products, which have no analogs and prototypes. Such problems are distinctive for their conceptual uncertainty. Conceptual uncertainty is understood as a unified system of uncertainty that includes: ambiguity and inconsistency of requirements for a product; inconsistency of purposes and ambiguity of application conditions of a product; uncertainty and unpredictability of probable competitors' actions; unboundedness and unpredictability of multiple risk situations at various stages of a product's life cycle. Disclosure of conceptual uncertainty is a system problem [115] since disclosure of all kinds of uncertainty is to be carried out on the basis of unified principles, criteria, and purposes. The complexity of problems and difficulties related to solving them are caused by the uncertainty of multiple alternative sets and choice criteria [29]. Therefore, the results of their solution fully depend on the researcher's abilities and skills in removing this drawback. The above determines the major requirements of the system researcher's abilities, which are:

- Self-estimation and self-adaptation—the researcher must know how to use what he or she knows, to understand what is necessary to know additionally that he does not know, and where and how to learn it.
- Estimation and formation of research apparatus—he or she must know how to use the resources of the available toolkit for achieving the research goals, to understand what is necessary to do additionally that cannot be done by the available toolkit, to identify what additional tools can perform those tasks that cannot be performed by the available toolkit.
- Problem estimation and formalization—he or she must know how to use what is known a priori for a problem formalization; to estimate the extent to which disclosure of a present uncertainty is necessary, possible, and expedient; and to reveal what must be done to make possible disclosure of an uncertainty during the formalization and solving of a problem.

To successfully meet the above requirements, it is not enough that a researcher has mastered formally the system methodology and reached a required level of its rational practical application. He must develop not only principally a new attitude toward his activities, but also change his mode of thinking, namely, to master system thinking.

Then, we should take into consideration that now, despite the truly unbounded application range of the term "system thinking", there is no generally accepted, detailed definition of this notion, although a many partial interpretations of its essential attributes can be found in the literature [35,65,157,181,188]. We shall understand the notion as follows: **system thinking** is a higher form of human cognition such that the processes of reflecting objective reality are based on the integrated representation of the studied object from the point of view of achieving the research goals, based on knowledge, experience, intuition, and foresight. An important feature of this definition is the dependence of the studied object boundaries on the research goals. Based on the above definition, the same physical object, for example, the airplane, can be the unique object of certain system research and also be one component of a system research object, for instance, a country's civil aviation as a whole. Moreover, in a physical object, for example, the airplane, it is possible to single out a certain functional element, in particular, a gas turbine engine, that can be the unique object of system research in the process of its construction, testing, and operational development. It should also be noted that system thinking becomes the major factor in achieving success in various spheres of practical activities [16,35]. The above-indicated circumstances call for the necessity and expediency of looking into the history, trends, and prospects of the methodology of system analysis and system research. This review seems useful taking into account the following important components of the given methodology development:

- Formation and development of systemness ideas as the basis of system thinking.
- Development of system representation of research objects.
- Development of means and methods of solving practical system problems.

It is necessary to point out that a limited number of system research directions offered for consideration and quite general statements are conditioned by fuzzy boundaries of system research, ambiguous definitions of basic notions, such as system research, system approach, system analysis, and system thinking.

The listed directions of system analysis are interconnected and reflect various aspects of the general process of system research, though their occurrence was not simultaneous, and their interconnection and interdependence became apparent only in the second half of the 20th century in the process of solving practical system problems.

Stages in the development of system analysis. An important role in the formation and development of system thinking has been played by fundamental scientific discoveries whose history goes back to the Aristotelian postulate [7, 161]: "the total is more than the sum of its parts", which in the modern understanding reflects the essence of the superadditive law of complex systems. Without going into the centuries-old history, we shall give a brief overview of the formation and development of systemness ideas and system thinking in the recent past. First, let us remember the fundamental discoveries that preconditioned system thinking. In the history of science of the last several centuries following N. Moiseyev's approach [100], it is possible to point out three basic steps in human understanding of the world.

The *first step* was laying the foundation of modern physics. This is connected with the discoveries of I. Newton, G. Galilei, M. Lomonosov, A. Poincare, and A. Einstein. The idea of motion that originated in ancient Greece was generalized and transformed into a harmonious mathematical theory of Newtonian mechanics. It was the beginning of a new stage in the development of the natural sciences and laid the foundation of the methodology of multifactor analysis of a state and the development of evolutionary physical processes and phenomena on the basis of which the present technosphere was created.

The *second step* is the creation of the fundamentals of Darwinism, carrying over the idea of motion and continuous changeability in the area of living matter. The idea of motion underlined a qualitatively new view of the development of nature. The major feature of this paradigm is a qualitative change in the properties of developing objects, which distinguishes fundamentally developmental processes from all other dynamic processes. The main achievements in understanding evolution and its basic processes and properties are heredity, changeability, and selection. They became the basis of the present conception of the development of living matter, defined the initial positions for biochemical methodology and physiological research, and laid the foundation of multifactorial analysis of the state and evolutionary processes of the present-day biosphere.

The *third step* is revelation of the unity of all evolutionary processes on the Earth: chemical, biological, physical, social, and others in the development of animate nature and society. V. Vernadsky made the first important step in this direction at the turn of the and 20th century. The fundamentals of his noosphere theory were formulated during this period [185]. Vernadsky revealed the systemness of interrelations of various evolutionary processes on the global scale, their system interrelation and interdependence with processes taking place in the Universe. As early as the 1920s, his system of ideas turned into a harmonious theory, which was called "the noosphere theory". According to this theory, all processes on the Earth are a constituent of the unified process of space evolution and are inseparable from the processes taking place in the Universe. The conclusions of this theory have an explicitly practical orientation. Thus, one of the main conclusions is the statement that at a certain stage in the development of civilization, the state of human interaction with Nature will be reached where human demand for resources will exceed the possibilities provided by Nature. In Vernadsky's opinion, to sort out such a contradiction, it is necessary that humankind should realize the practical necessity of taking responsibility for the further evolution of the planet; otherwise, humanity will not have a future.

Thus, the above-mentioned fundamental achievements in our understanding of Nature may be considered as initial steps to understanding the practical necessity of system perception of the world and system thinking. The former two achievements laid the foundation for the subsequent system understanding of Earth phenomena, as a constituent of the Universal processes. For the first time, the basic statements and conclusions of Vernadsky's theory on the noosphere are raising questions about the systemness of processes in the Universe, actually defining the need for system conceptualization of the development of civilization and the planet's evolution, and posing practical system problems that are global in scale. Therefore, it is possible to consider Vernadsky's theory as the original source of applied system thinking.

Below we briefly describe some milestones in the history of system analysis in the recent past, namely, during the 20th century. First, we do not intend to analyze in detail the works that laid the foundations of system analysis. Our purpose is to:

- Show the significance and role of those scientific results that have determined the origination and development of system analysis and those factors that have defined the content of the main directions and stages of its development.
- Show the state and trends of system problems and methodology development under present conditions.

To achieve our goal it is expedient to analyze the status and trend of development of the major directions of science and technology that influenced significantly the evolution of civilization during the considered period. Considering that in the 20th century many cardinal political, economic, structural, and other changes took place in the world as the result of various processes in various areas of life and human activity, it is hardly possible to embrace and analyze those changes in all their variety. We shall confine ourselves to considering processes that took place in the second half of the 20th century that influenced directly or indirectly the status and trends of development of system problems and methodology.

It is possible to distinguish four stages in the formation of system analysis as an applied scientific methodology [119].

The *first stage* is associated with the first half of the 20th century. It was a period when the main ideas of system thinking came into being and were formulated. Let us consider the basic historical sources of present-day system thinking. Taking into account the introduced definition of system thinking, we suggest using the list of the most important, basic works given in [161], but to add to that list the works of V.I. Vernadsky, whose importance for system analysis was shown above. Vernadsky's works should be placed at the top of the list, which is quite justified, in our opinion, not only from a chronological point of view, but also taking into account their great practical importance. As a result, we can compile the following list of fundamental theories (in brackets are indicated the years of execution of works in this area and dates of the first authors' publications): "The theory on the biosphere and its gradual transition in the noosphere" [185] by Vladimir Vernadsky (1893–1918); "General organizational science, or tectology" [11] by Alexander Bogdanov (1913–1929); "General theory of systems" [5] by Ludwig von Bertalanffy (1934–1949); "Cybernetics or control and communication in an animal and machine" [189] by Norbert Wiener (1948); and "Praxeology" [65] by Tadeusz Kotarbinsky (1930–1940).

The role, place, and practical importance of the works of Bogdanov, von Bertalanffy, Wiener, and Kotarbinsky in the development of ideas of systemness and system thinking are analyzed in detail in [161]. Therefore, here we point out only the most important aspects of the analysis and try to draw some conclusions. Bogdanov's monograph is distinguished by a deep system content, and its ideas and notions are based on the following postulate: the integral and most essential property of a universal organizational science is its systemness. Bogdanov can be considered the founder of system thinking with respect to the formation of such a concept of system research that was later on accepted as the general theory of systems. Kotarbinsky's praxeology was conceived and realized as a general theory of rational activities. Though the system orientation of the work does not raise any doubts, there are no proofs and confirmations of its influence on the formation of system thinking. There are no obvious proofs of mutual historical dependence of system tectology paradigms, cybernetics, the general theory of systems, and praxeology. However, the facts do not mean that Bogdanov's and Kotarbinsky's ideas are forgotten and have not found practical application in system analysis. Their monographs were republished in the second half of the 20th century, and the basic ideas have found application in practice in solving system problems of organizational management and in other fields.

The significant role and practical importance of the fundamental works of Ludwig von Bertalanffy [5] and Norbert Wiener [189] in the development of system thinking are generally recognized. It should be emphasized that in these works, independently of each other, these two scientists suggested a new idea, the core of which was transition to research of common properties characteristic of different types of objects. It is necessary to draw attention to the distinction of positions from which von Bertalanffy and Wiener define the necessity of research of common properties of various object types. Von Bertalanffy considers the question from the position of generality of principles of construction and structural properties of various types of systems, whereas Wiener does so from the position of generality of principles and properties of control of various types of complex objects, including living things and technical objects designed for various purposes. These approaches were developed independently for quite a long time. At the same time, both these approaches are directly related to system studies. For example, in the development of complex technical systems it is equally important to create a rational multilevel hierarchical structure of a product and to provide a coordinated control at the design, manufacture, and testing levels of a product and a rational control of the created system during its operation. Thus, the problem structure formation and product shape and the problem of substantiating the purposes and functions of control should be considered in a system statement that takes into account the interconnection, interdependence, and interaction in the closed structure of an integral object of research: "(person) $\leftrightarrow$ (product) $\leftrightarrow$ (environment) $\leftrightarrow$ (person)". The necessity of forming such a structure stems from the fact that environment determines the conditions of a product's operation and the person is a developer, manufacturer, and/or user of a product. This results in the practical necessity of coordinating the decisions of the corresponding system problems at the stage of conceptual design of a product. System coordination of the purposes, resources, terms, and expected results should be provided on the basis of the mutual, rational compromise of the inconsistent purposes of a product's development. Such integral representation of an object completely corresponds to Vernadsky's ideas about systemness of interactions, interconnections, and interdependence of diverse processes on the planet.

The fundamental research carried out independently by Vernadsky, von Bertalanffy, and Wiener have created a unified, ideological basis for forming an essentially new fundamental paradigm in science whose conceptual novelty consists in the following transitions:

- The transition from research of concrete properties of objects of a certain type (physical, chemical, biological, economic, etc.) to research of the common properties that are typical for various objects.
- The transition from research of properties and features of processes of a certain kind to research of the structure, properties, and features of interconnections, interdependence, and interactions of heterogeneous processes.
- The transition from research of properties of individual objects of a certain type to research of properties and structures of interconnections, interdependence, and interactions of polytypic objects.

Later on these attributes of novelty to some extent were realized in the form of the main principles of the theory of system analysis and, thus, the resulting paradigm may be characterized as "a theoretical paradigm of system methodology".

The performed analysis allows to make the following conclusions. The first stage in the system analysis development was the period of occurrence and formation of system thinking, the main theoretical sources of which were created independently from each other. It took place in the first half of the XXth century and was characterized by the independent occurrence of publications of philosophical and methodological ideas, principles, approaches which formed the basis of a new scientific direction which later received the name "the system analysis". It is considered that the major result of the first period was the creation of an ideological basis for formation of a new fundamental paradigm in science. The basic ideas of the given paradigm reflecting its various aspects, were independently presented in the works of V. I. Vernadsky, L. von Bertalanffy and Norbert Wiener. Thus, the first stage in the system analysis development was a stage of **forming of theoretical basis of system thinking**.

The *second stage* of development of system analysis was connected with the extreme conditions of increasing military threat of the 1930s and almost the global theatre of operations of the Second World War in the 1940s. Then many countries felt the need for operative decisions of the most complicated interdisciplinary problems to increase their defense capacity. This stage was characterized by qualitatively new properties of problems being solved and essentially important conditions under which they were solved. Among them are conceptual uncertainty, a lack of structure, NP-complexity, and the strategic importance of real problems; a high price of erroneous or ill-advised decisions leading to catastrophic consequences at a strategic level; the presence of ineradicable, a priori unknown threshold time restrictions on a cycle of formation and the realization of strategic decisions the violation of which may have catastrophic consequences on a national scale.

We would like to point out the importance of conceptual uncertainty in problem situations of development and batch production of new equipment in the presence of threshold time restrictions for a cycle of formation and realization of decisions. Such situations were typical in the initial stages of the Second World War. The notion of conceptual uncertainty in these conditions was supplemented with the following major factor: uncertainty and unpredictability of possible counteractions by the enemy. An extremely important factor of the problem situation is a threshold time restriction on the formation and realization of decisions, which is caused by the desire of each side to gain superiority in strategically important types of equipment. Superiority is reached by having better-quality equipment in greater quantity for a certain period of time.

Such factors have created the following important features and conditions for solving real problems:

- Determine the necessity of ensuring the system is coordinated with respect to purposes, terms, and expected results of formalization procedures and procedures for solving interdisciplinary problems at all stages of a product's life cycle given a set of interconnections, interdependences, and interactions of polytypic factors from qualitatively different areas of practical activities.
- Stiffen the contradiction between the necessity of investigating a great number of factors and the requirement of reducing the time required for making decisions at all stages of a product's life cycle.

- Greatly increase the degree and level of risk caused by making ill-advised or erroneous decisions at various stages of a product's life cycle.

As a result of the developed conditions, there arose a practical need to form system tools that would make it possible to solve real system problems within a practically acceptable time frame with a working accuracy under conceptual uncertainty conditions. To realize the above-mentioned possibility, the tools must be developed on the basis of a system of the following sets of methodological means, mutually coordinated with respect to purposes, terms, and expected results:

- A set of assumptions, conditions, approaches, techniques, and other means of problem formalization.
- A set of indicators, criteria, techniques, and other means for evaluating the quality and efficiency of the problem-solving approaches.
- A set of approaches, methods, equipment, algorithms, programs, and other means of problem solving.

The necessity of developing such tools resulted from the situation that was current at the time: approaches, methods, and techniques of solving various problems in the development of new equipment and the techniques, models, and methods operations of research that were being used did not meet the new requirements and conditions because of their restrictions. New approaches were required that would allow one to analyze, in line with the stated purpose and as a unified integral object, the totality of requirements, conditions, and possibilities of developing a piece of new equipment. On the basis of this analysis it is necessary to form a concept, plan, structure, and shape of a product; on the basis of the concept it is necessary to carry out formalization and solve the system-coordinated set of real system problems pertaining to a product development within a practically acceptable time frame with a working accuracy.

The conditions of the considered period excluded the possibility of developing the necessary tools on the basis of preliminary theoretical support. These conditions dictated more rigid requirements both for the building of new equipment and for organizing the activities of developers, testers, and production workers. In practice, as shown in [190], the chief designer of a certain type of military equipment was in charge not only of development, testing, and serial production of a new product and of the elimination of defects and drawbacks revealed in military operations, but also of organizing the joint activities of developers, testers, and production workers, ensuring the cooperation of enterprises involved in the manufacture of finished products and the production of raw materials and component parts.

The activities of the involved collectives were system-coordinated in regard to the purposes, problems, terms, and expected results. Each collective faced the necessity of efficiently developing its own methodology for solving real interdisciplinary problems, taking into account their specificity. This methodology was developed empirically on the basis of the experience, knowledge, intuition, and foresight of employees using collective system thinking and the method of individual generation of ideas and engineering solutions, which later came to be known as "brainstorming". The main goal of cooperative activities was to attain superiority in a strategically important type of military equipment.

It is obvious that rapidity of production and the unique quality of products of that period could be reached only by the system-coordinated, planned cooperative activities of scientists, developers, testers, and production workers of a certain kind of equipment, which demanded system-coordinated solutions to numerous real, multipurpose, interdisciplinary organizational, and technical problems. Specialists from different fields, including mathematicians, engineers, and scientists, later called system analysts, made a great contribution to the achievement of such results. In the military-industrial establishment the experience of a system-coordinated solution to real, complicated, interdisciplinary problems under limited time conditions had been achieved. Soon, following the outbreak of World War II the command of the US Air Force proposed that within 1 year the Harvard School of Business Administration should solve the problem of increasing Air Force personnel from 300,000 to 2,500,000 and the number of war planes from 4,000 to 80,000 on the condition that expenses would not exceed US$10 billion [105]. It was one of the first publications in the public press of real information about the development and application of a system approach to solving a problem connected with the development of the armed forces.

It is considered [105] that it was for the purpose of solving the above problem that for the first time some methods, principles, and approaches were stated that formed the basis of the "system analysis method". A great contribution to the widespread adoption and popularization of this method in the USA was made by the RAND Corporation, founded in 1947 [105]. In 1964 the Defense Department issued some manuals and guidelines regulating the procedures for system analysis.

The above brief review allows for the following conclusions. The second stage in the development of system analysis was formed in the extreme conditions from the beginning of the 1930s up to the end of the 1940s. It was a period when a practical necessity arose to solve efficiently real system problems of national significance, to create various engineering systems for military purposes within severe time constraints. The methodological apparatus was being developed empirically and independently in different countries. Wartime conditions excluded the possibility of exchanging experience and information. As a result, the empirical preconditions for forming a paradigm of system analysis as a methodology for solving real system problems within a practically permissible time frame and with a working accuracy under conceptual uncertainty conditions were created. Therefore, the given period may be considered a stage in the **empirical formation of system methodology**.

The *third stage* in the formation and development of system analysis was formed in the postwar period from the 1940s to the end of the 1970s. This period in principle differs from the previous ones and is characterized by problems of a new quality, fundamental sociopolitical changes taking place in the world after World War II, and outstanding scientific and technical achievements. During the first postwar decade the main goals and objectives of many countries consisted in finding an effective solution to interdisciplinary problems connected with the elimination of the consequences of the war and a fundamental reorientation of military economies for solving problems during peace time. The systemness and complexity of these problems were determined by many factors: fundamental differences and novelty of goals and objectives, expected results, limited financial and other resources, the shortage of qualified manpower, and other factors.

Among the most important changes in the world it is necessary to point out the establishment of the international organizations: the United Nations (UN) (1945) and the United Nations Educational, Scientific and Cultural Organization (UNESCO) (1946). These two institutions provided fundamentally new possibilities for international collaboration in such fields as education, science, and culture. In particular, these dates may be considered the beginning of the consolidation of science from different countries into unified world science. But a particular importance and practical significance of that process were truly appreciated only much later. Such a moment came when the world scientific community proved that the threat of global ecocatastrophe could become a reality as a result of nuclear tests or a nuclear war. The proof was based on system analysis and computer modeling of the results of nuclear weapons tests. And then the first political decision was made—the Moscow agreement on a partial nuclear test ban (1963).

That period was rich in scientific and technical achievements, some of which contributed significantly to the development of system analysis. Thus 1945, the year the war ended, was the first year nuclear energy was put into use. The next achievement of great importance was space exploration. Chronologically it happened after nuclear energy use, but it came first in terms of the level of uncertainty and unpredictability of problem situations. The pioneer of space exploration was the USSR, with its historic scientific and technical achievements. The world's first space satellite was launched on October 4, 1957, and the first manned space flight was made by Yurii Gagarin on April 12, 1961. These events spurred further achievements in space exploration. By the end of the 1970s more than 2,500 spacecraft had been launched and more than 60 manned space flights made. Space satellite systems were created, including those for radio communication and retransmission, meteorological observations, and remote probing for natural resource research. The exploration of the Solar System by unmanned space stations began—the Moon, the Sun, the planets Venus, Jupiter, Saturn, Mercury, and others all were studied. On June 21, 1969, US astronauts Neil Armstrong and "Buzz" Aldrin were the first humans to set foot on the Moon. The next important step was the creation and operation of a series of orbital, manned space stations "Salyut" built in the USSR during the period 1971–1977.

These stations gave their crews the possibility of carrying out different technical and biological experiments in space conditions; the duration of some of the missions exceeded 140 astronomical days. The stations provided the conditions for continuous space research, remote probing of the Earth, and the World Ocean observation. The orbital, manned space complex "Mir" became the next stage in the development of this type of spacecraft and extension of its possibilities. It operated continuously for more than 15 years. Then, in 2001, in the course of a few months it was withdrawn from orbit and sunk in a predetermined, uninhabited area of the Pacific Ocean.

The third unique achievement in space exploration is associated with two countries—the joint flight of the US "Apollo" and the Soviet "Soyuz". These spacecraft differ in many aspects including life support systems. Nevertheless, the crew of "Soyuz"—A. Leonov and V. Kubasov—and the crew of "Apollo"—T. Stafford, D. Slaton, and V. Brand—completely fulfilled the mission. They made two dockings and many transfers from one spaceship to the other. They carried out joint scientific research and technical experiments.

Chronologically the fourth achievement in space exploration was the creation of reusable spaceships, such as the "Space Shuttle" in the USA and the "Buran" in the Soviet Union. It should be pointed out that space exploration demanded solving problems unprecedented in their diversity, complexity, uncertainty, system-interconnection factors, and conditions. The fundamental and applied achievements and possibilities of practically all sciences were put to use, from astronomy to nuclear physics.

The creation of computer engineering is usually considered to be the next most important achievement of humanity. It contributed to the solution of real problems in different fields of vital activity. This direction originated from the first vacuum-tube computers, which were independently developed in many countries. The first computer in the world was created in the USA in 1948, the first computer in Europe was created in Great Britain in 1951, and the first one in continental Europe and the USSR was created in Ukraine in 1952. These computers were aimed at a wide field of applications and opened fundamentally new capabilities for solving complicated problems in different fields of practical activities.

At the same time, the need for solving qualitatively new problems arose in different fields of science. In particular, it was necessary to define the subject of investigation, choose the terminology, describe problems, develop a methodology, and, finally, create new sciences—the theory of algorithms, the theory of programming, the theory of computation systems, and others.

That is why the technical basis and major theoretical principles were being developed in parallel.

Thus in some countries simultaneously with general-purpose computers special-purpose computers were developed. In particular, in the USSR computers for controlling objects and technological processes; for scientific calculations and modeling of processes; for processing of measuring and radio-engineering information; for solving problems of accounting, statistics, planning, and economic modeling, and others were developed. In the USA scientists worked to develop a supercomputer. However, it should be noted that first-generation computers did not meet many practical requirements.

Fundamental changes in the technical basis and in computer hardware became possible thanks to a number of basic scientific and technical achievements, among must be mentioned the following. The invention of the transistor in 1948 was of the utmost importance. It provided not only for the creation of a fundamentally new elemental basis for computer logical schemes, but opened the door to technologies that continuously developed the micro-miniaturization of a computer's functional elements.

As a result, the amount of functional elements on the base plate doubled every one and a half years over the course of 30 years, and computer capabilities increased at virtually the same rate. In 1969 the invention of microprocessors and the development of technologies for the batch production of microprocessors were a new stage in the development of computer architecture and in the improvement of the technical characteristics of the computer.

Alongside developments in the technical basis, the rational use of computational capabilities of computers and the extension of their practical applications were ensured by continuous improvement in the theoretical basis of computer engineering, high rates of development of computer mathematical environments, and a rapid perfection of user tools. As a result, at a certain stage of the development of computer engineering, instead of having direct contact with technical facilities, users worked with applied software systems. Essentially, the conditions were created to provide the maximum ease of communication between people and computers, regardless of the field of application.

A high level of ease was realized in 1976 with the invention of the personal computer (PC). The PC has changed the means with which humanity makes calculations and records their results, while drawing up documents, preparing manuscripts, and sending and reading messages. Backup software for the PC was created in 1977. It was oriented toward a mass user audience and allowed for the batch production of operating systems and application software for the PC as well as for broad use of the PC for solving real tasks in all fields of practical activities. This circumstance opened the door to developing computer systems and networks.

The above survey shows that the situation during the considered period was characterized, on the one hand, by a continuously rising need to solve practically important system problems of an interdisciplinary nature, and, on the other hand, by the appearance of the qualitatively new possibilities for solving them provided by computer engineering. It determined the necessity of rational use of hardware-program capabilities of a computer and demanded the development of mathematical and methodological supporting facilities, adequate to emerging practical needs and available capabilities of computer engineering. These circumstances initiated the processes of the formation and theoretical motivation of methodologies of system analysis and related scientific directions and disciplines: the general theory of systems, system engineering, computer mathematics, applied mathematics, simulation equipment, computer science, designing computing machines, theory of programming, theory of automatic processing of digital information, and others.

Scientists from different countries made valuable contributions to the creation of the theories of system analysis and system methodology. In particular, we should mention the works of K. Boulding [12], G. Klir [53], M. Mesarovich [96], T. Saaty [156], G. Simon [174], A. Holl [49], and U.R. Eshby [3].

Scientific and technical achievements of this period are unique by many standards. For the first time, people could live and work outside the Earth and step on the Moon; for the first time, they had a fundamentally new source of energy; for the first time, the possibility of automating intellectual activities was realized. The most complicated system problems were solved and qualitatively new branches of industry and a new technosphere were created. System methodology and computer tools were developed simultaneously with growing practical needs and the increase in problem complexity. The balance of requirements and possibilities of problems solution was provided.

Complex and technical systems with different purposes were designed, constructed, and put into operation. The fundamental paradigm of system analysis was theoretically justified and practically realized by special mathematical and software products.Ideological basis of this paradigm was formed by the works of Vernadskiy, von Bertalanffy, and Wiener. This period is a unique stage of rapid development of civilization. It is characterized by the emergence of fundamentally new ideas, inventions, and discoveries, by the formation, assimilation, and broad application of results and possibilities of qualitatively new theoretical and applied sciences and scientific directions for the development and production of unique and complex systems, for the creation and development of key industries.

This stage was a period of **simultaneous development of system analysis theory and practice of system research**.

At the same time, by the end of the above-indicated period global system problems arose that could not be solved on the bases of the mathematical and methodological means available at the time.

Such a situation was connected with a number of general features of development of system analysis. First, in the course of evolution of civilization, complicated and important practical problems were emerging that could not be solved on the basis of existing theoretical and technical scientific facilities, including facilities of system analysis. And such a situation is quite explicable: if a problem situation is not anticipated by the scientific community, it is impossible to prepare in advance the facilities for the problem's solution. In such cases, preparation of facilities for the problem's solution begins only after the emergence of a problem. Second, it is impossible not to acknowledge the truth of the statement [52] that the development of system analysis did not at all look like a "triumphal procession" of successful solution of new system problems.

Along with significant successes some difficulties in the realization of the system approach became evident, first of all, in weakly structured subject domains: in the sphere of social management, in ecology, economy, etc. [52]. And a "critical mass" of unsuccessful attempts at applying the system approach and system methodology to a given set of problems had accumulated by the end of 1970s.

This opened the way for critics of system analysis to characterize it as a set of methods of a limited application and to speak about the groundlessness of its inefficiency to be a general scientific methodology. However, these difficulties were partly explained by the fact that the mathematical and methodological capabilities of system analysis, successfully applicable for solving problems pertaining to objects of one type, were mechanically applied to problems pertaining to objects of a qualitatively different type.

One more reason for the existing situation was that the mathematical, methodological, and computing capabilities of system analysis, which were developed and successfully used in defense, space exploration, and nuclear energy production, were inaccessible for wide application. At the same time, some important practical problems emerged that could not be solved by applying the available mathematical and methodological facilities of system analysis. Therefore, the situation in system methodology in the late 1970s was viewed as a "methodological crisis" [52].

The features of that crisis were determined by many factors. One of the main ones was a rapid growth in the complexity and range of real system problems preconditioned by globalization of all kinds of world processes. Interconnections, interdependencies, and interactions of economic, social, ecological, and other global and regional processes became the determining factors of the development of the world. As a result, a new consequence of globalization emerged that was distinctly and fully characterized by the French economist Gode in a short phrase: "The future ceased to look like the past" [39]. In these conditions global processes of the world system turned out to be under the influence of a complex-structured, multilevel, hierarchical set of practically unpredictable, continuously changing interconnections, interdependencies, and interactions. The results of this state of affairs were "consequences unforeseen and unpleasant" [39].

The *fourth stage* of the development of system analysis began in the early 1980s and continues to this day. This stage principally differs from the previous stage by the globalization of world processes and threats. On the one hand, globalization of economic, social, information, and other processes opens up new ways of applying the achievements of scientific and technical progress. In particular, the globalization of informational processes and telecommunications networks has created the conditions for a quick exchange of information, the emergence of distance learning, the information market, and electronic commerce, and a number of other innovations [114,164]. On the other hand, unequal access to information by the developed and developing countries has resulted in unfair competition and social inequality. A number of these emerging problems [31] are conditioned by the specific character of information spread over the Internet. Among them there are problems of information protection, intellectual property, transactional information, computer viruses, and other intentional unauthorized influences on informational systems.

In the early 1980s the scale of the globalization of economic, social, ecological, and technical threats took on special significance. Humanity entered a stage in its development where, as Vernadsky foresaw, human economic activity was bringing the planet to the verge of global ecological catastrophe [88]. Economic and social development came up against the finitness of Nature. This is evident from the exhaustion of natural resources of the land and oceans, an irrevocable loss of various plant and animal species, a technogenic breach of biogeochemical circulation of processes, pollution of all components of the natural environment, and the degradation of ecosystems.

Four categories of threats are becoming increasingly acute. They are connected with global changes in the world: (1) direct threats to the existence of humans (famine, diseases, radiation, terrorism, and others); (2) threats to large regions and territories of desertification, a rise in sea level, global warming, transboundary pollution, withdrawal of river water by countries located at river headwaters, etc.; (3) threats to systems of fresh water, forests, etc.; (4) threats to economic development (depletion of natural resources, increase in global inequality, including the digital divide, nonuniformity of economic prosperity, and development of countries, instability of financial systems and markets, etc.).

The interconnection, interdependence, and interaction of real problems in different fields of practical activities characteristic of that period were figuratively described by Pechchei—an economist and public figure and moving force behind the creation of the international scientific organization known as the Club of Rome (1972). His activities and role in the formulation of global problems are described in detail in [45]. In his book [144] Pechchei proved the following: "There are no more the economic, technical, or social problems existing separately, independently of each other, which could be discussed within the limits of one special terminology and be solved slowly, separately, one by one. In our artificially created world everything has literally reached unknown sizes and scale: speed, energy, complexity—and our problems, too. Now they are simultaneously psychological, and social, and economic, and technical, and, in addition, also political; moreover, closely intertwining and cooperating, they take root and sprout in adjacent and remote areas". Then, Pechchei lists the processes and factors that have led to the current situation. Among them: an inequality and heterogeneity of society; social injustice, famine, malnutrition and poverty; illiteracy, unemployment, a feeling of instability, and decline in moral values; a growth of criminality and violence; degradation of the environment and a potential or already existing lack of natural resources. This list is far from complete. Further, Pechchei comes to the conclusion that the major reason for the current situation is the lack of awareness of interconnections and interdependencies of these factors, processes, and problems that form a complex, intricate whole. And, therefore, it is necessary to undertake drastic measures aimed at estimating and explaining the essence of the problems before it is not too late.

The Club of Rome has accepted the paradigm of organic growth and holistic development. The following statements are worth repeating [45]:

- Development should be regular, multifold, and interdependent when any element of system cannot grow at the expense of others.
- The consistency of the world should be guaranteed by coordination of goals.
- The main aspect should be concentrated on the quality of development, with the purpose being growth of well-being of the human person.

The Club of Rome for 25 years of its existence has done much for the understanding of the state and development of global problems and their possible negative consequences [45,88,91–93,144,145]. During this period, as a result of the Club's activities, other international organizations were established, among them the International Institute for Applied System Analysis (Laxsenburg, Austria), which has carried out, in particular, a number of important programs on ecology and risks, some results and data about which are reflected in [44].

At the same time, continuing to investigate the modern state of the world, which has undergone fundamental changes during the last 15 years, the Club of Rome had to admit that the situation with global problems not only has not improved but also continues to worsen [45,91,93,173]. There are many different reasons for this, including political, economic, social, and others, that interfere with the development and realization of a rational strategy of joint actions of all humanity for the prevention of an approaching global catastrophe. We shall note one of the major methodological reasons for the declining situation in global cooperation—the modern methodology of system analysis does not correspond to the global, multilevel, hierarchical, multidisciplinary structure of diverse, multifactor, multipurpose interconnections, interdependencies, and interactions of objects under investigation. It does not use to the fullest extent the potential capabilities of the global, multilevel, hierarchical system of information computer systems and networks that are potential tools of research of global problems.

This drawback may be eliminated, in particular, by a consistent development of concepts, strategies, and programs of research of the most important current problems. Above all are the following problems: prediction of qualitative and quantitative changes in various spheres of practical activities [101, 192, 198, 200], control of risks and safety of complex technical systems, human-induced and ecologically dangerous processes [28, 134, 135, 168], development of intellectual information technologies and networks of scientific research support [106], relationships between Nature and society on the basis of global ecological monitoring, estimation of tendencies of development of the global ecosystem [51, 88, 100], and others.

Thus, the determining principle of system research of the fourth stage becomes a global vision of investigated problems accounting for increasing interconnections and interdependencies of all countries and nations of the world. An overall objective of research becomes the achievement of such a system-coordinated, interdependent development of all components of civilization in which any element of the world system cannot grow due to other elements. To this end it is necessary to concentrate efforts on quickly overcoming the methodological crisis, which became apparent in the late 1970s. It is expedient to expand the scope of system analysis methodology, having provided its systemic, functional coordination with the hierarchical structure of interconnections, interdependencies, and interactions of objects under investigation and in accordance with the hierarchical structure of information computer systems and networks, as the technical basis of its realization. Thus, the fourth stage is the **stage of globalization of system problems**.

The analysis of globalization processes. The current stage of globalization has essentially influenced the development of methodology and the solution of practical problems of system analysis. One example of the system approach to the analysis of the development of civilization at this stage is global modeling—a direction of system research whose foundations were laid down by the renowned work of J. Forrester, *Global Dynamics* [32]. Another typical example of an increase in interest in the specified problems is the research of scientific and technical innovations. In the 1970s, along with the problems of innovations of companies that were common in this area, there began a broad-based discussion of the influence of innovation on long-term trends of economic development and their role in the formation of so-called big cycles.

It is especially necessary to mention the transition from the industrial to the information society that occurred at a very fast rate and that was global in scale. This process is characterized, first of all, by an increase in the role of information technologies, telecommunications networks, in the world economy a whole and in every individual country. Currently the information sector of some economies around the world accounts for more than 20% of the gross domestic product. Estimates by leading world experts show the highest rates of development in information systems, technologies, and services sectors. The rates of growth are more than 11% annually, and the volume of the market doubles every 5 years. The rates of growth of the Internet—a growth of about 10% monthly over the last 5 years—are even more impressive. The experience of the development of the world economy shows that the world market of informatizaton by volume could surpass such highly profitable sectors as gas, oil, and power production in 2010.

The role of information telecommunications sector in the creation of new jobs and an increase in export opportunities of developed states is great. The most indicative will be the experience of the USA, the world leader in transitioning to the information society. Everything is learned in comparison to the USA. In particular, regarding the level of informatization there are the following comparative data. By the middle of the 1990s the USA had 43% of the world pool of computers. The share of the closest competitors from the most developed countries were: Japan – 7%, Germany – 6%, Great Britain – 5%, France – 4%, Canada – 3%, Italy – 2.5%, Spain – 1.8%, South Korea – 0.96%. Other countries had less than a 1% share. In the USA 52.2% of databases and knowledge bases are in exact and technical disciplines, 69.1% – in the social sciences. Of the ten largest software manufacturers, six are American.

The social and economic effects of the rapid development of the information sector is rather interesting. A paradoxical phenomenon is observed: informatization is likely to generate unemployment, but not to create new jobs. However, manual labor jobs are being eliminated and new jobs are being created in the telecommunications sector. This tendency was especially vivid in the USA in the mid-1990s, when that sector became the largest "employer" in the industry, having created in 1996 alone 4.3 million new jobs. According to studies conducted by the American Research Association information technology accounts for 6.2% of the gross domestic product of the USA, which has been one of the main driving forces behind national income growth. The wages of workers in this sector increased by 73% as compared with those of the average worker in other industries. It also dominated in expenditures for research and development, at US$40 billion annually, or 37% of all expenditures for science. The economic depression in the early part of the new century has slightly weakened this tendency, but it has not caused the USA and other developed countries to yield their position as leaders in this sector of the global economy.

As a result of redistribution of the labor force between the leading sectors of the economy in a number of countries in the telecommunications sphere more than 50% of the world labor force is concentrated in the USA—about 80%. By the estimates of experts, in the first decade of the 21st century a similar situation will characterize the majority of the developed countries of Europe, Asia, and North America. First, we should point out that for each type of society the labor force is usually concentrated in those areas where the main tasks of civilization are solved during the corresponding period. For example, agrarian societies saved humanity from starvation. And, consequently, the overwhelming part of the labor force was concentrated in agriculture. The industrial society solved another problem—providing the possibility of movement of individuals to any point on the planet, providing mechanization and automation of physical labor in various areas of activity. In such a society, the major part of the labor force was concentrated in industry. The information society sets an essentially new, complex task—to provide automation of intellectual work on the basis of a rational use of technical achievements of the industrial society, to release the individual from routine work of transferring, gathering, processing, and storing information, to create conditions of global access to information resources from any point on the planet, to ensure rational uses of the accumulated knowledge for solving various problems faced by society. The developed technological culture at the end of the 20th century and, first of all, arising from essentially new technological possibilities of processing, transferring, and storing information have created the necessary premises and conditions for practical realization of a given task.

Let us examine a number of other important properties and features of the information society that have largely determined the necessity and expediency of forming and developing ideas of systemness. Among them are the following:

- A sharp increase in the dynamism of economic, social, and political processes, both in the world at large and in separate regions.
- A continuous increase in information volume: currently the volume of information annually processed in the world is equal to the total amount of information accumulated by the humanity before the beginning of the First World War.
- In the information society the cost of labor is not paramount, but the cost of knowledge. The basic tendency of modern society lies in a considerable increase in the relative density of intellectualization of processes of control and production.
- An increase in the social and economic value of high technologies, an increase in rates of their development and introduction in various branches of the economy.

In modern understanding, **high technology** is a large-scale system technology technically realized in the form of a complex multilevel hierarchical system that unites structurally and connects functionally diverse physical, chemical, mechanical, and other technological processes in the unified production process for the achievement of practical purposes. Each such technology is, as a rule, unique, i.e., has no analogs or prototypes. It should be noted that the term "high technology" is predetermined by a significant contribution of scientific research results to its development and realization. High technologies have it possible to realize principally new ideas and decisions in various spheres of human activity. They allow one to gradually overcome technical barriers and constraints of the existing equipment and to develop products of various purposes hitherto unknown.

A prominent feature of modern scientific and technical progress is also a rapidly growing complexity of interconnections and interactions of various fields of activity of human beings with their environment. The global scale of positive and negative influences of various processes and their high dynamism have led to a change in the habitual stereotypes in understanding a degree and a level of influence of civilization on the environment. The possibility of global ecological catastrophes has become a reality. It confirms the fact that humanity's impact on the biosphere has already exceeded the real capabilities of the planet to compensate its consequences. Therefore, the main conclusion of Vernadsky's doctrine that at a certain stage of development humanity should take up the responsibility for the further evolution of our planet has become an absolute must. Otherwise, humanity will not have a future. The solution of this global, extremely important problem is impossible without carrying out system analysis of the existing global situation, without predicting possible scenarios of its development on the basis of system thinking, without planning necessary actions for the prevention of unfavorable outcomes on the basis of system methodology.

This brief analysis shows that the occurrence and development of system thinking, system methodology, and system analysis is caused by the objective necessity of perception of the world in its variety and integrity in the interests of the most rational use of material, power, and information resources of the planet for the welfare of humanity.

1.2 Place and Role of System Analysis in Human Practical Activities

The above analysis of the formation and evolution of systemness allows us to define the place and role of system analysis in human practical activities. To this end let us examine in detail the general structure of system interactions of a person, an object under investigation, and the environment. It may be easily shown that a system analysis problem pertaining to many types of complicated real systems of different purposes may be reduced to this type of interaction.

Here we introduce some definitions. The term **object under investigation** means an artificial or natural material object that is used or made in the course of human practical activities or a specific situation that is the result of natural processes or is formed in the process of practical work. The **environment** is the external vicinity of **a person** and **an object under consideration** (i.e., natural environment, external objects interconnected or interacting with an object under consideration). A **person** is a developer, producer, user, seller, or purchaser of an object or expert-researcher in the corresponding data domain the object under consideration belongs to.

Let us try to see the role of system analysis by comparing two problems: the traditional physical problem of gravitational interaction of three bodies and the problem of systemic interaction "**person** $\Leftrightarrow$ **object** $\Leftrightarrow$ **environment**". The systemic interaction of objects in these problems can be represented as a triangle in whose apexes three bodies—human, object, and environment—are placed (scheme 1). This physical problem is chosen because its scheme is maximally similar to scheme 1, which represents the interaction between objects in a system problem. However, the characteristics of interaction between objects in these problems are principally different. Indeed, for the problem of gravitational interaction of three bodies, the gravitational interference between them is the only true condition. Therefore, the systemic characteristics are defined only by the masses of the three bodies. There is a precise mathematic description for this type of interaction. The single complexity in the examination of this interaction is that the equations representing the dynamics of such a system do not have an analytical solution. The researcher only decides which numerical method to choose and, as a consequence, the accuracy of the obtained solution. In such a way, solutions obtained by different researchers will differ only in the values of an error and possibly the time spent on the calculations. The comparison of such solutions presents no mathematical or methodological difficulties.

The interconnection between three components of a complex system—**"person** $\Leftrightarrow$ **object** $\Leftrightarrow$ **environment"** is principally different. In this case incompleteness, indeterminacy, inaccuracy, and inconsistency of the initial information are characteristic properties of system problems. In fact, it is difficult to completely take into account all properties and influences of the environment during the development, use, or investigation of an object. The surrounding environment of an object usually does not have exact boundaries since any object involved in human practical activities has multiple connections and interactions with the environment, i.e., material, energy, informational, natural, etc. Furthermore, the system of universal constants that would allow one to evaluate all possible connections and interactions legibly and unambiguously does not exist in principle. For example, it is impossible to determine unambiguously quantitatively natural operation conditions, particularly for land-based vehicles. Thus, the conditions of the far north and those of deserts and semideserts differ; the latter, in turn, differ from the jungle. In addition, even within one climatic zone different groups of factors are opposite in many characteristics and set contrary requirements that an object (a building, technology, vehicle, etc.) should meet.

One more feature of the problem of system interaction "**person ⇔ object ⇔ environment**" is the indeterminacy and ambiguity of purposes. In the problem of interaction the purposes and development of an object cannot be precisely defined. In fact, any object should meet the often conflicting and opposite requirements, for example, achieving high strength at low weight or high reliability at a low cost, etc. In addition, many latent, intangible factors that may have primary importance during competition of alternative solutions should be taken into account. Among them are aesthetic, ergonomic, ethnic, and other factors, for example, habits, skills of a potential user, etc. Therefore, it is necessary to find a rational compromise both between different groups of factors characterizing "**person ⇔ object ⇔ environment**" interactions and inside each group. In particular, while defining the boundaries of an object and formulating a general statement of a problem connected with its investigation, it is necessary to provide a rational compromise between the determining generalized opposite criteria—maximization of solution quality and minimization of solution procedure complexity.

More examples could be given, but the above are enough to conclude that for solving many problems in practical activities a rational compromise is needed. But this procedure is subjective because comparison and preference criteria are chosen by *a decision maker*. This means that different decision makers, using the same initial data, can reach solutions that are different in practical terms by most criteria. There are many practical confirmations of this fact. The convincing proof is that in spite of very strict restrictions imposed by national and international standards, in the market we see a great variety of duplicate products.

With a diversity of products with similar purposes, the necessity naturally arises for estimating and comparing these products. First, it is important for a consumer. But it is no less important for the manufacturers of the respective products to have a substantiated and reliable solution to this problem if they want to be competitive. Moreover, for a manufacturer or researcher solving this problem is more important and complicated than for a user. A manufacturer should examine all aspects of product quality, taking into account both a competitor's actions and changes in customer needs. Thus, in practice a more general analysis of the interaction "**person ⇔ object ⇔ environment**" than that shown in scheme 1 is needed. This analysis essentially should be an investigation of the quality and efficiency of the accepted methodology and strategy of a person's actions in the system "**person ⇔ object ⇔ environment**".

Such problems should be solved by a system analyst. Thus, there is a new structural scheme 2 that includes an interaction between a system analyst and the system "**person ⇔ object ⇔ environment**". Thus we have a new system "**system analyst ⇔ person ⇔ object ⇔ environment**" (Fig. 1.1).

According to this scheme the investigation should include comparison, comparative assessment of the results obtained in the considered system with results similar to those achieved by competitors, as well as with general tendencies and prospects of production, market, and customer needs.

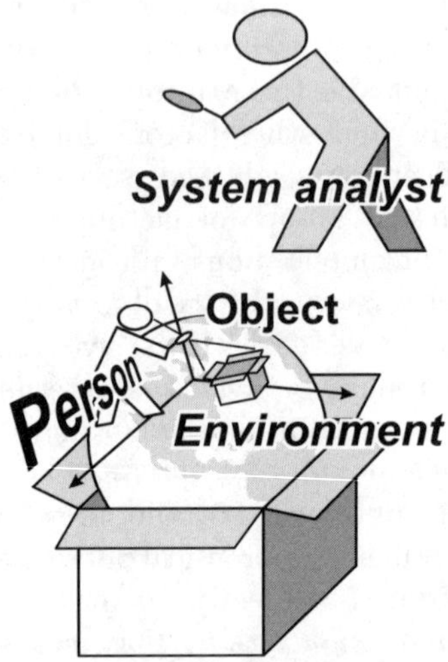

Fig. 1.1. Interaction between a **system analyst** and the system "**person ⇔ object ⇔ environment**"

At this point a natural question arises: why can this problem not be solved by a person (developer, user, researcher, or other), as is shown in scheme 1; why must the interaction data be analyzed from the point of view of a system analyst according to scheme 2. Indeed, an experienced developer or manufacturer analyzes his customer's needs, forecasts and estimates competitors' actions, estimates supply and demand, develops the corresponding strategy of actions, taking into account the above-mentioned factors and his own aims and capabilities. Nevertheless, there is a practical necessity of performing the interactions of system analysis according to scheme 2. It is determined by many factors. We note only two:

In the first place, any developer or manufacturer forms his own set of criteria, indices, assumptions, and suppositions while solving a problem. Thus, such a set of criteria is subjective, and therefore the corresponding evaluation is also subjective. Owing to this approach a developer or manufacturer dealing with a certain class of products often overestimates their positive features, potential, and advantages and underestimates defects as well as the potential and advantages of competing products. Numerous examples of business failures, unprofitability, and sales of companies in countries with a market economy are proving this conclusion.

The second factor, which is proving the necessity of system analyst in solving such problems, is described below. It is in principle impossible to evaluate the quality of the accepted methodology and strategy of problem solving, its credibility, and its validity in the context of the accepted set of criteria according to scheme 1. It directly follows from the mathematically proven principle of complementarity by Godel [182]. In accordance with this principle, it is necessary to use a more powerful set of criteria for evaluating the quality and efficiency of the accepted strategy and problem-solving methods based on a specified set of criteria. The more powerful set of criteria supplements the specified one with new characteristics and opens up possibilities for using additional evaluation tools. Such possibilities may ensue after performing a system analysis of the specified problem according to scheme 2. It follows that a solution to this sort of problem is fundamentally different in the methodological sense from the solution of a traditional physical problem.

The performed analysis allows us to draw the following conclusion. The systemness of human activity determines the necessity of carrying out system analysis at different levels of studying interactions, both inside the system "**person** $\Leftrightarrow$ **object** $\Leftrightarrow$ **environment**" and also at a higher level: "**system analyst** $\Leftrightarrow$ **person** $\Leftrightarrow$ **object** $\Leftrightarrow$ **environment**".

The considered example essentially determines the place and role of a system analyst in human activity. His role, as an expert who must evaluate the degree of goal achievement based on the results of a complex system operation, was described above. Obviously, the importance of the system analyst is not limited by this function only, and correspondingly the significance of system analysis, as a scientific methodology, as well. This conclusion directly follows from the principal features and differences between practical system problems and typical problems of axiomatic disciplines, such as those considered by the theory of operations research, optimization theory, etc. The main difference is that practical system problems are conceptually uncertain [115] since they contain principal uncertainties, ambiguities, risks, competing goals, and varied and opposing elements in the systems under examination and information of a different nature (qualitative and quantitative) that is circulating among these elements. It follows that the complexity of system problems is principally higher than the complexity of problems of axiomatic disciplines because they are principally nonformalizable in the initial statement. Thus, formalization is one of the most complicated procedures in stating and solving any kind of system problem. In practice, only separate parts of complex problems are formalizable. In particular, one of the most important procedures is goal setting and goal statement in the form of goal functions. The problem of goal function formation can be presented as a sequence of interconnected problems:

- A reduction of verbal, numerical, empirical, and other kinds of initial information to some standard information form, which makes it possible to state research goals; selection of a class and structure of approximating functions during goal setting.
- Selection of criteria, principles, approaches, and methods of goal function formation.
- Determination of goal functions in the accepted class of functions, which ensures formalization of the verbally formulated goals with an acceptable level of inaccuracy.

The final result of the new complex technical product development or complicated problem situation analysis depends on the validity and credibility of this initial procedure execution. At the same time, the work of large groups of front-line workers, research organizations, and enterprises depends on the final result. This is typical for the process of complex technical system development, i.e., airplanes and spacecraft, electric power plants, etc. Similar properties are peculiar to many local problems, which are generated upon decomposition of a general problem of system analysis. Obviously, solving these problems requires system thinking and system research.

It follows that to solve system problems at different stages of the life cycle of multilevel hierarchical complex systems or to conduct multilevel, multipurpose, multiple-factor problem situation research, it is necessary to create an adequate multilevel hierarchical organizational structure at every level of which there must be groups of system analysts who are able to solve complicated practical problems and analyze problem situations. A possible structure of such organizational systems will be described in the following sections.

1.3 System Analysis as an Applied Scientific Methodology

The history of the formation and evolution of system research and system analysis is closely connected with the problems and tendencies of a permanently increasing complexity of different natural and artificial systems, for example, technical, biological, social, economic, etc., that people encounter in different spheres of activities. The characteristic feature of such systems is their complexity, nonformalizability of a number of processes, occurrence in a system, uncertainty of external conditions, incompleteness, inaccuracy, fuzziness of the source information, and the possibility of occurrence of abnormal, critical, or extreme situations in the process of operation.

The emergence of large, extremely complicated systems typical of modern society vividly demonstrates these tendencies of increasing complexity.

On the one hand, these are organizational, nonphysical systems, for example, social and economic systems, on the other hand, large physical systems, for example, space, ecological, power, telecommunications, transport, etc. As a result, the complexity of analysis and control of such systems has greatly increased. This has necessitated the development of methods that enable one to perform effective planning and design in complicated situations, where no single traditional discipline can take into consideration all factors [144].

Modern scientific disciplines are traditionally divided into two large groups: the first group concerns itself with specified types of objects and objective laws describing their evolution and vital activity; the second group concerns itself with the principles and laws of relationships between objects. The first group includes those disciplines that deal with a certain type of object, for example, physical, chemical, biological, political, economic, social, etc., irrespective of their relationships. The second group includes those disciplines that investigate relationships between objects; the type of object may be arbitrary. This group includes such disciplines as cybernetics, the theory of control, decision-making theory, etc. These groups may be conditionally considered "orthogonal".

System analysis embraces both groups and is interdisciplinary by nature. It is predetermined by the following factors. First, the system principles, equipment, and methodology are applied to the investigation of problems typical for almost all disciplines of the first group. Second, system analysis has the ability to examine properties and explore characteristics of relationships between different objects in different conditions and situations, which is typical for the second group of disciplines. Third, system analysis allows one to consider an object under investigation as a whole, while at the same time providing for analysis of both the properties of elements of an object and the properties of relationships between these elements.

These features show that system analysis is a universal scientific methodology, which involves exploring both the characteristics of objects and relationships between them as a single whole. The formation and evolution of system analysis practice and theory confirms the correctness of the words of G. Klir [54]: "The main thing in the evolution of science in the second half of the twentieth century is the transfer from one-dimensional science to two-dimensional science, in which the science of systems, based mainly on relationships, is included as a second dimension". Surely, on the assumption of this definition the system analysis may be considered, as a two-dimensional science methodology.

The notion of one- and two-dimensional science methodologies is based on two fundamental criteria: the commonness of element types and the commonness of relationship types. It is presumed that these criteria form a mutually orthogonal coordinate system.

An example of the above-mentioned classification by the first criterion is the traditional division of science and engineering into disciplines, specialties, and branches. Every discipline (specialty, branch) deals with certain elements (physical, chemical, biological, political, economic, etc.). At the same time, relationship types are not registered, and relationships between elements of the same type may be different.

The second criterion leads to an absolutely different classification, in accordance with which a certain relationship type is set, while the type of element is not fixed. At the same time, the term "relationship" is used in the widest sense, including a whole set of related notions: organization, structure, interconnection, dependence, correlation, restriction, etc. The given classification is defined essentially by the results of processing information about relationships of a certain type for different object classes.

Examples of classification by the second criterion are different methods of grouping of knowledge sections accepted in cybernetics and based on the commonness of formalized elemental relationship descriptions, classification of mathematical programming problems, etc.

Current system research includes a wide complex of scientific, technical, and organizational methods and special computational approaches to system problem solving. Representing in this way a certain field of modern science and engineering, system research, as a form of scientific and technical activities, undoubtedly need to be theoretically well grounded.

It is quite natural that such a complicated and many-sided phenomenon as system research needs to establish its place among modern scientific methodologies.

The modern structure of scientific methodologies is not presented as a homogeneous knowledge unity. The rapid development of 20th-century science, the increasing complexity of its structure, the considerable growth in the role of theoretical and abstract thinking in it, the wide mathematization and formalization of modern science, and other factors were accompanied by an intensive process of differentiation of methodological research tools along with the development of general philosophical and methodological principles of cognition. This situation in the methodology of science makes it possible to distinguish its four fundamental levels [54]:

1. *The level of philosophic methodology* is the analysis of general principles of cognition and the categorical basis of science as a whole. This sphere of methodology represents a branch of philosophical knowledge and is developed by special philosophical methods.
2. *The level of general scientific methodology* is the development of the general scientific concepts of cognition and methodological theories (logic of science developed on the basis of mathematical logic tools, etc.). The general scientific nature of methodological concepts, in essence, refers to their interdisciplinary nature, i.e., their applicability for different areas of science, at interfaces between traditional disciplines and the principal possibility of transferring tools and methods of these concepts from one scientific knowledge area to another. As long as general scientific method-

ological concepts do not pretend to solve world view problems, or general philosophic problems, they are developed in the area of nonphilosophic knowledge, namely, in the context of modern logic and methodology of science.

3. *The level of specific scientific methodology* entails the analysis of the methods, principles, and procedures of development used in special scientific disciplines. The main task of this level of methodology is the revelation and description of the totality of the methodological tools and principles that are specific to some of these disciplines—physics, biology, chemistry, psychology, sociology, etc.
4. *The level of methodology of applied research* is the description of methods of obtaining initial information, experimental conditions, inaccurate accounting, methods of experimental data processing, etc. At this level methodological knowledge reduces to the requirements, regulations, and practical equipment in using given scientific results.

The introduction of different levels of methodology allows us to characterize system analysis as a general scientific interdisciplinary methodological approach. According to this classification, system analysis may be placed on the second level of general science methodology. Thus, system analysis methodology has a greater generality than methodological assertions formulated and accepted in specified areas of scientific knowledge.

The differentiation of methodological levels allows one not only to distinguish different types of methodological analysis, but also determine the interconnection between them. In particular, it presumes that philosophic methodology has the fundamental significance for any forms of methodological knowledge. The important consequence of such a representation is the influence of more general methodological levels on more special ones. Thus, for example, general scientific methodologies and system analysis, in particular, have an unquestionable influence on the methodological knowledge formed at the level of specific scientific methodology and at the level of research methods and apparatus.

Thus, system analysis, as a general scientific interdisciplinary methodology, is characterized by the following features: it is a coherent totality of methodological principles, procedures, and methods of research oriented toward the dialectical cognition of general properties and features of interdisciplinary system problems and methods of solving them in different areas of human practical activities.

System analysis is the next step in the development of modern science, namely, it involves the transfer from one-dimensional to multidimensional science, in which not only a system by itself is entirely explored, but also the initial conditions of its creation and conditions of its functioning and control in normal and abnormal situations. Therefore, there arises the necessity, first, for multidimensional system analysis of a system's functioning conditions, and, second, for multidimensional system analysis of an entire system proper.

Let us note one more principal difference between system analysis and traditional one-dimensional disciplines. This difference essentially boils down to the somewhat different investigation purposes. Thus, the heart of traditional disciplines is the axiomatic theory, based on a definite system of assumptions (axioms, postulates, theories). The main purpose of investigation is to prove that the formulated problem is solvable within the framework of the accepted theories, axioms, and constraints and to justify that these methods for solving the problem have definite properties (convergence, accuracy, correctness, existence of a solution, etc.). But the question of the existence of a practical interpretation of the formulated theory and the practical realizability of the introduced theoretical assumptions and constraints are not examined by these disciplines. "Applied" specialists search for answers to these questions and see their purpose finding practical interpretations of axiomatic theories and creating methodological tools for a specified usage of these theories in practice.

Proceeding from these positions, system analysis can be interpreted as an applied scientific methodology whose tools are based not so much on axioms, as on heuristic methods, equipment, and algorithms. System analysis, unlike other axiomatic disciplines, sets a quite opposite research goal—there is a practical problem with known physical, technological, economic, and other constraints. A system of means and methods must be found that will allow one to solve it with an acceptable level of accuracy within the allowed time with acceptable expenses of all types of resources. System analysis allows one to investigate the whole variety of properties of the object, from the point of view of the goal that has been set, as a single whole. At the same time, axiomatic disciplines allow one to investigate certain properties of an object from the point of view of only these properties, without taking into account the influence of other properties. For example, in the investigation of strength properties and characteristics of an object, such important properties as technological effectiveness, technoeconomic efficiency, and others are not taken into account. And, conversely, in the investigation of technoeconomic efficiency strength properties are ignored.

The second difference between axiomatic disciplines and applied system analysis is that their initial information platforms are different. In axiomatic disciplines it is expected that all initial information needed for the formalized description of an object under investigation will be assigned. In particular, the following is assumed to be known:

- Parameters characterizing the properties of the object being investigated.
- Dependence of these parameters on the value of external actions and control.
- Criteria by which it is possible to evaluate the degree to which goals have been achieved.

- Admissions and constraints on parameters and criteria.
- Other additional information determining the properties and features of the object under investigation.

Here the initial information is sufficient for the development of the problem description and mathematical statement of the problem of an object under investigation. Furthermore, a full mathematical problem statement with all necessary dependencies, criteria, constraints, and assumptions is often given as the initial information.

The amount and nature of the initial information specific to applied problems of system analysis are principally different from that of axiomatic disciplines. First of all, the initial information about an object under investigation is characterized by incompleteness, inaccuracy, fuzziness, and contradictoriness. At the same time, the most important information is usually missing, including information about criteria of evaluation of the goal achievement rate. Furthermore, in a number of cases, in designing particularly complex systems, designers set goals contradictory to the assigned technical, operation, and technological indices and characteristics. It is necessary to find a rational compromise; however, the conditions and criteria of a compromise should be selected and justified in the process of system investigation. The given list of features of the initial informational support of applied system problems may be expanded as needed. They all eventually lead to the conclusion that the problems of system analysis for real objects are principally nonformalizable because they contain essentially nonformalizable procedures, for example, criteria selection, selection of rational compromise conditions, etc. The problems of system analysis for real objects are solved on the basis of a rational combination of the capabilities of an expert and the possibilities of application of objective knowledge and scientific tools, above all computing machinery, heuristic equipment and procedures, and mathematical methods and algorithms.

Thus, with respect to traditional scientific disciplines system analysis may be considered as the next step in the evolution of modern scientific methodology, namely, the transfer from one-dimensional to a multidimensional science in which not only a system by itself is being thoroughly explored, but also the conditions of its creation and functioning and conditions of a system control in normal, abnormal, critical, and extreme situations, taking into account risk factors as well as incompleteness, inaccuracy, and uncertainty of information.

Taking into account the above-indicated features of system analysis, let us present its problem description contansive statement as a scientific methodology.

System analysis is an applied scientific methodology based on a diverse system of organized, structurally interconnected, and functionally interacting heuristic procedures, as well as methodological equipment, mathematical methods, and algorithmic, programmatic, and computing means that ensures the formation of complete interdisciplinary knowledge about an object under investigation, as the totality of various interconnected processes for subsequent decision making with regard to its further development and behavior, taking into account many conflicting criteria and goals, the presence of risk factors, as well as incompleteness and uncertainty of information.

Note that the classical definition of a scientific discipline cannot be applied to system analysis because of the impossibility of even more or less precisely describing an suite of methods and tools being applied, because the scientific approach of complex objects investigation is principally interdisciplinary and multidimensional.

The conditions of the transition of civilization to an information-oriented society and the penetration of new information technologies into different spheres of life and human activities change qualitatively the goals, tasks, and content of education as well as the role, place, and significance of individual disciplines in the general educational system. The practical task of educating a researcher capable of adapting to the fast changing conditions and norms of modern society and focused on the perfection of this society, determines the necessity of revising the educational paradigm and changing many aspects of the cognition process. Worldwide, the tendency toward radical changes in this process is clearly observed. This tendency consists in shifting to system thinking with the purpose of revealing deep-seated effects, interconnections, and influences of elements of the surrounding world. Therefore, the formation and development of the theoretical basis of system analysis as an applied scientific methodology should be coordinated with the goals, tasks, expected results, and development strategy of educational systems, as well as with the general trends in modern society. The necessity of such approach results from the general requirements of the anticipated educational system and, in particular, the requirements of mastering the scientific methodology and practical apparatus of system analysis application for the revealing the essence and determining the approaches to solving political, economic, social, scientific, technical, and other practically important problems.

2 Basic Notions of System Analysis

System analysis belongs to the category of scientific phenomena that cause numerous debates on a great number of problems. Nevertheless, till now there have been no generally recognized solutions to some of them. Still, you can find fundamental variances in the interpretation of many basic and practically important notions and definitions, such as subject of investigation/analysis, system, systemness, complexity, system approach, and system problem in particular. Thus, we consider it is necessary to analyze well-known interpretations of the concepts mentioned above and offer and justify our version of their definitions.

2.1 Objects of System Analysis

While solving real practical problems of system analysis one often uses such important notions as objects and models. They have sufficiently exact and unique definitions in traditional branches of science. Still, in system analysis these generally recognized notions have been significantly developed and expanded. This process is based on the modern specificity of problems and toolkit of system analysis, features, and characteristics of the subjects of system investigation, the conditions and features of formalization, and the solution of system problems.

Let us consider some factors and suppositions that define the diversity in approaches of different analysts to the selection of objects to investigate.Thus, following the approach of Klir [54] it is necessary to admit that in one's daily life one interacts with different objects from one's surroundings, but this interaction is usually limited by a number of representative properties that characterize the features of an object. As the necessity in investigating the object rises, such an interaction becomes more varied, but it is always nothing more than the limits of human perception, one's ability to evaluate, select, observe, and perform other actions.

We consider the following definition of an object a reasonable one. The **object of investigation** is a part of the world that is being investigated/analyzed, chosen/selected, formed, created, used, etc., by a person (specialist) as a single whole during a finite time period starting from when certain goals have been attained.

Let us now analyze the classification of objects following the approach of Klir [54]. Objects can be categorized as either **material** or **abstract**. Material objects can be further divided into natural, which do not depend on a person, and artificial, or man-made. Among natural objects are the Universe and the Solar System; flora and fauna of a particular region, continent, or the entire planet; mountains, caves, ravines, canyons, or other geological formations; herds of wild animals, cells of different organisms, etc. Examples of artificial objects are the technosphere of the planet, cities, airports, hospitals, and all the infrastructure of a given person's life style and activities. Abstract objects are man-made. But they are not material objects in a direct sense. Ideas, legends, folk songs, dances, and other abstract works can be given as examples of such objects. They can be passed on in oral or visual form from generation to generation, but they can die along with their authors and performers. Abstract objects may be embodied in such forms of material objects as books, magnetic and visual recording media, etc. In addition, these material objects are only the form in which abstract objects are embodied, the essence of which is the content of scientific, artistic, legal, and other types of information.

The properties of an object are its most important features. In many cases objects may have a practically unlimited number of properties, each of which can be conceived and studied, and as a result an object can be examined and evaluated with respect to a certain finite number of properties. Still, it is impossible to examine each object in its entirety, to investigate all its properties and features, including its aging and decomposition mechanisms. This statement follows directly from the first theorem of Godel [103]. Thus it is unrealistic to set oneself the goal of studying all properties and features of a material object. It is necessary to single out and study only those that refer to the set purpose or problem under consideration. It means that one should pick a limited number of properties, as a rule, a few that describe the given object with sufficient completeness from the point of view of the tasks set by an analyst. After that it is necessary to define for each property a limited class of indices, which are specifed by the vector of abstract variables. This vector is an image of the corresponding property.

Complex system as an object of investigation. A complex system is an area of system research that is chosen, formed, or created by a person as a single whole from the point of view of the goal he or she aims to attain. This system may be a material object or some model of a material or abstract object of investigation. In the first case, a system is a real material object. Its technological or organizational structure is a hierarchical, multilevel system of interconnected/interrelated, different- or single-type functional elements that are concentrated in a certain space. The government in different countries of the world may be an example of a real complex system with a hierarchical, multilevel organizational structure. Similar complex systems are widespread in industrial, economic, social, and other spheres. For example, real complex systems are a multiprofile industrial complex, railway network, telecommunications network, and other similar material objects.

The model of such objects is built on the basis of studying, describing, and formalizing processes in accordance with the formulated purposes and tasks, set indices, and parameters that define the properties of objects under investigation.

It is necessary to underline that **modeling** is one of the most universal methods of studying different processes and phenomena. Numerous methods and techniques of modeling are widely used in scientific research and engineering practice. At the same time we distinguish physical and mathematical modeling [15].

In *physical modeling*, a model repeats the behavior of the object under investigation preserving its physical character. Some similarity relationships between an object under investigation and the model must be maintained. They arise from the laws of physical character of phenomena and ensure the possibility of using the information obtained as a result of modeling for the evaluation of possibilities, properties, and characteristics of the object under investigation.

Physical modeling has a rather limited scope. Certainly, *mathematical modeling* has much broader applications. It is a method of examining objects based on studying phenomena having different physical contents but described by the same mathematical relationships. At the same time, similarity relationships between variables of a mathematical model and the most important properties and characteristics of the object under investigation must be preserved.

In practice different types of mathematical models are often used. They are based on integration of the possibilities of modern mathematics and computer engineering, e.g., graphic or simulation models. Thus, a *graphical model* deals with the system of interrelated layouts and images reflecting real interdependencies typical for the object under investigation. A *simulation model* is a system of interrelated computer programs for simulating an object's behavior.

The definition of a system given by Klir as the form of a real object model and the accepted approach to describing the object on the basis of multiple properties and multiple bases is of practical interest. It opens up possibilities for solving a wide class of real system problems. At the same time, real problems of managing complex organizational and technical objects do not always fit in the structure of the model conception given above. In particular, a multilevel hierarchical management system necessitates a rational distribution of functions between hierarchy levels and at each level. As a consequence, the hierarchical structure of purposes and tasks of control is formed. It calls for building such a hierarchical model on the basis of which it is possible to build a rational structure of control and ensure achievement of the set goals. Hence, for such an object a model should reflect not only its properties, but also its structure.

Let us introduce some definitions of a system, considering it first as a model of an object of system research and secondly as a real material object having a definite purpose. We shall characterize the system in the form of a model of an object of system research by the following series of multilevel enclosures. The characteristics of each enclosure are defined by the purposes of system research [119].

A **system** is an ordered set of structurally interrelated and functionally interdependent elements.

A **complex system** is an ordered set of structurally interrelated and functionally interdependent systems.

A **grand system** is an ordered set of structurally interrelated and functionally interdependent complex systems.

A **giant system** is an ordered set of structurally interrelated and functionally interdependent grand systems.

A **global system** is an ordered set of structurally interrelated and functionally interdependent giant systems.

A **global super system** is an ordered set of structurally interrelated and functionally interdependent global systems.

A system as a material object with a definite purpose may be characterized by the following series of multilevel enclosures:

An **engineering system** is a material, integral object designed for carrying out a certain function in specified conditions, technically realized on the basis of a nomenclature-ordered, finite set of structurally interconnected, functionally interdependent, and technologically interacting functional elements.

A **complex engineering system** is a material, integral object designed for carrying out a finite set of functions in specified conditions, technically realized on the basis of a nomenclature-ordered, finite set of structurally interconnected, functionally interdependent, and technically interacting engineering systems.

A **grand engineering system** is a material, spatially limited object designed for carrying out a certain set of functions in specified conditions, technically realized on the basis of a nomenclature-ordered, bounded set of spaced or concentrated complex engineering systems functionally interdependent, structurally and organizationally interrelated, and technically and resource interacting.

A **giant engineering system** is a material, spatially limited object designed for carrying out a certain set of functions in specified conditions, technically realized on the basis of a purpose and/or nomenclature, bounded by a set of grand/large and/or complex engineering systems of spaced and/or concentrated within the bounds of one region functionally and/or resource interdependent and technologically and organizationally interacting.

A **global engineering system** is a material object designed for carrying out practically needed functions in a certain sphere of activity under changing conditions, technically realized on the basis of an unlimited set of globally spaced giant, grand/large, and/or complex engineering systems coordinated by purposes, nomenclature of possibilities, and other indicators of activities, functionally and/or resource interdependent, and technologically and organizationally interacting.

A **global engineering supersystem** is a material object designed for carrying out practical functions under changing conditions, technically realized on the basis of an unlimited set of functionally and resource-interdependent and technologically and/or organizationally interacting global engineering systems.

A **global supersystem** is a material object designed for carrying out practical functions under changing conditions, technically realized on the basis of an unlimited set of global engineering, organizational, information, and other supersystems that are functionally and resource interdependent and technologically and/or organizationally interacting.

An infinite variety of complex multilevel hierarchical engineering and organizational systems and an infinite set of their functions determine the complexity of the formal presentation of system analysis problems. That is why it is necessary to formalize basic notions and describe assumptions and limitations in terms of system analysis problems. First, let us introduce the notion of a complex hierarchical system on the basis of the complex system notion given above.

A **complex hierarchical system** is an integral object, formed of functionally polytypic systems, structurally interrelated by hierarchical subordination, and functionally united in the interests of performing specified tasks in certain conditions.

Modern industrial objects, space systems of communication, navigation, remote probing, advanced systems of regions, corporations, multiprofile firm management, etc. are examples of complex hierarchical systems. It is necessary in practical terms to analyze such systems. It cannot be reduced only to the determination of types of elements or their relations. The presence of a hierarchical structure both in its own topology and in control systems is of principal importance.

First, such systems are characterized both by a variety of types of elements (different classes of physical, chemical, and mechanical types of elements and other) and by a variety of types of relations (starting with technological production interrelation and ending with information exchange and interaction).

Second, for such groups of objects the multilevel hierarchical structure of modern complex and grand/large systems is characterized by the following properties:

1. The significance and capabilities of functional elements (FEs) are different at different hierarchical levels.
2. The FEs of each level of the hierarchy act freely within the limits specified before or in the process of an object operation.

3. Higher-level FEs have the priority of actions or the right to interfere in the activities of lower-level FEs according to the actual execution of its functions.

Owing to the above-mentioned properties a complex hierarchical system has a number of peculiarities. These particularities determine both the main problems of investigation and the concrete purposes of analysis of its structure and functions, or **the structural-functional analysis** (SFA).

The most important feature of the objects under consideration is that many procedures of formation and analysis of a hierarchical structure are principally nonformalizable and their realization is of a subjective nature. First is the general structure of hierarchy. Such procedures as choosing the number of levels (strata, echelons, layers) in a hierarchy; separating elements at each level; choosing the description of elements at each level; choosing the apparatus for describing interconnections in a hierarchical structure—all this depends completely on decision maker (DM). In practice, when designing real complex objects such as space systems that have different purposes the chief designer determines the general organizational structure of an object [its division into first-level subsystems and identification of two upper echelons (object and subsystems) in accordance with the above terminology]. The chief engineer in charge of the first-level subsystem divides it into second-level subsystems and so on. Such an approach allows one to carry out system integration of organizational design problems: project planning, formation of developer teams, coordination of tasks, etc.

Let us introduce the notion of systems in terms of their directedness. A **purposeful system** is a system designed to achieve strictly defined purposes. It has a distinctly defined purpose for the specified conditions and is characterized by a set of purpose nomenclature limitations and by a specified range of admissible changes in operation conditions. Examples of such systems are very simple semiautomatic devices and automatic devices with a wired control program or very simple machinery. To this category belong automatic machine tools for manufacturing only some types of mechanical parts, domestic and industrial refrigerators, autonomous heating systems with semiautomatic temperature maintenance in a definite range, and motor vehicles for various purposes.

Principally different is the class of goal-seeking systems. A **goal-seeking** system is a system possessing the following properties:

1. It perceives the requirements of the environment with regard to the system and forms goals for fulfilling requirements under highly variable conditions.
2. It defines alternatives for all changes in the environment and rationally selects alternatives for achieving the goals under existing conditions.

The most important feature of goal-seeking systems is their ability to dynamically change the goals and methods of their achievement according to the changing situation. Goal-seeking systems principally differ from purposeful ones by their flexibility, dynamics, and ability to react to changes in the environment by adapting demands, goals, and actions in the existing conditions. Systems of this class can change their functions, properties, and even structure, both of FEs and of the system as a whole. The principal feature of goal-seeking systems is that they possess an intellect—natural, artificial, or combined. The majority of well-known goal-seeking systems belong to the class of organizational technical or expert systems. The main elements of such systems are human operators and various technical facilities for decision making with an intellectual component. Flexible automated production, capable of changing its nomenclature and output and the dispatcher service of large airports or seaports capable of attending to hundreds of airplanes or ships at the same time under changing weather conditions are examples of such systems. The actions of personnel at power plants, large metallurgical, mining, chemical, and other industries in critical and emergency conditions illustrate vividly the ability of goal-seeking systems to adapt to changing conditions.

A **structured system** is an object under investigation, presented in the form of a certain hierarchical structure of functional elements accounting for interconnections, interdependencies, and interactions between them. A **structured initial system** is a set of initial systems, systems of data, or generating systems with a common parameter set. The systems forming a structured system are usually called its elements. Some of their variables are shared. Shared variables are usually referred to as coupling variables. They present interactions between elements of a structured system.

Shared variables are used in both the investigation and design of complex systems. In designing a complex system, it is necessary to reach a rational compromise of opposing system requirements while choosing its structure and elements. However, the following problem arises: how does one ensure the fulfillment of requirements in real operating conditions on the basis of a rational control of functioning of a real object having a definite structure and containing concrete elements? The first stage in design is the determination of a so-called **generating system**. This presents the idea of the task that the given system should perform. Generally, this task is the transformation of the state of the corresponding input variables into the state of output variables. In this way, the obtained generating system is always a purposeful system. While compiling the initial information about the object of analysis, it is necessary to realize that all real objects are open systems.

Open systems are systems that deal with the exchange of different resources, including energy and information, with the environment. For this reason it is always important to determine the boundaries of the open system. This means that you must find out where the object ends and the environment begins. Here, the goal of the analysis is the determining factor. For example, if the goal of the system analysis is to determine the source of energy loss in a system, then the entire power energy network of this system must be analyzed. And if the goal of the analysis is to create an economical electrogenerator, then all the functional elements outside it should be the environment. But these examples are rather exceptions to the rule. In the majority of real problems of system analysis it is quite difficult to determine the physical boundaries of objects. The conflicting goals point to the different boundaries and the orientation toward the supremum of these boundaries leads not only to their excessive expansion, but, what is more important, toward a considerable complication of the problem. In addition, in a number of cases, in the early phases of development of a unique object, there is no information about the numerical data that describe the influence of different functional elements on the level of goal achievement and physical boundaries of the object. Moreover, the object under investigation cannot be said to be wholly determined. This happens because each functional element is, as a rule, described by a definite set of indicators that characterize the definite set of properties. But the influence of different properties on different goals of the system problem fundamentally differs. For this reason it is perfectly natural for analysts to try to exclude the properties of those functional elements that have only a slight influence. But in such cases the question arises as to what should be the criteria for determing the slightness of influence. The answer is not so simple, and it becomes more complicated when it is necessary to analyze the goal-seeking systems.

For the set structured system there exists another system, connected with the first one and determined by all the variables of its elements. This system is analyzed, as the **complete system**, that is, the system in the form of some combination of all included variables. From this point of view elements of every structured system are interpreted as subsystems of the corresponding complete system, and the complete system is a supersystem with respect to its elements. Thus, structured systems become the complete systems in the form of different subsystems. It is clear that the status of a complete system or subsystem is ambiguous. For example, a system in one case is an element of a structured system, and in another case it is a complete system with subsystems that form a structured system. Analogical ambiguity is typical for initial systems, data systems, or generating systems since it is possible to represent any complete system as a multiple enclosure of a structured system. For example, a complex structured system may have elements that, for their part, are complex structured systems, and so forth, down to the elements consisting of simple variables.

The necessity of representing a complete complex system as a set of its subsystems is determined by many factors. One of the main factors is its complexity, which in turn is caused by the visibility of the system. Another factor stems from observations and scope. When the parameters and characteristics of a complex system depend on time, it is often technically impossible or inexpedient to examine simultaneously all the variables that comprise the goal of investigation. In this case, it is possible to collect data only partially. In other cases, a researcher must use indirect data, which are collected by different indirect methods, but they are only a part of the variables required for work.

2.2 Properties of System Problems and System Methodology

Experience shows that the realization of new ideas and technical solutions depends on many factors and conditions, of which the following deserve special mention:

- The abilities of the specialist and possibilities of a given science to formalize and solve system design problems;
- Capabilities of industrial technologies for realizing a project and creating a product with the required quality indices.

It is necessary to point out the special significance of the problem formalization procedure for realizing innovative ideas and making technical decisions, and also for designing new equipment, which fundamentally has no analogs or prototypes. Such problems are distinguished by many contradictions and uncertainties, of which the major ones are:

- Ambiguity and contradictoriness of requirements of a product.
- Contradictoriness of goals and ambiguity of conditions of product application.
- Uncertainty and unpredictability of competitors' possible actions.
- Unboundedness and unpredictability of a set of risk situations at various stages of a product's life cycle.

Under these conditions and on the basis of various incomplete empirical, experimental, casual, and other initial information the developer must formalize and accomplish a product design task, including statement and justification of the product creation goal. The results of this problem solution are to prove the practical necessity, technological potential, and economic expediency of the manufacture of the developed product. Under market economic conditions the degrees and levels of risk at each stage of the product life cycle should also be estimated, and the decisions taken should be such that, accounting for all factors, the risk will be permissible. Similarly, under conditions of uncertainty, the procedures of formalization and solving of many other important system problems are carried out. In particular, the problems of analysis and control are solved when abnormal modes of complex engineering occur.

It is obvious that, for the best understanding of system problems, it is first necessary to establish basic concepts. Let us introduce the following definition.

The **system problem** is a problem involving the analysis of a certain set of properties of an object under investigation from the unified position achieving the assigned goals of an object under the real operating conditions.

System problems are infinitely diverse due to a large number and variety of goals, as well as to the complexity and uniqueness of an object under investigation, all of which is related to a diverse structure of its elements and their relationships (a real or designed product, a structure, production; a real or predicted situation, technology, etc.), the volume and quality of the initial information, the scope and level of requirements given the quality of the problem solution, and a number of other factors.

Nevertheless, all system problems may be united in a special class, as they have a number of common properties. In this class it is possible to distinguish a finite number of problem types. The difference in the problems can be characterized by a number of factors, in particular by:

1. The structure of ordering and interrelations between sets of the initial data of the problem and sets of final results of its solution.
2. Coordination and ordering of the requirements for the specified sets.
3. Level of computing complexity of the problem.
4. Degree of structurability and potential formalizability level of the problem.

The first two factors are characteristic both of system problems and problems of other classes studied in various axiomatic theories (theory of control, game theory, theory of decision making, etc.).

Therefore, following Klir [54], it is necessary to pay attention to the role and place of last two factors in system research. These factors are specific attributes of system problems and in many respects determine their basic properties and features. In particular, the credibility of real system problem formalization depends almost entirely on four factors, and the credibility of the calculations and results of the solution is determined by the credibility of problem formalization, the volume and credibility of the initial information, and the level of computational complexity of the formalized problem.

The procedure for solving formalizable problems is carried out in two steps:
Step 1: A problem is formalized and the appropriate materials are prepared for calculations.
Step 2: Calculations are carried out on a computer or in a computer network.

The above circumstance determines the necessity of the rational use of both human abilities and potential for modern computing methods and facilities.

System problems may arise in two basic contexts: upon investigating and upon designing complex systems. The objective of **investigating** systems is the accumulation of knowledge about the state of their internal elements and connections between investigated systems and the environment. The objective of complex system **design** is their creation based on the use of accumulated knowledge for attaining new properties and new goals of these systems when they interact with the environment.

The most important feature of system design is that parametrically invariant restrictions on some variables are determined by the user. Entirely different is the situation with system investigation, where these restrictions are unknown, and the task is to characterize them adequately in line with the concrete goal of investigation.

At the design phase, restrictions are determined explicitly either in the language of a specific generating system, as a rule a directional one, or in the data system language. In the first case, the design problem is reduced to the definition of sets of the structured systems that meet the assigned requirements. In the second case, it is necessary to determine certain generating systems that adequately describe the restrictions contained in the data.

In system analysis problems, as a rule, only some requirements of given properties and certain factors are known, and they must be taken into account to achieve these properties (the maximum speed, carrying capacity of a car, time of acceleration up to a certain speed, maximum fuel consumption, climatic conditions, etc.). The criteria by which the quality of the system performance is estimated, as a rule, are unknown and nonformalizable. Thus, in designing a new aircraft, the designer, without knowing its shape, wants it to possess the maximum strength, reliability, durability, performance, and minimum fuel consumption and be cheap enough. It is clear that these conditions, in principle, cannot be simultaneously met.

In solving system analysis problems the level and quality of mathematical and informational support are of great importance. The volume of useful information should be determined taking into account the necessity of decision making in the process of control at every stage of the system life cycle. The researcher faces problems both in the case of the lack of information and in the case of its redundancy. Therefore, the problem of informational support, in a broad sense, is decisive at the early stages of the system life cycle, in particular at the stage of development, testing, and revision of a production prototype. At these early phases, most problems are nonformalizable, and therefore these problems are very complicated and important. The importance of finding of a correct solution is so high because mistakes made at the early phases may lead to costs that are several times higher than those incurred later.

Properties and features of system problems. Let us examine the specificity of system analysis problems using as an example the development of a product of new engineering. To make properties and features of system problems more comprehensible, let us introduce a number of restrictions and assumptions that, in practice, are not always fulfilled and will be further excluded. Let us assume that the following information about the problem is known a priori:

1. The problem is formalized on the basis of the initial data, which are presented in one of the following forms: 1.1. Technical proposal formulated by the developer and subject to a competition. 1.2. Requirements specification stated and approved by customer.
2. The goal of the problem solution is the creation of a material object in the form of a complex hierarchical system, according to customer requirements.

From the above it follows that there are two basic principles of creation and operation of complex engineering objects:

- The principle of work division.
- The principle of functional division.

Work division consists in the following: the customer develops the requirements specification and the developer creates the product design. In a number of cases the functions of the developer and the customer can demand more labor and time inputs. For example, the developer may carry out all production functions at all stages of a product life cycle: design; development, manufacture, and operation of a product, and recycling of a product the life cycle of which is ower. In this case the customer may carry out all control functions: control over the quality of materials, development, and operation of a product at the corresponding stages of its life cycle. The division of functions between the components of a product is due to the product's being created as a complex hierarchical system, each level of which carries out only functions defined in advance. During operation, such a realization of a product allows one to simplify control and improve operability and efficiency. It should be emphasized that the terms "developer" and "customer" are general notions of organizations, which perform the above-indicated functions.

Contensive statement of a system problem. The basic data for the development of a new engineering product are **known**. These data presented in the form of the requirements specification characterize:

- The purpose, operating conditions, and dimensions of a product.
- The requirements for technical, technological, constructive, operational, economic, and other indices of product quality.

A product **must** be developed that completely satisfies the conditions and restrictions of the requirements specification.

Main procedures of a problem solution. Let us examine the contents of the main procedures, which for clarity will be presented in the form of a requirements sequence:

1. To determine a number of hierarchical levels of a complex system.
2. To determine a rational number of functional elements for each hierarchical level of a complex system.
3. To state the requirements of each hierarchical level of a complex system on the basis of restrictions and requirements of the specifications.
4. To state the requirements for each functional element of each hierarchical level on the basis of restrictions and requirements for this level.
5. To choose the types of functional elements of each hierarchical level in conformity with the restrictions and requirements of each functional element.
6. To ensure the system coordination of all functional elements with respect to goals, tasks, and expected results for all hierarchical levels.

7. On the basis of the results of procedures 1 to 5, to carry out an expert estimation of the possibility of practical realization of the assigned restrictions and requirements for a product.
8. Using the results of procedures 1 to 7, to carry out an expert estimation of the practical necessity, technological possibility, and technical-economic efficiency of production of the developed product, taking into account restrictions and requirements of the specifications and the condition and dynamics of competing products.
9. To produce results of the expert estimation in order to present the generalized estimations of advantages and disadvantages of the developed product.
10. To develop and establish the basis for conclusions regarding the expediency of further actions with respect to the developed product project.
11. To develop and establish recommendations for the customer about possible variants of further actions with the developed product.
12. To realize the decision taken by the customer about measures and actions regarding the developed product.

Analysis of system problem characteristics. The content of the basic procedures makes it possible to reveal a number of important characteristics of a system problem. The main feature is that procedures governing a problem's solution are executed by different performers whose purposes differ considerably. From this it follows that the results of the problem solution should reflect the basic aspects of the performers' activities. The results of the developer's activities, which determine the properties and capabilities of a developed product, are of major importance. The developer must execute procedures 1 to 5 to achieve the primary goal determined by procedure 6. The complexity of these procedures is connected with the fact that the developer is forced to execute them under conditions of incompleteness and uncertainty of the initial information, since in the requirements specification the requirements are set only for a product as a whole. In such conditions the problems are characterized by a conceptual uncertainty and are nonformalizable [120]. Therefore, the choice of the number of hierarchical levels and the quantity and types of each functional element of each hierarchical level depend on the skills, experience, knowledge, intuition, prediction of the developer. From this it follows that, when the same type of product is developed by different developers, the final product may differ in many aspects. The airplanes of the chief designers Antonov, Tupolev, Yakovlev are typical examples.

The indicated properties and characteristics distinguish in principle the procedures for solving a system problem from those used to solve routine determinate problems, in which the numerical values of all initial data are known a priori. Therefore, methodological approaches and mathematical methods of estimation of accuracy and credibility of routine determinate tasks cannot be applied directly to system problems, which are characterized by incompleteness and uncertainty of the initial information.

It is rational to apply the methods of expert estimation to such problems [204]. The expert estimation of advantages and disadvantages of a developed product is performed by procedures 7 and 8. It should be pointed out that the credibility of expert estimation under the conditions of incompleteness and uncertainty of the initial information depends on many factors, which are considered in detail in Chap. 9. Therefore, besides the expert estimation of the advantages and disadvantages of a product, it is necessary to develop recommendations about the possible variants of the customer's actions on the basis of which he forms and makes decisions about the further measures and actions regarding the developed product.

The performed analysis shows that the developer of a new technique must choose every functional element of every hierarchical level and form the complex hierarchical system under conditions of incompleteness and noncredibility of the initial information. In such conditions, it is necessary to present the developer's actions 1–6 in the form of the following functionally interdependent and structurally interconnected system problems:

1. The problem of system-coordinated decomposition of a specified product into a complex hierarchical system based on the rational choice of the number of hierarchical levels and functional elements at every level.
2. The problem of system-coordinated decomposition of specified general requirements for a product into the rational requirements for each hierarchical level of the developed system.
3. The problem of system-coordinated decomposition of specified general requirements for each hierarchical level of the developed system into the rational requirements for each functional element of each hierarchical level.
4. The problem of the rational aggregation of functional elemental types chosen for each hierarchical level into a unified system-coordinated structure of each hierarchical level.
5. The problem of the rational aggregation of the formed system-coordinated structures of each hierarchical level into a unified complex multilevel hierarchical system, meeting the specified restrictions and requirements for a product.

In the practice of developing, testing, and finishing real complex engineering systems (CESs), the requirements, criteria, main research, and operating conditions are determined at the corresponding stages of implementation at each phase of the system life cycle. The life cycle of different types of complex systems may differ in variants of phases. In particular, for complex unique engineering products, the life cycle includes the following stages: designing developing a prototype, series of tests, trial operation, full-scale production, and operation. As a rule, the life cycle of the engineering product is terminated by the stage of its recycling.

Attention should be draw to several of the initial conditions and situations. For example, in the CES development process different situations occur, where a number of factors, conditions, and limitations are not completely set or are completely indefinite. Therefore, the problem of the optimum variant searching of the CES structure is not directly reduced to a routine mathematical optimization problem. As a rule, in the process of similar problem solving, it is necessary to include human experience, knowledge, skills, intuition, and foresight.

Thus, there is no exact method or algorithm for searching a rational system structure and its division into hierarchic levels. Problems like this are usually referred to as **weakly structured**. The limit case of weakly structured problems are nonformalizable problems.

Thus, the majority of system analysis problems may be interpreted as problems including nonformalizable and formalizable constituents. In addition, these problems are characterized by a weak structure, algorithmic insolubility, incompleteness, uncertainty, inaccuracy, fuzziness, and contradictoriness of the initial information. Nonformalizable and formalizable parts of such problems are indivisible, and at every phase of the analysis there are both formalizable and nonformalizable parts that are also indivisible.

As a rule, the solution to a formalizable constituent of the problem is determined by the completeness and quality of the general characteristics of a nonformalizable part. The nonformalizable constituents are formed on the basis of human experience and intuition using the principles and methods of expert estimation and also computer mathematics and simulation techniques. These methods may also be used for verification of the credibility and possibile realization of new ideas, hypotheses, suppositions, and foresight in different spheres, for example, in innovative activities.

In this way, in solving many problems of system analysis the principle of the initial problem decomposition into a sequence of more simple ones, with subsequent aggregation of separate problem solution, results in the final solution of the initial general problem. In its turn, each partial problem of system analysis is decomposed into a sequence of simpler problems. The process of decomposition takes place until the obtained partial problem has a formalizable part, which is solvable by known or specially developed accurate and approximate methods and algorithms. Aggregation of the obtained results is carried out in the same manner.

The problems of system analysis possess such properties as multicriterionness, multifactoriness, and multiparametricity. Let us assume that the given properties are described as follows:

- A **criterion** is an indicator or functional allowing one to estimate quantitative or qualitative characteristics of a particular system property.
- A **factor** is an external or internal influence affecting certain properties and complex systems behavior.
- A **parameter** is a quantitative indicator determining certain properties of the system or a degree of the factor influence on its property.

Note that in many real system analysis problems, as a rule, only certain requirements for given properties and certain factors to be taken into account upon obtaining these properties are known. For example, the maximum speed, maximum and minimum fuel consumption, maximum carrying capacity, quantitative values of other transport-vehicle indices, maximum values of humidity, temperature, or other environmental and climatic conditions, and indices of other external conditions and factors are assigned. The criteria by which it is necessary to estimate qualitative characteristics of the system frequently are not formalizable, and they are formed on the basis of knowledge, skills, experience, intuition, and foresight of a decision maker or a group of experts.

Research and design of complex systems are carried out using system analysis methodology, which is an essential tool of a system analyst. Obviously, the methodology, as a system tool, must possess functional and logical completeness and system-coordinated interconnection of techniques, principles, and methods. To achieve these qualities, the methodology must correspond to certain principles, possess certain properties, and use the system approach potential. In addition, the system approach must be meant for solving real system problems. With this purpose it seems expedient to use simultaneously deductive and inductive methods: the first—for understanding the probable result in the analyzed situation, and the second—for realization of expedient actions in such a way that a desired outcome will be achieved with the highest probability.

Let us consider the properties and principles of system methodology. The fundamental properties of system methodology are resultivenes, efficiency and scope.

Resultiveness is the capacity of the methodology, as a user's working tool, to make it possible for the user to obtain a practically acceptable solution to applied system problems in real conditions characterized by incompleteness, inaccuracy, contradictoriness, and uncertainty of the initial information. A practically acceptable solution is a result that satisfies a researcher with regard to the required accuracy, credibility, and validity.

Efficiency is the capacity to achieve the final result in the form of a solution to real system problems in a practically acceptable time frame with practically acceptable computational, financial, and other resource expenses. The significance of this property is determined by the practical necessity of comparing the effect of using the obtained result and expenses for obtaining it. This means that economic, social, ecological, defensive, or other expected effects resulting from the solving of a system problem should be commensurable with the expenditures made to obtain the solution. Thus, the solution will fall into one of two categories with respect to costs: pessimistic—the obtained effect compensates expenditures related to obtaining a solution; or optimistic—the obtained effect considerably exceeds such expenditures.

The **scope** is the capacity of the methodology to be used for solving a wide range of applied problems, which differ considerably in a number of respects, namely, the nature of objects, their application domain, the researcher's informedness level, problem content, and other features. The meaning and importance of this property are obvious from its definition.

We will not deal here with other practically important properties of the methodology, which are more expedient to consider in the process of its practical application for certain classes of system problems. Let us point out only a number of fundamental principles that must be followed in forming a system methodology and its practical realization in the form of an aggregate of concrete approaches, methods, algorithms, application packages, and so on.

1. **Principle of system consistency**: methods, procedures, approaches, algorithms, and application packages must be functionally and structurally interdependent and interconnected in a unified system methodology.
2. **Principle of procedural completeness**: the system methodology must ensure implementation of all procedures from the formalization of a system problem to the verification of the obtained results.
3. **Principle of functional orthogonality**: every procedure in the system methodology is realized in the form of a set of functions that are independent of other procedures' functions.
4. **Principle of information interdependence**: the initial information and results of every procedure implementation must be informatively consistent with other interdependent procedures of this methodology.
5. **Principle of purposeful conformity**: the procedures and approaches of the methodology must be mutually coordinated and correspond to achievement of the unified purpose of ensuring the required credibility and validity of the obtained results of the problem solution.
6. **Principle of functional rationality**: in the system methodology, no functions should be duplicated.
7. **Principle of multipurpose generality**: methods and approaches of the methodology must possess a sufficient level of generality and ensure solving different-type classes of system problems, differing in functions, purposes, area of application, nature of objects, and other factors.
8. **Principle of multifactor adaptability**: procedures and approaches of the methodology should be adapted both to peculiarities and properties of system problems differing with respect to complication, degree of completeness of the initial information, and a number of other factors, as well as to the requirements of a decision maker in whose interests a system problem is being solved.
9. **Principle of procedural openness**: methods and approaches must preserve a structural interconnection and functional interaction and ensure general resultiveness of the methodology both with respect to replacement of some procedures by others and in their structural or functional aggregation.

10. **Principle of rational supplementation**: the methodology must ensure the possibility of introducing additional methods, procedures, and principles in the interests of extending the applicability field under the condition of consistency between them and in regard to the initial methodology.

The indicated principles are realized only on the basis of a rational combination of possibilities of an expert and modern computing facilities and methods. Therefore, in system methodology the heuristic approaches based on the use of intuition and experience of the researcher play an important part.

It is also important to consider factors that are essential for ensuring the credibility and validity of the solutions of system analysis problems. First of all, it should be noted that the general methodology of system analysis is open. In other words, applying this methodology, it is possible to exchange information with the external environment in regard to the problem-solving space. Therefore, the system methodology should make it possible to use the information coming from the environment for checking the validity of decisions made at various stages of the system analysis of an object or a situation. In this sense, it may be considered that the methodology has a feedback mechanism. Such a mechanism is an important part of the methodology, as it opens up the possibilities of comparing calculated and real results. The comparison of results allows one, for example, to correct control processes. It is especially important in controlling complex ecologically dangerous objects under critical or emergency conditions. However, it is necessary to take into account that the feedback mechanism has a sluggishness. Therefore, a certain time interval is needed to check the validity of the system problem solutions. Under abnormal, critical, or emergency conditions a decision maker might have no slack time to carry out a validity check of a given solution. In that case the method of multifactor simulation modeling has proved to be very useful to determine the reaction of a system to various alternative variants of the decisions being made.

It should be noted that the concept of "simulation modeling" is interpreted in system analysis more broadly than its traditional meaning. The principal distinction is the necessity to carry out simulation modeling from the position of **a the principle of integrity**. The essence of the principle consists in the system coordination of the goals, tasks, conditions, properties of an object under investigation in the process problem solving. It is necessary to realize the principle of integrity from the position of the theory of risks. Namely, it is required to take into account a degree and level of risk of the decision which is being taken in real-life environment. For such kind of conditions both incompleteness, uncertainty, inaccuracy of the initial information and contradictoriness of the purposes of research are characteristic. Therefore, the traditional method of simulation modeling that is used to solve problems of system analysis should be supplemented by principles and methods by both situation modeling and risk theory. For this reason it is possible to create rational tools capable of working in an interactive mode using the rational potential capabilities of computer facilities and software, as well as knowledge, skills, experience, intuition, and human abilities of prediction.

Thus, a general system analysis methodology should possess the following major properties:

1. Integrity of the methodology from the point of view of achieving the set research goals.
2. Systemness in accounting for the factors inherent in the real initial conditions of operation object—incompleteness, uncertainty, inaccuracy of the initial information, and contradictoriness and ambiguity of the research goals.
3. The possibility of explanation, substantiation, and realization of a rational compromise in solving a system problem from the point of view of obtaining the best results in achieving the goals and decreasing risks.
4. The possibility of preliminarily forecasting the various alternative solutions' efficiency and subsequent verification of the decision taken in a real situation.

In addition to the above-indicated properties, it is necessary to formulate a number of heuristic hypotheses whose correctness is intuitively obvious and results from the experience of practical research.

Hypothesis 1. An object of system analysis is characterized by a common purpose, defined by the decision-maker's point of view and by his understanding of goals and objectives of the research.

Hypothesis 2. An object of system analysis is studied only in the process of its interaction with the environment.

Hypothesis 3. The goals and objectives of system analysis in regard to an object are defined from the point of view of the subjective understanding of the decision maker of a rational compromise between these contradictory goals (or requirements) and the degree and level of possible risk.

Hypothesis 4. It is assumed that the mode of steady functioning of an object is the normal mode for that object.

Hypothesis 5. An unforeseen exit from the mode of steady functioning requires the application of special methods of system analysis used for such situations.

The necessity of introducing these hypotheses is dictated by a number of factors. First, it is necessary to point out the possibility of different subjective approaches to the choice of different methods and procedures of system analysis for the same objects by different experts.

Thus, for engineers in the field of radio electronics and electrical engineering, it is convenient to analyze complex systems in the language of circuit theory. Specialists in the field of process and systems control distinguish in such connections the circuitry of direct and feedback couplings, with the purpose of formalization and synthesis of control. Specialists in operational research analyze properties of the social, ecological, economic, and other complex systems on the basis of their reduction to optimization problems or a multipurpose sequence of such problems. This list can be extended indefinitely.

Often, the main goal of formalization of a system research problem of a concrete object consists only in its reduction to some sort of formal mathematical statement. Here, a number of important factors and characteristics of an object may remain beyond the researcher's vision, for example, factors of risk that an object will fall out of the mode of steady functioning, etc. This drawback is quite essential in carrying out system analysis of organizational and organizational-engineering systems, whose efficiency and functioning stability depend, to a great extent, on human preferences, judgements, and errors.

In addition, for a number of complex systems the presence of principally unavoidable risk factors is typical. For example, in the conditions of a market economy, for entrepreneurs risks conditioned by competition are largely unavoidable. Such risks are especially dangerous when modern transport vehicles are used, at chemical and nuclear power plants, and in different critical and extreme situations. Consideration of risks in different system researches also frequently is of a subjective nature.

Let us define a subjective approach of a researcher to the solution of a real system problem in the form of his methodological paradigm. A **methodological paradigm** is a system-coordinated set of ideas, approaches, methods, suppositions, and limitations, selected by a researcher for solving a concrete system problem.

If the problem is solved by using a certain methodological paradigm, the found solution should not contain features incompatible with this paradigm. It is expedient to consider a particular case, where one paradigm is a subset of another.

We shall define also the term "**contensive paradigm**", which is understood as a set of ideas, approaches, methods, and suppositions, ensuring the possibility of solving all concrete problems of a given type.

In recent years there has been a tendency to generalize paradigms, stimulated by new achievements in the development of system mathematics and computational engineering. Any generalization of a paradigm extends the class of the solved problems and in many cases allows for the best solution.

The connections between possible methodological paradigms and classes of system problems are the subject of the **metamethodology of systems**, which is a new area of research that has not yet been much explored.

An important aspect of the system metamethodology is the development of paradigms that, for different classes of problems and the modern state of system mathematics and computational engineering, would provide the best compromise between two groups of contradictory goals—the quality and complexity of system problem solving.

The basic difficulty of this research is that a set of alternative procedures for solving a given problem may be developed using the same methodological paradigm.

At the same time, any mathematical theory that makes sense from the point of view of solving system problems is a methodological paradigm. It is connected with a type of problem and presents a local system that may be used for development methods for solving specific problems of a given type. One of the problems of system methodology is the compilation (drawing up) of mathematical approaches and definitions of their place in solving a complete set of a given class of problems. Another problem is the construction of new contensive paradigms for solving each type of problem. Since each new paradigm serves as an impetus for the creation of a new mathematical approach or a set of methods, comprehensive research of system metamethodology will become a powerful stimulus for getting fundamental and applied mathematical results in the field of system mathematics, the theory of complex systems, and other subdisciplines of system knowledge.

2.3 Classification of System Analysis Problems and Procedures

We shall move on to a consideration of the basic classes of system analysis problems. First, we shall define the contensive statement of the general problem of system analysis. For simplification of understanding system problems properties, and features we shall assume that:

1. A problem is formalized on the basis of some requirements specification formulated by the developer.
2. A problem consists in creating a material object in the form of a multilevel hierarchical complex system.

Let the initial data, presented in formalized form, and the requirements specification that determine the purpose, general characteristics, and properties of a system be known; and let the basic requirements of its technical, technological, constructive, operational, and economic parameters be assigned.

The structure and properties must be determined in the form of a multilevel hierarchical system to justify the requirements for each functional element (FE) of all hierarchical levels so as to choose and justify the functions of each functional element of all hierarchical levels proceeding from the condition of meeting the specified requirements for achieving the purpose in the specified set conditions.

Thus, in the given variant of a system problem on the basis of general conditions and general characteristics of a developed product, its structure must be determined and justified as a complex multilevel hierarchical system and all functional elements of the system chosen so that they are coordinated with respect to goals, objectives, and expected results. In addition, indices must be determined from the point of view of achieving the required functioning efficiency in known and predicted situations. At the same time, the designer must work in conditions of incompleteness, incredibility, and fuzziness of information.

The problems of system analysis are considered from the point of view of the **principle of integrity** (i.e., research of an object, the environment, and a control system, as a single whole that ensures the achievement of the set goal with the assigned quality indicators).

This problem of system analysis is divided into two stages:

1. **Rational decomposition** of general requirements for a system into requirements for functional subsystems and functional elements of all hierarchical levels.
2. **Aggregation of results** of the analysis and realizability of requirements at each hierarchical level for each functional subsystem and each functional element into a final result, which determines the technical characteristics of a system and rational parameters of its functioning under set requirements in assigned and predicted situations.

The analysis of structurization processes for systems and problems. We shall consider some important features of decomposition and aggregation processes and their role and place in system research and system problem solving. First, it is appropriate to recall the statement of two known experts in the field of system theory. Gauguin and Varel in one of their works have noted that the world for us is not mostly divided into systems, subsystems, environment, etc. We ourselves subdivide it for different reasons, as a rule leading to the only common reason: for convenience [54]. This phrase precisely defines the main feature of the formation processes of system problem structure and complex systems—the creation of conveniences for the performance by people of their production functions. These processes are of various types, which differ in purpose, function, and field of application. And, as a consequence, they have different names: decomposition, aggregation, structurization, and others. For example, the basic purpose of structurization for many types of complex engineering systems (CESs) is providing convenience for maintenance and repair. One of the main reasons for the structurization of complex engineering systems is the convenience of realizing and controlling processes at all stages of the life cycle of new products. It is possible to list several factors and examples proving this thesis. For example, structurization allows one to considerably simplify the solutions of some practically important problems. In particular, for manufacturing, structurization makes it possible to limit the nomenclature of parts to a defined set of standard modules and other elements. During operation of complex engineering systems it allows one to simplify diagnostics, to raise the level of maintainability and improve repair, to reduce the nomenclature of required spare parts and devices, and, finally, to ensure a sufficiently high level of product serviceability.

However, in the process of structurization we have not only advantages, but also face some difficulties. One of the main difficulties is to create a product's system structure and its formalized description in such a way that in the process of operation the flaws of the constructive and functional elements of the duplicate products will be revealed with sufficient reliability and make it possible to compare their operational resources comprehensively and objectively.

During the design, testing, and operational development of real CESs the requirements, criteria, and basic conditions of research and operation are determined at the corresponding phases of each stage of the system life cycle. The life cycle of different complex systems may differ in variants of stages. In particular, for unique complex engineering products the life cycle includes the following stages: design, advanced development, a series of sample tests, preproduction operation of a sample, batch production, and operation of products. The life cycle of engineering products is finished, as a rule, with utilization of each product. Let us move on to the analysis of goals, properties, features, and possibilities of system research apparatus. We need to take into account that the processes of globalization determine the necessity of developing new notions, approaches, and methods for solving the complicated problems of modern life, requiring one to take a fundamentally new approach to stating and solving such problems. When formalizing and solving real system problems, the initial information about an object should be represented constructively and conveniently in terms of conceptual functional spaces of conditions and properties of complex systems [119]. Under modern conditions, these spaces must provide a new vision of interactions in the following structure: **system analyst** $\Leftrightarrow$ **person** $\Leftrightarrow$ **object** $\Leftrightarrow$ **environment**. This method should create the clearness of coordination of the main factors: the properties of the object under investigation, the initial information, and the conditions under which this object functions while taking into account the various uncertainties and multifactorial risks. The coordination should be system defined and take into account the goals, problems, and expected results of an object's function; the complexity of situations in which an object functions; and the level of information deficiency about the complexity of the goals and conditions under which an object functions (Fig. 2.1). The limiting factors are the resources required by computations and empirical procedures of system research.

It is expedient to present the conditions of an object functioning in the form of the conceptual functional space of conditions in the following system of coordinates $\alpha_1\beta_1\gamma_1$, where:

the α_1 axis determines the level of goal complexity of the object's operation; with an increase of α_1 the goal complexity increases.

the β_1 axis determines the level of situational complexity in which the object functions; with an increase of β_1 the complexity of a situation increases.

the γ_1 axis determines the entropy of information as the level of an information deficit on the complexity of goals and conditions of the object's operation, which is characterized by points of the functional space above the surface $O\alpha_1\beta_1$.

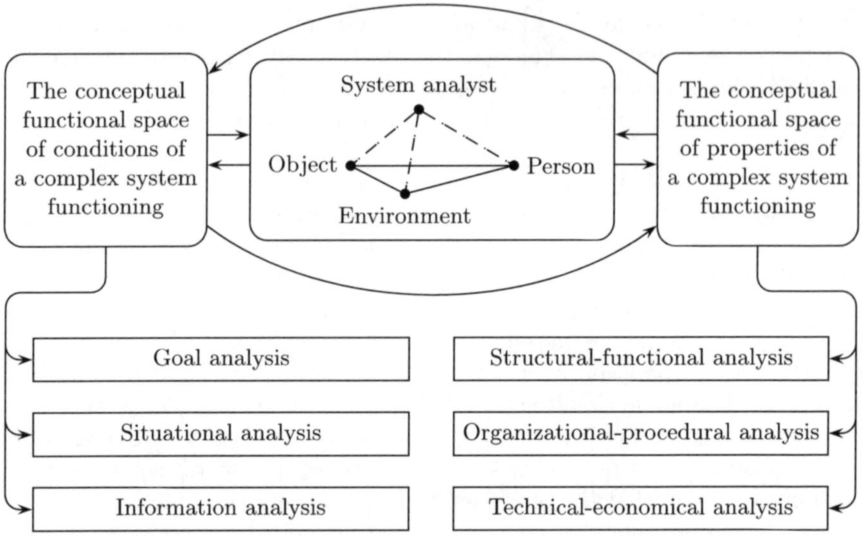

Fig. 2.1. Formalization of system analysis problems

The conceptual functional space of conditions of a system's operation (Fig. 2.2) is formed as a result of performing computing procedures of three types of analysis—goal, situational, and information:

- Goal analysis allows one to determine a set of points on the α_1 axis that quantitatively characterize all kinds of goal properties and features.
- Situational analysis allows one to determine a set of points on the β_1 axis that quantitatively characterize the properties and features of all kinds of normal and abnormal situations.
- Information analysis allows one to determine a set of points on the γ_1 axis that quantitatively characterize the level of information deficit on situations and goals of an object's operation.

The functional space of conditions of an object's operation is based on the above-indicated sets of points.

It is obvious that as the complexity of functioning conditions increases, the complexity of creation and functioning of the corresponding complex system also grows. In fact, the more complicated the goals and situations, and the greater the deficit of the information, the more difficult it is to create a system that meets these conditions and control it in these conditions. Depending on the level of condition complexity, it is possible to introduce some classification series in which with an increase in the number of series elements, the complexity of the system functioning conditions grows in a certain way.

Similarly, it is possible to present a multidimensional conceptual functional space of the complex-system properties (Fig. 2.3). This functional space is formed as a result of performing computing procedures of three kinds

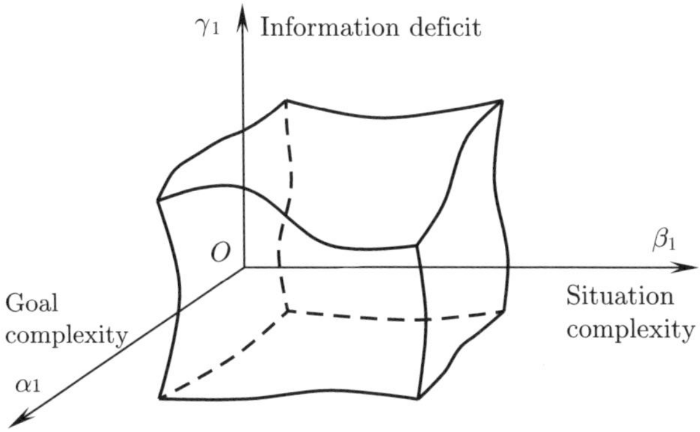

Fig. 2.2. Conceptual functional space of conditions of complex system operation

of analysis: structural-functional, organizational-procedural, and technical-economic.

- A structural-functional analysis allows one to determine a set of points on the α_2 axis that quantitatively characterize the allowable values of parameters of the structure and function of the object that are required for the achievement of the assigned goals in the assigned situations, which in turn are determined by the functional space of conditions.
- An organizational-procedural analysis determines a set of points on the β_2 axis that quantitatively characterize the properties, structure, and functions of a control system for the assigned conditions.
- A technical-economical analysis determines a set of points on the γ_2 axis that characterize the obtained effect and expenses for its achievement in the assigned conditions, as well as the technical-economic efficiency of an object.

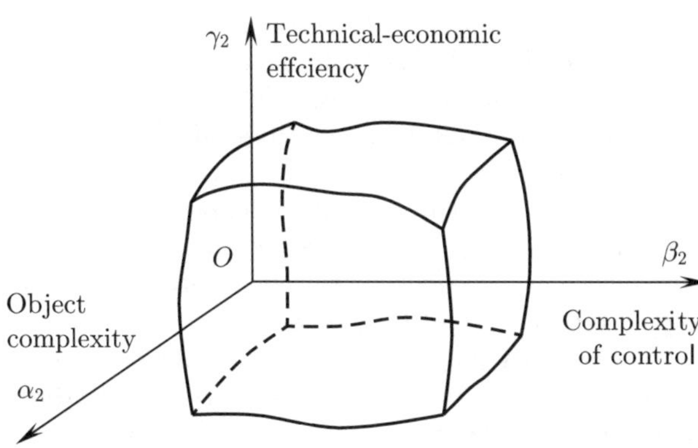

Fig. 2.3. Conceptual functional space of properties of complex system operation

The interconnection of the conceptual functional spaces (CFSs). Two conceptual functional spaces introduced above are interconnected. Thus, two variants of interconnections are possible.

The first variant: initial is the functional space of conditions, on the informational basis of which the functional space of properties and characteristics of the object is formed.

The second variant: initial is the functional space of properties and characteristics of an object for which the functional space of conditions is found.

The first variant is used in designing complex systems based on an a priori known assignment. The structure, properties, and characteristics of a developed system are determined on the basis of the assigned functional space of conditions. Such situations are conditioned by the fact any complex system is characterized by the uncertainty and contradictoriness of goals. For example, for the aircraft designer the contradictoriness of goals is manifested in the natural desire to design the most economic, most reliable, most load-lifting plane with the maximal range ability. But it is obvious that it is in principle impossible to achieve simultaneously these goals. From here the problem arises as to how to find their rational combination. However, the designer does not know beforehand what combination is most expedient, economical, and technologically acceptable. In this connection one of the primary problems of system analysis consists in disclosing the uncertainty of goals.

The second variant arises in the testing of newly created systems designed for operation under a wide range of changing conditions. For example, a newly created plane is tested under various conditions with the purpose of revealing the most expedient modes and spheres of its application. Designers study at what height and with what speed it is possible to ensure the maximal carrying capacity of the plane, or with what carrying capacity, speed, and height of flight its maximal range is possible.

The necessity of carrying out such an analysis follows from the circumstance that at the design stage it is impossible to take into account the whole set of factors present in real operating conditions. In particular, it is practically impossible to reveal the levels of influence of any destabilizing factors (design, technological and operational miscalculations, unpredictable errors, deterioration and aging of plane parts, regulation of its components in the process of operation, etc.).

In connection with the influence of the indicated factors a practical problem arises: during testing of a pilot model it is necessary to improve the system in such a way as to ensure attainment of the set goals, to reveal possible abnormal and critical situations, and to introduce certain technical corrections and restrictions that will allow one to decrease considerably the probability of occurrence of these abnormal and critical situations in the process of operation. As a result of such actions, there may occur updating of the project and updating of the goals and properties of the developed system. This updating is in essence the adjusted result of disclosing the uncertainty of goals. Thus, the problem of disclosing the uncertainty of goals is real in both variants of interconnection.

The conceptual spaces suggested above are based on properties and features of the introduced procedures of complex systems' analysis. We shall consider their purposes and basic functions.

1. **Goal analysis** is used for revealing the individual purposes of a complex system behavior for achievement of its main goal.
2. **Situational analysis** is used for revealing situations and their characteristics that determine the basic conditions of a complex system's operation.
3. **Information analysis** is used for defining the scope, completeness, and other parameters of the information about a complex system and its environment that are necessary for quantitative determination of a degree of the system's achievement of the set goals in the given situation.
4. **Structural-functional analysis** is used for determination of a required level of potential possibilities of a complex system's functional elements and the degree of interconnections and interdependence of its functional elements for achievement of the set goals of the system functioning in an a priori developing situation.
5. **Organizational-procedural analysis** is used to reveal the rational ways of organizing control and a rational choice of procedures that are necessary for achieving the set goals in a developing situation.
6. **Technical-economical analysis** is used for determination of the resources necessary for a complex system to achieve set goal, taking into consideration the assigned quality parameters.

Let us consider the interconnections of the introduced procedures of the purposeful analysis for the first variant of interconnections of conceptual spaces. In this variant it is expedient to perform the following sequence of the specified procedures of the purposeful analysis.

Goal analysis (the definition of the goals of functional elements based on the set goals of a system) $\Rightarrow$ **situational analysis** (a selection of rational conditions of functioning) $\Rightarrow$ **information analysis** (compilation of the basic data on a system for achieving the set goals) $\Rightarrow$ **structural-functional analysis** (determination of the structure and functions of system elements necessary for achieving the set goals) $\Rightarrow$ **organizational-procedural analysis** (organization and realization of the control procedures in the conditions of a changing environment) $\Rightarrow$ **technical-economical analysis** (determination of the resources necessary for achieving the set goals with the required quality parameters).

The interconnection of the above procedures is determined by the goals and features of a complex system's operation and the features of the problem being solved (a problem of system design, optimization and operation, forecasting of potential capabilities of the developed system in new abnormal situations, problems of technical diagnosis of the system serviceability, etc.).

The interconnection and degree of application of the specified procedures may change depending on the features and statement of the problem.

We shall consider in more detail the properties of the system analysis procedures, assuming that the sequence of the basic processes is fixed.

1. **Goal analysis procedure**
   a) Consecutive multilevel decomposition of an assigned set of goals into the goals elements for each of the assigned hierarchical levels.
   b) Formation of the quantity indicators that determine the degree and level of achievement of the general system goal and the local goals of elements.
   c) Establishment of functional interconnection of goal indicators of elements for different hierarchical levels with the indicators of the general goal of an object (system).
   d) Establishment of allowable intervals of changes in goal indicators of hierarchical levels of functional elements in accordance with allowable intervals of changes of goal function indicators of an object (system).

   In the development of complex engineering systems the procedure for goal analysis consists in verification of the correctness of the assignment specification in such a way that different parameters should comprehensively define the goals of the system being developed and the allowable intervals between changes to them. The values and allowable intervals of changes should be mutually coordinated and technically realizable. At this stage the decomposition of requirements of the system is carried out in accordance with the requirements of the basic functional elements (FEs) comprising the complex system.

   For example, when a thermal power plant is designed, the requirements of the system as a whole extend to the requirements of the capacity and quantity of electric generators and their other basic and auxiliary subsystems.
2. **The situational analysis procedure**
   a) Formation of a set of controlled normal situations of a complex system and forecasting the most probable abnormal and critical situations during its functioning.

      Here, **situation** is understood as a certain state of the system under consideration and the environment in which it functions, characterized by a priori established intervals of the values of the system parameters and the functional characteristics of the environment.

      Normal situations are situations where the system parameters and functional characteristics of the environment are in a priori set intervals, for example, operation of motor vehicles in certain climatic conditions, in particular, in conditions of the Far North.

      Abnormal situations are situations where separate parameters of a system or separate functional characteristics of the environment go beyond the allowable intervals but do not create conditions of complete violation of functioning or destruction of an object.

Critical situations are situations in which a number of system parameters or functional characteristics of the environment go beyond the allowable intervals and create conditions for the object's operation is violated, which results in its partial or complete destruction, creates dangerous conditions for the attending personnel, or results in ecologically dangerous consequences.

Then, the following sequence is carried out:

b) Definition of quantitative characteristics of normal situations and intervals of their changes.
c) Forecasting the possible set of abnormal situations and identifying the most probable among them.
d) Revealing features and definitions of characteristics of the most probable abnormal situations.
e) Revealing a set of critical situations, conditions of their occurrence, and their characteristics.
f) Revealing conditions of a possible transition from a normal to a abnormal or critical situation.
g) Revealing conditions of a possible transition from a abnormal to a normal situation.
h) Revealing conditions for the prevention of critical situations.

3. **Information analysis procedure**
   a) Determination of completeness, credibility, and timeliness of getting the information necessary to regulate the system with the purpose of attaining the set goals in normal and abnormal situations.
   b) Determination of the information system characteristics corresponding to the assigned level of information support (completeness, credibility, timeliness of information) for their regulation in normal and abnormal situations.
   c) A selection and analysis of procedures for obtaining, storing, and processing the information for ensuring the system controllability in normal and abnormal situations.
   d) A selection and analysis of procedures for formation, justification, and decision-making in regulating the system in normal situations and during a transition from a abnormal situation to a normal one.
   e) Determination of information support indicators of the forecasting procedures of abnormal and critical situations and their consequences.
4. **Structural-functional analysis procedure**
   a) Determination of the complete structure of the functions ensuring attainment of the set goals in the assigned conditions of functioning (for example, the list of all control functions required for electric power generation).

   b) Determination of a rational hierarchical structure of the system ensuring attainment of the set goals with the assigned restrictions on resources.
   c) Determination of a functionally complete number of elements for each hierarchical level.
   d) Determination of rational characteristics of functional elements for each hierarchical level.
   e) Determination of conditions for attainment of the assigned characteristics of functional elements for each hierarchical level.
5. **Organizational-procedural analysis procedure**
   a) Determination of a functionally complete structure of control procedures in normal and abnormal situations.
   b) Determination of a rational distribution of control procedures between the person and a complex of technical means.
   c) Determination of a rational organizational structure of the object control system in normal and abnormal situations.
   d) Determination of a rational structure of the technical system of an object control in normal and abnormal situations.
   e) Determination of a functionally complete number of elements of system technical control.
   f) Determination of the characteristics of functional elements of technical control systems in normal and abnormal situations.
   g) Justification of the degree and level of intellectualization of technical means for decision support in normal and abnormal situations.
   h) Determination of a rational structure of control in critical situations.
   i) Determination of the intellectual level of technical means for decision support in a control system in critical situations.
6. **Technical-economical analysis procedure**
   a) Determination of all expenses incurred to carry out the basic functions of an object.
   b) Determination of all expenses to carry out object control procedures.
   c) Determination of all expenses for the prevention of abnormal and critical situations.
   d) Determination of the socioeconomic and technical-economic efficiency of a system.

2.4 Notions of Complexity of System Problems, Complexity Spectrums, Transcomputational Complexity

The modern world is marked by an increasing complexity and interdependence of its components, which was underlined on several occasions by Pechey [144], Saaty [151], Klir [53], and other scientists, as well as by public figures and politicians at the second half of the last century. Problems, whether social, political, or economic, do not exist in isolation. They cannot be extracted from a certain integral environment, solved and explained separately, and then integrated again for an explanation of this integral environment. The environment itself in which these problems arise does not possess constant properties of integrity. It is dynamic, since it always changes, being exposed to both external and internal influences. The environment changes together with its problems and their solutions in physical and conceptual spaces. For example, as a result of technological processes, relations between elements of these processes also change and relationships between the elements of these processes are selected by means of specific interpreting structures. The environment also changes over time, since it is noticeably influenced by various events and changes in conditions.

Let us introduce a definition of complexity from the standpoint of system methodology. Such a necessity is caused by a number of factors. As Klir [54] points out, complexity is many sided and leads to a number of definitions. Ashby [3] also confirms this idea: "The term 'complexity', as applied to systems, has many meanings". From the standpoint of system analysis, it is expedient to consider the notion of complexity based on estimating costs for solving and investigating system problems and situations. We shall use the concept of complexity that is used in general system theory. According to this concept, complexity is a general property of some set of various objects that are structurally interconnected and functionally interacting. For system analysis problems this definition should be expanded and presented in the following form. Complexity is a general property of a unified set of various objects that are structurally interrelated, functionally interdependent, and interacting within the existing parameters and characteristics of their environment in the presence of uncontrollable external influences, risk factors, and other conditions typical of system problems.

Following the approach of Klir [54], we shall distinguish the notions of complexity and difficulty. Many problems are difficult, but simple, having a single or a finite set of solutions. For example, the problem to undo a knot in a cord may be difficult, but it has one solution. Complex problems usually have a set of possible solutions, which correspond to different purposes. For example, designing or modifying a complex engineering system is realized proceeding from technical, economic, organizational, and other kinds of purposes for each of which there are solutions.

Complexity is connected with a manifestation of interaction and interdependence of processes in interconnected systems, which is determined by the degree of influence of one or several elements on the behavior of other elements.

For example, it is possible to draw a sequence of the following relationships: the economy depends on power engineering and other industries, the availability of energy resources depends on politics, politics depends on the power of the state, and the power of the state depends on its military potential and economic stability. Note that these relationships are symmetric: politics depends on the economy; however, the economy also depends on politics.

The interpretation and understanding of complexity depends on many factors, frequently subjective, that are used when the system analyst unites the considered parts into a single whole.

This subjectivity represents a dilemma of a higher degree, which dispels any latent suspicion that there is an objective interpretation of reality surpassing our intellect and cognitive ability.

People who decide what actions to take for complex problems and those people whom these decisions influence usually have different interests and contradictory goals. Within these two groups of people or between them there can be no consensus regarding desirable goals or strategies developed for their achievement.

The other difficulty in defining the notion of complexity is connected with the fact that the solution of a number of problems, or their set as a whole, cannot be obtained by breaking them up into subproblems with their subsequent solution and synthesis of the general solution to the initial problem.

In system problems one must consider the notion of "complexity" in various aspects, depending on the properties of systems and features of these problems. As a matter of fact, the notion of complexity "is as fundamental a notion in systems science as the notion of energy in the natural sciences" [54].

The notion of complexity is many-sided: complexity = "to have many different interconnected parts, structures or elements and, hence, to be difficult to understand"; or complexity = "to include many parts, aspects, details, notions, demanding serious investigation or consideration for understanding or mastering", an explanatory definition; or complexity = "to have a complex nature or state". All these definitions may be applied to any type of system—material and abstract, natural and artificial, to creations of science or art, as well as to problems, methods, theories, laws, games, languages, machines, organisms, and any other systems. Regardless of what is considered complex or simple, in the general case, the degree of complexity is connected with a number of differentiated parts and a measure of their interaction and interrelation. In addition, the notion of complexity has subjective conditionality, since it is connected with the abilities of understanding or using the considered system. Thus, what is complex for one person may be simple for another.

The notion of complexity expresses the interaction between the investigator and the object of investigation, which results in the study of the initial system with a certain level of detail. In this sense, complexity is not the immanent property of the investigated system, but rather it is determined in such a way that the investigator interacts with the system from the viewpoint of the set goals.

At the level of initial systems in certain conditions system complexity is expressed [54] only as the cardinality of considered sets—sets of variables, sets of parameters, sets of states, and parametric sets—since there are no interconnections among these sets. On higher epistemological levels the notion of system complexity becomes more important. It is different for different types of systems. The same initial system on different higher epistemological levels may be described in very different ways.

In some cases, a certain degree of complexity is a necessary condition for getting certain system properties, which are usually named by revealing properties (self-reproduction, training, development).

In other, more widespread cases, in solving system problems, we build either a simple system or an attempt is made to simplify an already existing one.

Complex problems never exist separately and are seldom characterized by unilateral causal relationships. Rather, complexity links problems together and forms a picture of interrelations and plural causality. The exact character of causality is difficult to describe fully: the interrelation of problems is too often found out only after making decisions that cause secondary problems.

As an illustration, it is possible to cite problems related to preventing the proliferation of weapons of mass destruction and problems of struggle against international terrorism. These complex problems cause many others that are no less complex. For example, the problem of the reduction of arms leads to fundamental changes in the structure of military expenditures and influences on any economic system, in which it is an integral part. A decrease in military expenditures cuts down research activities, which often leads to obtaining results, for both military and peaceful purposes [52]. A given prediction was logically proved using the following reasoning. A reduction in arms demands solving the contradiction: the reduction of expenditures on arms guarantees the required level of national security. In addition, it is necessary to consider that defense expenditures considerably influence the infrastructure of industry, transport, power engineering, which are the basis of any economic system.

Note that in a given problem an important role is played by the complexity and interdependence of various components of a social and economic system. To deal with such problems one should use an approach that would allow for the use of any kind of information, including precise data, quantitative information, and fuzzy data obtained intuitively, from experience, taking into account the values of inferences and figurative guesses. It is obvious that, in addition to the information necessary for arriving at solutions to the considered problems, it is necessary to use certain techniques, principles, and approaches that allow one to find solutions with the required accuracy, validity, and credibility. At the same time, formal methods cannot always be well adapted for solving practical problems. For example, in operations research and science of control many models and methods have been developed that are frequently mechanically used for solving a complex problem. As a result, the initial data to be used in a model are generated before a problem really arises. However, no problem exists exactly in the form in which people expect and realize it. These peculiarities are encountered in the study of conflicts. It is possible to try to prevent conflicts, but when they occur, it is necessary to apply another procedure for their neutralization, investigation, and interruption.

The history of science and engineering shows that up until the 20th century science dealt basically with relatively simple systems. From the 17th to the 20th centuries, the list of the major events in science mainly consisted of variations of the same theme: revealing the latent simplicity in a seemingly complex situation. A characteristic feature of such situations is that some essential factors are distinguished, and many others are considered insignificant. It allows the researcher to introduce essential, experimentally justified simplifications and, hence, to consider the investigated characteristics "isolated" from all others.

Many situations in which it is possible to extract a few essential factors from a large number of factors are typical for physics, which explains the significant successes of this science in enriching the knowledge other disciplines. Newton began this development, having shown that in physics considerable simplifications are possible. His law of gravity is a consequence of very strong simplifications. However, using correct calculations, it allows one to calculate very precisely the orbits of planetary motion.

Up until almost 1900 science was under the influence of Newton's achievements. His powerful simplifications had been applied in most areas, which in investigating some physical phenomena (electricity, magnetism, hydromechanics) had produced good results; however, in other sciences, especially biology and medicine, they had not been so applied. The problems science had been dealing with and learned to solve belonged to research of determinate systems with two or three variables. They were presented analytically, generally in the form of differential equation systems. Similar problems with a small number of variables and a high degree of determinism that are solved analytically are usually called **organized simplicity problems**.

In the late 19th century some physicists began to investigate systems of gas molecule motion in closed space. Such systems usually have ca. 10^{23} molecules. Moving molecules have huge velocities, and their trajectories have a chaotic nature because of constant, repeated collisions. Obviously, Newton's law with its strong simplifications of investigated processes cannot be applied in research of such complex systems. Therefore, it is absolutely hopeless trying to solve the problem of the analysis of gas molecule motion in closed space (i.e., the problem of a very complex and unorganized system) by means and methods used for the problems of organized simplicity. An absolutely different approach is required in this case. Some scientists, primarily Boltzmann and Gibbs, have created powerful statistical methods for solving system problems with a great number of variables, which appear in very random manner. Such problems have been called **unorganized complexity problems**.

Statistical methods do not describe separate variables (e.g., motion of a molecule). They allow one to define general characteristics of investigated processes. Developed in the early part of the last century, statistical methods were successfully applied to solving the many unorganized complexity problems that were arising, both in theoretical science and in its many applications. There are well-known examples of the successful application of these methods in statistical mechanics, thermodynamics, statistical physics, and statistical genetics. In engineering these methods play an important part in creating large telephone networks and computer systems with time sharing, in solving problems of technical reliability control, etc. In business these methods are widely used for solving problems of marketing, insurance, etc.

Unlike analytical methods, which are used for a class of organized simplicity problems and which turn out to be inapplicable even with a relatively small number of variables (e.g., five), the efficiency and expediency of statistical methods increase with an increase in the number of variables. Thus, these two classes of methods are complementary. They correspond to two opposite parts of the complexity spectrum and, despite their augmentability, cover a very small part of this spectrum. Thus, besides the two specified parts, the complexity spectrum remains unsupported methodologically in the sense that neither statistical nor analytical methods are suitable for a wide class of problems. The problems, connected with the middle part of the complexity spectrum, are called **organized complexity problems**.

There are many examples of problems with properties of organized complexity, especially in sciences studying life on the earth, behavior of animals, the environment, social sciences, and high-tech scientific applications.

Systems with organized complexity possess many properties that cannot be neglected in principle. However, in most cases it is impossible to avoid simplifications. Since both Newton's and statistical simplification strategies are inapplicable for such problems, it is necessary to find new ways to solve them. One of them is based on an assumption regarding the inaccuracy in describing the studied systems. In this case, the inaccuracy is not statistical, but of a more general nature. The mathematical tool underlying this approach was developed in the mid-1960s and is known as "the theory of fuzzy sets". The essence and importance of this problem are perfectly described by its founder, Zadeh.

Note that traditional quantitative methods of system analysis are not applicable in studying, for example, humanitarian problems and in general any systems whose complexity exceeds the complexity of simple mechanical and physical systems, described analytically by means of differential, integrated, algebraic, and other types of equations. This statement is based on a principle called the principle of incompatibility. The essence of this principle is that a growth in the complexity of systems is accompanied by an exponential decrease in the ability of a person to perform multifactorial system analysis efficiently, with sufficient validity and credibility and to make sufficiently precise and valid statements about the conditions and behavior of an object under investigation. It is possible that a person's ability to perform this function will drop below some critical limit. In this case, some characteristics of the analysis results, for example accuracy and meaningfulness, credibility, and efficiency, become mutually exclusive. From here it may be concluded: "The more detail a real problem is considered in, the less efficient and less justified its solution may be" [80]. Hence, the quantitative approach to solving real system problems based on a detailed description of the properties and behavior of complex systems cannot always provide the required levels of justification and efficiency of solutions that are required in the real conditions of control and decision making. At the same time, in practice the need for efficient formation and reliable justification of decisions arises often enough. For example, such a need arises when systems are controlled in conditions of augmented risk. In particular, it occurs in the following cases of:

- Studying the modes of functioning of complex engineering systems, machines, and mechanisms.
- Critical or emergency situations, arising in interconnected technological processes and industrial complexes.
- Critical or emergency human-induced ecological and other abnormal modes of system behavior.

The most important characteristic of the indicated situations is the incompleteness, uncertainty, and inaccuracy of the initial information. In such conditions it is necessary to apply a nontraditional and, in some ways, alternative approach to arriving at a solution of the studied problems. The idea of such an approach is based on rational human thinking. It is characterized by the ability to operate not using numbers, but images of various objects of observation, investigation, and control for whose elements the transition from belonging to a certain class to nonbelonging is not sharp, but fuzzy and gradual. In fact, the "omnipresent" fuzziness of human thinking suggests that the logic of human reasoning is not ordinary, single-valued, or even many-valued; it is a logic with fuzzy truth, fuzzy relations, and fuzzy rules of inference. Such fuzzy logic is the major component of one of the main features of human thinking, namely, the ability to generalize information—to extract from the huge data arrays only those that are necessary for solving a specific problem.

The considered factor of complexity is most important, but not the only feature that allows one to single out system problems into a special class. The second major factor is the potential formalizability of the initial practical problem.

From the above brief overview of complexity it follows that formalization of the organized simplicity problems does not cause difficulties. The mentioned problem is getting simplified to such a degree that such formalization becomes possible. Unorganized complexity problems, where laws of large numbers predominate, are considered similarly.

Organized complexity problems are considered in a principally different way. For example, modern engineering systems and technologies are characterized not only by a large number of interacting and interconnected functional elements, but, first of all, by the complexity of their interaction, by the complexity of influence of the environment on internal processes of the studied object, and by a number of other factors. Therefore, the problem of analysis of almost any field of human activity consists not in defining its "inputs" and "outputs", but in simultaneous tracking of all interactions of object functional elements with such a level of accuracy and credibility that will allow one to obtain practically acceptable solutions and recommendations.

The complexity of this problem is well illustrated by the well-known system mathematician Saaty [156] using the economy as an example. Frequent failures in forecasting economic fluctuations confirm that the complexity peculiar to social and economic behavior can exceed the limits of our intellectual potential. Even using complex economic theories and models, it is impossible to cope with a complex system of interdependencies of all components of the economy. We face the difficulty of forecasting the short-term, annual, or even monthly state of the economy. What appears to be a symptom in one sector of the economy appears in another sector to be the result of some event. For example, productivity depends on capital investments, which in turn depend on interest rates and on credits and taxes. The state of the economy also depends on the mobility of manpower, which in turn depends on many local and regional factors, such as employment opportunities, cost of living, and the normal of living. In addition, the state of the economy is determined by the confidence by the electorate in their elected leaders.

From the given example it is obvious that a complete formalization of practical system problems is not only a complex but, in some cases, an unsolvable problem. The fundamental nonformalizability of some system analysis problems is determined not only by their complexity. There are also other factors. For example, a decision maker subjectively understands the purposes of specific activities and degrees of their importance, has the individual ability to deal with risks, and fulfills other functions. Similarly, while designing a complex engineering system or technology, many problems are solved only by the person, the designer. For example, for the structure of an object, a possible set of alternative technical solutions is formed only by a person. Only a person can suggest a new way, a new technical solution, that will become the basis for a new object or new technology.

Thus, the class of system problems differs from other problems in two basic properties:

1. The organized complexity of a problem.
2. The principal impossibility of full unambiguous formalization of a problem.

It is necessary to explain how we understand the property of "the principal impossibility of a full unambiguous formalization of a problem". This property of a problem does not mean that the problem is in principle unsolvable. Methodological approaches, principles, and methods of system problem solving allow one to present the initial system problem in the form of some sequence of nonformalizable and formalizable problems. Thus, the solving of nonformalizable problems is carried out by a person on the basis of his or her knowledge, experience, skills, intuition, and potential of foresight. Owing to the subjectivity of this approach, formalization of the same system problems by several experts of the same profession will differ. Thus, the difference may be principal, since in conditions of uncertainty one expert may be inclined to take risks, while the other may be extremely cautious.

This circumstance determines the necessity of a rational use of both the potentialities of a person for empirical solving of nonformalizable problems and the potentialities of modern computing methods and means for solving the formalized problems.

In solving system problems, complexity is understood in two ways: as a property of investigated systems and as a feature of system problems solutions. We shall call them the **complexity of systems** and the **complexity of problems**, respectively. In the literature the difficulties connected with the complexity of problems are known as the computing complexity.

Regardless of the type of studied system, it is possible to specify two general principles for estimating their complexity, which may be the basis for comparative study of these systems' complexity.

First principle: the complexity of a system increases proportionally to the volume of information required for this system description.

One way to describe such descriptive complexity is to estimate the number of elements contained in the system (variables, states, components) and their various interdependencies. In fact, with an increase in the number of elements or their various interconnections, the difficulty of working with such a system also increases.

Second principle: the complexity of a system should be proportional to the volume of information required for solving any fuzziness that is characteristic of the studied system.

The complexity of systems is studied, first of all, to create methods by means of which it can be lowered to an acceptable level. System simplification is realized in two ways: simplification based on descriptive information and simplification based on fuzzy information.

The problem of transcomputional complexity arises in the most varied cases, for example, in image recognition problems. A similar problem arises in such areas as testing of highly integrated chips. They present complex electronic schemes with a large number of inputs and outputs.

According to Bremermann's limit, solving transcomputional complexity problems requires processing more than 10^{93} bytes of information. It has been shown that the problem for a system containing n elements with k different states exceeds the value of the specified limit at $n \geq 308$ and $k \geq 2$. Taking into account that in real systems, for example in computer processors, we have $n \geq 10^6$ at $k \geq 2$, it is obvious that the computational complexity of applied problems of system analysis exceeds Bremermann's limit by several orders of magnitude.

If the problem is a transcomputional complexity problem, then to solve we must restate it. The most natural way to restate it is by making the conditions weaker. Making the conditions weaker allows one to apply heuristic methods, which allow one to reject a large number of unpromising variants, and approximate (fuzzy) methods, which allow one to solve these problems taking into consideration a set of variants.

Bremermann's limit leads to a too-simple breaking-up of system problems by their complexity. It does not reflect real, practical computational restrictions. Nevertheless, it is a useful characteristic at the preliminary assessment of a situation.

One of the most obvious consequences, and one that is virtually entirely neglected now, is that before starting the investigation of any complex system, it is necessary, at least approximately, to estimate the required volume of information.

2.5 Principles of Overcoming the Transcomputational Complexity of System Problems

The purpose of this section is the analysis of known and development of new approaches, methods, and principles of solving transcomputational complexity problems for real systems. Let us assume that the complexity of solving real problems is determined by three fundamental factors:

- The complexity of the system problem formalization under conditions of uncertainty, incompleteness, and inaccuracy of the initial information and contradictoriness of research goals.
- The complexity of forming a system describing the object of system research.
- The computational complexity of a system research problem.

From the above list a new methodological problem arises—to define what techniques, principles, and methods are necessary for explaining the possibility of solving system analysis problems for real objects with a practically acceptable level of accuracy and within a technically and economically acceptable period of time.

First, we should take into account that Bremmerman's limit gives only a general estimation of limiting a problem's complexity; it does not reflect real, actual computational limitations. Nevertheless, it is a useful characteristic for the preliminary estimation of the information resource capacity of an applied problem. Since real system analysis problems require a huge computational capacity considerably exceeding that of modern computers, this estimation is very important. Consider possible ways of overcoming the transcomputational complexity of system analysis problems for which a determining principle is the rationality principle. It is quite justified owing to contradictory factors, efforts, and purposes typical of applied problems. In such situations the rationality principle is the only real basis for obtaining a practically acceptable solution. That is why this principle is one of the underlying principles of system analysis methodology.

Let us recall that **rationality** is a property of alternative procedures, allowing one to achieve a compromise of goals without excessive resource expenses.

Let us propose the following rationality types:

1. **Technical rationality** is used in search of a compromise between contradictory requirements for the technical characteristics of an object, as well as in other cases involving the choice of technical parameters, characteristics, and properties. Thus, determining factors in the selection are technical characteristics of an analysis object.
2. **Social rationality** is used in search of a compromise between contradictory requirements for the social properties of an object, and also in other cases of choices connected with social parameters, characteristics, and properties of an object. The determining factors in such a choice are the properties of an object determining its social importance.
3. **Economic rationality** is a type of rationality for which the economic characteristics and parameters of an object are of prime importance, including economic efficiency.
4. **Technological rationality** is a type of rationality for which parameters, characteristics, and properties describing manufacturability of an object are of prime importance.
5. **Operational rationality** is a type of rationality for which parameters describing operational properties of an object are of prime importance.
6. **Ergonomic rationality** is a type of rationality for which the parameters, characteristics, and properties determining the quality of interaction of an object and a person, as the user or participant in the process of object control are determining factors. It is important to note that at first sight it seems that ergonomic rationality and social rationality overlap each other. But this is not so. Social rationality determines the contribution of an object in the social environment as a whole (employment, improvement of the ecological conditions in a certain region due to the replacement of out-of-date production by a more advanced technology, etc.). Ergonomic rationality determines principally different object properties, namely, the

ability to create for an individual the necessary conditions for his or her direct interaction with an object.

7. **Aesthetic rationality** is a type of rationality for which ethical and aesthetic properties of an object are of prime importance. It is used in searching for a compromise in the expert procedure of forming ethical and aesthetic requirements for an object and estimation of the corresponding properties of an object.
8. **Legal rationality** is a type of rationality for which degrees of conformity of properties, characteristics, and parameters of a system analysis object to laws currently in force, government standards, standard acts, and other norms (standards) and regulations of society—customs, traditions, etc.—are determining factors. It is used in expert evaluation of novelty competitiveness on some foreign market, characteristics and parameters of competition, and also in the development of legislation and as a support mechanism for national producers in solving other system problems in the social and economic spheres. In particular, this type of rationality is used in the development and application of new laws, expert evaluation of the degree of conformity and consistency with the laws currently in force, and the degree of conformity to the goals and objectives of the new laws.
9. **Informational rationality** is a type of rationality essentially consisting in searching for a rational compromise between the levels of completeness, credibility, and timeliness of the information regarding an object under investigation and resource consumption, including time margins, for their achievement in the process of creation of dataware of system analysis problems.
10. **Methodological rationality** is a type of rationality used in an operative choice of system analysis tools in specific conditions. Characteristics and properties of methods, algorithms, application packages that are analyzed from principally different sides are determining factors. First,they are considered from the position of attaining the necessary computation accuracy, second—from the position of the realized level of velocity and acceptable spending of computing facility memory and, third—from the position of their compatibility in a single computational process. The practical necessity of this approach is dictated by the fact that in the formation and solving of applied problems of system analysis, especially in the social, economic, and ecological spheres, it is necessary to ensure multiple-factor compatibility of both computational methods and algorithms, and empirical approaches, heuristic methods, and expert procedures in single purposefully organized processes, in many fragments of which interactive procedures are used.
11. **Coalition rationality** is a type of rationality that is used in forming a single strategy of actions of a number of subjects that simultaneously have common and individual purposes. The search of a compromise of

contradictory individual purposes from the position of attaining the required level of common purposes is a determining condition. This type of rationality is basic in the search for collective coordinated decisions of group expert evaluation, in the formation of multilateral treaties, and in the creation of military and economic coalitions.

12. **Commonsense rationality** is a type of rationality that is used generally in conditions of extreme time deficit available for making decisions or for the preliminary selection of a solution set. Here, the determining factor is the use of intuition, experience, and foresight of experts and analysts or a decision maker. The distinctive feature of this type of rationality is almost a complete absence of quantitative justification of a solution. In system analysis it is used for different applications. First, it is used in the interactive mode for fast decision making in the process of a computing experiment, and second, for solving problems of preselection. In particular, for preliminary formation of alternative solution sets regarding the structure and functions of an object, for the preliminary formation of a possible situation set for predicting abnormal and critical modes of a designed object, and in other similar situations. The sets of solutions generated in this way are further selected by expert evaluation or other methods—with the use of simulation modeling, computational algorithms, and other approaches. In addition, this type of rationality is widely used by decision makers in practice in critical and extreme situations, for example, when great fires, earthquakes, shipwrecks, and other natural disasters occur, and also in military operations.
13. **Game rationality** is a type of rationality for which the principles of rational game strategy determine the principles on which choices are based. Step-by-step balance of risk and caution, comparability of alternative outcome values, commensurability of expected profit and possible damage to the final result are the principles of game rationality. As in commonsense rationality, experience, intuition, and foresight are basic in the procedure of forming alternative decisions. But unlike commonsense rationality, the choice of decision is made on the basis of multicriterion alternative evaluation based on the specified principles of game strategy. This type of rationality is widely used in conditions of high uncertainty of an expected result, particularly in strategic planning of large and innovative projects. **Venture project**: the research project of fundamentally new ideas allowing to obtain, with the same probability, both a big profit and a new practically important result, or a complete failure. A venture capitalist takes risk and deposits funds as a charitable contribution to support research. **Innovation project**: a project of new-idea realization; an executor takes all the risk and takes credit for this project. In the process planning a business game of two expert groups supporting opposite opinions is organized. During the game strengths and weaknesses of the project are revealed and a general conclusion is drawn. This type of

rationality is also used in the planning of great strategies in the economy, industry, and military science and in the development of new weapons, new technologies, etc.

14. **Conceptual rationality** is a type of rationality that is basically a purposeful integration of all the above-examined types of rationality. This type of rationality is most important in solving complicated applied problems of a global character, dealing with, for example, national interests, relations of states, large transnational companies, and others.
This type of rationality plays an important role in the market strategy of large companies. For example, a typical method for removing a competitor from the computer market has giant companies buying up small and medium-sized firms, advancing in the market, and then removing their products. This is especially characteristic of system and network software products. Consequently, the balance is disturbed in the market interaction of informal associations of companies opposing the giants in the industry. As a result, the successful small and medium firms incur big losses in a short time.
It is necessary especially to underline that this type of rationality is not a mere totality of the concepts considered above. Conceptual rationality is a unified complex of different types of rationalities, system coordinated for achievement of a common goal.
Unity in conceptual rationality is achieved by the unity of an idea and a mechanism of its realization, which is ensured by an agreement of problems, terms, resources, and expected results of interaction of all types of rationality. From this position, conceptual rationality may be treated as multirationality and metarationality. This type of rationality is used in the development of strategy of crediting certain countries, conceptions of activities of a company, conceptions of a rational strategy in a competitive struggle over the allocation of free capital and in other similar cases connected with solving strategic problems for companies, corporations, and nations.

Techniques for solving transcomputational problems. Consider the techniques for solving transcomputational complexity on the basis of rational decomposition of a general problem of system analysis. The above list of principles for overcoming transcomputational complexity of such problems has defined the possibilities of different types of rationality and indicated appropriate areas of their application with quite evident fundamental methods, namely:

- The rational limitation of a set of analyzed properties, factors, states, and situations.
- The rational limitation of a set of alternative variants of solutions.
- The rational choice of system analysis tools.

The main question of overcoming transcomputational complexity remains open, namely, how and by what means are the mentioned techniques to be realized and how to use potentials of different types of rationality. Answering this question is rather difficult for several reasons, among which the most essential are:

- Principal nonformalizability of a general system analysis problem, owing to the necessity of a nonformalized choice of system analysis goals, criteria of their attainment, and impact of some other factors.
- Uniqueness of each practical problem of system analysis, owing to the uniqueness of its goals, the structure of each complex system, conditions of development, production, operation, and also the uniqueness of abnormal, critical, and emergency situations.

Therefore, approaches to overcoming transcomputational complexity for the various practical problems differ in conditions and ways of realization, properties, and possibilities of the different types of rationality.

The basic idea consists in implementation of a definite sequence of procedures of a purposeful reduction in the power of the set of analyzed variants by selecting, based on various criteria and limitations, only a subset of rational variants and exclusion from the subsequent analysis of unpromising and nonrational ones.

Realization of this idea is reduced to a sequence of three basic stages:

1. Rational formalization of system analysis (SA) problem.
2. Rational choice of SA computational process.
3. Rational choice of SA computational facilities.

The necessity of finding a rational approach at each stage is determined by the presence of contradictions at all SA stages and in solving each problem that is included in the functional space of SA conditions and directly in the functional space of an object's properties.

For instance, in the formalization of the general SA problem it is necessary to solve the **contradiction between the level of credibility and justification of the solution and the level of complexity of SA problem**. Here, on the one hand, to raise the level of credibility and justification, it is necessary to increase the level of completeness and credibility of the initial information used for SA, particularly to increase the quantity of factors taken into account and the number of parameters describing each factor. On the other hand, with a growth in the number of factors and indices, the scope of a concrete variant of an SA problem statement increases, and, as consequence, the computational complexity of the problem increases, and by the polynomial or exponential law, the complexity of the algorithm realizing the accepted variant of the SA problem increases. Correspondingly, the time spent for the algorithm realization increases. Therefore, for a reduction in problem complexity it is necessary to decrease the quantity of factors and indices. Consequently, the rational formalization of a problem must ensure a rational resolution of this contradiction. Here, there arises an informal problem of estimating the quality of this contradiction resolution or, in other words, estimation of the efficiency of the accepted variant of the problem formalization rationality.

In this problem a substantial nonformalizable constituent is present, connected with an uncertainty: on what parameters to estimate the level of rationality of the accepted variant of problem formalization and by what criteria to optimize the level of its effectiveness. This problem is one example of a nonformalizable problem.

Let us point out that the notions "parameter" and "criterion" are not synonyms. A parameter is an index by which the level of the problem solution is estimated, for example, the time spent on computation. A criterion is a formalized expression of quality (effectiveness) of attainment of the set goal, for example, Chebyshev's criterion, mean-square, or some other criterion of computation time minimization for a concrete problem or for a definite class of SA problems.

The choice of a rationality type for each of the indicated stages in the practical realization of the above-stated fundamental idea of overcoming transcomputational complexity has its specific features and peculiarities and depends on the goals and features of a concrete system analysis problem. That is why we shall give only general recommendations. These questions will be considered in more detail when we study concrete classes of SA problems.

At the **first stage** it is expedient to use almost all types of rationalities described above, above all, conceptual, technical, economic, and operational rationalities. In addition, in some applications such types of rationality as social and legal rationality are important. At this stage a reduction in the problem dimensionality is achieved by excluding from analysis such factors as the influence of what may be neglected or the influence of what is known a priori, and also the influence of what may be taken into account in the final phase of analysis by introducing corrections, accurate definitions, or in other ways. This problem can be solved using various criteria that characterize the efficiency of achievement of certain goals, in particular, minimizing the influence of risk factors.

At the **second stage** it is expedient to use such types of rationality as conceptual, technological, methodological, and informational rationality. At this stage a reduction in problem dimensionality is achieved in specific ways based on the principle of system decomposition of the initial SA problem into a sequence of more simple partial problems and the principle of aggregation of solution of the partial problems and solution of the initial SA problem.

The **third stage**, to some extent, is provisional in relation to the above-mentioned first and second stages. At this stage, the principles of rationality are used for the purpose of choosing those computational facilities that are rational from the point of view of attaining SA goals with the available possibilities. Therefore, at this stage it is expedient to use the following types of rationality: economic, technological, ergonomic, and others.

Let us consider the methods and possibilities of the second stage in more detail, because it of prime importance both in the development of the mathematical statement of a problem and in the choice of computing facilities. It is known that the choice of the method for solving an applied problem depends on its mathematical statement, and the choice of an algorithm and method for software realization determines requirements of the computing facilities. In algorithmization of the fundamental idea of overcoming transcomputational complexity of a system analysis problem, the basic type of rationality is conceptual rationality, on which basis the principle of system decomposition of a general system analysis problem is realized. The system decomposition is executed in two aspects—in conceptual-functional and in structural-functional. Their realization has its peculiarities for different classes of system analysis problems. But the essence of methods is common. Therefore, in this section we shall deal only with the general scheme of system decomposition using the example of problems connected with the development of a complex engineering system (CES).

The primary is the conceptual-functional decomposition of the general system analysis problem into problems of functional space of the conditions in which the CES functions and the conceptual space of the properties of the CES structure. Each space consists of three partial system analysis problems, based on (a) goal, situation, and information analysis and (b) structural-functional, organizational-procedural, and technical-economical analysis, respectively.

The decomposition procedure is executed individually for each functional space.

Let us explain the essence of structural-functional decomposition using as an example the conceptual space of a CES structure. In this case, the CES system analysis problem may be reduced to the following sequence of partial problems:

- Formation and analysis of a permissible set of principles of CES operation, the choice of the functioning principle on the basis of conceptual rationality.
- Formation and analysis of a permissible set of CES structures for the accepted principle of functioning.
- Choice of a rational structure.

The following partial problems are of the same type, proceeding from the methods of realization. Therefore, we shall state them in a generalized form, as a problem for a functional element (FE) of the structure, and then we shall show the concrete value of the functional element on the different hierarchical levels of the system:

1st level: formation and analysis of a permissible set of functional elements;

2nd level: choice of the rational FE.

The following elements are gradually taken as functional in the course of transition to a lower level of the hierarchical structure of the CES: "functional system" – "functional unit" – "functional block" – "functional module". A CES is the highest level of hierarchical structure and presents an integrated object that is purposefully made of structurally interconnected and functionally interactive functional systems.

In its turn, a "functional system" presents a FE, which is purposefully made of structurally interconnected and functionally interactive "functional units". Similarly, a "functional unit" is made of "functional blocks" and a "functional block" – of "functional modules". In the process of transition to a lower level, the number of FEs composing FEs of a higher level increases, but at the same time the number of functions performed by one element decreases. For example, a CES may consist of three to ten "functional systems", each performing up to several dozen functions. A "functional module", as a rule, performs one to two functions. Attention should be drawn to one important tendency of current technology. A widespread introduction of high technologies on the basis of modern computer engineering, based on microelectronics, makes possible a constructive realization of FEs of several levels in the form of a unified very-large-scale integration circuit. For example, most PC functions are performed by one very-large-scale integration circuit—a processor. Current very-large-scale integration circuits may contain more than 10^7 active and passive elements (transistors, diodes, inductive elements). It follows that, if each element is characterized by only one parameter and each parameter has only two values, then to analyze a very-large-scale integration circuit as a unified functional element (without decomposition), it is necessary to process the following information volume (bit):

$$k^n = 2^{10^7} .$$

Here, n is the number of parameters and k the number of values (states) of each parameter. Note that at $k = 2$ and $n = 308$, we have $k^n \geq 10^{93}$, which is greater than Bremermann's limit.

The structural and functional decomposition allows one to carry out system analysis subsequently for each hierarchy and to ensure problem solvability on each level and for the one very-large-scale integration circuit as a whole. It is reached, on the one hand, by simplifying the screening procedure of unpromising alternative solutions for FEs of each level using expert estimations. On the basis of their intuition, experience, and knowledge, experts can rather quickly and without detailed calculations form a small set of possible alternative solutions (not big) at the stage of choosing the principle of CES's operation.

On the other hand, it becomes possible to carry out a comparative analysis on the basis of quantitative calculations with an application of computational procedures. Such a possibility appears thanks to a drastic reduction in the required volume of information processing. If, for example, each alternative is characterized by a group from ten parameters and each is constrained by only two values (min and max), which can be ensured by each alternative, then we get:

$$k^n = 2^{10} = 1024\ (\text{bit})\,.$$

If we take into account that, in practice, the number of analyzed alternatives does not exceed ten, and each of them may be analyzed separately, then approx. 10^4 bits of information must be processed, i.e.,

$$10 \cdot k^n = 10 \cdot 2^{10} \approx 10^4\,.$$

Certainly, this estimation is too simplified. In practice, for this case it is necessary to process about 10^4–10^5 bytes (1 byte = 8 bits), i.e., 10–100 KB of information. An increase in processed information is conditioned by the necessity to describe and identify each alternative, each parameter, a variation interval of each parameter, and so on. But the problem is, in principle, solvable. For each of the above-mentioned sets we can rank alternatives in accordance with the degree of their usefulness based on the generalized parameters of a CES, the level of execution of the assigned requirements, or other criteria. For example, at present there are packages of applied programs that allow one to choose the structure and all functional elements of a corporate network on a number of generalized indices (volume of processed information, volume of database, number of jobs, etc.).

A choice of alternatives on the hierarchical levels such as "functional unit", "functional block", and "functional module" is carried out on the basis of more exact estimations. On these levels, the choice and a draft design of alternatives virtually coincide, since for the comparative estimation and the choice of an alternative it is necessary to know the basic technical characteristics of the corresponding functional elements. Therefore, on these levels the methods and facilities of artificial intelligence are widely used, above all as expert systems and as intelligence facilities supporting the developer's decisions. In practice, as experience shows, each CES is unique, with respect to purpose, conditions of operation and application, or the base of elements used. Therefore, in carrying out system analysis on different hierarchical levels, such as "functional system"—"functional unit" computing experiments are widely used, and on the hierarchical levels "functional block"—"functional module", both computational and natural experiments are used. The general idea of system analysis methodological techniques in using computational and natural experiments consists in the construction of such a sequence of iterations that allows one to rationally distribute requirements for an object, as a whole, as a system of requirements for functional elements of all levels, using the principle "top-down", and then to aggregate the system analysis results using the principle "bottom-up". Upon execution of the draft design on the lower level, such as "functional module" for all FEs, it is possible on the basis of a computing experiment, in particular using simulation methods, to determine the characteristics of FEs on the higher levels and the CES as a whole, to estimate the extent of fulfillment of the given requirements and limitations, and also to define such generalized indices as technical and economic efficiency. If the obtained solutions satisfy a decision maker, then engineering design is executed consecutively, beginning with the lowest level. Otherwise, another iteration of a choice is executed, beginning with a hierarchical level for which the given requirements, restrictions, and other conditions are unrealizable. As a whole, this approach makes it possible to reduce considerably the number of analyzed alternatives and decrease the volume of processed information.

Thus, the main methodological technique of overcoming transcomputational complexity of system analysis problems is the system decomposition of the initial problem into a sequence of simpler problems. These problems comprise:

- **Rational use of the possibilities of the system "person-expert and computing procedures" (including those based on intellectual systems of decision supporting).**
- **Rational organization of the iteration process of system analysis on the basis of a rational decomposition of the "top-down" requirements and the "bottom-up" aggregation of results.**

The major feature of the considered process of system decomposition is the difference in accuracy, credibility, and justification of the solution on different levels of structural and functional decomposition. On levels such as "functional system" the indicated indices are much lower than on the lowest levels such as "functional unit" or "functional module". This approach is quite justified and practically useful, since the choice of a structure, for example, for the "functional systems" level is made in conditions of goal uncertainty and, what is especially important, in conditions of incompleteness, inaccuracy, and uncertainty of the initial information. Therefore, at these stages an optimal approach may be a rational combination of guessing, intuition, expert experience, and available information. This would make it possible to provide the initial justification for a structure and for functional systems of a CES and proceed to the next stages of system analysis, the last of which is the draft design of functional blocks and functional modules. Realization of the final system analysis stages allows one to increase considerably the knowledge of the capabilities of a CES as a whole on the basis of aggregating these results by the methods of a computing experiment or simulation modeling. As a result, it allows one to obtain the final justification of the decisions taken, for example, on the level of "functional systems", or justification that these decisions are unsuitable, after which the iteration process is to be repeated, starting from the highest levels.

3 Disclosing Uncertainties in System Analysis Problems

Disclosure of uncertainties is an important problem for the class of system analysis formalizable problems. First, we will notice that uncertainty is a typical property of system analysis practical problems. It is connected with the diversity of goals, properties, and system analysis object features. Applied problems without uncertainties are rather the exception than the rule. An adequate problem statement usually has various types of uncertainties, which reflect the natural state of a researcher. His knowledge is always rather incomplete and inaccurate. It directly follows from the Gųdel theorem about incompleteness [103] and the evolution of human cognition. If everything were known with certainty, completely, and exactly about the processes, factors, and evolution of the Universe, then civilization would stop developing.

In practice, the most widespread uncertainties are those related to goals, situations, and conflicts. In the brief contensive formulation, the essence of these uncertainties consists in the following:

1. Goal uncertainty is the uncertainty in the choice of goals in multicriterion decision-making problems.
2. An uncertainty of knowledge about possible situations is the uncertainty of uncontrolled factors influencing practical activities (situational uncertainty).
3. An uncertainty of conflicts is the uncertainty of a choice of intentions, goals, and plans during partners' interaction and competitors' or opponents' counteraction (information uncertainty of conflicts).

3.1 Problems and Methods of Disclosing Goal Uncertainty

Firstall, we shall consider problems of disclosing goal uncertainty. In the general case, when we are studying an integral object, the necessity coordinating its goals arises. For some goals the optimal solutions correspond to the minimum value of the corresponding criterion, and for others—to the maximum value. But, with the help of a certain replacement of variables, these problems are easily reduced to a single type of criteria and to a single type of optimization problems. They may be considered as a multicriterion optimization problem [99]:

$$f_1(\overline{x}) \to \max_{\overline{x}\in D}, \quad f_2(\overline{x}) \to \max_{\overline{x}\in D}, \quad \ldots, \quad f_m(\overline{x}) \to \max_{\overline{x}\in D}. \tag{3.1}$$

It is obvious that the best solution of problem goal uncertainty disclosure will be a value $\bar{x}$ at which conditions (3.1) are fulfilled simultaneously for all goal functions. However, in practice, this is not feasible, as the functions $f_k(\bar{x})$, $k = 1, 2, \ldots, m$ are different by nature (generally, they are determined by criteria of a different nature—various physical, technical, economic, or other contents). Therefore, each function extremum is achieved at the value $\bar{x}$, and it is practically impossible to find a value $\bar{x}^0$ at which conditions (3.1) are fulfilled simultaneously for all goal functions. From here it follows that the problem is reduced to finding a value $\bar{x}^0$ at which a rational compromise of the set goals will be ensured. Two basic approaches are applied to searching for a rational compromise. The essence of the **first approach** is to exclude from the analysis a priori unacceptable variants of a solution. The essence of the **second approach** is to use methods and means of multipurpose problem reduction to an ordinary optimization problem with one criterion and to solve it.

First, we shall consider the first approach based on the **idea suggested by Pareto**: try to reduce a set of initial variants of solutions by excepting from the analysis those variants that are a priori unacceptable [99]. This idea is realized as follows. Let us assume that some vector x is chosen; we shall designate its value as x^*. Now, we make some other choice of $\hat{x}$ such that for all goal functions

$$f_i(\hat{x}) \geq f_i(x^*), \quad i = \overline{1, m}, \tag{3.2}$$

and at least one of the inequalities is strict. It is obvious that the choice $\hat{x}$ is more preferable than x^*. It is necessary to exclude from the given analysis all vectors with value x^* for which condition (3.2) is fulfilled. It is necessary to put to informal analysis, for the purpose of comparing among themselves only those vectors x^* for which there is no such value $\hat{x}$ that satisfies inequality (3.2) for all criteria.

A set of all such values x^* is called the **Pareto set**, and vector x^* is called the **unimprovable vector of results (Pareto's vector)**.

Let us consider in more detail the method of finding the Pareto set. Let vector $\bar{f}$ of goal functions $f_i(x)$, given on set D, be defined as

$$\overline{f} = \left\{ f_i(x) \,|\, i = \overline{1, m};\ x \in D \right\}, \quad D = \left\{ x | x^- \leq x \leq x^+ \right\}.$$

There is such a set Γ of limit values $\tilde{x} \in D$ that divides initial set D into two sets: set Π and set $\overline{D}$. Set D is to satisfy the following condition:

$$\Pi \cup \overline{D} = D, \quad \Pi \cap \overline{D} = 0. \tag{3.3}$$

Set Π consists of such values $x_{k_1} \in D$ for which condition $f_i(x_{k_1}) \geq f_i(x^*)$ is fulfilled for all $i = \overline{1,m}$. Set Π is defined by the following relationship:

$$\Pi = \left\{x | x = x_{k_1};\ x_{k_1} \in D;\ f_i(x_{k_1}) \geq f_i(x^*);\ i = \overline{1,m}\right\}. \tag{3.4}$$

Set $\overline{D}$ consists of such $x_{k_2} \in D$ for which condition $f_i(x_{k_2}) < f_i(x^*)$ is fulfilled at least for one function $f_i(x_{k_2})$. Set $\overline{D}$ is described by the following relationship:

$$\overline{D} = \{x | x = x_{k_2};\ x_{k_2} \in D;\ f_i(x_{k_2}) < f_i(x^*);\ i \in [1,m]\}.$$

Thus, vector $x^* \in D$ is an unimproved vector of results, and set Π, which satisfies condition (3.4), is a Pareto set.

Owing to condition (3.4) set Γ is a boundary that separates the Pareto set from set D. According to (3.3) set $\overline{D}$ is a subset of the initial set D, from which Pareto's set is singled out $\Pi : \overline{D} = D \setminus \Pi$. Therefore, all alternative solutions, which belong to $\overline{D}$, are excluded from consideration.

Example. Pareto's set must be singled out in area $D \in [x^-, x^*]$ (Fig. 3.1). We shall break the given set D into three subsets. Area D_1 is the area x from x^- up to x_1^0, where x_1^0 is the value of x, at which $f_1(x)$ reaches the maximum:

$$f_1\left(x_1^0\right) = \max_{x \in D} f_1(x),\ x \in \left[x^-, x_1^0\right).$$

Area D_2 is the area x from x_1^0 up to x_2^0, where x_2^0 is the value of x, at which $f_2(x)$ reaches the maximum:

$$f_2\left(x_2^0\right) = \max_{x \in D} f_2(x),\ x \in \left[x_1^0, x_2^0\right].$$

Area D_3 is the area x from x_2^0 up to x^+. We shall compare areas D_1 and D_2. For these areas $f_2(x)|_{x \in D_1} < f_2(x)|_{x \in D_2}$, i.e., the values of function $f_2(x)$ for any $x \in D_1$ are less than the values of function $f_2(x)$ for any $x \in D_2$. For $f_1(x)$ we have $f_1(x)|_{x \in D_1} \leq f_1(x)|_{x \in D_2}$, i.e., in a certain interval of area D_1 the values of $f_1(x)$ are commensurable with its values in a certain interval

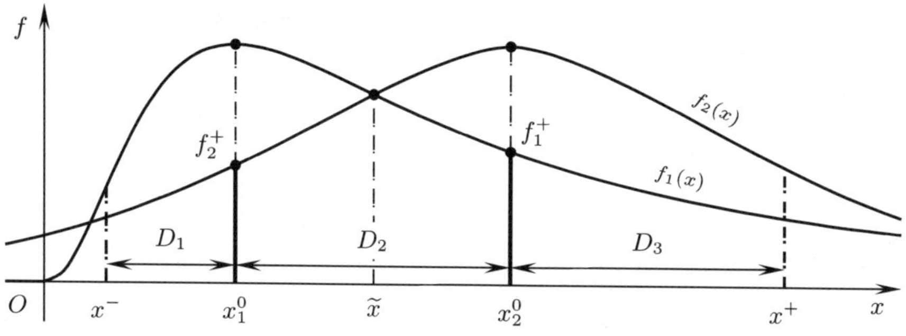

Fig. 3.1. Singling out the Pareto set

of area D_2. Hence, area D_1 obviously yields to area D_2 in the sense of the goal function $f_2(x)$. Similarly, for area D_3 of any $x \in D_3$ the value of the function $f_1(x)$ is less than the value of the function $f_1(x)$ for any $x \in D_2$ $f_1(x)|_{x \in D_3} < f_1(x)|_{x \in D_2}$. But for $f_2(x)$ we have $f_2(x)|_{x \in D_3} \leq f_2(x)|_{x \in D_2}$, i.e., the values of $f_2(x)$ are commensurable in a certain interval of area D_3 with its values in a certain interval of area D_2, hence area D_3 obviously yields to area D_2 in the sense of the goal function $f_1(x)$. Thus, it is necessary to exclude from area D areas D_1 and D_3, as condition (3.2) is not fulfilled in them. According to (3.2) area D_2 is a Pareto set. For the given area the following restrictions are fulfilled

$$f_1(x)|_{x \in D_2} \geq f_1^+; \quad f_2(x)|_{x \in D_2} \geq f_2^+ .$$

In decision-making theory there is a principle called "***the Pareto principle***", which states that a rational solution of a multicriterion problem or a rational compromise in a multipurpose problem is among x, belonging to the Pareto set. But the Pareto principle does not allow one to select a single solution. It only allows one to restrict a set of possible alternative solutions. In the considered example it is necessary to find a rational solution in area D_2.

Then it is necessary to point out that the question of what single solution (or what value $x \in D_2$) is rational remains open. In fact, the problem of finding a rational compromise has an informal character due to a number of factors. First, the importance of various goal functions may depend on a number of objective factors. For example, a plane designed for operation in Arctic conditions will essentially differ in many properties from a plane aimed at functioning in hot climate conditions. Second, the importance of goal functions may depend also on subjective factors. For example, in one design bureau, due to its established traditions, plane speed is of primary importance, and in another, its carrying capacity. Third, a choice of goal priority may be determined by additional conditions or restrictions. For example, restrictions may be imposed on plane production cost or on total cost of production and operation.

In all considered cases the construction of a Pareto set allows one to obtain additional information that gives a qualitative estimation in a comparison of different variants. The decision maker on the basis of the Pareto set analysis may estimate how an increase of one goal function may affect changes in others.

From the considered example it follows that the $f_1(x)$ maximum and $f_2(x)$ minimum are at point x_1^0, the $f_2(x)$ maximum and $f_1(x)$ minimum are achieved at point x_2^0, and at point $\tilde{x}$ the equality $f_1(\tilde{x}) = f_2(\tilde{x})$ is fulfilled. The decision maker defines which of these variants is preferable. If the decision maker believes that the criteria are equally preferable, so the variant $\tilde{x}$ is rational, when $f_1(\tilde{x}) = f_2(\tilde{x})$. If the goal $f_1(x)$ is more important, then a rational solution lies in the interval $\left[x_1^0, \tilde{x}\right)$. If the goal $f_2(x)$ is more important, then a rational solution is in the interval $\left(\tilde{x}, x_2^0\right]$.

However, in the two latter cases, a degree of preference of one goal to the other remains a subjective measure of the decision maker.

Quantitatively, a measure of goal preference is usually characterized by ***the importance coefficient*** of each of the considered goals. The introduction of such coefficients allows one to apply the second approach of disclosing goal uncertainties based on a reduction of a multipurpose problem to an ordinary one with one criterion. In this case, a rational compromise of goals is defined by the decision maker's choice of importance coefficients.

We shall consider some of the most frequently used ways of reducing a multipurpose problem to a one-criterion problem.

Linear convolution. The essence of this method is that instead of m given goals described by functions

$$f_i\left(x\right), \quad i = \overline{1, m}\,,$$

one generalized goal is introduced, which is described by the function

$$F\left(x\right) = \sum_{i=1}^{m} c_i f_i\left(x\right) . \tag{3.5}$$

Here, c_i, $i = \overline{1, m}$ are the initial goal-importance coefficients, which reflect the preference measure of the decision maker. Usually it is assumed that the coefficients c_i are normalized in some way. As a rule, normalization is applied in the following form:

$$\sum_{i=1}^{m} c_i = 1 . \tag{3.6}$$

Note that, for positively defined functions $f_i(x) > 0$, $i = \overline{1, m}$ is fulfilled, which corresponds to our practical tasks. Instead of the additive function $F(x)$, it is possible to use the multiplicative function $F_{\mathrm{M}}(x)$ of the form

$$F_{\mathrm{M}}\left(x\right) = \prod_{i=1}^{m} \left(f_i\left(x\right)\right)^{c_i} . \tag{3.7}$$

Taking the logarithm of the left and right sides of equation (3.7), and having introduced the designation

$$\lg F_{\mathrm{M}}\left(x\right) = \hat{F}\left(x\right) ,$$
$$\lg f_i\left(x\right) = \hat{f}_i\left(x\right) ,$$

we get the expression of an additive kind $\hat{F}(x) = \sum\limits_{i=1}^{m} c_i \cdot \hat{f}_i(x)$. As a result of convolutions (3.5) and (3.7), the problem of disclosing goal uncertainties, presented as a multipurpose problem of optimization $f_i(x) \to \max$, $i = \overline{1, m}$ is reduced to a single-purpose standard problem of mathematical programming $F(x) \to \max$ with restrictions determined by the initial data or requirements specification.

This method is widely used in practice. But in applying it in real problem solving the following principal drawbacks are to be taken into consideration:

- The choice of importance coefficients c_i is subjective to a considerable degree, and therefore principal errors and miscalculations are possible.
- Solving an optimization problem for $F(x)$ does not mean that the rational values for all given goals are achieved. In convolutions (3.5) and (3.7) the deficient value of one goal function can be compensated by an increase in values of other goal functions. Moreover, by changing the values of the importance coefficients, it is possible to obtain a series of different values x, for which the function $F(x)$ has the same value.

Hence, in essence, goal uncertainty in such a statement of the problem is not disclosed to the extent required for practical purposes, since the convolution does not give an unambiguous rational variant of values x. In essence, goal uncertainty is reduced to an uncertainty of importance coefficients, since solving the obtained one-purpose problem is many-valued, determined, at best, by the Pareto set.

Application of technical restrictions. A number of methods used for disclosing goal uncertainties are based on the application of a priori information about the set goals. For example, in design and planning problems certain normative restrictions are frequently imposed from above (total cost, permissible dimensions, weight, and others), or bottom limits exist of certain characteristics and parameters (for example, parameters of reliability, strength, service life etc.). Both upper and lower restrictions may be reduced to one of the forms. Let the following restrictions be given:

$$\begin{cases} f_i(x) \geq f_i^*(x), & i = \overline{1, m_0}\,, \\ f_i(x) \leq f_i^*(x), & i = \overline{(m_0+1)\,, m}\,. \end{cases} \tag{3.8}$$

To simplify the problem solving, it is possible to write down (3.8), reasoning from one type of restrictions: upper or lower restriction only. For example, the upper restriction:

$$\begin{cases} -f_i(x) \leq -f_i^*, & i = \overline{1, m_0}\,, \\ f_i(x) \leq f_i^*, & i = \overline{(m_0+1)\,, m}\,, \end{cases}$$

or the lower restriction:

$$\begin{cases} f_i(x) \geq f_i^*, & i = \overline{1, m_0}\,, \\ -f_i(x) \geq -f_i^*, & i = \overline{(m_0+1)\,, m}\,. \end{cases}$$

Assume that a priori (for example, in the requirements specifications) restrictions are imposed on the goal functions

$$f_i\,(x) \leq f_i^*, \quad i = \overline{1, m} \tag{3.9}$$

or

$$f_i(x) \geq f_i^*, \quad i = \overline{1,m}. \tag{3.10}$$

With these restrictions, it is necessary to ensure

$$f_i(x) \to \max, \quad i = \overline{1,m}. \tag{3.11}$$

For such a statement, various options of disclosing goal uncertainties are possible on the basis of a reduction of a multipurpose problem to a standard one-criterion problem.

Variant 1. For each x value we introduce a function

$$F_1(x) = \min_{i\in[1,m]} \frac{f_i(x)}{f_i^*} \tag{3.12}$$

and define such values x^0 that correspond to the condition $F_1(x^0) = \max_{x\in D} F_1(x)$. Here, D is the permissible multidimensional area of vector x variation, assigned, for example, by means of constructive or technological restrictions. With such a statement it is guaranteed that in the worst case, which corresponds to $\min_{i\in[1,m]} f_i(x)/f_i^*$, the maximum value $F_1(x)$ is ensured. Such a problem of provision is a ***maximin*** optimization problem.

Variant 2. For each x value we introduce a function

$$F_2(x) = \max_{i\in[1,m]} \frac{f_i(x)}{f_i^*} \tag{3.13}$$

and define such a value x^0 at which the function $F_2(x)$ has the minimum value

$$F_2(x^0) = \min_{x\in D} F_2(x). \tag{3.14}$$

With the given statement of the problem it is guaranteed that its solution, which, in the worst case, corresponds to $\max_{i\in[1,m]} f_i(x)/f_i^*$, the minimum value of $F_2(x)$, will be ensured. Such a problem is a ***minimax*** optimization problem.

The difference between variants 1 and 2 consists in their belonging to different optimality conditions. Variant 1 provides the maximum possible deviation among all $f_i(x)$ from their assigned values f_i^*, since such a deviation is provided for the worst case characterized by the following relationship:

$$F_1(x^0) = \max_{x\in D} \min_{i\in[1,m]} \frac{f_i(x)}{f_i^*}. \tag{3.15}$$

Variant 2 pertains to the inverse problem—a problem providing for possible minimum deviation of all $f_i(x)$ from the assigned values f_i^*. Such a deviation is reached for the worst case under the condition of

$$F_2(x^0) = \min_{x\in D} \max_{i\in[1,m]} \frac{f_i(x)}{f_i^*}. \tag{3.16}$$

Let us consider an example of the peculiarities of realizing the given variants. Assume that an optimization problem must be solved under restrictions (3.9). In this case, it is convenient to use convolution in the form (3.11) and to solve an uncertainty-disclosure problem as an optimization problem in the form (3.12). The solution will be accomplished in two stages. At the first stage, it is necessary to define the Pareto set on the basis of graphically presented initial data (Fig. 3.2). Area I, corresponding to $x \in \left[x^-, x_2^0\right)$, contains unacceptable values of x, since the assigned conditions for $f_2(x)$ are violated, because here $f_2(x) > f_2^*$. Area III, corresponding to $x \in \left(x_1^0, x^+\right]$, contains unacceptable values of x, since the conditions for $f_1(x)$ are violated, because here $f_1(x) > f_1^*$. From here it follows that the Pareto set lies between areas I and III, and its boundaries are defined by the condition $x \in \left[x_2^0, x_1^0\right]$. On the boundaries of this set the following conditions are fulfilled:

$$f_2\left(x_2^0\right) = f_2^*, \quad f_1\left(x_2^0\right) < f_1^*; \quad f_1\left(x_1^0\right) = f_1^*, \quad f_2\left(x_1^0\right) < f_2^*,$$

and at $x_2^0 < x < x_1^0$, $x \in \left(x_2^0, x_1^0\right)$:

$$f_1(x) \leq f_1^*, \quad f_2(x) \leq f_2^*.$$

Thus, the first stage of the problem solving is completed.

The second stage consists in defining the rational compromise conditions on the Pareto set.

In the given area it is necessary to define a value x^0 for which condition (3.9) is fulfilled, and value (3.15) is ensured. Let us choose some values x' and x'' and compare the relationships $f_i(x)/f_i^*$ for different values $i = 1, 2$.

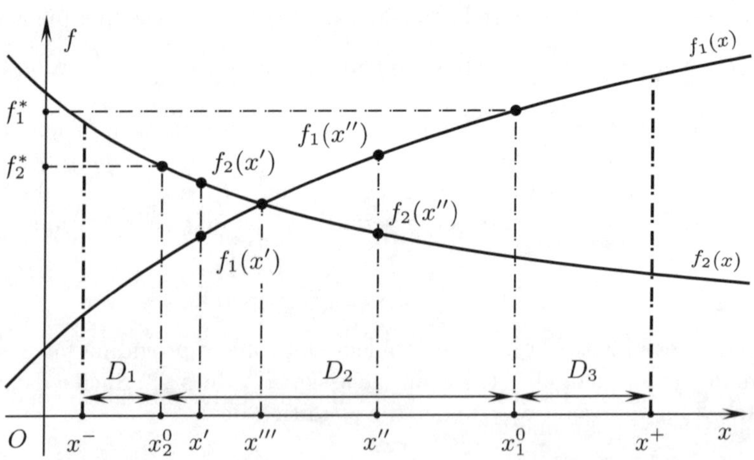

Fig. 3.2. Initial data pertaining to problems of disclosing goal uncertainties $f_1(x)$, $f_2(x)$

As seen from Fig. 3.2, at point $x = x'$ we have

$$\frac{f_2(x')}{f_2^*} > \frac{f_1(x')}{f_1^*},$$

and at point x''

$$\frac{f_2(x'')}{f_2^*} < \frac{f_1(x'')}{f_1^*}.$$

From the given inequalities it follows that it is necessary to find a rational compromise between two points x' and x''.

With the same goal preference such a compromise is reached at point $x = x'''$, where

$$\frac{f_2(x''')}{f_2^*} = \frac{f_1(x''')}{f_1^*}.$$

In the case of differences in goal preferences that are characterized by the importance coefficients k_1 and k_2, a compromise is defined by the following relationship:

$$k_2\frac{f_2\left(x^{(4)}\right)}{f_2^*} = k_1\frac{f_1\left(x^{(4)}\right)}{f_1^*},$$

where the value at point $x^{(4)}$ is a compromise point.

It is necessary to draw attention to the fact that the problem solving equation (3.15) with restrictions (3.10) does not exist, since the condition $f_2(x) > f_2^*$ is fulfilled in area I, and the condition $f_1(x) > f_1^*$ is fulfilled in area III, but these areas do not cross, i.e., it is impossible to fulfill both conditions simultaneously.

The example above allows us to reveal some drawbacks of the considered approach to disclosing goal uncertainty.

First, both variants 1 and 2 give us an answer only to one question—what will be ensured in the worst case—but they leave open some other questions. In particular, it is unknown how one is to realize conditions of a rational compromise when for some values $i_1 \in [1, m_1]$ conditions (3.15) are fulfilled and for others $i_2 \in [1, m_2]$ $(m = m_1 + m_2)$ they are not fulfilled.

Second, the considered approach does not allow us to compare quantitatively various options of permissible solutions from the Pareto set. For example, it is impossible to tell which of the variants is better at points x' and x''.

If on solving a system of inequalities we get several intervals of the Pareto set, then the process of interval narrowing is fulfilled separately for each of them.

Example. Find the Pareto area and define rational compromise conditions for the assigned goal functions:

$$f_1(x) = 10 + 5x + 7x^2,$$
$$f_2(x) = 35 - 10x, \quad x \in [0; 2], \tag{3.17}$$

with the following restrictions:

$$f_1(x) \geq f_1^*, \quad f_2(x) \geq f_2^*; \quad f_1^* = 22, \quad f_2^* = 20\,.$$

First, we define the Pareto set on the interval $[x^-, x^+]$, where the following inequalities are fulfilled:

$$\frac{f_1(x)}{22} \geq 1, \quad \frac{f_2(x)}{20} \geq 1 \tag{3.18}$$

or

$$\begin{aligned} 7x^2 + 5x + 10 &\geq 22\,, \\ 35 - 10x &\geq 20\,. \end{aligned} \tag{3.19}$$

The analytical solution to inequality system (3.18) shows that the required Pareto set is in the interval $x \in [1; 1.5]$ (Fig. 3.3).

For narrowing the Pareto area and reducing the initial two-criteria problem to a one-criterion problem, we shall use technical restrictions based on the minmax principles of $\min_x \max_i f_i(x)/f_i^*$ and the maxmin principles $\max_x \min_i f_i(x)/f_i^*$. The values of the relationships $f_1(x)/f_1^*$ and $f_2(x)/f_2^*$ on interval $x \in [1; 1.5]$ with the step of grid 0.1 are given in Table 3.1.

We can see from the table that at point $x = 1,2$, the conditions $\max\limits_x \min\limits_i f_i(x)/f_i^* = \min\limits_x \max\limits_i f_i(x)/f_i^*$ are fulfilled. Thus, the strategy $x = 1,2$ should be chosen as the rational compromise for two investigated functions.

Disclosing goal uncertainty on the basis of reduction of a multipurpose problem to a system of equations.

The above considered methods of disclosing goal uncertainty are based on the initial multipurpose problem reduction to a single-purpose optimization

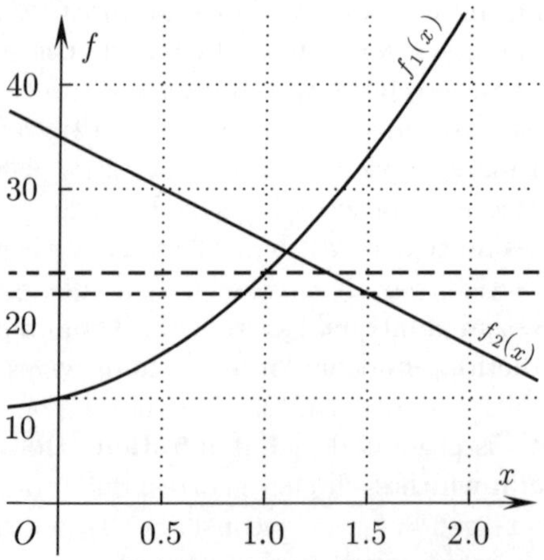

Fig. 3.3. Definition of the Pareto set for system (3.17)

Table 3.1. Values of relationships $f_1(x)/f_1^*$ and $f_2(x)/f_2^*$

x	1	1.1	1.2	1.3	1.4	1.5
$\dfrac{f_1(x)}{f_1^*}$	1	1.089	1.185	1.288	1.396	1.511
$\dfrac{f_2(x)}{f_2^*}$	1.25	1.2	1.15	1.1	1.05	1

problem of the same type, in particular, to a single functional minimization or maximization problem.

Now we consider another approach, which is based on reducing a problem of disclosing goal uncertainty to an optimization problem described by a system of equations. First, consider a problem, in which the restrictions of the form (3.9) or (3.10) are imposed. Assume that it is required for each goal to fulfill the following conditions:

$$f_i\,(x) = f_i^*, \quad i = \overline{1, m}\,. \tag{3.20}$$

However, as follows from Fig. 3.4, these conditions for various functions $f_i(x)$ are fulfilled not at the same values of x, but when they differ considerably. As a result, the necessity arises to analyze jointly all goals from the position of estimating difference in realized restrictions from those assigned for each $f_i(x)$. We present a set of the initial restrictions in the form of a system of equations

$$f_i\,(x) - f_i^* = 0, \quad i = \overline{1, m}\,. \tag{3.21}$$

In practical problems x represents a vector

$$x = \{x_1, x_2, \ldots, x_j, \ldots, x_n\}\,, \quad j = \overline{1, n}\,.$$

Therefore, (3.21) may be considered as a system of m equations with n unknown quantities x_1, $x_2, \ldots, x_n$. It is obvious that the method of solving this system depends on the relationship between m and n.

The following cases are possible: (1) $m = n$; (2) $m < n$; (3) $m > n$. The simplest case is $m = n$, for which it is possible, in a number of situations, to find the goal uncertainty solving system (3.21) by one of the known methods [42]. Here, condition (3.20) will be precisely ensured for each ith goal. However, in practice, the given case is, as a rule, an exception. Moreover, even with $m = n$, a solution of system (3.21) cannot always be found, owing to the contradiction of goals. An example of when it is impossible to find a value x at which condition (3.20) is simultaneously fulfilled for two two-variable functions is presented in Fig. 3.5. Here, the set A is formed by the family of curves, in which each curve corresponds to a certain numerical value of function $f_1\,(x_1, x_2) = c_1$, $c_1 = \text{const}$. Set B is formed by a family of curves, in which each curve corresponds to a certain value $f_2\,(x_1, x_2) = c_2$, $c_2 = \text{const}$.

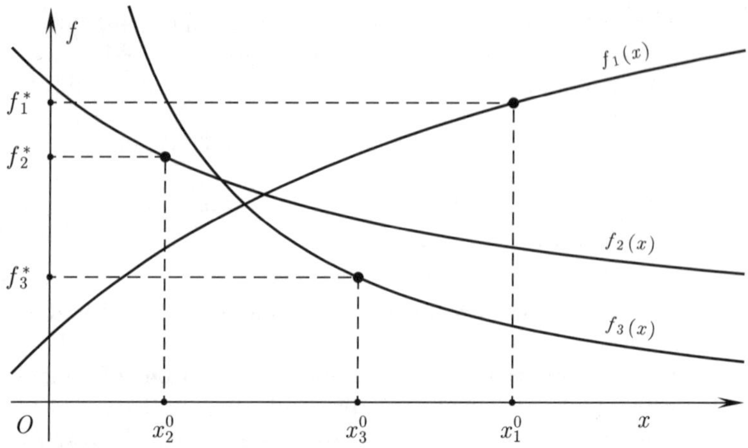

Fig. 3.4. Uncertainty disclosure on the basis of a system of equations

Curves f_1^* and f_2^* define the given restrictions for $f_1(x_1, x_2)$ and $f_2(x_1, x_2)$, respectively. As is seen from Fig. 3.5 there are no values x_1 and x_2 for which the conditions $f_1(x_1, x_2) = f_1^*$, $f_2(x_1, x_2) = f_2^*$ are fulfilled simultaneously, since curves f_1^* and f_2^* do not cross.

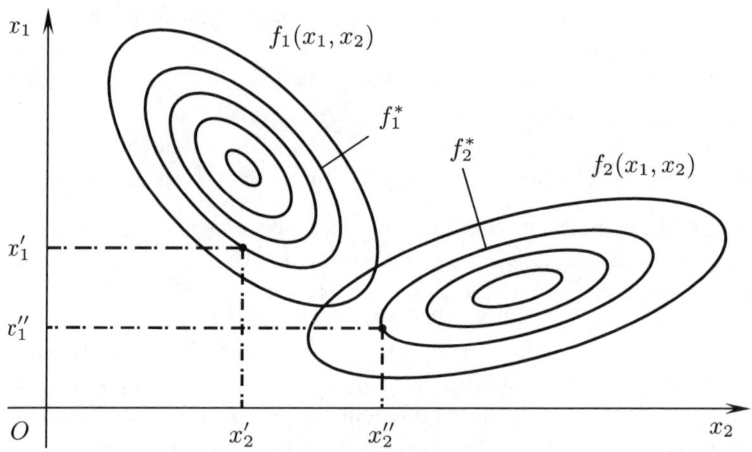

Fig. 3.5. Solving system of equations (3.21) for $m = n$

In the case $m < n$, system (3.21) is overdetermined, which allows one to change arbitrarily some components of vector x. In the general case, the maximum number of such components may be equal to $n - m$. The simplest way of solving the problem at $m < n$ may consist in a priori assignment of some variable values on the basis of the decision maker's intuition and experience, or on the basis of the analysis of the corresponding indices of the known product and project prototypes. As a result, the system passes to the previously considered variant $m = n$. Changing vector x's components it is possible to obtain various conditions for disclosing goal uncertainty. Then, rational solutions are chosen from the obtained variants. However, in practice, this case occurs very seldom, since, as a rule, the number of requirements (i.e., the number of goals $f_i(x)$) exceeds the number of parameters (i.e., the number of components of vector x).

Now, proceed to a consideration of the third case, where $m > n$. System (3.21) is incompatible (underdetermined system), i.e., it is impossible to find values $x_1, x_2, \ldots x_n$ at which conditions (3.20) are fulfilled for all $i = \overline{1, m}$. That is why it is said that in the algebraic sense an incompatible system of equations cannot be solved. Also, it is necessary to note that the functions $f_i(x)$, $i = \overline{1, m}$ are nonlinear in the general case, and system (3.21) is an incompatible system of nonlinear equations.

The methods of solving incompatible systems, such as (3.21), are developed in the function approximation theory [77, 150]. Since the number of equations m exceeds the number of variables n, for such systems it is in principle impossible to obtain a solution what will ensure fulfillment of condition (3.21). Therefore, solving the system is reduced to minimizing the functions

$$\tilde{f}_i(x) = f_i(x) - f_i^*, \quad i = \overline{1, m}.$$

The quality of solving the presented system is characterized by a quadratic criterion, root-mean-square, Chebyshev, medium-scale criterion, and a number of other characteristics. The most expedient criterion from the practical point of view is the ***Chebyshev criterion*** of function approximation. For this criterion the Chebyshev problem of system approximation (3.21) consists in finding a value of $x^0 = \left(x_1^0, x_2^0, \ldots x_n^0\right)$ for which the maximum residual

$$\Delta = \max_{i \in [1, m]} \left| f_i(x) - f_i^* \right|, \tag{3.22}$$

taken as the Chebyshev approximation measure of system (3.21) (or, otherwise, as the Chebyshev approximation measure of the function system $\tilde{f}_i(x)$, $i = \overline{1, m}$), would be minimally possible

$$\Delta^0 = \min_x \Delta. \tag{3.23}$$

Fulfillment of conditions (3.23) means that at $x = x^0$ the function system $\tilde{f}_i(x)$ maximum deviation from zero values will be minimally possible. The choice of any value $x \neq x^0$ will result in the inequality $\Delta > \Delta^0$. The value x^0 is an argument that provides minimization of the maximum residual, i.e.,

$$x^0 = \arg \min_{x \in [x^-, x^+]} \max_{i \in [1, m]} \left| f_i(x) - f_i^* \right|.$$

The advantage of such an approach to disclosing goal uncertainty is that along with finding x^0, it makes it possible to define the values of absolute deviations of all functions $f_i(x)$ from their assigned values f_i^*, i.e., to define deviations not only for the worst case determined by condition (3.22), but also for all other goals. This possibility is, in fact, an intermediate result of solving the Chebyshev approximation problem. In fact, if the value x^0 is known, then the value of the absolute deviation for all goals is found directly from (3.22), where, we assume, $x = x^0$. It is possible to solve problems (3.22) and (3.23) for system (3.21) by various methods.

- Methods of direct solution of the Chebyshev approximation problem.
- Methods of reduction of the Chebyshev approximation problem to a problem of linear or nonlinear mathematical programming.

Concerning the given approach, some essential remarks must be made:

1. In the general case, where $f_i(x)$ are nonlinear, problems (3.22) and (3.23) for system (3.21) may have a solution set, and each solution is referred to as local. Thus, the local solution, which ensures a minimally possible Δ^0 among all local solutions, is referred to as global. The global solution of the given system can be written in the form

$$x_{\text{opt}}^0 = \arg \min_k \min_x \max_i |f_i(x) - f_i^*| ,$$

where $\min(\cdot)$ means that the minimum among all local solutions $k = \overline{1, k_0}$ is defined, i.e., for any local solution $x_k^0 \neq x_{\text{opt}}^0$ we have

$$\Delta_k^0 \geq \Delta_{\text{opt}}^0 \quad \text{and} \quad \Delta_{\text{opt}}^0 = \min_k \Delta_k^0 .$$

2. Besides the Chebyshev criterion, it is possible to use other criteria. In particular, the most common one is a medium-scale criterion. We designate

$$\Delta f_i = (f_i(x) - f_i^*) . \tag{3.24}$$

Also, we assume that

$$\Delta f_{(q)} = \left[\sum_{i=1}^{m} (\Delta f_i)^q \right]^{1/q} , \tag{3.25}$$

where Δf_i, $\Delta f_{(q)}$ are, respectively, residual measure and approximation medium-scale criteria. Here, q is a degree of the medium-scale criterion. In practice, the most widespread value is $q = 2$ and the corresponding criterion usually is referred to as the root-mean-square criterion

$$\Delta f_{(2)} = \left[\sum_{i=1}^{m} (\Delta f_i)^2 \right]^{1/2} = \sqrt{\sum_{i=1}^{m} (\Delta f_i)^2} . \tag{3.26}$$

When criteria (3.25) and (3.26) are used, the approximation problem for system (3.21) consists in definition of such value x^0, for which:

$$\Delta f^0_{(q)} = \left\{ \sum_{i=1}^{m} \left[\Delta f_i \left(x^0 \right) \right]^q \right\}^{1/q} = \min_{x \in [x^-,\, x^+]} \Delta f_q \,,$$

$$\Delta f^0_{(2)} = \left\{ \sum_{i=1}^{m} \left[\Delta f_i \left(x^0 \right) \right]^2 \right\}^{1/2} = \min_{x \in [x^-,\, x^+]} \Delta f_2 \,.$$

It is necessary to draw attention to one principal difference between the Chebyshev criterion and the medium-scale criterion (at $q < \infty$). The Chebyshev criterion guarantees that the absolute deviation of any function $f_i(x)$ from the assigned value f_i^* will not exceed Δ^0 at the chosen $x = x^0$. The medium-scale criterion guarantees that the average deviation (for the corresponding degree) of any function will not exceed $\Delta^0_{(q)}$. But it does not guarantee that the separate functions will not deviate by a certain value Δ^0_i assigned beforehand. Therefore, the medium-scale criterion is not applicable in problems where the absolute deviation of certain parameters and characteristics is critical. For example, such problems are characteristic of chemical production, where a deviation of chemical-process parameters from the absolute value may lead to catastrophic consequences.

Finally, note that the medium-scale criterion is most general in the sense that at $q \to \infty$ it coincides with the Chebyshev criterion, and at $q = 2$ it is a root-mean-square criterion.

3. The criteria presented above in the form (3.23), for the Chebyshev approximation, and in the form (3.24) or (3.25), for the medium-scale criterion, are oriented toward the condition at which all goals are equally preferable. In practice, this condition is rarely fulfilled, i.e., there is goal preference in most cases.

Therefore, we consider a case where **the goals have different importance degrees**, which we shall characterize by the importance coefficients k_i. Then, the residual value (3.22) for the Chebyshev criterion will be defined by the relationship

$$\Delta^k = \max_{i \in [1,m]} \left\{ k_i \left| f_i \left(x \right) - f_i^* \right| \right\} . \tag{3.27}$$

The problem consists in finding a value x^0 at which the residual Δ^k will be minimally possible, i.e.,

$$\Delta^k |_{x=x^0} = \Delta_{\text{opt}} = \min_x \Delta^k \,. \tag{3.28}$$

For the medium-scale criterion, instead of (3.24) and (3.25), we shall obtain, respectively,

$$\Delta f_i^k = k_i \cdot (f_i(x) - f_i^*) , \tag{3.29}$$

$$\Delta f_{(q)}^k = \left(\sum_{i=1}^{m} (\Delta f_i^k)^q \right)^{1/q} . \tag{3.30}$$

The coefficients k_i are chosen taking into account the following normalization condition

$$\sum_{i=1}^{m} k_i = 1 . \tag{3.31}$$

In practice, the coefficients k_i are defined on the basis of expert estimation or assigned by the decision maker on the basis of his or her own experience and intuition. Condition (3.31) brings in certain difficulties in carrying out expert evaluation. Therefore, to simplify the process of introducing importance coefficients, the estimation of importance degree may be carried out according to the score system, for example, by accepting $k_{\max} = 10$ as the maximum value. Then, the importance coefficient values will possess any value in the interval (0, 10]. In this case, normalization of the form (3.31) can be carried out by defining

$$k_i = \frac{k_i'}{\sum\limits_{i=1}^{m} k_i'} \tag{3.32}$$

for all $i = \overline{1, m}$, where k_i' is a numerical score of the ith importance coefficient from the given interval.

It is necessary to pay attention to some difference in the way we introduce importance coefficients in linear convolution and in the considered case. In linear convolution, the importance coefficients mutually connect the values of the corresponding functions, but there is an open question: What will the deviation of the corresponding function from the assigned value be? In this case, the importance coefficient directly takes into account a degree of deviation of the corresponding function from the assigned value (3.21) and (3.24), since, by changing the value of one or several coefficients k_i, we know beforehand how it will affect the deviation of the corresponding function from its assigned value. Other conditions being equal, the less the value k_i, the greater the difference $|f_i(x) - f_i^*|$ a deviation will for the Chebyshev criterion and a difference $(f_i(x) - f_i^*)$ for the medium-scale criterion.

For example, as a result of solving the problem using the Chebyshev criterion, it was determined that $\Delta_{\text{opt}} = 0.2$. Let two functions $f_1(x)$ and $f_2(x)$ and their importance coefficients $k_1 = 0.25$, $k_2 = 0.75$ be assigned. Then, in conformity with (3.27), we have

$$k_1 |f_1(x) - f_1^*| = \Delta_{\text{opt}}, \quad k_2 |f_2(x) - f_2^*| = \Delta_{\text{opt}} , \tag{3.33}$$

$$0.25|f_1(x) - f_1^*| = 0.2; \quad 0.75|f_2(x) - f_2^*| = 0.2\,. \tag{3.34}$$

From here, taking into account (3.33) and (3.34), it follows that

$$|f_1(x) - f_1^*| = \frac{0.2}{0.25} = 0.8, \quad |f_2(x) - f_2^*| = \frac{0.2}{0.75} = 0.267\,, \tag{3.35}$$

which confirms the conclusion drawn above.

From the given example it also follows that

$$\frac{|f_1(x) - f_1^*|}{|f_2(x) - f_2^*|} = \frac{k_2}{k_1}\,, \tag{3.36}$$

i.e., the value of one function deviation from the assigned value differs from the corresponding value of another function deviation in inverse proportion to their importance coefficients (3.36).

4. The above considered methods of disclosing goal uncertainty correspond to the case where the restrictions on function, describing the corresponding goals, are assigned. It is possible to reduce the general case to a given variant when x^0 must be defined based on the condition

$$f_1(x) \to \max, \ldots, f_m(x) \to \max\,. \tag{3.37}$$

For this it is enough to find the maximum $f_k(x)$ value in the assigned interval for $i = \overline{1,m}$ and to define $\max_{x \in D} f_i(x) = f_k(x)$.

3.2 Disclosure of Situation Uncertainty

First, we shall define two types of uncertainty: situational and natural.

A situational uncertainty is characterized by unpredictable actions of different uncontrolled factors (human activities, natural disasters, influences of noosphere, etc.). All these factors could cause an unpredictable behavior of the explored system.

A natural uncertainty is a result of random actions of the difficult-to-predict and difficult-to-control factors of nature (flood, drought, etc.).

A problem of disclosing situational uncertainty will be considered using a concrete example. Assume it is required to establish an aviation route from Kiev to Paris that must be rational in terms of time; correspond to the international flights corridors; ensure the acceptable rate of fuel consumption; create the required conditions for passengers and crew; and conform to international, European, regional, and national requirements and terms.

Obviously, such a solution to this problem requires carrying out system analysis of a great number of factors and conditions with the purpose of defining the necessity of choosing rational options of actions at different stages of route development, realization, and exploitation. Let us consider only one factor that has a considerable impact on many aspects of the route usage. This factor is flight time. Obviously, flight time depends on many a priori known factors, in particular, technical aircraft performance characteristics, the conditions and capabilities of airports, etc. For a better understanding of the specific features of this problem we assume that all these factors are characterized by the a priori known vector x. In addition, there are factors that act in an unknown way or may change in the course of flight. Among them, for example, are such factors as weather conditions along the flight route and at the airfields during landing and takeoff, and many other factors. We will characterize all these factors by the generalized parameter of uncertainty α.

Taking into consideration that total flight time is limited, although the takeoff and landing times have certain permissible variations, actions of the crew must be directed at minimizing the difference between the assigned time T_0 and the permissible time T^* of flight duration in real conditions of situational uncertainty. The indicated difference is determined by the relationship

$$\Delta T^* = |T_0 - T^*|; \quad T^* = f\left(x^*, \alpha\right). \tag{3.38}$$

During the flight the objective of the crew is to make a decision x on choosing a speed that will ensure arrival at the destination in the existing weather conditions at time $T = f(x, \alpha)$.

Thus, time T deviation from the assigned value T_0 must satisfy the condition

$$\Delta T = |T_0 - T| \le \Delta T^* .$$

For fulfillment of this condition, it is necessary that

$$\begin{aligned} \Delta T = F\left(x, \alpha\right) = \left|T_0 - f\left(x, \alpha\right)\right| \to \min_{\alpha}, \\ \min_{\alpha} F\left(x, \alpha\right) = F\left(x^*, \tilde{\alpha}\right). \end{aligned} \tag{3.39}$$

Here, x^* is an index of the effectiveness of the crew's actions during the flight under conditions of uncertainty, which determines the value of vector x, if condition (3.39) is fulfilled. Vector x depends on the uncertainty factors and, thus, is the function of the generalized index α:

$$x = \tilde{x}(\alpha). \tag{3.40}$$

The value of the situational uncertainty generalized index α at the moment of making a decision x^* is determined by a variable $\tilde{\alpha} \in G_\alpha$, where G_α is some set of typical situation uncertainties, in particular, weather conditions typical for different seasons in the vicinity of the considered flight route. The important feature of such situational uncertainties is their high dynamic quality and, as a result, the complex qualitative changes in a situation on route within a limited time. Therefore, in real situations, as a rule, information about an index α in the form $\tilde{\alpha} \in G_\alpha$ is not sufficient for making a single-valued decision for a long period. As a result, the necessity arises to correct it in the process of realization.

Thus, disclosing situational uncertainty in this problem is realized as a choice of actions in conditions where a goal function is known but contains the generalized index of uncertainty α, which may appreciably change over time.

To solve the initial problem it is possible to choose an approach that allows one to obtain an estimate sufficiently justified, though one-sided. Such an approach is based on the principle of the guaranteed result. We consider this approach supposing that the situational uncertainty factors worsen the flight conditions.

Let the following condition be fulfilled for any x:

$$\min_{\alpha \in G_\alpha} F(x, \alpha) \leq F(x, \alpha^*) , \tag{3.41}$$

and for any $\alpha > \alpha^*$, $\alpha \in G_\alpha$

$$F^* = \max_x \min_{\alpha \in G_\alpha} F(x, \alpha) \leq \max_x F(x, \alpha^*) . \tag{3.42}$$

Here, F^* is a guaranteed estimation, and the corresponding value $x = x^*$ is a guaranteeing strategy in the sense that, whatever the value of uncertainty parameter α, the choice of $x = x^*$ in accordance with (3.42) guarantees that at any α the value of goal function $F(x, \alpha)$ will not be less than F^*.

To arrive at the guaranteeing strategy $x = x^*$, it is necessary to solve the following optimization problems:

- To calculate the value $\min\limits_{\alpha \in G_\alpha} F(x, \alpha)$ for any x, the following will be found as a result:
$$\alpha = \alpha^*(x); \quad F(x, \alpha) = F(x, \alpha^*(x)) .$$
- To calculate $\max\limits_x F(x, \alpha^*(x))$, the following will be determined as a result:
$$x = x^*, \quad F^*(x^*, \alpha^*) = F^* .$$

Note that the guaranteed estimation may be improved if some information about a parameter α is available before a decision is made. Information that the value of an uncertain factor, for example weather conditions, is known allows one to obtain a new, more exact guaranteed estimation. In this case, the guaranteeing strategy will not be a vector $x = x^*$, but some function $\tilde{x} = x(\alpha(t), t)$, where t is the current time.

Thus, the choice of the guaranteeing strategy is a rational method of decision making. Using this strategy we get the following guaranteed result: whatever uncontrollable factors may exist, the value of the goal function is ensured to be not less than F^*. Such a guaranteed result, presented in the form (3.42), allows us to find the best solution for the worst case.

But one question remains open: What is the probability of such an outcome, and how do we find the solution for the most likely outcome? For this it is necessary to make a decision connected with some risk. We shall characterize the risk using two indicators:

1. Degree of risk, as the probability of undesirable events occurrence.
2. Level of risk, expressed in the form of the quantitative estimation of possible damage.

Here two extreme cases are singled out: the choice of a strategy is made repeatedly, and the choice is a single operation. In both cases it is assumed that α is a random variable whose distribution law is unknown. As α is a random variable, the value of function $F(x, \alpha)$ will be a random variable, too. Therefore, in case of repetitive operations, the goal function of the initial problem may be presented as some probabilistic characteristics, for example, the mathematical expectation value $F_1 = \max_x \overline{F(x, \alpha)}$ or in some other form of designation $F_1 = \max_x MF(x, \alpha)$.

Certainly, other criteria may be chosen. If we denote $\tilde{\alpha}$ as a mean value of the random variable α, then the function $F_3 = F(x, \tilde{\alpha})$, whose maximum value may be used for estimation, will correspond to it.

Assume that the parameter α takes on the discrete values α_1, α_2, ... Then, the condition $F(x, \alpha) \to \max$ is identical to maximization of a set of the following criteria:

$$\begin{aligned} &F_1(x, \alpha_1) \to \max, \quad F_2(x, \alpha_2) \to \max, \quad \ldots, \\ &\ldots, \quad F_i(x, \alpha_i) \to \max, \quad \ldots, \quad F_m(x, \alpha_n) \to \max . \end{aligned} \tag{3.43}$$

Consequently, a decision-making problem in the presence of a situational uncertainty, when the parameter characterizing this uncertainty is random, has much in common with a problem of taking a decision under the conditions of goal uncertainty.

If we consider solving a problem involving the disclosure of a situational uncertainty on using the Chebyshev approximation, i.e., reducing the initial problem to a Chebyshev approximation problem for an incompatible system of nonlinear equations, then the solution can be characterized by:

1. The value of probability P_k of the kth situation occurrence characterized by α_k.
2. The level of possible damage $\Delta F_\alpha = \left|F\left(x^0\right) - F\left(\tilde{x}^0, \alpha_k\right)\right|$.

We need to find $\tilde{x}^0$ at each $\Delta F_\alpha \to \min_\alpha$. In the general case, $\tilde{x}^0 \neq x^0$. Owing to a situational uncertainty action, a variable ΔF_α takes into account the extent to which the characteristics of the ith goal $F_i(x)$ differ from the level of a rational compromise.

3.3 Disclosing the Uncertainty of Partners' or Opponents' Actions

The third type of uncertainty is specific to active practical activities of people during the development and realization of strategies for attaining certain goals. It is characterized by the information uncertainty that springs up in the course of active actions or counteractions of participants and is completely uncontrolled. This type of uncertainty is caused by strategies in which two or more partners cooperate or competitors or opponents counteract in a certain field of activities. Examples of such problems are industrial planning and forecasting of company activities, taking into account partners' actions or competitors' counteractions; objectives of national security under conditions of conflicting goals and interests; planning of measures and actions aimed at preventing illegal migration; maintenance of military and economic security; etc.

The easiest case of **interaction of two partners** may be formulated as follows: two partners are interacting, each of the two partners having his own goal, but the achievement level degree to which each partner attains his or her goals depends on the actions of the partner.

Let $f_1(x_1, x_2)$ and $f_1(x_1, x_2)$ be, respectivley, the goal functions of the first and second partner and x_1, x_2 be vectors of parameters whose values may be changed by the first and second partner, respectively. In the course of active interaction, the partners may exchange information regarding their actions. Here, two variants of information exchange are possible:

- Variant A: complete information exchange about goals, actions, parameters of activities, etc., which is typical, for example, for a production association (the parent enterprise and its branches).
- Variant B: partial information exchange, for example, only regarding the volume of output of some products or other parameters that are characterized by vector x_1 or x_2. But the goal functions are not communicated, which is typical of partners' activities under conditions of fair competition.

In variant A uncertainty may be caused by an incompleteness of information regarding a developing situation and predicted by market forces of supply and demand. In these conditions each partner may act independently, and disclosing goal uncertainty is reduced to disclosing situational uncertainty with known $f_1(x_1, x_2, \alpha_1)$, $f_2(x_1, x_2, \alpha_2)$, where α_1 and α_2 are parameters of situational uncertainty.

In variant B uncertainty may be caused by two factors: uncertainty of a developing situation and uncoordinated actions of partners.

First assume that there is no situational uncertainty. In this case disclosing goal uncertainty of partners' actions is carried out sequentially in the following steps:

Let partner 1 consider it necessary for attaining his goal to have the value $x_1 = x_1'$; he informs partner 2 of this. Partner 2 maximizes his own goal, taking into consideration this information, i.e., he finds a value $x_2 = x_2'$ at which:

$$f_2(x_2', x_1') = \max_{x_2} f_2(x_2, x_1') \,. \tag{3.44}$$

Partner 2 informs partner 1 about the value $x_2 = x_2'$ he desires. Partner 1 solves his problem of goal optimization by determining $\max f_1(x_1, x_2)$ under the condition that $x_2 = x_2'$, keeping the x_1' value or choosing a value x_1' where

$$f_1(x_1'', x_2') = \max_{x_1} f_1(x_1, x_2') \,. \tag{3.45}$$

If this condition satisfies both partners, the problem is solved. But, usually, the value x_1'' at which condition (3.45) is fulfilled is not equal to the initial value $x_1 \neq x_1'$. Therefore, partner 1 informs partner 2 about a new value $x_1 = x_1''$ that is desirable for him. After that, partner 2 solves problem (3.44) with a new value. This process stops when a rational compromise is found for both partners.

Now, let us consider the case where two factors operate simultaneously, namely, there is a **situational uncertainty** and **noncoordinated actions of partners**. Suppose that for partner 1 a situational uncertainty is characterized by parameter α_1, and for partner 2 by α_2, where

$$\alpha_1 \in \left[\alpha_1^-, \alpha_1^+\right], \quad \alpha_2 \in \left[\alpha_2^-, \alpha_2^+\right] .$$

Assume that the partners know the values x_1^0, x_2^0 that ensure a rational compromise in the absence of a situational uncertainty. Then, in the conditions of a situational uncertainty partner 1 determines the value $\hat{x}_1'$ by maximization of the mathematical expectation of the function $f_1(x_1, x_2, \alpha_1)$ at the known values $x_2 = x_2^0$, i.e.,

$$\hat{x}_1' \to \max_{x_1} M f_1(x_1, x_2^0, \alpha_1) \,. \tag{3.46}$$

And partner 2 determines the value $\hat{x}_2'$ by maximization of the mathematical expectation of the function $f_2(x_1, x_2, \alpha_2)$ at the known value $x_1 = x_1^0$, i.e.,

$$\hat{x}_2' \to \max_{x_2} M f_2(x_1^0, x_2, \alpha_2) \,. \tag{3.47}$$

Then, the values x_1^0 and $\hat{x}_1'$, x_2^0 and $\hat{x}_2'$ are compared, i.e., the following values are found:

$$\Delta x_1' = |x_1^0 - \hat{x}_1'|, \quad \Delta x_2' = |x_2^0 - \hat{x}_2'| \,. \tag{3.48}$$

If $\Delta x_1'$ and $\Delta x_2'$ do not exceed the assigned values

$$\Delta x_1' \leq \varepsilon_1, \quad \Delta x_2' \leq \varepsilon_2 \,, \tag{3.49}$$

then it is assumed that the values x_1^0 and x_2^0 may be taken as a rational compromise. If condition (3.49) is not fulfilled, the search for a rational compromise continues by the algorithm considered above. But instead of the corresponding functions, their mathematical expectations are taken. And if $x_2 = \hat{x}'_2$, then $\hat{x}''_1$ can be found from the condition

$$\hat{x}''_1 \to \max_{x_1} Mf_1(x_1, \hat{x}'_2, \alpha_1).$$

Then, a value $\hat{x}''_2$ is found from the condition $x_1 = \hat{x}''_1$:

$$\hat{x}''_2 \to \max_{x_2} Mf_2(\hat{x}''_1, x_2, \alpha_2).$$

If the obtained values $\hat{x}''_1$ and $\hat{x}''_2$ satisfy the partners, then the calculations stop, and the values $\hat{x}''_1$ and $\hat{x}''_2$ are accepted as a rational compromise. Otherwise, the process continues until a compromise is reached. In each index of the compromise the following values may be taken:

$$\Delta\hat{x}''_1 = |\hat{x}'_1 - \hat{x}''_1|, \quad \Delta\hat{x}''_2 = |\hat{x}'_2 - \hat{x}''_2|$$

or

$$\Delta f'_1 = |Mf_1(\hat{x}''_1, \hat{x}''_2, \alpha_1) - Mf_1(\hat{x}'_1, \hat{x}'_2, \alpha_1)|,$$
$$\Delta f'_2 = |Mf_2(\hat{x}''_1, \hat{x}''_2, \alpha_2) - Mf_2(\hat{x}'_1, \hat{x}'_2, \alpha_2)|.$$

The conditions of the type (3.49) may be used as criteria of a rational compromise.

The considered method is oriented toward the averaged parameters and is of practical interest in cases where:

1. Various situations have equal probability.
2. Values of the goal function differ insignificantly.

In practice, these conditions rarely occur. Therefore, more general is an approach to uncertainty disclosure taking into account risk factors.

We replace the assigned interval $\left[\alpha_1^-, \alpha_2^+\right]$ of changing α_1 with a discrete set $\alpha_1^{(1)}, \ldots, \alpha_1^{(q)}, \ldots, \alpha_1^{(q_{01})}$. The probability of occurrence of different values $\alpha_1^{(q)}$ is not the same and is characterized by the set $p_1^{(1)}, \ldots, p_1^{(q)}, \ldots, p_1^{(q_{01})}$.

Similarly, we can build the sets for α_2 and p_2. For each value $q_1 = \overline{1, q_{01}}$, $q_2 = \overline{1, q_{02}}$ we define values of the goal functions of each partner, assuming the conditions of a rational compromise are known in the absence of situational uncertainty factors, i.e., assuming that $x_1 = x_1^0$, $x_2 = x_2^0$ are known. Then, for any $q_1 = \overline{1, q_{01}}$, $q_2 = \overline{1, q_{02}}$, we have the following values:

$$f_1^{(q_1)} = f_1\left(x_1^0, x_2^0, \alpha_1^{(q_1)}\right), \quad f_2^{(q_2)} = f_2\left(x_1^0, x_2^0, \alpha_2^{(q_2)}\right).$$

In the general case $q_{01} \neq q_{02}$.

A problem of uncertainty disclosure may be solved by using different optimality criteria. If we maximize the income or minimize the loss, the best solution is to use the Chebyshev criterion, which allows one to directly estimate the attainment of specified goals. In this case the main problem is finding values $\hat{x}_1^0, \hat{x}_2^0$ whereby the maximum deviation of goal functions from the rational compromise should be minimally possible taking into account the probabilities of the corresponding situations.

For partner 1 the problem is to find $\hat{x}_1^0$ at the known value $x_2 = x_2^0$ such that the residual value

$$\Delta_1 = \max_{x_1} \left[p_1^{(q_1)} \left| f_1^{(q_1)} - f_1 \left(x_1, x_2^0, \alpha_1^{(q_1)} \right) \right| \right] \tag{3.50}$$

is minimal:

$$\Delta_1 |_{x=\hat{x}_1^0} = \Delta_1^0 = \min_{x_1} \Delta_1 \,. \tag{3.51}$$

Partner 2 must find a value $\hat{x}_2^0$ at the known $x_1 = x_1^0$ such that the residual value

$$\Delta_2 = \max_{x_2} \left[p_2^{(q_2)} \left| f_2^{(q_2)} - f_2 \left(x_1^0, x_2, \alpha_2^{(q_2)} \right) \right| \right] \tag{3.52}$$

would be minimally possible

$$\Delta_2 |_{x=\hat{x}_2^0} = \Delta_2^0 = \min_{x_2} \Delta_2 \,. \tag{3.53}$$

If conditions (3.50) and (3.51) are fulfilled, then the value $\hat{x}_1^0$ is defined from the following system of equations:

$$f_1 \left(x_1, x_2^0, \alpha_1^{(q_1)} \right) - f_1^{(q_1)} = 0, \quad q_1 = \overline{1, q_{01}} \,. \tag{3.54}$$

In a similar manner, provided that conditions (3.52) and (3.53) are fulfilled, the value $\hat{x}_2^0$ is defined from the following system of equations:

$$f_2 \left(x_1^0, x_2, \alpha_2^{(q_2)} \right) - f_2^{(q_2)} = 0, \quad q_2 = \overline{1, q_{02}} \,. \tag{3.55}$$

In the general case, the solution of each of these problems may be reduced to the Chebyshev approximation problem for an incompatible system of nonlinear equations (3.54) or (3.55), since usually the number of equations q_1 or q_2 is, as a rule, greater than the number of variables (a component of vector x_1 and x_2). As was mentioned before, the search for a rational compromise for both partners entails a sequence of iterations. And this process is finished completed when conditions (3.49) are satisfied.

The problems considered above are among the simplest problems of uncertainty disclosure with two partners interacting.

More complicated is the case of **mutliple partner interaction**. Consider the first type. Let k_0 partners be interacting, each having a goal that can be described by a known goal function of the following form:

$$f_k\left(x_1, \ldots, x_k, \ldots, x_{k_0}\right); \quad k = \overline{1, k_0},$$

where x_k is the vector of parameters for the kth partner.

The partners exchange information about the values of parameters $x_k = x'_k$. Each partner knows all the values x'_k of the other partners. In this case the problem of uncertainty disclosure may be solved taking into account these two variants:

1. Each kth partner finds his own solution individually but informs others of the chosen values of parameters x_k and on the degree of satisfaction with the solutions accepted by the other partners $\left(x'_1, \ldots, x'_{k-1}, x'_{k+1}, \ldots, x'_{k_0}\right)$.
2. The decision is made together and a rational compromise is found regarding certain criteria and importance of each partner's goal.

The first variant is typical for situations where partners are negotiating, for example, while establishing production quotas for the countries of the European Union. The second variant is more appropriate for development of strategy of transnational corporations on the international market, where each partner (company division) makes its contribution to the aggregate profit of the corporation, taking into account the market features of a given country. A measure of efficiency of such strategy is the size of income, including the value of the relative level of income, if a new market is explored.

Formalization and solving these problems is carried out in accordance with the scheme given above for variant B, with the only difference being that, instead of two functions, k_0 functions are optimized. Therefore, we consider only variant 2 for two cases:

1. Degrees of importance for all goals are assigned (for example, the importance of the relative contribution of each division in the aggregate profit of a corporation).
2. Additional conditions are assigned, for example, to maximize a certain criterion (for example, aggregate profit) given certain restrictions, for example, under the condition that the income in some regions or from the manufacture of certain products should not be below the assigned level.

Let us consider the formalization of these problems using the approach of uncertainty disclosure by reduction to the Chebyshev approximation for an incompatible system of equations [77, 150].

For the first case the problem of uncertainty disclosure is formulated as follows: the values $x_1^0, \ldots, x_k^0, \ldots, x_{k_0}^0$ must be found such that the residual value

$$\Delta = \max_k \left[V_k \left| f_k(x) - f_k^* \right|\right] , \tag{3.56}$$

taken as a measure of the Chebyshev approximation of a system of equations

$$f_k(x) - f_k^* = 0, \quad k = \overline{1, k_0} , \tag{3.57}$$

will be minimally possible

$$\Delta|_{x=x^0} = \Delta^0 = \min_x \Delta . \tag{3.58}$$

Here, $x = \{x_1, \ldots, x_k, \ldots, x_{k_0}\}$, V_k is an importance coefficient of the goal of the kth partner and f_k^* is the assigned (desirable) value of the goal function of the kth partner.

Formally, this problem is similar to the problem considered earlier of disclosing goal uncertainty when each goal is characterized by its own importance coefficient k_i.

The problem of uncertainty disclosure for the second case may be formulated as follows: certain values $x_1^0, \ldots, x_k^0, \ldots, x_{k_0}^0$ must be found to provide the maximum of the assigned criterion

$$F(x_1, \ldots, x_k, \ldots, x_{k_0}) \to \max_{x_k} \tag{3.59}$$

at the restrictions

$$f_k(x_1, \ldots, x_k, \ldots, x_{k_0}) \geq f_k^* . \tag{3.60}$$

We define the function $F(x_1, \ldots, x_k, \ldots, x_{k_0})$ as follows:

$$F(x) = \sum_{k=1}^{k_0} f_k(x) , \tag{3.61}$$

or

$$F(x) = \frac{\sum\limits_{k=1}^{k_0} V_k f_k(x)}{\sum\limits_{k=1}^{k_0} V_k} . \tag{3.62}$$

Here, it is accepted that in criterion (3.61) the goal functions $f_k(x)$ of all partners have an identical degree of importance, and in criterion (3.62) the degree of importance of each partner's goal is taken into account by a parameter V_k at $k = \overline{1, k_0}$. Generally, the maximization of function (3.61) or (3.62) at restrictions (3.60) is reduced to a typical nonlinear programming problem.

From the considered examples it follows that problems of uncertainty disclosure for several partners are reduced to the problems considered earlier for two partners but are more complicated from a computational point of view.

3.4 Uncertainty Disclosure in Problems of Strategies Conflict

We consider the peculiarity of the information uncertainty disclosure in the case, when strategies of participants of some common process **counteract** each other. For example, a problem of choosing a rational strategy for competitors on the common market at the absence of contracts between them, and some restrictions imposed on their behavior. Basic difference of strategy counteraction problems from the interaction of partners previously considered consists in the following:

1. The parties not merely do not inform each other about their actions, but even misinform about the goals and the appropriate parameters.
2. Situations in which events are developing depend not only on external conditions, but also on strategy of the parties.
3. Actions of the parties result in the necessity of mutual changing not only parameters, but also the goals in the course of situation development.
4. The goals of the parties do not coincide and are, as a rule, opposite and frequently conflicting.

The discrepancy of interests or goals of the participants in some common process will be referred to as a **conflict of strategies**. Studying of properties and features of conflict situations is one of the primary objectives of such discipline, as the theory of games. In the given case we consider problems of uncertainty disclosure, when opponents are acting under additional conditions and assumptions.

First, we study the simplest case of **counteractions of two opponents**. This problem in its statement is to some extent similar to the problem of two interacting partners but differs in the character and degree of information exchange. The situation can be described as follows. Each party has the goal function: participant 1—$f_1(x_1, x_2)$, participant 2—$f_2(x_1, x_2)$. The parties are acting independently, and neither knows either the goal function or the parameters of the other party.

Two different approaches may be used for solving the given problem:

1. Aim at achievement of a guaranteed result in the worst conditions.
2. Aim at the most probable variant of the other party's behavior and to ensure for oneself the best outcome in these conditions.

In the first case each participant tries to ensure getting certain guaranteed results in the worst conditions created by an active counteraction of the competitor or opponent. For achieving a guaranteed result each participant proceeds from two conditions:

1. The opponent has chosen action parameters that will cause the greatest damage to the other party.

2. Taking into account the first condition, the opponent chooses parameters of actions that ensure getting the maximum possible values of the goal function in the worst situation.

With such an approach the guaranteed result f_1^* for the first participant will be expressed by the relationship

$$f_1^* = \max_{x_1} \min_{x_2} f_1(x_1, x_2) , \tag{3.63}$$

and for the second participant

$$f_2^* = \max_{x_2} \min_{x_1} f_2(x_1, x_2) . \tag{3.64}$$

Let condition (3.63) be fulfilled at $x_1 = x_1^*$ and condition (3.64) at $x_2 = x_2^*$, i.e.,

$$x_1^* = \operatorname{Arg} \max_{x_1} \min_{x_2} f_1(x_1, x_2) , \tag{3.65}$$

$$x_2^* = \operatorname{Arg} \max_{x_2} \min_{x_1} f_2(x_1, x_2) . \tag{3.66}$$

Choosing $x_1 = x_1^*$ for the first participant it is guaranteed that for any value x_2 the following condition is fulfilled:

$$f_1(x_1^*, x_2) \geq f_1^* . \tag{3.67}$$

For the second participant it is similarly guaranteed that at $x_2 = x_2^*$ and any value x_1 the following requirement is satisfied:

$$f_2(x_1, x_2^*) \geq f_2^* . \tag{3.68}$$

Obviously, in the general case, the value x_2^* chosen from condition (3.66) does not coincide with the value at $\hat{x}_2$, with which the minimum value of the goal function of the first participant is ensured, i.e.,

$$x_2^* \neq \hat{x}_2, \quad \hat{x}_2 = \operatorname{Arg} \min_{x_2} f_1(x_1, x_2) . \tag{3.69}$$

Similarly, for the second participant

$$x_1^* \neq \hat{x}_1, \quad \hat{x}_1 = \operatorname{Arg} \min_{x_1} f_2(x_1, x_2) . \tag{3.70}$$

Hence, in the general case, these conditions of maximization of one's own results and minimization of one's opponent's results are not the same. This circumstance forces each participant to incur a certain risk. For example, it may be assumed that the opponent will choose tactics that will yield the greatest income, instead of reducing the income of the competitor. Such a choice is the most effective in cases where demand essentially exceeds the opportunities for supply. Therefore, the competitor will choose paramter values that will ensure a greater income rather than a guaranteed result. Then, the first participant may assume $x_2 = x_2^*$ and find x_1^{**} by maximization of the goal function $f_1(x_1, x_2)$ at $x_2 = x_2^*$. In this case we have

$$x_1^{**} = \operatorname{Arg} \max_{x_1} f_1(x_1, x_2^*) ,$$
$$f_1^{**} = \max_{x_1} f_1(x_1, x_2^*) .$$

Due to inequality (3.67), we have

$$f_1^{**} > f_1^* . \tag{3.71}$$

Similarly, for the second participant we assume that $x_1 = x_1^*$ and find

$$x_2^{**} = \operatorname{Arg} \max_{x_2} f_2(x_1^*, x_2) ,$$
$$f_2^{**} = \max_{x_2} f_2(x_1^*, x_2) .$$

Here, due to inequality (3.68),

$$f_2^{**} > f_2^* . \tag{3.72}$$

Inequality (3.71) is strict, owing to the fact that inequality (3.67) becomes an equality in accordance with (3.69) only at $x_2 = \hat{x}_2$, but f_2^{**} is determined at $x_2 = x_2^*$, where $x_2^* \neq \hat{x}_2$.

Similarly, from (3.68) and (3.70) it follows that inequality (3.72) also is strict.

It should be pointed out that expressions (3.71) and (3.72) are true with certain restrictions on properties of the functions $f_1(x_1, x_2)$ and $f_2(x_1, x_2)$, namely, these functions should be convex or concave on each variable at the same time. If a function is convex on one variable and concave on another, then another situation is possible in which inequalities (3.71) and (3.72) will have opposite meanings:

$$f_1^{**} < f_1^* , \tag{3.73}$$
$$f_2^{**} < f_2^* . \tag{3.74}$$

In the general case, when the functions $f_1(x_1, x_2)$ and $f_2(x_1, x_2)$ have a set of extremums on every or only one variable, it may happen that conditions (3.71) and (3.72) are fulfilled at some values x_1 and x_2, and at other values, conditions (3.73) and (3.74) are fulfilled.

Therefore, in the general case, the above considered conditions of choosing the action strategies by the participants may not be optimal. In each choice there is a certain risk: if the first participant plans the actions proceeding on the assumption that the opponent has chosen $x_2 = x_2^*$, and the opponent chooses $x_2 = x_2^{**}$, then it may happen that

$$\max_{x_1} f_1(x_1, x_2^{**}) < f_1^* .$$

Similarly, for the second participant;

$$\max_{x_2} f_2\left(x_1^{**}, x_2\right) < f_2^* .$$

Therefore, in disclosing the uncertainty of counteractions there arises a problem of system evaluation of the degree and level of risk. In solving this problem one should pay attention to the fact that the considered approach does not take into account one important thing: the probability that the opponent will choose this or that action strategy, i.e., the probability of choosing $x_2 = x_2^*$, or $x_2 = x_2^{**}$, or any other $x_2 = x_2^{***}$.

Moreover, in real conditions of counteraction mutual misinformation usually takes place. On accepting a certain strategy each party pretends that a completely different strategy has been accepted. Thus, in counteraction the degree of risk depends both on the probability of the opponent's choosing a certain strategy and on the probability of recognizing the decision taken. The level of risk will be defined by the extent of damage caused to each participant.

We assume that the second participant has made the decision $x_2 = x_2^{**}$ and the first has built up a strategy proceeding from the condition $x_2 = x_2^*$. Then, the level of damage U_1 of the first participant can be defined as

$$U_1 = \Delta f_1 = \max_{x_1} f_1\left(x_1, x_2^{**}\right) - \max_{x_1} f_1\left(x_1, x_2^*\right) . \quad (3.75)$$

Similarly, if the first participant has made the decision $x_1 = x_1^{**}$ and the second thinks that $x_1 = x_1^*$, then the extent of damage to the second participant U_2 is

$$U_2 = \Delta f_2 = \max_{x_2} f_2\left(x_1^{**}, x_2\right) - \max_{x_2} f_2\left(x_1^*, x_2\right) . \quad (3.76)$$

The accepted estimations (3.75) and (3.76) show the extent to which the obtained result exceeds the guaranteed result, i.e., these estimations are lower estimations. But another approach is possible—compare the obtained result with the potentially attainable results, i.e., define an approximation of the obtained result to the maximum possible value.

The maximum possible value of the goal function of the first participant is equal to

$$f_1^+ = \max_{x_1} \max_{x_2} f_1\left(x_1, x_2\right) . \quad (3.77)$$

Then, the extent of the damage to the participant will define a deviation of the obtained result from the potentially achievable one.

The extent of the damage caused to the first participant U_1^+ under the condition that he assumes the choice x_2 made by the second participant to be in the form $x_2 = x_2^*$ will be defined by the relationship

$$U_1^* = f_1^+ - \max_{x_1} f_1\left(x_1, x_2^{**}\right) . \quad (3.78)$$

Similarly, for the second participant

$$U_2^* = f_2^+ - \max_{x_2} f_2\left(x_1^{**}, x_2\right) , \tag{3.79}$$

where

$$f_2^+ = \max_{x_2} \max_{x_1} f_2\left(x_1, x_2\right) . \tag{3.80}$$

It should be pointed out that in a number of cases it is interesting to compare the obtained value of the goal function with the absolute minimum.

Such a comparative estimation is useful in the analysis of critical situations, for example, if an opponent causes irreparable damage to another party. In this case, the given estimation shows the extent to which the developed situation approaches the critical situation, which is characterized by the absolute minimum. The value of the absolute minimum for the first participant is defined as

$$f_1^- = \min_{x_1} \min_{x_2} f_1\left(x_1, x_2\right) . \tag{3.81}$$

Then a variable U_1^- defines a difference of the goal functions in critical situations:

$$U_1^- = f_1\left(x_1', x_2'\right) - f_1^- ,$$

where x_1', x_2' are numerical values of x_1, x_2 in a situation developing for the first participant.

Similarly, for the second participant we have

$$U_2^- = f_2\left(x_1'', x_2''\right) - f_2^- ,$$

where

$$f_2^- = \min_{x_1} \min_{x_2} f_2\left(x_1, x_2\right) , \tag{3.82}$$

where x_1'', x_2'' are numerical values in a situation developing for the second participant.

In conclusion we should note that the above indicated characteristics allow one to obtain quantitative estimations of the results of one's own decisions in real situations with assumed actions by one's opponent. But at the same time these characteristics have some drawbacks, namely, they give one-sided estimations—the extent to which the possible result differs from some quite definite value. Such estimations are usually called **point estimations**, in the sense that they show the distance between two points—preliminarily given (or calculated) and a point determining the situation. For example, they show how close the result is to the maximum possible value, but the question remains open as to the extent to which it exceeds the minimally possible value. And conversely, they make it possible to make a comparison with minimally possible values, not answering the question of how this result differs from the maximum possible value.

This drawback can be eliminated if, instead of the point principle of comparison, we use the **interval** comparison. The meaning of interval comparison is that a developing situation is estimated by the value of the goal function in a numerical interval whose minimum value corresponds to the worst situation, and the maximum to the best.

As was shown above, the worst and best situations for the first participant are described respectively by the values f_1^- and f_1^+, which are defined by (3.81) and (3.77).

For the first participant the interval estimaton for arbitrary values x_1 and x_2 may be presented in the form

$$I_1^- (x_1, x_2) = \frac{f_1 (x_1, x_2) - f_1^-}{f_1^+ - f_1^-} . \tag{3.83}$$

Taking into consideration that, owing to the definition f_1^+ and f_1^-, the function $f_1 (x_1, x_2)$ satisfies the condition

$$f_1^- \leq f_1 (x_1, x_2) \leq f_1^+ ,$$

from (3.83) it follows that the following inequality is true:

$$0 \leq I_1^- (x_1, x_2) \leq 1 .$$

Similarly, for the second participant we have the interval estimation for arbitrary values x_1 and x_2:

$$I_2^- (x_1, x_2) = \frac{f_2 (x_1, x_2) - f_2^-}{f_2^+ - f_2^-} , \tag{3.84}$$

where f_2^+, f_2^- are defined respectively by (3.80) and (3.82).

In view of the fact that the following is true

$$f_2^- \leq f_2 (x_1, x_2) \leq f_2^+ ,$$

as a result of (3.84) the following inequality is true:

$$0 \leq I_2^- (x_1, x_2) \leq 1 .$$

Estimations (3.83) and (3.84) allow one to define (in relative units) the degree to which the functions $f_1 (x_1, x_2)$ and $f_2 (x_1, x_2)$ exceed their minimum values, since at $f_1 (x_1, x_2) = f_1^-$ and $f_2 (x_1, x_2) = f_2^-$ we have, respectively, $I_1^- (x_1, x_2) = 0$ and $I_2^- (x_1, x_2) = 0$.

Let us introduce the estimates that define a relative level of difference of the goal functions from their maximum values:

$$I_1^+ (x_1, x_2) = \frac{f_1^+ - f_1 (x_1, x_2)}{f_1^+ - f_1^-} , \tag{3.85}$$

$$I_2^+ (x_1, x_2) = \frac{f_2^+ - f_2 (x_1, x_2)}{f_2^+ - f_2^-} . \tag{3.86}$$

It is easy to verify that the following expressions are true:

$$I_1^+(x_1, x_2) + I_1^-(x_1, x_2) = 1,$$
$$I_2^+(x_1, x_2) + I_2^-(x_1, x_2) = 1.$$

Thus, it is enough to define only one estimation for each participant, for example, $I_1^-(x_1, x_2)$ or $I_1^+(x_1, x_2)$ for the first participant and $I_2^-(x_1, x_2)$ or $I_2^+(x_1, x_2)$ for the second participant.

The practical importance of these estimates is that it is possible beforehand to define or assign the lower boundary of the maximum permissible damage, regardless of the absolute values f_1^+, f_1^- and $f_1(x_1, x_2)$.

Such an estimation of the allowable damage is necessary in many fields, in particular, in various business operations connected with competition-related risks, crises, and other peculiar features of the market. In practical terms, in all countries allowable risk levels of banking operations are established. This level, for example, may be expressed as a percentage of bank asset costs.

Example. For two opposing participants the following goal functions are assigned

$$f_{12}(x_1, x_2) = -x_1^2 + 2x_1 + x_2^2 - 4x_2 + 8,$$
$$f_{21}(x_1, x_2) = x_1^2 - 6x_1 - x_2^2 + 2x_2 + 2,$$

and ranges of their strategic definitions are

$$x_1 \in [0; 4], \quad x_2 \in [0; 4].$$

The optimum solutions must be found for the opposing participants, taking into account certain restrictions:

$$\begin{aligned} f_{12}(x_1, x_2) &\geq f_{12}^*, \quad (a) \\ f_{21}(x_2, x_1) &\geq f_{21}^*, \quad (b) \end{aligned} \tag{3.87}$$

where $f_{12}^* = \max_{x_1} \min_{x_2} f_{12}$, $f_{21}^* = \max_{x_2} \min_{x_1} f_{21}$.

First, for finding optimum solutions it is necessary to find rational solutions from the Pareto set. To this end the guaranteed results f_{12}^*, f_{21}^* should be defined. To define the guaranteed results various techniques may be used, in particular, the classical analysis of a function on an extremum, the tabular method, the graphic method, etc. In this case we will use the graphic method. To find f_{12}^* fixing a number of values x_1, we build the diagrams of the function $f_{12}(x_1, x_2)$ using x_2 (Fig. 3.6). This figure shows that the maximum minimum is reached at $x_2 = 2$ and $f_{12}^* = 5$.

Similarly, to find f_{21}^* we fix a number of x_2 values and build diagrams of the function $f_{21}(x_1, x_2)$ using x_1 (Fig. 3.7). The maximum minimum is reached at $x_1 = 3$ and $x_2 = 1$, so $f_{21}^* = -6$. Thus, the guaranteed results are $f_{12}^* = 5$ and $f_{21}^* = -6$.

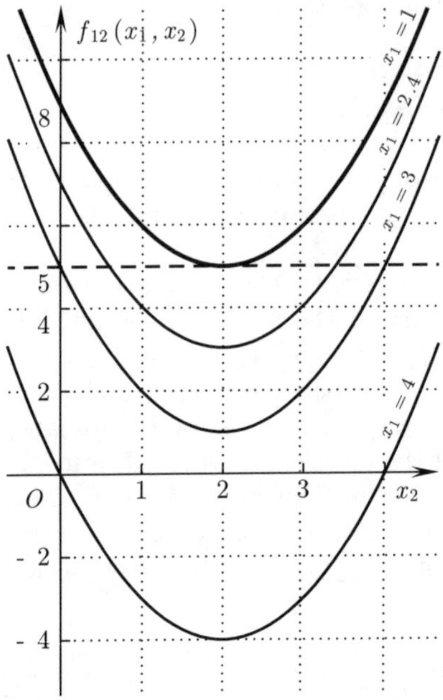

Fig. 3.6. Finding function f_{12}^* using diagram method

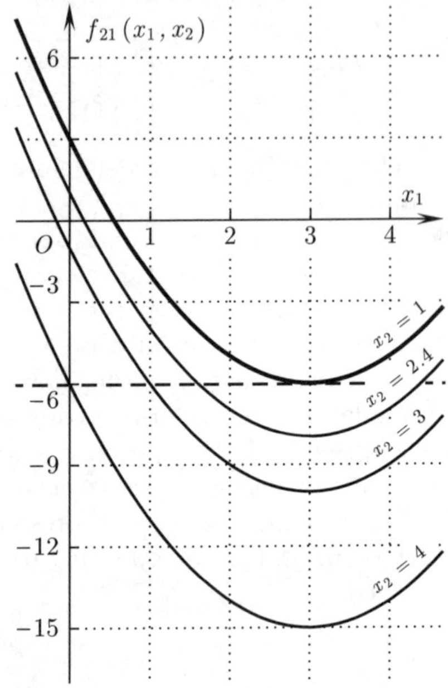

Fig. 3.7. Finding function f_{21}^* using diagram method

The Pareto set is found proceeding from restrictions $f_{12}^*(x) \geq 5$; $f_{21}^*(x) \geq -6$. The system of inequalities (3.87) is rather simple for solving graphically. Figure 3.8 shows the Pareto area in the form of an intersection of the areas given above where there are rational solutions for counteracting participants. Also, here the optimum values x_1^* and x_2^* are shown, at which $\Delta = \min_{x_1,x_2} \max_i \Delta_i = 0$, where $\Delta_i = |f_i(x_1, x_2) - f_i^*|$, $i = 1, 2$.

3.5 Problems and Methods of System Uncertainty Disclosure

In the preceding paragraphs, approaches and methods of disclosing certain types of uncertainties, namely, goal uncertainties, situational uncertainties, partners' interaction or parties' counteraction uncertainties, were considered. This means that when considering one type of uncertainty it was a priori assumed that there were no other types of uncertainties. However, in real system problems various types of uncertainties usually operate simultaneously. Therefore, a number of practical questions arise:

1. Is it expedient to carry out separate analysis of various types of uncertainties?

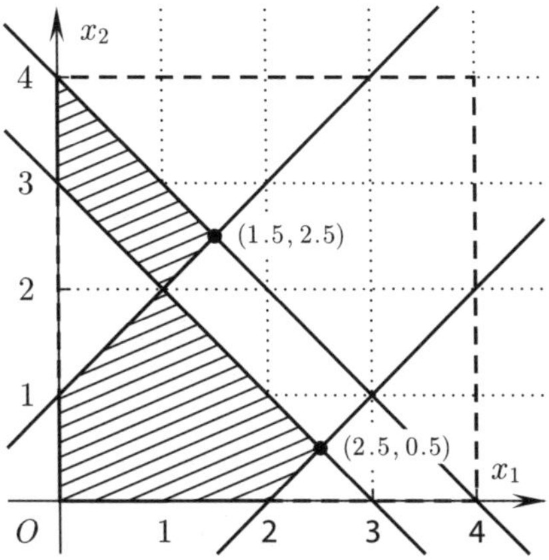

Fig. 3.8. Definition of Pareto area using graphic method

2. Is it possible to aggregate the results of separate analyses in a cumulative result with the sufficient degree of credibility?
3. Are there any effective methods and approaches that disclose simultaneously several types of uncertainties in one system problem?
4. What restrictions and assumptions make it possible to solve several types of uncertainties simultaneously?

To answer these questions it is first necessary to formulate the problem of disclosing system uncertainty.

Contensive statement of problem. Suppose it is known that there is a need to solve two participants' goal uncertainties in two possible situations: (1) when participants are interacting as partners and (2) when participants are solving contradictions in conditions of counteraction.

It is known that the goals of participants 1 and 2 are described by the following vector functions:

$$f_1(x_1, \widehat{x}_2, \alpha_1) = \left\{ f_{1i_1}(x_1, \widehat{x}_2, \alpha_1) \mid i_1 = \overline{1, m_1} \right\}, \tag{3.88}$$

$$f_2(x_2, \widehat{x}_1, \alpha_2) = \left\{ f_{2i_2}(x_2, \widehat{x}_1, \alpha_2) \mid i_2 = \overline{1, m_2} \right\}. \tag{3.89}$$

The vector function of each participant's goals will consist of components $f_{1i_1}(x_1, \widehat{x}_2, \alpha_1)$, $i_1 = \overline{1, m_1}$ and $f_{2i_1}\left(x_2, \widehat{x}_1, \alpha_2\right)$, $i_2 = \overline{1, m_2}$, respectively. Every component depends on the series of parameters and conditions simultaneously, which differs essentially in the properties and capabilities of participants to control them. Thus, first, participants can decide themselves to change certain parameters in permissible limits: participant 1 can change the parameters of vector x_1 in vector function (3.88), and participant 2 can change the parameters of vector x_2 in vector function (3.89). Second, in uncertainty conditions no participant knows a priori the other participant's actions; therefore, each participant predicts the partner's or opponent's possible actions based on the estimation of possible values and boundaries of intervals for the following parameters: participant 1 predicts the parameters of vector $\widehat{x}_2$ of vector function (3.88), and participant 2 predicts the parameters of vector $\widehat{x}_1$ of vector function (3.89). Third, each participant knows a priori that external natural, ecological, and other factors influence which properties will not depend on participants, and, consequently, no party can change their parameters. Each participant can independently only describe and characterize possible influences of various factors. In particular, participant 1 predicts the influences by selecting the quantity and values of parameters of vector α_1, and participant 2 predicts the influences by selecting the quantity and values of parameters of vector α_2. At the same time participant 1 believes that basic external influence is determined by climatic, ecological, and other external conditions and factors, as well as by market forces of supply and demand of certain types of multipurpose products. Vector α_1 characterizes the range of the corresponding changing external conditions and factors. Participant 2 can predict possible influences, taking into account other external conditions and factors, and has other components and values of vector α_2. Therefore, vectors α_1 and α_2, describing the external influences in the understanding of participants 1 and 2, respectively, may differ both in their structure and quantity of components.

Therefore, it is known a priori, that in the general case participants form their goal functions independently. Thus, f_{1i_1} and f_{2i_2} are functions in which the total argument quantity N_1 and N_2 differs:

$$N_1 = n_1' + n_2' + n_\alpha', \quad N_2 = n_1'' + n_2'' + n_\alpha''.$$

Here, participants 1 and 2 define the following values: n_1' and n_1'' are the component quantity vectors x_1 and $\widehat{x}_1$, n_2' and n_2'' are the component quantity vectors $\widehat{x}_2$ and x_2, and n_α' and n_α'' are the component quantity vectors α_1 and α_2, respectively.

System uncertainty ***must*** be disclosed from the point of view of each participant's perception of it.

Mathematical statement of problem. The general problem of disclosing system uncertainty in its mathematical statement is reduced to a ***simultaneous research of different types of uncertainties***, which is based on the following conditions, properties, and restrictions:

- Conditions for disclosing uncertainties of each of two participants' goals in two possible situations: participants' interactions as partners or in conditions of their counteraction.
- Properties and restrictions of goals described by the assigned vector functions (3.88) and (3.89) and differ in their parameter compositions, number of components, and possible influences.
- Conditions of forming the vector function components that are independently defined by the participants.

- Ranges of possible variation of the corresponding external conditions and factors, which are independently predicted by participants.

For disclosing goal uncertainties the following procedures ***must*** be carried out:

- A mathematical model of the conditions, properties, and possible external influences must be developed for each participant, taking into account the uncertainty, incompleteness, and inauthenticity of the initial information about external influences.
- A mathematical model of each participant's strategy must be developed for two variants: (1) when the participants are interacting as partners and (2) when the participants are resolving contradictions in the conditions of their counteraction, taking into account the uncertainty, incompleteness, and nonncredibility of information about the other participant's actions either as a partner or as an opponent.
- Participant 1 must define a rational relationship between the components $f_{1i_1}\left(x_1, \frown{x}_2, \alpha_1\right)$, $i_1 = \overline{1, m_1}$ of the goal vector function (3.88), the rational values of the components $x_{11}, \ldots, x_{1j_1}, \ldots, x_{1n'_1}$ of the vector x_1, and the rational values of components $\frown{x}_{21}, \ldots, \frown{x}_{2j_2}, \ldots, \frown{x}_{2n'_2}$ of the vector $\frown{x}_2$.
- Participant 2 must define a rational relationship between the components $f_{2i_1}\left(x_2, \frown{x}_1, \alpha_2\right)$, $i_2 = \overline{1, m_2}$ of the goal vector function (3.89), the rational values of the components $x_{21}, \ldots, x_{2j_2}, \ldots, x_{2n''_2}$ of the vector x_2, and the components $\frown{x}_{11}, \ldots, \frown{x}_{1j_1}, \ldots, \frown{x}_{1n''_1}$ of the vector $\frown{x}_1$.

Solving the problem of system uncertainty disclosure. First, it should be pointed out that the problem-solving procedures have a number of practically important features. First, a mathematical model of external influences on the participants and a mathematical model of the participants' action strategy should be developed in the conditions of various types of uncertainty, namely, uncertainties of goals, situations, factors, and conditions of the participants' interaction or counteraction. Therefore, such models are developed on the basis of nonformalizable procedures that are realized for each specific problem by a group of developers based on their experience, knowledge, intuition, and foresight. Here, only the necessity of developing such models is underlined, and the basic questions of their development and application are considered in subsequent chapters.

Second, all the previously listed procedures are one type of uncertainty in terms of their formal statement for the participants, but they are different in their structural realization. In particular, each participant forms vector functions on the basis of his own interpretation of goal and action strategies. Therefore, in the general case the number of the participants' goal functions m_1 and m_2, and the number of components of the vectors x_1, x_2 and α_1, α_2 are different for each participant. For example, participant 1 estimates partner 2's goals and actions only on the basis of macroeconomic indices, and participant 2 can characterize his own actions more accurately and completely taking into account certain additionally defined economic production indices and nationwide demand. Therefore, the number n_2' of components of vector $\widehat{x}_2$ in the goal function $f_1\left(x_1, \widehat{x}_2, \alpha_1\right)$ cannot coincide with the number n_2'' of components of vector x_2 in the goal function $f_2\left(\widehat{x}_1, x_2, \alpha_2\right)$. Similarly, the participants may characterize a predictable situation α by a different number of components n_α'' and n_α''. Moreover, they may characterize the situation not only by a different number of components, but also by choosing different meanings of the same components. Therefore, vectors α_1 and α_2, describing the situation in understanding of participants 1 and 2 respectively, may differ both in their structure and quantity and in the contensive meaning of the components.

Third, each participant should take into account the uncertainty, incompleteness, and noncredibility of the information about the other participant's actions. In particular, it is necessary to define beforehand the role each participant is to play with respect to the other participant and to predict his role as partner or opponent. Then, in practical activities each participant should specify the information; in particular, it is necessary to explore the market dynamics and analyze tendencies of the other participant's actions. Moreover, each participant should take into account the influence of external factors. In addition, there is a need to pay special attention to the uncertainty, incompleteness, and noncredibility of the information about the ranges of possible changes in predictable and occurring external conditions and factors. The performed analysis will prove a practical necessity in specifying information and correct the action strategy taking into consideration the other participant's actions and dynamics of changes in external conditions and factors. And on this basis, each participant should correct his goals, objectives, and actions in a timely manner.

Then, one should pay attention to peculiarities and possibilities of disclosing system uncertainty. In the beginning, we will find out under what conditions it is possible in principle **to analyze independently various types of uncertainties**. It is obvious that the most favorable condition is independence of the influence of the analyzed factors, conditions, and restrictions. In such situations, each goal function can be expressed in an additive form:

$$f_{1i_1} = f_{11i_1}(x_1) + f_{12i_1}\left(\widehat{x}_2\right) + f_{13i_1}(\alpha_1), \quad i_1 = \overline{1, m_1}, \tag{3.90}$$

$$f_{2i_2} = f_{21i_2}\left(\widehat{x}_1\right) + f_{22i_2}(x_2) + f_{23i_2}(\alpha_2), \quad i_2 = \overline{1, m_2}. \tag{3.91}$$

Here for the function f_{1i_1} summands have the following meaning: $f_{11i_1}(x_1)$ describes the i_1th goal component as a function of vector x_1, which may be changed by participant 1; $f_{12i_1}\left(\widehat{x}_2\right)$ is the i_1th goal component as predicted by participant 1 the function of vector $\widehat{x}_2$ of participant 2's probable actions; $f_{13i_1}(\alpha_1)$ is the i_1th goal component determined by a developing situation for participant 1 that depends on the values of vector α_1 quantitatively characterizing this situation. Similar meanings for function f_{2i_2} have corresponding summands $f_{21i_2}\left(\widehat{x}_1\right)$; $f_{22i_2}(x_2)$; $f_{23i_2}(\alpha_2)$.

It should be noted that the function $f_{1i_1} = f_{1i_1}\left(x_1; \widehat{x}_2; \alpha_1\right)$ characterizes the resulting dependence of the i_1th component on all vectors x_1 $\widehat{x}_2$, α_1 as understood by participant 1. Based on this function a system of $i_1 = \overline{1; m_1}$ equations is formed; the solution to this system defines a result of the system uncertainty disclosure for participant 1. A similar result is defined by the function f_{2i_2} for participant 2. Then we should pay attention to the fact that for (3.90) it is possible to determine the i_1th component separately for each vector x_1, $\widehat{x}_2$, α_1, since components of each vector do not depend on other vectors and the general result for function f_{1i_1} is defined by summation. Such an approach is completely applicable for the system (3.91).

In this case, ***the problem of disclosing goal uncertainty*** can be solved by using the approaches previously considered for functions

$$f_1' = \left\{ f_{11i_1}(x_1) \,\middle|\, i_1 = \overline{1, m_1} \right\} . \tag{3.92}$$

The problem of simultaneous uncertainty disclosure for several vectors can be solved on the basis of various approaches and methods. The simplest is the convolution method [99]. It is realized on the basis of the transformation of the system of functions

$$f_1'' = \left\{ f_{11i_1}(x_1), f_{13i_1}(\alpha_1) \,\middle|\, i_1 = \overline{1, m_1} \right\} \tag{3.93}$$

into the generalized function of the following form:

$$F(x_1, \alpha_1) = \sum_{i_1=1}^{m_1} \left[c'_{i_1} f_{11i_1}(x_1) + c''_{i_1} f_{13i_1}(\alpha_1) \right] \tag{3.94}$$

and subsequent disclosure of goal uncertainty for the generalized function $F(x_1, \alpha_1)$ on the basis of the approaches considered above.

One more method is the reduction of the problem of goal uncertainty disclosure to the following system of equations:

$$f_{11i_1}(x_1) + f_{13i_1}(\alpha_1) = f^*_{11i_1}, \quad i_1 = \overline{1, m_1} . \tag{3.95}$$

Here $f^*_{11i_1}$ defines the value of the function $f_{11i_1}(x_1)$, corresponding to the solution of the Chebyshev approximation problem for the system of equations

$$f_{11i_1}(x_1) - b_{i_1} = 0, \quad i_1 = \overline{1, m_1} .$$

Such a system is formed in disclosing goal uncertainty in cases where there are no other uncertainty factors, in particular $\widehat{x}_2$ and α_1, or they are not taken into consideration. This approach possesses two practical important features. First, this approach makes it possible to separately reveal in the problem a quantitative influence of each uncertainty factor $\widehat{x}_2$ and α_1 on the general result of goal uncertainty disclosure. Second, there is no need to convolute the system of equations (3.95) to the form of the generalized function $F(x_1, \alpha_1)$. It excludes the necessity of introducing factors c'_{i_1} and c''_{i_1}, the choice of which is nonformalizable in principle and, consequently, is carried out by the decision macer or by experts, which brings elements of subjectivity into the structure of function $F(x_1, \alpha_1)$ structure (3.94).

Let us proceed to consideration of the ***uncertainty of a partner's or opponent's actions***. Initially we consider the problem of disclosing the uncertainty of two competing participants' counteraction, each having strictly defined goals. Suppose that for each participant a functional dependence determining the degree of influence of the competitor's actions on his goals and the influence of external factors on the developing situation are known.

The counteraction is analyzed with respect to the influence of one of the following conditions:

1. The situation properties are completely determined, and their influence on each participant's attaining his goal may be ignored.
2. The situation' basic properties are determined, and the degree of their influence on each participant's attaining his goal is determined by a single probabilistic indicator.

First, we consider the methods of disclosing the ***uncertainty of two partners' interaction*** when the situation is clearly defined. Suppose that participant 1 has m_1 goals, and participant 2 has m_2 goals. If condition 1 is fulfilled for all $i_1 = \overline{1, m_1}$, then we have $f_{13i_1}(\alpha_1) = 0$ and for all $i_2 = \overline{1, m_2}$ we have $f_{23i_2}(\alpha_2) = 0$ in accordance with (3.90) and (3.91). Hence, the participants' goals are described by the following relationships:

For participant 1

$$f_{1i_1}\left(x_1, \widehat{x}_2\right) = f_{11i_1}(x_1) + f_{12i_1}(\widehat{x}_2), \quad i_1 = \overline{1, m_1}. \tag{3.96}$$

For participant 2

$$f_{2i_2}\left(\widehat{x}_1, x_2\right) = f_{21i_2}\left(\widehat{x}_1\right) + f_{22i_2}(x_2), \quad i_2 = \overline{1, m_2}. \tag{3.97}$$

The problem consists in finding values x_1 for participant 1 that provide a guaranteed result for each goal function. A guaranteed result value is determined by the following relationship:

$$f^*_{1i_1} = \max_{x_1} \min_{\widehat{x}_2} f_{1i_1}\left(x_1, \widehat{x}_2\right), \quad i_1 = \overline{1, m_1},$$

or, taking into account (3.96),

$$f_{1i_1}^* = \max_{x_1} f_{11i_1}(x_1) + \min_{\widehat{x}_2} f_{12i_1}\left(\widehat{x}_2\right) . \tag{3.98}$$

The required value x_1^* can be obtained from the following formula:

$$x_1^* = \operatorname{Arg} \max_{x_1} \min_{\widehat{x}_2} f_1\left(x_1, \widehat{x}_2\right) , \tag{3.99}$$

where

$$f_1\left(x_1, \widehat{x}_2\right) = \left\{ f_{1i_1}\left(x_1, \widehat{x}_2\right) \mid i_1 = \overline{1, m_1} \right\} .$$

It should be noted that relationship (3.99) only formally determines the required value x_1^* of vector x_1, since in the general case the number m_1 of components of the goal function $f_1\left(x_1, \widehat{x}_2\right)$ is not equal to the number n_1 of components of vector x_1 and the number n_2 of components of vector $\widehat{x}_2$.

In this case, the problem of disclosing interaction uncertainty for goal function systems (3.96) and (3.97) may be reduced to the following two Chebyshev approximation problems

Problem A1. Determine such values x_2^* at which for all $i_1 = \overline{1, m_1}$ the following conditions are fulfilled:

$$f_{12i_1}^* = \min_{\widehat{x}_2} f_{12i_1}\left(\widehat{x}_2\right) , \quad x_2^* = \operatorname{Arg} \min_{\widehat{x}_2} f_{12i_1}\left(\widehat{x}_2\right) .$$

Problem A2. Determine those values x_1^* at which for all $i_1 = \overline{1, m_1}$ the following conditions are fulfilled:

$$f_{11i_1}^* = \max_{x_1} f_{11i_1}(x_1) , \quad x_1^* = \operatorname{Arg} \max_{x_1} f_{11i_1}(x_1) .$$

Problem A1 will consist in determining value x_2^* from the incompatible system of equations

$$\begin{cases} f_{121}\left(\widehat{x}_2\right) - b_{121} = 0 , \\ \dots\dots\dots\dots\dots\dots \\ f_{12i_1}\left(\widehat{x}_2\right) - b_{12i_1} = 0 , \\ \dots\dots\dots\dots\dots\dots \\ f_{12m_1}\left(\widehat{x}_2\right) - b_{12m_1} = 0 , \end{cases}$$

under the condition that value

$$\Delta_{12} = \max_{i_1} \left| f_{12i_1} - b_{12i_1} \right| , \tag{3.100}$$

taken as the Chebyshev approximation measure, should be minimally possible

$$\Delta_{12}\left(x_2^*\right) = \min_{\widehat{x}_2} \Delta_{12} = \Delta_{12}^0 .$$

In this problem the choice of b_{12i_1} for all $i_1 = \overline{1, m_1}$ remains an open question. Such a choice is possible on the basis of utility theory, according to which for value b_{12i_1} it is possible to take the most desirable value. From the physical meaning of the problem it is obvious that the most desirable values are $b_{12i_1} = 0$, for every $i_1 = \overline{1, m_1}$, which corresponds to a complete absence of the competitor's influence on the activities of participant 1. Then, problem A1 is reduced to the Chebyshev approximation for an incompatible system of equations:

$$f_{12i_1}\left(\widehat{x}_2\right) = 0, \quad i = \overline{1, m_1}. \tag{3.101}$$

As a result of solving system (3.101), we obtain the value x_2^* and values of functions

$$f_{12i_1}\left(x_2^*\right) = \Delta_{12i_1}^*, \tag{3.102}$$

where, according to formulas (3.99) and (3.100),

$$\Delta_{12i_1}^* \leq \Delta_{12}^0.$$

Similarly, problem A2 will consist in finding the following values x_1^* from the incompatible system of equations:

$$f_{11i_1}\left(x_1\right) - b_{11i_1} = 0, \quad i_1 = \overline{1, m_1}, \tag{3.103}$$

for which

$$\Delta_{11} = \max_i \left|f_{11i_1} - b_{11i_1}\right|, \tag{3.104}$$

taken as the Chebyshev approximation measure, should be minimally possible

$$\Delta_{11}\left(x_1^*\right) = \min_{x_1} \Delta_{11} = \Delta_{11}^0. \tag{3.105}$$

Here value b_{11i_1} for $i_1 = \overline{1, m_1}$ is chosen similarly. In particular, it is possible to determine

$$b_{11i_1} = \max_{x_1 \in D} f_{11i_1}\left(x_1\right),$$

where D is the definition domain of the vector x_1 components assigned from physical considerations in the form

$$d_{1j_1}^- \leq x_{1j_1} \leq d_{1j_1}^+.$$

As a result of solving the given problem, we obtain x_1^* values and values of the functions

$$f_{11i_1}^* = f_{11i_1}\left(x_1^*\right) = b_{11i_1} \pm \Delta_{11i_1}^*, \tag{3.106}$$

where, owing to conditions (3.104) and (3.105), the following inequality is true:

$$\Delta_{11i_1}^* \leq \Delta_{11}^0.$$

As a result of solving problems A1 and A2, we have the required values x_2^* and x_1^*, respectively; in addition the values of the guaranteed result for each function on the basis of (3.98), (3.102), and (3.106) are determined by the relationship

$$f_{1i_1}^* = f_{11i_1}^* + f_{12i_1}^* .$$

Also, the following inequality is true for each $i_1 = \overline{1, m_1}$:

$$f_{1i_1}^- \leq f_{1i_1}^* \leq f_{1i_1}^+ .$$

Here $f_{1i_1}^-$ and $f_{1i_1}^+$ define respectively the exact bottom and top boundaries of the function $f_{1i}\left(x_1, \widehat{x}_2\right)$ in the vicinity of points x_1^* and x_2^* in the form

$$\begin{aligned} f_{1i_1}^- &= \left(b_{11i_1} - \left|\Delta_{11i_1}^*\right|\right) - \left|\Delta_{12i_1}^*\right| , \\ f_{1i_1}^+ &= \left(b_{11i_1} + \left|\Delta_{11i_1}^*\right|\right) + \left|\Delta_{12i_1}^*\right| . \end{aligned} \qquad (3.107)$$

Relationship (3.107) follows from the properties of Chebyshev approximation functions [77, 150].

We now consider the problem of **disclosing the uncertainty of competitors' (opponents') counteractions taking into account both the competitor's actions and the situational uncertainty**. Owing to (3.90 and 3.91) this problem can be reduced to problems B1 and B2.

Problem B1. Disclosing goal and situational uncertainty for participant 1.

Problem B2. Disclosing uncertainty of opponents' counteractions, when the situation is completely determined, and its influence on each participant's attaining his goals may be neglected (the given problem was considered in detail above).

Problem A1 differs from the earlier considered problem of disclosing goals and situational uncertainty by imposing an additional condition: the situation is described not by one probabilistic parameter, but by vector α_1. On the basis of the above considered approaches the given problem can be reduced to the following Chebyshev approximation problem: to determine values x_1^{**} for the system of equations

$$f_{11i_1}(x_1) + f_{13i_1}(\alpha_1) = b_{11i_1}, \quad i_1 = \overline{1, m_1} ,$$

at which

$$\Delta_{11\alpha} = \max_i \left| f_{11i_1}(x_1) + f_{13i_1}(\alpha_1) - b_{11i_1} \right| ,$$

taken as the Chebyshev approximation measure, should be minimally possible

$$\Delta_{11\alpha}(x_1^{**}) = \min_{x_1} \Delta_{11\alpha} = \Delta_{11\alpha}^0 .$$

In this statement the problem is practically similar to problem A1 considered above in the part of relationships (3.103)–(3.105); however, it has its own peculiarities. In particular, this statement of the problem is expedient if a concrete situation is known for which the numerical values of vector α_1 and function $f_{13i_1}(\alpha_1)$ are determined.

In practice, a consecutive influence of several situations is possible. We assume that k_0 situations are known, for each of which probability coefficients p_k and $k = \overline{1; k_0}$, functions $f_{13i_1}(\alpha_{1k})$ and $k = \overline{1; k_0}$, and numerical values of vector α_{1k} and $k = \overline{1; k_0}$ are predicted. The problem should be solved separately for each situation $k = \overline{1; k_0}$ and is reduced to the following Chebyshev approximation problem: to determine such values x_1^{**} for the system of equations

$$f_{11i_1}(x_1) + f_{13i_1}(\alpha_{1k}) = b_{11i_1}; \quad i = \overline{1, m_1}; \quad k \in [1, k_0] , \tag{3.108}$$

for which

$$\Delta_{11} = \max_i \max_k \left[p_k \cdot |f_{11i_1}(x_1) + f_{13i_1}(\alpha_{1k}) - b_{11i_1}| \right] , \tag{3.109}$$

taken as the Chebyshev approximation measure, should be minimally possible

$$\Delta_{11\alpha}^k (x_1^{**}) = \min_{x_1} \Delta_{11\alpha}^k = \Delta_{11}^{0k} .$$

Here we should point out some peculiar features of the given problem. First, system (3.108) contains m_1^0 equations, where $m_1^0 = m_1 \cdot k_0$, since each i_1th equation is written down for each kth situation. Second, relationship (3.109) takes into account the probability of the kth situation occurrence, which is determined by the value p_k.

Thus, solving problem B1 allows one to disclose goals and situational uncertainty, and solving problem B2 allows one to disclose the uncertainty of two opponents' (competitors') interactions. Hence, these problems encompass all three types of uncertainty. The considered problems can be solved independently owing to assumptions (3.90) and (3.91). But independently solving these problems gives two values of vector x_1: the first, as the solution of problem B1, the second, of problem B2. It is obvious that these solutions may coincide only in exceptional cases. Therefore, there is an open question: how to solve the initial general problem of uncertainty disclosure if solutions to problems B1 and B2 are known. To answer this question one should pay attention to a peculiarity of solving problem B2—the required values x_1 and x_2 are found independently. Therefore, it is possible to consider other types of uncertainties, irrespective of the level of participants' counteraction uncertainty. This is exactly what was done in problem B1. Therefore, the totality of solutions—value x_1 as the solution to problem B1 and value x_2 as the solution to problem B2, determine the solution to the initial general problem of disclosing all considered types of uncertainty. The value x_1 obtained in solving problem B2 provides disclosure only of two uncertainty types—counteraction (interaction) uncertainty and goal uncertainty, which is a special case of the general problem of uncertainty disclosure. Thus, under conditions (3.90) and (3.91) it is possible to aggregate separate problem solutions in the final solution of the general problem of uncertainty disclosure.

Solving the problem of **disclosing the uncertainty of partners' actions under conditions** (3.90) and (3.91) is much simpler owing to the fact that there is no disagreement of interests. To solve this problem it is enough in all the above considered relationships, in determining partners' interaction conditions, to assume that $f_1\left(x_1, \widehat{x}_2\right)$ and $f_2\left(\widehat{x}_1, x_2\right)$ are vector functions and to consider the problem separately for their components, supposing that their components are defined by relationships (3.96) and (3.97), respectively. It is expedient to solve the corresponding problems by reducing them to the Chebyshev problem, using an approximation similar to the approaches considered above.

The performed analysis shows that the system problem of disclosing various types of uncertainty can be reduced to separate solutions to problems of disclosing each type of uncertainty, and then the obtained solutions are aggregated in the final solution of the initial problem. But this can be done under the strict condition that the partners' goals are described by additive functions of the form (3.90) and (3.91). At the same time, it is obvious that in practice such a representation of goal functions is not always possible.

We will show that the system problem of uncertainty disclosure can be solved not only for the type of additive function considered above (3.90) and (3.91). A similar procedure is also admissible if the participants' goals are interdependent and, therefore, described by multiplicative functions of the following form:

$$\begin{aligned}\left[1+f_{1i_1}\left(x_1;\widehat{x}_2;\alpha_1\right)\right] &= \left[1+f_{11i_1}\left(x_1\right)\right]^{\gamma_{11i_1}}\left[1+f_{12i_1}\left(\widehat{x}_2\right)\right]^{\gamma_{12i_1}}\\ &\quad\times\left[1+f_{13i_1}\left(\alpha_1\right)\right]^{\gamma_{13i_1}},\\ f_{1i_1}\left(x_1;\widehat{x}_2;\alpha_1\right) &= \left\{\left[1+f_{11i_1}\left(x_1\right)\right]^{\gamma_{11i_1}}\left[1+f_{12i_1}\left(\widehat{x}_2\right)\right]^{\gamma_{12i_1}}\right.\\ &\quad\left.\times\left[1+f_{13i_1}\left(\alpha_1\right)\right]^{\gamma_{13i_1}}-1\right\},\\ i_1 &= \overline{1,m_1}\,. \end{aligned} \tag{3.110}$$

Taking into account that the physical meaning of functions, arguments, and parameters included in relationship (3.110) meets the following conditions:

$$\begin{aligned}&f_{1i_1}\left(x_1;\widehat{x}_2;\alpha_1\right)\geq 0;\quad f_{11i_1}\left(x_1\right)\geq 0\,,\quad x_1=\left(x_{1j_1}|x_{1j_1}\geq 0,\ j_1=\overline{1,n_{1j_1}}\right),\\ &\widehat{x}_2=\left(x_{2j_2}|\,\widehat{x}_{2j_2}\geq 0,\ j_2=\overline{1,n_{2j_2}}\right),\qquad \alpha_1=\left(\alpha_{1k_1}|x_{1k_1}\geq 0,\ k_1=\overline{1,k_{01}}\right),\\ &\gamma_{11}=\left(\gamma_{11i_1}|\gamma_{11i_1}\geq 0,\ i_1=\overline{1,m_1}\right),\qquad \gamma_{12}=\left(\gamma_{12i_1}|\gamma_{12i_1}\geq 0,\ i_1=\overline{1,m_1}\right),\\ &\gamma_{13}=\left(\gamma_{13i_1}|\gamma_{13i_1}\geq 0,\ i_1=\overline{1,m_1}\right),\quad \gamma_{11i_1}\in\gamma_{11},\ \gamma_{12i_1}\in\gamma_{12},\ \gamma_{13i_1}\in\gamma_{13}\,,\end{aligned} \tag{3.111}$$

for participants 1 and 2 expression (3.110) may be transformed into an additive form:

$$f^n_{1i_1}\left(x_1,x_2,\alpha_1\right)=\gamma_{11i_1}\cdot f^n_{11i_1}\left(x_1\right)+\gamma_{12i_1}\cdot f^n_{12i_1}\left(x_2\right)+\gamma_{13i_1}f^n_{13i_1}\left(\alpha_1\right)\,. \tag{3.112}$$

Thus, multiplicative function (3.110) under condition (3.111) can be reduced to an additive function (3.112).

It may be shown that other types of functions can be represented in a multiplicative form and then reduced to an additive type. For example, it may be done with functions of the following form:

$$f_{i_1}\left(x_1, \widehat{x}_2, \alpha_1\right) = a^{f_{11i_1}(x_1)} \cdot b^{f_{12i_1}\left(\widehat{x}_2\right)} \cdot c^{f_{13i_1}(\alpha_1)} . \tag{3.113}$$

At $a > 0$, $b > 0$, and $c > 0$ function (3.113) is reduced to an additive form by taking the logarithm at any radix.

At the same time, it is possible to suggest a number of more complicated functions for which the solution to the initial problem may be reduced to a sequence of various types of uncertainty disclosure problems. Generally, such functions may be presented in the form of the following convolution:

$$f_{1i_1}\left(x_1, \widehat{x}_2, \alpha_1\right) = f_{13i_1}\left(\alpha_2, f_{12i_1}\left(\widehat{x}_2, f_{11i_1}(x_1)\right)\right) .$$

For the given type of function it is expedient to reduce problems of uncertainty disclosure to the following sequence of problems:

1. The problem of disclosing goal uncertainty, which is reduced to finding the value x_1^* of the assigned goal functions and their numerical values from the system of equations

$$f_{11i_1}(x_1) - b_{11i_1} = 0 .$$

2. The problem of disclosing uncertainty of interaction or counteraction, which is reduced to finding values x_2^* when the value x_1 is known:

$$x_2^* = \operatorname{Arg}\max_{\widehat{x}_2} f_{12i}\left(\widehat{x}_2, y_1^*\right) , \tag{3.114}$$

   where

$$y_1^* = \left\{y_{1i_1}^* \mid i_1 = \overline{1, m_1}\right\} , \quad y_{1i_1}^* = f_{11i_1}(x_1^*) .$$

3. The problem of disclosing situational uncertainty with known x_1^* and x_2^* on the basis of an analysis of the following function

$$y_{13i} = f_{13i}(\alpha_1, x_2^*) ,$$

   where x_2^* is defined by relationship (3.114).

The above problems can be solved using the methods of the Chebyshev function approximation theory [77, 150].

Thus, in this chapter the general structure of system analysis problems was considered and formalizable problems were singled out and their peculiarities were pointed out. It was shown that it is important in system analysis problems to disclose various types of uncertainty, provided that it is not known a priori whether various components of functions

$f_{1i_1}(x_1; \widehat{x}_2; \alpha_1)$, $i_1 = \overline{1, m_1}$ and $f_{2i_2}(\widehat{x}_1; x_2; \alpha_2)$, $i_2 = \overline{1, m_2}$, and various components $(x_{1j_{11}}; j_{11} = \overline{1, n_{11}})$, $(\widehat{x}_{2j_{22}}; j_{12} = \overline{1, n_{12}})$ $(\widehat{x}_{1j_{21}}; j_{21} = \overline{1, n_{21}})$, $(x_{2j_{22}}; j_{22} = \overline{1, n_{22}})$, $(\alpha_{1k_1};\ k_1 = \overline{1, k_{01}})$, $(\alpha_{2k_2};\ k_2 = \overline{1, k_{02}})$ of vectors x_1, x_2 and α are mutually independent or interconnected. The introduction multiplicative functions of the form (3.110) and (3.113) allow one to solve an uncertainty disclosure problem irrespective of whether or not components are interrelated [139]. On this basis the main problems of uncertainty disclosure were formulated and solved for the cases of participants' strategies for interaction and counteraction.

3.6 Recovery of Functional Dependences in Conceptual Uncertainty Disclosure Problems

In this chapter an approach to recovering functional dependences using an experimentally obtained discrete sample for system problems of conceptual uncertainty disclosure is considered. In particular, the given problem is common in the formation of a concept of new engineering equipment [176], in testing automated aircraft systems [186], in systems of automatic control of complex dynamic objects functioning in real time [50], in systems of technical diagnostics [55], and in a number of other applications.

Problems of functional dependence recovery and problems related to revealing regularities using empirical data are widespread in practice. Therefore, ways and methods of their solving are being continuously improved and adapted for the specificity of a concrete subject domain and to peculiarities of real problems [19, 34, 184, 187, 208].

Let us consider a series of such peculiarities. Their mathematical statement has a number of principal differences from classical statements of interpolation problems and statistical processing of a final sample. In the classical interpolation problem, one must find a function that provides a recovery of its meanings at the assigned points. In problems related to revealing regularities, it is necessary to find a function that characterizes most accurately the true dependence of this function on the most important factors on all intervals of the initial data. This difference results in a number of important peculiarities of the considered problem, such as the determination of a rational set of attributes and rational volume of a sample, construction of a regularized functional in solving incorrect problems of indirect experimental interpretation, problems of informal choices of recovered function structure, and a number of other peculiar features [1, 97, 169, 183, 184, 208, 209].

The above indicated factors have determined the formation and development of a special mathematical apparatus on which basis lie heuristic procedures and algorithms, inductive methods, and heuristic information models based to a considerable extent on intuition, experience, hypotheses, and assumptions. These problems are very important for certain kinds of applications. These applications are characterized by conditions of incompleteness, uncertainty, inaccuracy, and contradictoriness of the initial information, for example, for weakly structured and weakly formalized areas of application (medicine, sociology, technical diagnostics of abnormal, critical situations of complex objects, etc.).

Recall that the peculiarity of disclosing a conceptual uncertainty problem is determined by the necessity of finding a rational compromise between contradictory goals, for example, those set when a new product is being created, revealing its advantages and disadvantages in relation to the competing offers, and evaluating and forecasting possible risk factors [110, 112, 134, 135]. The ***problem of conceptual uncertainty disclosure*** in its contensive statement is reduced to a problem of system-coordinated disclosure of a set of diverse uncertainties on the basis of unified principles, techniques, and criteria. This set includes the uncertainty of product development goals and the uncertainty of its competitiveness prospects, the uncertainty of variation in supply and demand and the uncertainty of a competitor's active counteraction, and also the situational uncertainty of risks in the course of development, manufacturing, selling, and operation of a product. Such an uncertainty refers to the conceptual one in the sense that, as distinct from information uncertainty, it represents a unified complex of the lack of information, ambiguity, and contradictoriness of interconnected and interdependent elements of a specified set of polytypic uncertainties [115].

The problem of disclosing conceptual uncertainty may be presented as a kind of modification of a system optimization problem in the treatment of Glushkov [36]. However, the considered problem has a number of essentially important features. These features exclude a direct usage of the known system optimization methods and methods of uncertainty disclosure, since their application is possible if the corresponding goal functions are assigned [36, 38]. But in real problem solving, for example, at the initial stage of designing a concept and of complex engineering products, only incomplete, diverse initial information is available. This information may include empirical data, expert evaluations, a priori information about analogs and prototypes, some information about purposes and qualitative parameters of a product, standard restrictions and information, describing conditions of manufacture and operation, etc. On the basis of such information, it is necessary to generate goal functions of a new product creation. In these conditions, a choice of the number of goal functions, their analytical forms, and the substantiation of their contents and purposes is a nonformalizable procedure, and only the researcher can carry it out. The result depends on the competence, skill, experience, intuition, and other individual qualities of an actual researcher who is carrying out the given procedure.

The formation of functional dependences will be presented as the following sequence of interconnected problems [120]:

- Reducing the initial information to some standard type, which makes it possible to form functional dependences.

- Choosing a class and structure of approximating functions in forming functional dependences.
- Choosing criteria, principles, approaches, and methods of approximating function construction.
- Finding approximating functions in the accepted class, which ensures their best approximation according to the accepted criterion.

The problem of choosing a class and structure of approximating functions is basic and determines the requirements to other problems. In particular, the required functions do not only have to be as much as possible approximated to empirical data in the sense some criterion, but also have some extreme properties [64]. The specificity of the extreme properties of the required approximating functions results from the resrictiveness of the interval of the initial data assignment and consists in a considerable impact exerted by the disturbing actions on the extreme properties of the function on the interval boundaries. This peculiarity is of prime importance and results in a more complicated structure of approximating functions than that in interpolation problems. From the above the urgency and practical importance of the problem of rationally choosing a class of approximating functions is evident. An important feature of this problem is the necessity of choosing a rational compromise between contradictory requirements—maximization of the credibility level of the procedure for revealing the required regularity, which leads to the necessity of increasing the complexity of a class of approximating functions, and the requirement of minimizing the complexity of the procedure of forming the required functional dependence, which causes the necessity to simplify approximating functions. An unsuccessful choice of approximating functions may lead to a situation where a restored function will be good in approximating the concrete initial data for most of the assigned interval, but it will be bad in describing the true functional dependence in general.

Taking into account the above explanations, we state the problem of forming approaching functions on the assigned empirical data [120, 128].

Statement of problem of approximating function recovery

Let in the general case the vector $y = \left(y_i | i = \overline{1,m}\right)$ determine the required or desirable values of the required functions. These functions characterize quantitatively the basic properties of an object. The vector x_1 forms indices of the design decisions of a decision maker, in particular, for an engineering product they are its constructive, technical, technological, and other indices. Vector x_2 forms controllable indices of an external influence. In particular, they are indices of static and dynamic loading and others. Vector x_3 forms indices of random and uncontrollable factors having an external influence, for example, indices of influence of the external environment and of operating conditions, various risk factors, predictable actions of partners and competitors, predictable abnormal situations, and predictable market indices of demand and supply.

Let the initial information be assigned in the form of the following discrete array

$$
\begin{aligned}
M_0 &= \langle\, Y_0, X_1, X_2, X_3 \rangle\,,\\
Y_0 &= \left(Y_i | i = \overline{1,m}\right), \quad Y_i = \left(Y_i\,[q_0]\,| q_0 = \overline{1,k_0}\right),\\
X_1 &= \left(X_{1j_1} | j_1 = \overline{1,n_1}\right), \quad X_{1j_1} = \left(X_{1j_1}\,[q_1]\,| q_1 = \overline{1,k_1}\right),\\
X_2 &= \left(X_{2j_2} | j_2 = \overline{1,n_2}\right), \quad X_{2j_2} = \left(X_{2j_2}\,[q_2]\,| q_2 = \overline{1,k_2}\right),\\
X_3 &= \left(X_{3j_3} | j_3 = \overline{1,n_3}\right), \quad X_{3j_3} = \left(X_{3j_3}\,[q_3]\,| q_3 = \overline{1,k_3}\right),
\end{aligned}
$$

where the set Y_0 determines numerical values $Y_i\,[q_0] \Rightarrow \langle X_{1j_1}\,[q_1],\ X_{2j_2}\,[q_2],\ X_{3j_3}\,[q_3]\rangle$ of the required continual functions $y_i = f_i\,(x_1, x_2, x_3)$, $i = \overline{1,m}$; $x_1 = \left(x_{1j_1} | j_1 = \overline{1,n_1}\right)$, $x_2 = \left(x_{2j_2} | j_2 = \overline{1,n_2}\right)$, $x_3 = \left(x_{3j_3} | j_3 = \overline{1,n_3}\right)$. Each value $q_0 \in [1, k_0]$ corresponds to a certain set $q_0 \Leftrightarrow \langle q_1, q_2, q_3\rangle$ of the values $q_1 \in [1, k_1]$, $q_2 \in [1, k_2]$, $q_3 \in [1, k_3]$. The set Y_0 consists of k_0 various values $Y_i[q_0]$. In the sets X_1, X_2, X_3 a certain part of the values is $X_{1j_1}\,[q_1]$, $X_{2j_2}\,[q_2]$, $X_{3j_3}\,[q_3]$, while some values are $q_1 = \hat{q}_1 \in \hat{Q}_1 \subset [1, k_1]$, $q_2 = \hat{q}_2 \in \hat{Q}_2 \subset [1, k_2]$, $q_3 = \hat{q}_3 \in \hat{Q}_3 \subset [1, k_3]$ and separately repeat, but for different $q_0 \in [1, k_0]$ there are no completely coincident sets $\langle X_{1j_1}\,[q_1]\,, X_{2j_2}\,[q_2]\,, X_{3j_3}\,[q_3]\rangle$. Here $n_1 + n_2 + n_3 = n_0$, $n_0 \le k_0$. It is known that $x_1 \in D_1$, $x_2 \in D_2$, $x_3 \in D_3$, $X_1 \in \hat{D}_1$, $X_2 \in \hat{D}_2$, $X_3 \in \hat{D}_3$, where

$$
\begin{aligned}
D_s &= \left\langle x_{sj_s} | d^-_{sj_s} \le x_{sj_s} \le d^+_{sj_s},\ j_s = \overline{1,n_s} \right\rangle, \quad s = \overline{1,3}\,,\\
\hat{D}_s &= \left\langle X_{sj_s} | \hat{d}^-_{sj_s} \le X_{sj_s} \le \hat{d}^+_{sj_s},\ j_s = \overline{1,n_s} \right\rangle, \quad s = \overline{1,3}\,,\\
& d^-_{sj_s} \le \hat{d}^-_{sj_s}, \quad d^+_{sj_s} \ge \hat{d}^+_{sj_s}\,.
\end{aligned}
$$

Approximating functions $\Phi_i\,(x_1, x_2, x_3)$, $i = \overline{1,m}$, must be found that, with a practically acceptable error, in the sense of the Chebyshev problem of approximation, characterize the true functional dependences $y_i = f_i\,(x_1, x_2, x_3)$, $i = \overline{1,m}$, on the set D_s.

In a real problem the meaning of variables y, x_1, x_2, and x_3 is concretized. For example, in designing and/or testing a product, vector x_1 determines the external indices of the product. These indices characterize technical, operational, economic, and other parameters of its quality. The components of vector x_2 are the internal parameters of the product. They characterize constructive, technological, and other indices of the product. The components of vector x_2 are controllable parameters of an external influence. In particular, there are the indices of carrying capacity (maximum weight, dimensions, types of cargo) and general parameters of allowable climatic zones of operation (the moderate, polar, or tropical climate). The components of vector x_3 are uncontrollable indices of an external influence, in particular, concrete indices of the environment (a permissible range of variations of temperature, humidity, and other operation conditions).

The given problem is principally distinct from a typical problem of functional dependence recovery by its complexity, which is preconditioned not only by the diversity of initial information, but also by the diversity of properties of the considered groups of factors, which are determined by vectors x_1, x_2, and x_3, respectively. In fact, the values of components of a vector x_1 are the developer's own decisions and, consequently, may be changed in the process of designing. The values of components of vector x_2 are requirements that are determined by the purpose of a product, and, if it changes, they may be corrected by the customer. In any case the developer is obliged to fulfill the requirements of the customer. The values of components of vector x_3 are requirements that are determined by the standards for operating conditions of a product, and, consequently, they are to be fulfilled by the developer.

Thus, in practice the necessity arises of estimating separately the degree of influence of each group of factors on the properties of the approximating functions. For this we will form the approximating functions in the form of a hierarchical multilevel system of models. At the top level a model is realized that determines the dependence of approximating functions on variables x_1, x_2, x_3. The required functions are formed in a class of additive functions and presented as a superposition of functions of variables x_1, x_2, x_3. The possibility of such a presentation follows from the theorem of Kolmohorov [57]. Thus, the required functions $\Phi_i(x_1, x_2, x_3)$ will be presented in the following form [115]:

$$\Phi_i(x_1, x_2, x_3) = c_{i1}\Phi_{i1}(x_1) + c_{i2}\Phi_{i2}(x_2) + c_{i3}\Phi_{i3}(x_3), \quad i = \overline{1, m}. \tag{3.115}$$

At the second hierarchical level models are formed that determine separately the dependence on approximating functions on the components of variables x_1, x_2, x_3, respectivly. For this one must proceed from the vector functions to superpositions of functions of these vector components. Taking into account that components of each vector x_1, x_2, x_3 are different in their physical composition, for adding functions (3.115) it is expedient to choose a class of generalized polynomials and to present them in the following form:

$$\Phi_{i1}(x_1) = \sum_{j_1=1}^{n_1} a_{ij_1}^{(1)} \Psi_{1j_1}(x_{1j_1}), \quad \Phi_{i2}(x_2) = \sum_{j_2=1}^{n_2} a_{ij_2}^{(2)} \Psi_{2j_2}(x_{2j_2}),$$
$$\Phi_{i3}(x_3) = \sum_{j_3=1}^{n_3} a_{ij_3}^{(3)} \Psi_{3j_3}(x_{3j_3}). \tag{3.116}$$

It is suggested for all $i = \overline{1, m}$ on each variable to choose the one-type functions Ψ_{1j_1}, Ψ_{2j_2}, Ψ_{3j_3}, which allows one to simplify further the solution to the problem.

At the third hierarchical level models are formed that determine functions Ψ_{1j_1}, Ψ_{2j_2}, Ψ_{3j_3}. Here the main problem consists in choosing a structure and components of functions Ψ_{1j_1}, Ψ_{2j_2}, Ψ_{3j_3}. Structures of these functions are chosen similar to (3.116). Let us present the functions in the form of the following generalized polynomials:

$$\Psi_{sj_s}(x_{j_s}) = \sum_{p=0}^{P_{j_s}} \lambda_{j_s p} \varphi_{j_s p}(x_{sj_s}), \quad s = 1, 2, 3 . \tag{3.117}$$

In choosing the functions a number of requirements must be taken into account. First, these functions are the basic structure forming the functional elements of all models. Therefore, they should have such extreme properties at the assigned intervals for the corresponding variables x_{sj_s}, $j_s = \overline{1, n_s}$, $s = \overline{1, 3}$, which are fuzzily defined by the array M_0. Second, they should ensure the possibility of realizing the uniform approximation of the true functional dependences on set D_s and the conformity of the extreme properties of functions $\Phi_i(x_1, x_2, x_3)$ and $f_i(x_1, x_2, x_3)$ $\forall i = \overline{1, m}$. Taking into account the fact that, for the majority of variables, the condition $x_{sj_s} \geq 0$ is physically fulfilled and variables x_{sj_s}, $\forall j_s = \overline{1, n_s}$, $s = \overline{1, 3}$ can be normalized to the interval [0,1]. Then, shifted Chebyshev polynomials can ensure fulfilment of these requirements [81]. We choose them as functions $\varphi_{j_s p}$.

The next practically important problem is a choice of criteria and methods of approximating function construction. As a rule, while choosing criteria and methods, the convenience of computing process realization is considered to be a determining condition. From this position the most convenient is the root-mean-square criterion, which is widely used in practice. At the same time, in solving real problems a more important approach may become the requirement that the chosen criterion conform to the specificity of the studied objects' properties. For example, in controlling high-dynamic objects in real time it is important to find out in a timely manner the separate, sufficiently large deviations in a sample of technical diagnostics, and on this basis reveal regularities of transition to an abnormal mode. In particular, in aircraft gas-turbine engines such an approach allows one to reveal in a timely manner the tendency of transition to a surge, which may cause engine destruction.

Similar conditions are characteristic of other real problems. For example, the absolute deviation of the dimensions is a determining condition in problems involving the matching of tolerances of conjugate constructive elements in the design, manufacture, and operation of a product. However, the root-mean-square criterion carries out a smoothable action on separate and sufficiently large spikes, and, consequently, it does not allow one to reveal such deviations. The Chebyshev criterion does not have this disadvantage, which is why it is chosen as the criterion in all problems considered in this work.

Solving the criterion choice problem and other problems allows one to move on to the procedure of finding approximating functions in the accepted class of functions. Since the initial procedure involves choosing functions $\varphi_{j_s p}$, which are the basic elements of all models, it is expedient to start the formation of a system of models from the lowest hierarchical level. Then the approximating functions should be found on the basis of the following sequence [142]:

$$\Psi_1, \Psi_2, \Psi_3 \rightarrow \Phi_{i1}, \Phi_{i2}, \Phi_{i3} \rightarrow \Phi_i \,,$$

which will yield the final result by aggregation of the corresponding solutions. Such an approach allows one to reduce the formation procedure of approximating functions to a sequence of Chebyshev problems of approximation for incompatible systems of linear equations, the methods of their solving are well known [77, 81, 150]. In particular, Chebyshev problems are reduced to a problem of linear programming [210]. Let us proceed to the formalization and solving of problems.

The problem of forming the function Ψ_s. This problem is the most important and most complicated. It is important because some flaws, for example an inappropriate choice of the number and degrees of Chebyshev polynomials, cannot be eliminated completely at the subsequent levels of a system of models. Moreover, they may be aggravated. The given problem is also complicated because there is of contradictory requirements for the required functions. First, the functions should reflect with sufficient accuracy the extreme properties characteristic of a set of approximating functions. Second, they should, to a sufficient extent, take into account individual features of the extreme properties of each function and provide the possibility of adapting to them at subsequent levels. From here it follows that the functions $\Psi_{1j_1}(x_{j_1})$, $\Psi_{2j_2}(x_{j_2})$, $\Psi_{3j_3}(x_{j_3})$ are to be formed in conformity with the conditions

$$\Psi_1 \rightarrow \left\langle \Phi_{i1} | i = \overline{1, m} \right\rangle, \quad \Psi_2 \rightarrow \left\langle \Phi_{i2} | i = \overline{1, m} \right\rangle, \quad \Psi_3 \rightarrow \left\langle \Phi_{i3} | i = \overline{1, m} \right\rangle,$$
$$\Psi_s = \left\langle \Psi_{sj_s} | j_s = \overline{1, n_s} \right\rangle, \quad s = \overline{1, 3}.$$

Further, while forming a system of equations it is necessary to take into account the properties and features of the recovered functional dependences determined by the initial data, and also the structure of the functions $\Psi_{1j_1}(x_{j_1})$, $\Psi_{2j_2}(x_{j_2})$, $\Psi_{3j_3}(x_{j_3})$, which are described by (3.117). In addition, we assume that the degree of influence of functions Ψ_1, Ψ_2, Ψ_3 on the general properties of a set of recovered functions is the same. This assumption is conditioned by two factors. On the one hand, there is no a priori information on mutual influence of these functions, and on the other hand, there is the necessity to take into account with sufficient accuracy the influence of these functions on the extreme properties characteristic of this set as a whole.

At other levels of a system of models individual features of extreme properties of each function will be taken into account. From here it follows that the problem of forming is reduced to the Chebyshev problem of approximation for the following system of equations:

$$F_{i1}\left(\hat{X}[q_0]\right) - b_{q_0} = 0, \quad q_0 = \overline{1, k_0}, \tag{3.118}$$

$$F_{i1}\left(\hat{X}\left[q_0\right]\right)=\sum_{j_1=1}^{n_1}\sum_{p_1=0}^{P_{j1}}\lambda_{j_1p_1}T^*_{p_1}\left(\hat{X}_{1j_1}\left[q_1\right]\right)+\sum_{j_2=1}^{n_2}\sum_{p_2=0}^{P_{j2}}\lambda_{j_2p_2}T^*_{p_2}\left(\hat{X}_{2j_2}\left[q_2\right]\right)$$
$$+\sum_{j_3=1}^{n_3}\sum_{p_3=0}^{P_{j3}}\lambda_{j_3p_3}T^*_{p_3}\left(\hat{X}_{3j_3}\left[q_3\right]\right);$$
$$\hat{X}\left[q_0\right]=\left(\hat{X}_{1j_1}\left[q_1\right],\ \hat{X}_{2j_2}\left[q_2\right],\ \hat{X}_{3j_3}\left[q_3\right]\left|q_0\Leftrightarrow\langle q_1,q_2,q_3\rangle\right.\right).$$

Here $T^*_{p_1}$, $T^*_{p_2}$, $T^*_{p_3}$ are Chebyshev shifted polynomials, b_{q_0} is the value determined by the relationship

$$b_{q_0}=\left\{\max_{i\in[1,m]}\hat{Y}_i\left[q_0\right]+\min_{i\in[1,m]}\hat{Y}_i\left[q_0\right]\right\}/2,\quad q_0=\overline{1,k_0},$$

and $\hat{X}_{1j_1}\left[q_1\right]$, $\hat{X}_{2j_2}\left[q_2\right]$, $\hat{X}_{3j_3}\left[q_3\right]$, $\hat{Y}_i\left[q_0\right]$ are the meanings of the values $X_{1j_1}\left[q_1\right]$, $X_{2j_2}\left[q_2\right]$, $X_{3j_3}\left[q_3\right]$, $Y_i\left[q_0\right]$, respectively, normalized to the interval [0,1].

Solving the system consists in finding those matrices $\left\|\lambda^0_{j_1p_1}\right\|$, $\left\|\lambda^0_{j_2p_2}\right\|$, $\left\|\lambda^0_{j_3p_3}\right\|$ that ensure the best approximation

$$\Delta^0_\lambda=\min_{\|\lambda\|}\Delta_\lambda$$

taking into account the maximum residual

$$\Delta_\lambda=\max_{q_0\in[1,k_0]}\left|F_1\left(\hat{X}\left[q_0\right]\right)-b_{q_0}\right|, \tag{3.119}$$

accepted as a measure of the Chebyshev approximation of system (3.118).

Here the value of the best approximation Δ^0_λ and of the required matrices are characterized by the relationship

$$\Delta^0_\lambda=\min_{\|\lambda\|}\max_{q_0\in[1,k_0]}\left|F_1\left(\hat{X}\left[q_0\right]\right)-b_{q_0}\right|,$$
$$\lambda^0=\operatorname{Arg}\min_{\|\lambda\|}\max_{q_0\in[1,k_0]}\left|F_1\left(\hat{X}\left[q_0\right]\right)-b_{q_0}\right|.$$

Here

$$\left\|\lambda^0\right\|=\left\langle\left\|\lambda^0_{j_1p_1}\right\|,\left\|\lambda^0_{j_2p_2}\right\|,\left\|\lambda^0_{j_3p_3}\right\|\right\rangle,\quad\|\lambda\|=\left\langle\left\|\lambda_{j_1p_1}\right\|,\left\|\lambda_{j_2p_2}\right\|,\left\|\lambda_{j_3p_3}\right\|\right\rangle.$$

The problem of forming the functions Φ_{is}. In this problem we assume that $\forall i\in[1,m]$, the influence degree of functions $\Phi_{i1}(x_1)$, $\Phi_{i2}(x_2)$, $\Phi_{i3}(x_3)$ on the properties of the corresponding approximating function $\Phi_i(x_1,x_2,x_3)$, is the same. This assumption is conditioned by the absence of a priori information. At the same time, the assumption allows one to form separately the functions $\Phi_{i1}(x_1)$, $\Phi_{i2}(x_2)$, $\Phi_{i3}(x_3)$ and to determine the degree of influence of each of them on the following higher level of the model's hierarchy. As a result, the problem consists in finding matrices $\left\|a_{ij_1}^{(1)}\right\|$, $\left\|a_{ij_2}^{(2)}\right\|$, $\left\|a_{ij_3}^{(3)}\right\|$ $\forall i \in [1, m]$ and is reduced to the Chebyshev problem of approximation for the following three systems of equations:

$$\begin{aligned} &F_{i21}\left(\hat{X}_1[q_0]\right) - \hat{Y}_i[q_0] = 0, \quad F_{i22}\left(\hat{X}_2[q_0]\right) - \hat{Y}_i[q_0] = 0\,, \\ &F_{i23}\left(\hat{X}_3[q_0]\right) - \hat{Y}_i[q_0] = 0, \quad q_0 = \overline{1, k_0}\,, \end{aligned} \tag{3.120}$$

where

$$\begin{aligned} F_{i21}\left(\hat{X}_1[q_0]\right) &= \sum_{j_1=1}^{n_1} a_{ij_1}^{(1)} \Psi_{1j_1}\left(\hat{X}_{1j_1}[q_1]\right), \\ F_{i22}\left(\hat{X}_2[q_0]\right) &= \sum_{j_2=1}^{n_2} a_{ij_2}^{(2)} \Psi_{2j_2}\left(\hat{X}_{2j_2}[q_2]\right), \\ F_{i23}\left(\hat{X}_3[q_0]\right) &= \sum_{j_3=1}^{n_3} a_{ij_3}^{(3)} \Psi_{3j_3}\left(\hat{X}_{3j_3}[q_3]\right). \end{aligned}$$

Solving each system consists in finding those matrices $\left\|a_s^0\right\| = \left\|\hat{a}_{ij_s}^{(s)}\right\|$, $s = \overline{1, 3}$ respectively, that ensure the best approximation

$$\Delta_s^0 = \min_{\|a_s\|} \Delta_{a_s}$$

for the maximum residual value

$$\Delta_{a_s} = \max_{q_0 \in [1, k_0]} \left| F_{i2s}\left(\hat{X}_s[q_0]\right) - \hat{Y}_i[q_0]\right|$$

that is accepted as a measure of the Chebyshev approximation of system (3.120). At the same time, the best approximation values and the required matrices are characterized by the following relationships:

$$\begin{aligned} \Delta_1^0 &= \min_{\|a_1\|} \max_{q_0 \in [1, k_0]} \left| F_{i21}\left(\hat{X}_1[q_0]\right) - \hat{Y}_i[q_0]\right|, \\ \left\|a_1^0\right\| &= \operatorname{Arg} \min_{\|a_1\|} \max_{q_0 \in [1, k_0]} \left| F_{i21}\left(\hat{X}_1[q_0]\right) - \hat{Y}_i[q_0]\right|, \\ \Delta_2^0 &= \min_{\|a_2\|} \max_{q_0 \in [1, k_0]} \left| F_{i22}\left(\hat{X}_2[q_0]\right) - \hat{Y}_i[q_0]\right|, \\ \left\|a_2^0\right\| &= \operatorname{Arg} \min_{\|a_2\|} \max_{q_0 \in [1, k_0]} \left| F_{i22}\left(\hat{X}_2[q_0]\right) - \hat{Y}_i[q_0]\right|, \\ \Delta_3^0 &= \min_{\|a_3\|} \max_{q_0 \in [1, k_0]} \left| F_{i23}\left(\hat{X}_3[q_0]\right) - \hat{Y}_i[q_0]\right|, \\ \left\|a_3^0\right\| &= \operatorname{Arg} \min_{\|a_3\|} \max_{q_0 \in [1, k_0]} \left| F_{i23}\left(\hat{X}_3[q_0]\right) - \hat{Y}_i[q_0]\right|, \end{aligned}$$

where

$$\hat{X}_1[q_0] = \left\langle \hat{X}_{1j_1}[q_1] \,|j_1 = \overline{1, n_1},\ q_1 \in [1, k_1] \right\rangle ,$$
$$\hat{X}_2[q_0] = \left\langle \hat{X}_{2j_2}[q_2] \,|j_2 = \overline{1, n_2},\ q_2 \in [1, k_2] \right\rangle ,$$
$$\hat{X}_3[q_0] = \left\langle \hat{X}_{3j_3}[q_3] \,|j_3 = \overline{1, n_3},\ q_3 \in [1, k_3] \right\rangle ,$$
$$q_0 \Leftrightarrow \langle q_1, q_2, q_3 \rangle , \quad \|a_1\| = \left\| a_{ij_1}^{(1)} \right\| , \quad \|a_2\| = \left\| a_{ij_2}^{(2)} \right\| , \quad \|a_3\| = \left\| a_{ij_3}^{(3)} \right\| .$$

The problem of forming functions Φ_i. This problem consists in finding the set of required approximating functions $\Phi = \langle \Phi_i(x_1, x_2, x_3) \,|i = \overline{1, m} \rangle$ and is realized at the final stage of forming a system of models. The results of the previous stages and the initial discrete values of functions $\hat{Y}_i[q_0]$ are taken as the initial data. The formation of each function $\Phi_i(x_1, x_2, x_3)$ is independent. Thus, $\forall\ i \in [1, m]$, all processes of calculating $\Phi_i(x_1, x_2, x_3)$ may be carried out simultaneously and in parallel. Solving the problem for $\forall\, i \in [1, m]$ consists in finding matrices $\|c_{i1}\|, \|c_{i2}\|, \|c_{i3}\|$ and is reduced to the Chebyshev problem of approximation for the following system of equations:

$$F_{i3}\left(\hat{X}[q_0]\right) - \hat{Y}_i[q_0] = 0, \quad q_0 = \overline{1, k_0}, \quad i \in [1, m] ,$$

where

$$F_{i3}\left(\hat{X}[q_0]\right) = c_{i1}\Phi_{i1}\left(\hat{X}_1[q_1]\right) + c_{i2}\Phi_{i2}\left(\hat{X}_2[q_2]\right) + c_{i3}\Phi_{i3}\left(\hat{X}_3[q_3]\right) ,$$
$$q_0 \Leftrightarrow \langle q_1, q_2, q_3 \rangle .$$

The Chebyshev criterion of evaluating the quality of the solution is formalized similarly to the criteria of the previous problems. The results of the solution are characterized by the following relationships:

$$\Delta_c^0 = \min_{\|c\|} \max_{i \in [1,m]} \left| F_{i3}\left(\hat{X}[q_0]\right) - \hat{Y}_i[q_0] \right| ,$$
$$\left\| c^0 \right\| = \operatorname{Arg} \min_{\|c\|} \max_{i \in [1,m]} \left| F_{i3}\left(\hat{X}[q_0]\right) - \hat{Y}_i[q_0] \right| ,$$
$$\left\| c^0 \right\| = \left\langle \left\| c_{i1}^0 \right\| , \left\| c_{i2}^0 \right\| , \left\| c_{i3}^0 \right\| \right\rangle , \quad \|c\| = \langle \|c_{i1}\| , \|c_2\| , \|c_{i3}\| \rangle .$$

Thus, the consecutive solving of the formulated Chebyshev problems allows one to find all unknown quantities in the structure of the approximating functions. As a result, the following hierarchical system of approximating functions is obtained:

$$\Phi_i(x_1, x_2, x_3) = c_{i3}^0 \Phi_{i1}(x_1) + c_{i2}^0 \Phi_{i2}(x_2) + c_{i3}^0 \Phi_{i3}(x_3) , \quad i = \overline{1, m} ; \quad (3.121)$$

$$\Phi_{i1}(x_1) = \sum_{j_1=1}^{n_1} \hat{a}_{ij_1}^{(1)} \Psi_{1j_1}(x_{1j_1}) ,$$

$$\Phi_{i2}(x_2) = \sum_{j_2=1}^{n_2} \hat{a}_{ij_2}^{(2)} \Psi_{2j_2}(x_{2j_2}) ,$$

$$\Phi_{i3}(x_3) = \sum_{j_3=1}^{n_3} \hat{a}_{ij_3}^{(3)} \Psi_{3j_3}(x_{3j_3}) ; \qquad (3.122)$$

$$\Psi_{sj_s}(x_{j_s}) = \sum_{p_s=0}^{P_{j_s}} \lambda_{j_s p_s}^{0} T_{p_s}^{*}(x_{sj_s}), \quad s = \overline{1,3}. \qquad (3.123)$$

Further, according to the problem statement it is necessary to estimate errors of the functions $\Phi_i(x_1, x_2, x_3)$, $i = \overline{1,m}$, in relation to the true functional dependence $y_i = f_i(x_1, x_2, x_3)$, $i = \overline{1,m}$. If an error turns out practically unacceptable, it is necessary to eliminate the flaw. In practice the given problem is very complicated since the true functional dependence is determined by many variables $n = n_1 + n_2 + n_3$; $n \gg 10$ and is characterized by a multidimensional discrete array with irregular counts, but its analytical form $y_i = f_i(x_1, x_2, x_3)$ is absent. These features make it impossible to apply typical methods of analysis and evaluate an error of the empirical data. It is suggested that the method of a repetitive use of the initial array should be applied. Briefly its essence is as follows. On the basis M_0 several (for example, from 3 to 6) samples are formed. One of them is complete (i.e., it coincides with M_0), while others have noncoincident absences of data. On the basis of each sample, the functions $\Phi_i(x_1, x_2, x_3)$ are determined. The comparison of these functions between themselves and the functions' values with the missed data allows one to obtain the information necessary for error estimation and for making a decision on the measures to be taken to reduce errors.

The suggested approach to forming functional dependences, due to the properties of Chebyshev polynomials, makes it possible to extrapolate the approximating functions constructed for segments $\left[\hat{d}_{j_s}^{-}, \hat{d}_{j_s}^{+}\right]$ on wider segments $\left[\hat{d}_{j_s}^{-}, \hat{d}_{j_s}^{+}\right]$, which allows one to predict the properties of a product beyond the testing intervals.

Consider some examples.

Example 1. Let it be necessary to solve a problem of recovering functions $y_i(x_1, x_2, x_3)$, $i = \overline{1,4}$, on the assigned discrete meanings X_s, $s = \overline{1,3}$ and Y_i, $i = \overline{1,4}$ of the sample given in Table 3.2. The dimensions of vectors X_1, X_2, X_3 are $n_1 = 2$, $n_2 = 2$, $n_3 = 3$, respectively; the volume of a sample $q_0 = \overline{1,45}$; and the number of functions is $m = 4$.

Table 3.2. Initial discrete data $X_1\left[X_{11}, X_{12}\right],\ X_2\left[X_{21}, X_{22}\right],\ X_3\left[X_{31}, X_{32}, X_{33}\right]$ and $Y_i\left[X_1, X_2, X_3\right],\ i=\overline{1,4}$

q_0	X_{11}	X_{12}	X_{21}	X_{22}	X_{31}	X_{32}	X_{33}	Y_1	Y_2	Y_3	Y_4
1	5.050	2.015	7.050	8.015	10.000	1.000	5.100	254.621	98.145	119.406	117.683
2	5.150	2.100	7.150	9.109	15.800	2.100	4.200	198.163	73.368	92.651	90.123
3	5.200	2.120	7.192	9.125	22.500	2.500	3.500	187.411	71.084	87.691	83.576
4	5.250	2.170	7.250	9.175	25.000	3.510	2.720	167.197	63.567	78.793	74.789
5	5.325	2.200	7.325	9.198	32.500	4.200	2.530	166.547	63.813	79.497	74.316
6	5.350	2.250	7.350	9.251	35.000	5.020	2.100	153.789	61.378	77.082	72.817
7	5.400	2.400	7.411	9.395	40.700	8.200	1.150	110.926	55.579	67.758	77.425
8	5.500	2.500	7.505	9.498	51.800	10.100	0.720	151.381	60.432	71.956	89.519
9	5.600	2.600	7.610	9.598	65.000	12.800	0.540	187.364	76.283	91.123	121.374
10	5.700	2.700	7.695	9.699	72.000	14.400	0.120	236.123	93.657	112.859	149.173
11	5.750	2.750	7.750	9.748	75.400	14.700	1.250	292.341	118.624	153.717	184.136
12	5.800	2.775	7.804	9.775	82.800	15.500	1.760	288.324	114.324	117.965	179.152
13	5.850	2.800	7.850	9.798	85.000	16.300	2.230	326.939	128.926	155.912	201.239
14	5.907	2.850	8.050	9.850	90.780	16.700	2.610	377.128	148.675	169.359	225.482
15	5.910	2.855	7.910	9.855	91.000	16.900	4.160	405.327	159.367	192.924	240.976
16	5.925	2.865	7.925	9.865	92.500	17.500	5.250	458.386	180.567	218.549	275.846
17	5.929	2.885	8.011	9.875	92.900	17.700	6.370	518.859	183.932	247.354	316.124
18	5.933	2.915	7.933	9.899	93.500	18.200	7.260	595.737	235.124	284.167	363.928
19	5.935	2.950	7.935	9.951	94.580	19.100	7.510	506.168	261.946	316.375	403.153
20	5.950	2.975	7.950	9.975	95.400	19.500	7.740	685.761	281.387	341.326	431.195
21	5.010	1.995	6.950	9.015	11.500	21.000	8.140	790.639	310.519	375.651	471.588
22	5.050	2.975	7.108	9.975	10.500	19.560	8.350	723.784	285.142	344.856	436.847
23	5.150	2.950	7.151	9.950	15.800	19.300	8.580	731.438	288.125	348.314	441.842
24	5.200	2.900	7.204	9.915	21.500	18.700	8.740	721.321	283.435	344.716	439.425
25	5.250	2.875	7.248	9.875	26.400	17.560	8.850	691.845	272.834	329.942	422.147
26	5.325	2.865	7.325	9.865	32.500	17.100	9.210	708.614	280.562	349.316	435.954

27	5.350	2.855	7.351	9.855	35.300	16.700	19.520	729.956	287.987	348.231	450.492
28	5.400	2.850	7.408	9.850	41.700	16.200	9.750	730.129	288.951	347.987	454.897
29	5.500	2.775	7.495	9.775	50.200	15.700	10.100	717.152	285.494	342.967	458.289
30	5.600	2.750	7.607	9.750	62.700	15.360	0.100	278.654	111.209	132.856	172.164
31	5.700	2.710	7.697	9.697	69.800	14.700	1.150	242.145	96.197	115.632	153.356
32	5.750	2.603	7.750	9.605	75.100	13.340	1.360	186.243	77.325	93.135	127.168
33	5.800	2.495	7.798	9.495	80.520	11.720	1.750	162.345	64.615	77.824	106.123
34	5.850	2.394	7.850	9.415	85.200	8.900	2.130	132.879	52.534	63.453	82.659
35	5.907	2.245	7.913	9.255	90.760	7.740	2.570	167.156	65.178	79.167	93.834
36	5.910	2.192	7.910	9.205	91.100	6.360	2.750	170.531	66.176	80.836	91.345
37	5.925	2.175	7.925	9.175	92.500	5.700	3.260	184.243	70.364	87.192	96.841
38	5.929	2.125	7.929	9.125	92.900	3.750	3.790	181.956	70.428	85.834	93.952
39	6.010	2.105	7.933	9.091	93.300	3.650	4.120	216.829	83.475	101.985	109.463
40	5.935	2.010	7.935	8.985	94.500	3.520	4.360	273.329	104.924	128.591	133.415
41	5.950	2.110	7.950	9.115	98.600	2.720	3.850	219.421	84.183	102.861	108.613
42	5.020	2.115	6.995	9.115	110.000	2.340	2.340	225.356	86.324	105.817	107.319
43	6.050	2.128	7.950	9.120	95.260	2.560	1.680	176.578	66.457	78.473	82.263
44	5.935	2.131	7.935	9.130	93.520	2.760	1.320	170.948	65.814	81.417	84.132
45	5.925	2.135	7.925	9.135	92.800	2.980	1.160	168.334	64.549	78.653	81.953

Finding approximating functions $\Phi_i(x_1, x_2, x_3)$, $i = \overline{1, m}$, which with a practically acceptable error in the sense of the Chebyshev approximation characterize the true functional dependences $y_i = f_i(x_1, x_2, x_3)$, $i = \overline{1, m}$, is carried out on the basis of the following sequence of models:

$$\Psi_1, \Psi_2, \Psi_3 \rightarrow \Phi_{i1}, \Phi_{i2}, \Phi_{i3} \rightarrow \Phi_i .$$

Let us give results of the formation of these models. We find matrices $\left\|\lambda^0_{j_1 p_1}\right\|$ $\left\|\lambda^0_{j_2 p_2}\right\|$, $\left\|\lambda^0_{j_3 p_3}\right\|$ from equation system (3.118). And on the basis of (3.123) the following expressions are obtained for $\Psi_{sj_s}(x_{j_s})$, $s = \overline{1, 3}$, $j_1 = 1; 2$, $j_2 = 1; 2$, $j_3 = \overline{1, 3}$:

$$\begin{aligned}
\Psi_{11}(x_{11}) &= 0.171T_0^*(x_{11}) - 0.108T_1^*(x_{11}) - 0.089T_2^*(x_{11}) , \\
\Psi_{12}(x_{12}) &= 0.171T_0^*(x_{12}) + 0.278T_1^*(x_{12}) + 0.379T_2^*(x_{12}) , \\
\Psi_{21}(x_{21}) &= 0.123T_0^*(x_{21}) - 0.257T_1^*(x_{21}) + 0.204T_2^*(x_{21}) , \\
\Psi_{22}(x_{22}) &= 0.123T_0^*(x_{22}) + 0.434T_1^*(x_{22}) + 0.175T_2^*(x_{22}) , \\
\Psi_{31}(x_{31}) &= 0.150T_0^*(x_{31}) + 0.025T_1^*(x_{31}) + 0.018T_2^*(x_{31}) , \\
\Psi_{32}(x_{32}) &= 0.150T_0^*(x_{32}) + 0.238T_1^*(x_{32}) + 0.113T_2^*(x_{32}) , \\
\Psi_{33}(x_{33}) &= 0.150T_0^*(x_{33}) + 0.355T_1^*(x_{33}) + 0.117T_2^*(x_{33}) .
\end{aligned}$$

Then, using (3.122) we find matrices $\left\|a^{(1)}_{ij_1}\right\|$, $\left\|a^{(2)}_{ij_2}\right\|$, $\left\|a^{(3)}_{ij_3}\right\|$ $\forall i \in [1, m]$ from equation systems (3.120). As a result, for Φ_{is}, $i = \overline{1, 4}$; $s = \overline{1, 3}$ we get the following relationships:

$$\begin{aligned}
\Phi_{11}(x_1) &= 1.073\Psi_{11}(x_{11}) + 0.962\Psi_{12}(x_{12}) , \\
\Phi_{21}(x_1) &= 0.957\Psi_{11}(x_{11}) + 1.014\Psi_{12}(x_{12}) , \\
\Phi_{31}(x_1) &= 0.964\Psi_{11}(x_{11}) + 1.018\Psi_{12}(x_{12}) , \\
\Phi_{41}(x_1) &= 0.960\Psi_{11}(x_{11}) + 1.048\Psi_{12}(x_{12}) , \\
\Phi_{12}(x_2) &= 1.040\Psi_{21}(x_{21}) + 0.989\Psi_{22}(x_{22}) , \\
\Phi_{22}(x_2) &= 1.012\Psi_{21}(x_{21}) + 0.991\Psi_{22}(x_{22}) , \\
\Phi_{32}(x_2) &= 1.025\Psi_{21}(x_{21}) + 0.994\Psi_{22}(x_{22}) , \\
\Phi_{42}(x_2) &= 0.940\Psi_{21}(x_{21}) + 1.038\Psi_{22}(x_{22}) , \\
\Phi_{13}(x_3) &= 1.158\Psi_{31}(x_{31}) + 0.925\Psi_{32}(x_{32}) + 0.960\Psi_{33}(x_{33}) , \\
\Phi_{23}(x_3) &= 0.946\Psi_{31}(x_{31}) + 1.031\Psi_{32}(x_{32}) + 0.983\Psi_{33}(x_{33}) , \\
\Phi_{33}(x_3) &= 0.971\Psi_{31}(x_{31}) + 1.014\Psi_{32}(x_{32}) + 0.999\Psi_{33}(x_{33}) , \\
\Phi_{43}(x_3) &= 0.858\Psi_{31}(x_{31}) + 1.156\Psi_{32}(x_{32}) + 0.981\Psi_{33}(x_{33}) .
\end{aligned}$$

The final result of formation of the approximating functions $\Phi_i(x_1, x_2, x_3)$, $i = \overline{1, 4}$, is obtained by aggregation of the corresponding solutions on the basis of relationships (3.121):

$$\begin{aligned}
\Phi_1(x_1, x_2, x_3) &= 0.073\Phi_{11}(x_1) - 0.022\Phi_{12}(x_2) + 0.953\Phi_{13}(x_3) , \\
\Phi_2(x_1, x_2, x_3) &= 0.120\Phi_{21}(x_1) - 0.078\Phi_{22}(x_2) + 0.959\Phi_{23}(x_3) , \\
\Phi_3(x_1, x_2, x_3) &= 0.127\Phi_{31}(x_1) - 0.081\Phi_{32}(x_2) + 0.954\Phi_{33}(x_3) , \\
\Phi_4(x_1, x_2, x_3) &= -0.023\Phi_{41}(x_1) + 0.018\Phi_{42}(x_2) + 1.005\Phi_{43}(x_3) .
\end{aligned}$$

The recovered functions $\Phi_i(x_1, x_2, x_3)$, $i = \overline{1,4}$, expressed through shifted Chebyshev polynomials, have the form

$$\begin{aligned}
\Phi_1(x_1,x_2,x_3) = {} & 0.013T_0^*(x_{11}) - 0.008T_1^*(x_{11}) - 0.007T_2^*(x_{11}) + \\
& + 0.012T_0^*(x_{12}) + 0.019T_1^*(x_{12}) + 0.027T_2^*(x_{12}) - \\
& - 0.003T_0^*(x_{21}) + 0.006T_1^*(x_{21}) - 0.005T_2^*(x_{21}) - \\
& - 0.003T_0^*(x_{22}) - 0.010T_1^*(x_{22}) - 0.004T_2^*(x_{22}) + \\
& + 0.166T_0^*(x_{31}) + 0.028T_1^*(x_{31}) + 0.020T_2^*(x_{31}) + \\
& + 0.132T_0^*(x_{32}) + 0.210T_1^*(x_{32}) + 0.099T_2^*(x_{32}) + \\
& + 0.137T_0^*(x_{33}) + 0.325T_1^*(x_{33}) + 0.107T_2^*(x_{33}) ; \\
\Phi_2(x_1,x_2,x_3) = {} & 0.020T_0^*(x_{11}) - 0.012T_1^*(x_{11}) - 0.010T_2^*(x_{11}) + \\
& + 0.021T_0^*(x_{12}) + 0.034T_1^*(x_{12}) + 0.046T_2^*(x_{12}) - \\
& - 0.010T_0^*(x_{21}) + 0.020T_1^*(x_{21}) - 0.016T_2^*(x_{21}) - \\
& - 0.009T_0^*(x_{22}) - 0.033T_1^*(x_{22}) - 0.013T_2^*(x_{22}) + \\
& + 0.136T_0^*(x_{31}) + 0.023T_1^*(x_{31}) + 0.016T_2^*(x_{31}) + \\
& + 0.148T_0^*(x_{32}) + 0.236T_1^*(x_{32}) + 0.111T_2^*(x_{32}) + \\
& + 0.141T_0^*(x_{33}) + 0.334T_1^*(x_{33}) + 0.110T_2^*(x_{33}) ; \\
\Phi_3(x_1,x_2,x_3) = {} & 0.021T_0^*(x_{11}) - 0.013T_1^*(x_{11}) - 0.011T_2^*(x_{11}) + \\
& + 0.022T_0^*(x_{12}) + 0.036T_1^*(x_{12}) + 0.049T_2^*(x_{12}) - \\
& - 0.010T_0^*(x_{21}) + 0.021T_1^*(x_{21}) - 0.017T_2^*(x_{21}) - \\
& - 0.010T_0^*(x_{22}) - 0.035T_1^*(x_{22}) - 0.014T_2^*(x_{22}) + \\
& + 0.139T_0^*(x_{31}) + 0.023T_1^*(x_{31}) + 0.016T_2^*(x_{31}) + \\
& + 0.145T_0^*(x_{32}) + 0.231T_1^*(x_{32}) + 0.109T_2^*(x_{32}) + \\
& + 0.143T_0^*(x_{33}) + 0.338T_1^*(x_{33}) + 0.112T_2^*(x_{33}) ; \\
\Phi_4(x_1,x_2,x_3) = {} & -0.004T_0^*(x_{11}) + 0.002T_1^*(x_{11}) + 0.002T_2^*(x_{11}) - \\
& - 0.004T_0^*(x_{12}) - 0.007T_1^*(x_{12}) - 0.009T_2^*(x_{12}) - \\
& + 0.002T_0^*(x_{21}) - 0.004T_1^*(x_{21}) + 0.003T_2^*(x_{21}) + \\
& + 0.002T_0^*(x_{22}) + 0.008T_1^*(x_{22}) + 0.003T_2^*(x_{22}) + \\
& + 0.129T_0^*(x_{31}) + 0.022T_1^*(x_{31}) + 0.015T_2^*(x_{31}) + \\
& + 0.174T_0^*(x_{32}) + 0.277T_1^*(x_{32}) + 0.131T_2^*(x_{32}) + \\
& + 0.148T_0^*(x_{33}) + 0.350T_1^*(x_{33}) + 0.116T_2^*(x_{33}) .
\end{aligned}$$

The same functions $\Phi_i(x_1, x_2, x_3)$, $i = \overline{1,4}$, presented as polynomials, for normalized variables are described by the relationships:

$$\begin{aligned}
\Phi_1(x_1,x_2,x_3) = {} & 0.039x_{11} - 0.056x_{11}^2 - 0.173x_{12} + 0.212x_{12}^2 + \\
& + 0.050x_{21} - 0.038x_{21}^2 + 0.012x_{22} - 0.031x_{22}^2 - \\
& - 0.101x_{31} + 0.157x_{31}^2 - 0.375x_{32} + 0.795x_{32}^2 - \\
& - 0.209x_{33} + 0.858x_{33}^2 + 0.123 ; \\
\Phi_2(x_1,x_2,x_3) = {} & 0.057x_{11} - 0.082x_{11}^2 - 0.300x_{12} + 0.368x_{12}^2 + \\
& + 0.168x_{21} - 0.128x_{21}^2 + 0.041x_{22} - 0.108x_{22}^2 - \\
& - 0.083x_{31} + 0.129x_{31}^2 - 0.421x_{32} + 0.892x_{32}^2 - \\
& - 0.215x_{33} + 0.883x_{33}^2 + 0.091 ;
\end{aligned}$$

$$\begin{aligned}\Phi_3(x_1,x_2,x_3) = {} & 0.061x_{11} - 0.088x_{11}^2 - 0.321x_{12} + 0.393x_{12}^2 + \\ & + 0.178x_{21} - 0.135x_{21}^2 + 0.043x_{22} - 0.113x_{22}^2 - \\ & - 0.085x_{31} + 0.131x_{31}^2 - 0.412x_{32} + 0.873x_{32}^2 - \\ & - 0.218x_{33} + 0.894x_{33}^2 + 0.094\,;\end{aligned}$$

$$\begin{aligned}\Phi_4(x_1,x_2,x_3) = {} & 0.011x_{11} + 0.016x_{11}^2 + 0.060x_{12} - 0.073x_{12}^2 - \\ & - 0.036x_{21} + 0.027x_{21}^2 - 0.010x_{22} + 0.026x_{22}^2 - \\ & - 0.079x_{31} + 0.122x_{31}^2 - 0.495x_{32} + 1.049x_{32}^2 - \\ & - 0.225x_{33} + 0.925x_{33}^2 + 0.062\,.\end{aligned}$$

Finally, the functions in the form of polynomials at unnormalized variables have the form

$$\begin{aligned}\Phi_1(x_1,x_2,x_3) = {} & 376.152x_{11} - 35.010x_{11}^2 - 718.833x_{12} + 150.055x_{12}^2 + \\ & + 325.590x_{21} - 21.216x_{21}^2 + 91.772x_{22} - 5.469x_{22}^2 - \\ & - 0.902x_{31} + 0.011x_{31}^2 - 15.456x_{32} + 1.351x_{32}^2 - \\ & - 15.375x_{33} + 5.833x_{33}^2 - 1573.261\,;\end{aligned}$$

$$\begin{aligned}\Phi_2(x_1,x_2,x_3) = {} & 209.199x_{11} - 19.471x_{11}^2 - 472.927x_{12} + 98.723x_{12}^2 + \\ & + 419.049x_{21} - 27.307x_{21}^2 + 121.617x_{22} - 7.248x_{22}^2 - \\ & - 0.281x_{31} + 0.003x_{31}^2 - 6.578x_{32} + 0.575x_{32}^2 - \\ & - 6.008x_{33} + 2.279x_{33}^2 - 2026.330\,;\end{aligned}$$

$$\begin{aligned}\Phi_3(x_1,x_2,x_3) = {} & 271.605x_{11} - 25.279x_{11}^2 - 611.813x_{12} + 127.715x_{12}^2 + \\ & + 536.092x_{21} - 34.933x_{21}^2 + 154.137x_{22} - 9.186x_{22}^2 - \\ & - 0.348x_{31} + 0.004x_{31}^2 - 7.793x_{32} + 0.681x_{32}^2 - \\ & - 7.360x_{33} + 2.792x_{33}^2 - 2594.071\,;\end{aligned}$$

$$\begin{aligned}\Phi_4(x_1,x_2,x_3) = {} & -62.649x_{11} + 5.831x_{11}^2 + 145.776x_{12} - 30.430x_{12}^2 - \\ & - 137.746x_{21} + 8.976x_{21}^2 + 45.103x_{22} + 2.688x_{22}^2 - \\ & - 0.414x_{31} + 0.005x_{31}^2 - 11.955x_{32} + 1.045x_{32}^2 - \\ & - 9.727x_{33} + 3.690x_{33}^2 + 823.432\,.\end{aligned}$$

Recovered in a class of additive functions, the functional dependence $\Phi_2(x_1, x_2, x_3)$ and a diagram of function $Y_2[X_1, X_2, X_3]$ were constructed using discrete meanings of its sample. They are given in Fig. 3.9.

Example 2. The problem of recovering functions in a multiplicative form on a discrete sample. In the above-mentioned example the functional dependences are formed in a class of additive functions and presented as a superposition of functions that depend on variables x_1, x_2, x_3. This choice is quite justified since it is accepted that the components of vectors x_1, x_2, x_3 are considered to be independent. However, in a number of practical problems such a choice is inadmissible because it is unknown whether the components of vectors x_1, x_2, x_3 are dependent or not. The most complex is the condition according to which the components of vectors x_1, x_2, x_3 are dependent. In this case forming the structure $\Phi_i(x_1, x_2, x_3)$, $i = \overline{1,m}$, in a class of additive functions will lead to considerable deviations of the obtained dependences from the true multifactor regularities; mutual influences

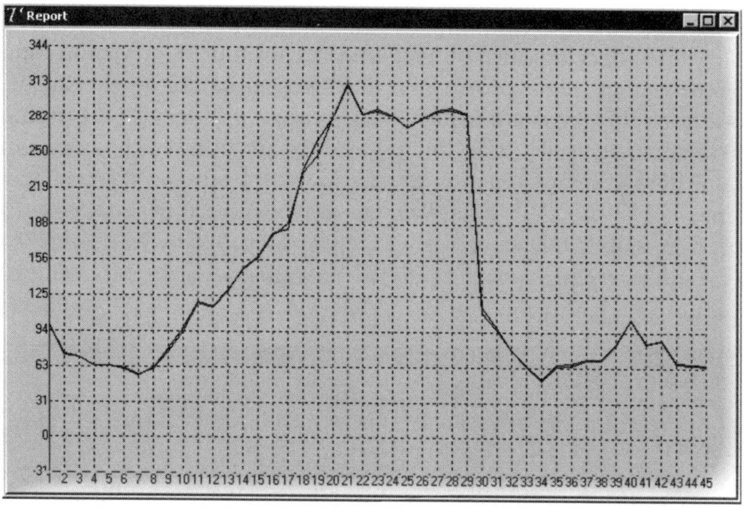

Fig. 3.9. Recovered functional dependence $\Phi_2(x_1, x_2, x_3)$ and diagram of function $Y_2[X_1, X_2, X_3]$

of the components of vectors x_1, x_2, x_3 on the properties $\Phi_i(x_1, x_2, x_3)$ are not taken into consideration.

Here, while forming the model's structure we take into account the influences on properties of the required functions $\Phi_i(x_1, x_2, x_3)$, $i = \overline{1,m}$, not only by a group of components of each vector x_1, x_2, x_3, but also by mutual influences of components of different vectors x_1, x_2, x_3. Therefore, for revealing multifactor regularities it is suggested to form a hierarchical multilevel system of models in a class of multiplicative functions [139]. Let us present the system of models in a sequence of the following levels:

$$[1 + \Phi_i(x)] = \prod_{k=1}^{K_0} [1 + \Phi_{ik}(x_k)]^{c_{ik}} \; ; \tag{3.124}$$

$$[1 + \Phi_{ik}(x_k)] = \prod_{j_k=1}^{n_k} [1 + \Psi_{kj_k}(x_{kj_k})]^{a_{ikj_k}} \; ; \tag{3.125}$$

$$[1 + \Psi_{kj_k}(x_{kj_k})] = \prod_{p_{j_k}=1}^{P_{kj_k}} \left[1 + \Phi_{p_{j_k}}(x_{kj_k})\right]^{\lambda_{kj_k}} , \tag{3.126}$$

where the functions are constructed on the basis of Chebyshev polynomials.

Let us recover the functions $\Phi_i(x_1, x_2, x_3) = \overline{1,4}$ according to the assigned sample volume of the discrete data X_s, $s = \overline{1,3}$ and Y_i, $i = \overline{1,4}$ given in Table 3.2.

In Fig. 3.10 a window from the initial data set software is presented: a volume of the sample; number of goal functions; number of vectors and their dimensions; meanings of polynomial degrees; evaluating conditions of initial approximation of functions; solution to equation system (3.116) (unified system or three systems separately); and also the meaning of the maximum and minimum deviations of function $\Phi_2(x_1, x_2, x_3)$ from the discrete meanings of sample $Y_2[X_1, X_2, X_3]$. The diagram of recovered function $\Phi_2(x_1, x_2, x_3)$ presented in a multiplicative form is given here. The diagram of function $Y_2[X_1, X_2, X_3]$ is constructed using the meanings of a discrete sample.

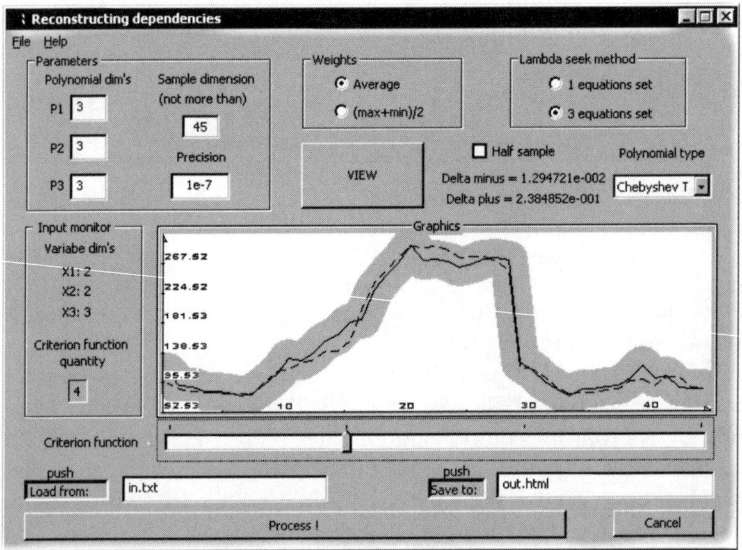

Fig. 3.10. Window of initial data set for Example 2. The recovered functional dependence $\Phi_2(x_1, x_2, x_3)$ and a diagram of function $Y_2[X_1, X_2, X_3]$ are shown

The recovered functions $\Phi_i(x_1, x_2, x_3) = \overline{1, 4}$ are obtained by aggregation of the corresponding solutions on the basis of relationships (3.124)–(3.126) and are presented in the following form:

$$\begin{aligned}\Psi_{11}(x_{11}) = {} & (1.5)^{0.0777295} (3.12x_{11} + 15.03)^{0.114941} (9.7344x_{11}^2 + 86.5072x_{11} + \\ & + 192.831)^{0.146933} (37.1205x_{11}^3 + 485.628x_{11}^2 + 2114.29x_{11} + \\ & + 3065,28)^{0,118541} - 1\,; \\ \Psi_{12}(x_{12}) = {} & (1.5)^{0.0777295} (2.94x_{12} + 5.985)^{0.602141} (8.6436x_{12}^2 + 28.3318x_{12} + \\ & + 23.8552)^{0.329136} (31.0593x_{12}^3 + 144.545x_{12}^2 + 220.983x_{12} + \\ & + 112.869)^{-0.079337} - 1\,;\end{aligned}$$

$$\begin{aligned}
\Psi_{21}(x_{21}) = {}& (1.5)^{0.103405}(3.3x_{21}+20.85)^{-0.32523}\left(10.89x_{21}^2+129.91x_{21}+\right.\\
&\left.+388.073\right)^{0.389109}\left(43,923x_{21}^3+775.671x_{21}^2+4562.41x_{21}+\right.\\
&\left.+8940.01\right)^{-0.0834892}-1;\\
\Psi_{22}(x_{22}) = {}& (1.5)^{0.103405}(5.88x_{22}+24.045)^{0.392262}\left(34.5744x_{22}^2+269.049x_{22}+\right.\\
&\left.+524.057\right)^{0.370097}\left(248.475x_{22}^3+2867.7x_{22}^2+11025.7x_{22}+\right.\\
&\left.+14124.2\right)^{0.197783}-1;\\
\Psi_{31}(x_{31}) = {}& (1.5)^{0.0351869}(300x_{31}+30)^{0.085386}\left(90000x_{31}^2+17300x_{31}+\right.\\
&\left.+832\right)^{0.0502533}\left(0,07x_{31}^3+0.06x_{31}^2+897900x_{31}+\right.\\
&\left.+28490\right)^{0.0280712}-1;\\
\Psi_{32}(x_{32}) = {}& (1.5)^{0.0351869}(60x_{32}+3)^{0.47717}\left(3600x_{32}^2+220x_{32}+\right.\\
&\left.+4\right)^{0.0268206}\left(264000x_{32}^3+20800x_{32}^2+408x_{32}+\right.\\
&\left.+5\right)^{-0.00681945}-1;\\
\Psi_{33}(x_{33}) = {}& (1.5)^{0.0351869}(30x_{33}+0.3)^{0.582053}\left(900x_{33}^2-52x_{33}+\right.\\
&\left.+1.39\right)^{-0.0204128}\left(33000x_{33}^3-3710x_{33}^2+105.9x_{33}+\right.\\
&\left.+1.463\right)^{-0.0215009}-1;\\
\Phi_{11}(x_1) = {}& (\Psi_{11}+1)^{0.637338}(\Psi_{12}+1)^{0.0930169}-1;\\
\Phi_{12}(x_2) = {}& (\Psi_{21}+1)^{0.107742}(\Psi_{22}+1)^{-0.066333}-1;\\
\Phi_{13}(x_3) = {}& (\Psi_{31}+1)^{0.106661}(\Psi_{32}+1)^{0.5654043}(\Psi_{33}+1)^{0.74188}-1;\\
\Phi_{21}(x_1) = {}& (\Psi_{11}+1)^{0.4387406}(\Psi_{12}+1)^{0.173249}-1;\\
\Phi_{22}(x_2) = {}& (\Psi_{21}+1)^{0.0599518}(\Psi_{22}+1)^{-0.448568}-1;\\
\Phi_{23}(x_3) = {}& (\Psi_{31}+1)^{0.132208}(\Psi_{32}+1)^{0.566331}(\Psi_{33}+1)^{0.792533}-1;\\
\Phi_{31}(x_1) = {}& (\Psi_{11}+1)^{0.46741}(\Psi_{12}+1)^{0.188974}-1;\\
\Phi_{32}(x_2) = {}& (\Psi_{21}+1)^{0.074937}(\Psi_{22}+1)^{-0.46837}-1;\\
\Phi_{33}(x_3) = {}& (\Psi_{31}+1)^{0.115089}(\Psi_{32}+1)^{0.529878}(\Psi_{33}+1)^{0.806946}-1;\\
\Phi_{41}(x_1) = {}& (\Psi_{11}+1)^{0.125671}(\Psi_{12}+1)^{0.0591671}-1;\\
\Phi_{42}(x_2) = {}& (\Psi_{21}+1)^{0.0508546}(\Psi_{22}+1)^{-0.00753406}-1;\\
\Phi_{43}(x_3) = {}& (\Psi_{31}+1)^{0.0850168}(\Psi_{32}+1)^{0.707902}(\Psi_{33}+1)^{0.838657}-1;
\end{aligned}$$

$$\begin{aligned}
\Phi_1(x_1,x_2,x_3) = {}& 679.713\,[\Phi_{11}(x_1)+1]^{1.06018}[\Phi_{12}(x_2)+1]^{1.46359}\times\\
&\times[\Phi_{13}(x_3)+1]^{0.961271}+109.926;\\
\Phi_2(x_1,x_2,x_3) = {}& 257.985\,[\Phi_{21}(x_1)+1]^{1.28673}[\Phi_{22}(x_2)+1]^{1.4494}\times\\
&\times[\Phi_{23}(x_3)+1]^{0.946762}+51.534;\\
\Phi_3(x_1,x_2,x_3) = {}& 312.198\,[\Phi_{31}(x_1)+1]^{1.28158}[\Phi_{32}(x_2)+1]^{1.49631}\times\\
&\times[\Phi_{33}(x_3)+1]^{0.951132}+62.453;\\
\Phi_4(x_1,x_2,x_3) = {}& 398.771\,[\Phi_{41}(x_1)+1]^{1.56375}[\Phi_{42}(x_2)+1]^{1.75381}\times\\
&\times[\Phi_{43}(x_3)+1]^{0.889233}+71.817.
\end{aligned}$$

A comparison of the functional dependences obtained in the class of additive functions and in the class of multiplicative functions shows an inconsiderable difference in the results. This difference is preconditioned by the independence of vectors x_1, x_2, x_3.

The above examples show that the suggested approach to recovering functional regularities on the basis of a model system in a class of additive and multiplicative functions allows one to estimate and correct the credibility of recovery on discrete data independently of properties of indices of the required functions' definition domain.

3.7 System Coordination of Contradictory Goals in the Search for Rational Compromises

In practical activities problems related to resolving conflicts and searching for rational compromises of contradictory goals occur frequently enough. Typical examples are various problems of coordinating a person's needs and capabilities, including coordination of his demands and available resources. In industry, to such problems belong rather complicated, multicriterion problems of system coordination of requirements set for a product's quality parameters and technological capacities of production, where there are restrictions on material, financial, and other types of resources, as well as restrictions on operating conditions. Here we suggest an approach to forming Pareto sets on the basis of a systemwide, joint coordination of the definition domain and a set of goal function values, provided that it is a priori unknown whether there exists or does not exist a nonempty set of possible compromises of contradictory goals [117, 120]. At the same time, in a typical statement of a search compromise problem, it is assumed a priori that a Pareto set exists [99, 147].

The essence of this approach will be considered using as an example a practically important problem of complex system conceptual design, namely, problems related to coordinating requirements for external and internal indices of a product, with a priori known restrictions on external-influence indices. Goal functions are assumed to be known in the form (3.121)–(3.123).

Mathematical statement of problem. For a conditional product let the following requirements to external indices y be assigned:

$$y \in B^{\pm}, \quad B^{\pm} = \{B_i^{\pm},\, i = \overline{1,m}\}, \quad B_i^{\pm} = \{y_i | b_i^- \leq y_i \leq b_i^+,\, i \in [1,m]\} \tag{3.127}$$

and the requirements for internal indices x_1 be

$$x_1 \in D_1^{\pm}, \quad D_1^{\pm} = \{x_1 | x_1 = \langle x_{1j_1},\, j_1 = \overline{1,n_1} \rangle,\, d_{1j_1}^- \leq x_{1j_1} \leq d_{1j_1}^+\}. \tag{3.128}$$

In addition, restrictions are imposed on external-influence indices in the form

$$x_2 \in D_2^*, \quad D_2^* = \{x_2 | x_2 = \langle x_{2j_2},\, j_2 = \overline{1,n_2} \rangle,\, d_{2j_2}^- \leq x_{2j_2} \leq d_{2j_2}^+\}, \tag{3.129}$$

$$x_3 \in D_3^*, \quad D_3^* = \{x_3 | x_3 = \langle x_{3j_3},\, j_3 = \overline{1,n_3} \rangle,\, d_{3j_3}^- \leq x_{3j_3} \leq d_{3j_3}^+\}. \tag{3.130}$$

Let functional dependences of external indices y on variables $x = (x_1, x_2, x_3)$ also be known in the form (3.121)–(3.123).

For functions $\Phi = \{\Phi_i(x),\ i = \overline{1,m}\}$, a Pareto set $P_{D,B}(\Phi)$ must be generated in which, on the basis of the initial data (3.127)–(3.130), the rational mutual coordination of definition domain D^* of functions Φ and a set of these functions values B^* is ensured to take into account the following conditions:

$$(\forall x_1 \in D_1^*) \wedge (\forall x_2 \in D_2^*) \wedge (\forall x_3 \in D_3^*) \Rightarrow \exists\ y \in B^*\ ; \qquad (3.131)$$

$$\forall y \in B^* \Rightarrow (\exists x_1 \in D_1^*) \wedge (\exists x_2 \in D_2^*) \wedge (\exists x_3 \in D_3^*)\ ; \qquad (3.132)$$

$$y \in B^* \Leftrightarrow (y_1 \in B_1^*) \wedge \ldots \wedge (y_i \in B_i^*) \wedge \ldots \wedge (y_m \in B_m^*)\ . \qquad (3.133)$$

In relationships (3.127)–(3.133) it is assumed that the signs $\pm$ and $*$ define, respectively, those values that can be corrected and those values that are exempt from correction. These values are concerned either with a priori assigned restrictions or the final results of the solution. An important feature of the used formalization consists in choosing the rationality principle instead of the typical optimality principle. This choice is conditioned by a number of factors, the most important of which is the presence of contradiction of goals. For example, in the development of a complex engineering system there is a contradiction between economic, technological, constructive, and other goals. In addition, in a market economy time is of great importance. In these conditions, it is necessary to reach a rational, reasonably justified compromise in attaining contradictory goals on a set of contradictory restrictions, and the restriction on the time frame for attaining the set goals is considered a priority.

Let us note the practical necessity and expediency of simultaneous fulfillment of conditions (3.131) and (3.132) while forming a Pareto set. These conditions present different approaches to a choice of a rational solution determining the required Pareto set. Condition (3.131) presupposes that internal indices are taken as the initial data with the presence of unadjustable restrictions on external-influence indices. In practice, such restrictions are set by the standards for various types of products. Here, the quality indices are determinate values. The given approach allows one to obtain an answer to the following question: What quality indices can be determined with the chosen internal indices and in the presence of assigned restrictions on external-influence indices? Condition (3.132) presupposes the opposite approach: one takes the requirements for external-quality indices as the initial data, and the determinate values will be the product's internal indices. From here the following question can be answered: What internal indices are necessary for the realization of the assigned quality indices in the presence of the assigned restrictions on the external-influence indices.

Consequently, given invariable restrictions on the external-influence indices the first approach ensures the formation of product "appearance" depending on the assigned internal indices used to define product quality. The second approach realizes the opposite sequence: from assignment of product quality to a choice of its structure and internal indices. The mathematical essence of the first approach consists in constructing a Pareto set as a set of values of function Φ, optimal by Pareto on an a priori assigned invariable definition domain D^*. Such an approach is typical for a search of compromises on the basis of a Pareto set [99, 147]. The second approach realizes the formation of area D^* by correcting it in such a way that the optimum values of function Φ are reached. Such an approach was proposed by Glushkov for system optimization problems [36]. Generalization of this approach for multicriterion linear optimization problems with an interval-preference setting is given in [38], and for human-machine optimization procedures, in [206].

It is necessary to point out that in a system problem of conceptual uncertainty the given approaches are not interchangeable in the sense that it is possible to choose any of them and on that basis to form the structure of a complex system and the requirements for its functioning. In conceptual uncertainty conditions a separate use of any approach does not ensure a rational compromise of goals for the following reasons. First, contradictions of goals have a multilevel, hierarchical structure. At the highest level there are conceptual contradictions pertaining to product design, system contradictions in the form of essential needs and potential capabilities at various stages of the product life cycle, etc. At the second level are intersystem contradictions of product realizability in the form of contradictions between a system of requirements for the external parameters describing the product properties and quality and a system of requirements for the internal parameters describing constructive, technological, and other properties of product realizability. At the next level are intrasystem contradictions between various groups of requirements (technical, operational, economic, and others) in a system of external parameters and between different groups of requirements (constructive, technological, and others) in a system of internal parameters. The given contradictions reflect, in essence, the complexity of only one conceptual stage in the product design stage. The complexity of the indicated contradictions is quite obvious, and it is aggravated by other factors: first, all contradictions are interconnected and interdependent; second, the requirements for constructive, technological, and other groups of product parameters are formed by experts in various fields.

From here follows a practical necessity of developing and perfecting methodological and mathematical apparatus of resolving contradictions on the basis of a rational compromise.

We consider the cases of requirements for system coordination between various groups and complexes of external and internal parameters of a conditional product. System coordination is understood as a rational compromise between various groups of requirements for a product for ensuring its profitability and competitiveness by adapting the design to existing technological, resource, and other production capacities. Thus, it is necessary to take into account both the direct and the inverse interconnection of parameters. Thus, for example, replacement of a material may result not only in a change in technology of constructive-element treatment, but also in a change in certain quality indices, in particular, technical, operational, economic, and/or other quality indices. And conversely, a change in requirements for a certain quality index may lead not only to a change in material treatment technology but also to a complete change in the design structure and constructive elements and the product form as a whole.

The indicated factors determine the practical need and expediency of forming a Pareto set on the basis of simultaneous fulfillment of conditions (3.131) and (3.132). The main idea of this approach to forming such a set consists in consecutive, iterative correction of the initial sets (3.127) and (3.128) with invariable restrictions (3.129) and (3.130) on the basis of computational and interactive procedures for the purpose of simultaneous fulfillment of (3.131) and (3.132).

The sequence of checking (3.131) or (3.132) may be determined by objective and subjective factors, and the procedure may begin by checking any of these conditions. We consider an algorithm of Pareto set formation in which the realization of computational procedures (CPs) begins by checking condition (3.131).

CP-1. With the initial data (3.127)–(3.128) for all $i = \overline{1,m}$ on the basis of (3.121)–(3.123) it is required:

1.1 To define:

$$y_i^- = \min_{x \in D^\pm} \Phi_i(x); \quad y_i^+ = \max_{x \in D^\pm} \Phi_i(x) \,. \tag{3.134}$$

1.2 To form:

$$B^* = \left\langle B_i^*,\, i = \overline{1,m} \right\rangle; \quad B_i^* = \left\{ y_i | y_i^- \leq y_i \leq y_i^+ \right\} . \tag{3.135}$$

1.3 To check the fulfillment of the following conditions:

$$\left(\forall i = \overline{1,m} \to y_i^- \leq b_i^-\right) \wedge \left(\forall i = \overline{1,m} \to y_i^+ \geq b_i^+\right) \Rightarrow B^\pm \supseteq B^* \,, \tag{3.136}$$

$$\left(\exists i \in [1,m] \to y_i^- > b_i^-\right) \vee \left(\exists i \in [1,m] \to y_i^+ < b_i^+\right) \Rightarrow B^\pm \neq B^* \,. \tag{3.137}$$

1.4 To analyze the results on the basis of condition checks:

- If (3.136) is fulfilled, then (3.127) and (3.128) are compatible, and (3.135) is the required result.
- If (3.137) is fulfilled, then (3.127) and (3.128) are incompatible, and (3.127) or (3.128) is to be corrected.

Comments

C1.1. If (3.136) is fulfilled, then requirements for the external parameters (3.127) are realized under the initial data (3.128), and the relationship (3.135) defines the required result.

C1.2. If (3.137) is fulfilled, then the decision maker makes a correction: either in (3.127) of those values B_i, $i \in [1, m]$ that correspond to condition (3.132), and the procedure **CP-1** is carried out once again, or in the corresponding initial data in (3.128) and proceeds to the following computational procedure.

CP-2. With the initial data for B_i^*, $i = \overline{1, m}$, D_2^*, D_3^* it is required:

2.1 To form discrete analogs:

$$\hat{B}^* = \left\{ y_i | y_i = y_i \left[q_0 \right]; \; y_i \left[q_0 \right] \in B_i^*; \; q_0 = \overline{1, k_0}; \; y_i[1] = b_i^-; \; y_i \left[k_0 \right] = b_i^+ \right\}, \tag{3.138}$$

$$\hat{D}_2^* = \left\{ x_2 | x_2 = x_2 \left[q_2 \right]; \; x_2 \left[q_2 \right] \in D_2^*; \; q_2 = \overline{1, k_2}; \; x_2[1] = d_2^-; \; x_2 \left[k_2 \right] = d_2^+ \right\}, \tag{3.139}$$

$$\hat{D}_3^* = \left\{ x_3 | x_3 = x_3 \left[q_3 \right]; \; x_3 \left[q_3 \right] \in D_3^*; \; q_3 = \overline{1, k_3}; \; x_3[1] = d_3^-; \; x_3 \left[k_3 \right] = d_3^+ \right\}. \tag{3.140}$$

2.2 To form a system of equations

$$y_i \left[q_0 \right] - \Phi_i \left(x_1, x_2 \left[q_2 \right], x_3 \left[q_3 \right] \right) = 0; \quad i = \overline{1, m}; \quad q_0 = \overline{1, k_0}. \tag{3.141}$$

2.3 To solve (3.141) and with the results to define:

$$x_1^- = \operatorname{Arg} \min_{i \in [1,m]} \Phi_i(x); \quad x_1^+ = \operatorname{Arg} \max_{i \in [1,m]} \Phi_i(x),$$
$$x_1^- = \left\{ x_{1j_1}^- | j_1 = \overline{1, n_1} \right\}; \quad x_1^+ = \left\{ x_{1j_1}^+ | j_1 = \overline{1, n_1} \right\}.$$

2.4 To check conditions:

$$\left(\forall j_1 = \overline{1, n_1} \to x_{1j_1}^- \geq d_{1j_1}^- \right) \wedge \left(\forall j_1 = \overline{1, n_1} \to x_{1j_1}^+ \leq d_{1j_1}^+ \right) \Rightarrow D_1^* \supseteq D_1^\pm, \tag{3.142}$$

$$\left(\exists j_1 \in [1, n_1] \to x_{j_1}^- < d_{j_1}^- \right) \vee \left(\exists j_1 \in [1, n_1] \to x_{j_1}^+ > d_{j_1}^+ \right) \Rightarrow D_1^* \neq D_1^\pm, \tag{3.143}$$

$$D_1^* = \left\{ x_1 | x_1 = \left\langle x_{1j_1}, \; j_1 = \overline{1, n_1} \right\rangle, \; x_{1j_1}^- \leq x_{1j_1} \leq x_{1j_1}^+ \right\}. \tag{3.144}$$

2.5 To analyze results on the basis of the condition checks:

- If (3.142) is fulfilled, then (3.127) and (3.128) are compatible, and (3.144) is the required result.
- If (3.143) is fulfilled, then (3.127) and (3.128) are incompatible, and (3.127) or (3.128) is to be corrected.

Comments

C2.1. The system of equations (3.141) contains $N = m \times k_0$ equations and is formed from the condition:

$$y_i\left[q_0\right] = \Phi_i\left(x_1, x_2\left[q_2\right], x_3\left[q_3\right]\right); \quad i = \overline{1, m}; \quad q_0 = \overline{1, k_0}; \quad q_0 \Leftrightarrow \langle q_2, q_3 \rangle \,.$$

C2.2. In the general case $N > n_1$ and 3.141) is an incompatible system of nonlinear equations, where components x_{1j_1}, $j_1 = \overline{1, n_1}$ of vector x_1 are unknown.

C2.3. If (3.142) is fulfilled, then the initial values $x_1 \in D_1^{\pm}$ ensure the fulfillment of requirements $y \in B^*$, and instead of $x_1 \in D_1^{\pm}$ it is possible to take $x_1 \in D_1^*$. If (3.143) is fulfilled, then the initial values $x_1 \in D_1^{\pm}$ do not allow one to realize the assigned requirements $y \in B^*$. Therefore, a correction is necessary: either of the values x_{1j_1} for which (3.143) is fulfilled, or of the requirements for those $y_i \in B^*$ for which condition (3.142) is not fulfilled. The decision maker chooses variants of a solution, and, depending on the accepted variant, procedure **CP-1** or procedure **CP-2** is carried out once again. The condition for completion of calculations is the simultaneous fulfillment of conditions (3.136) and (3.142). Thus, a consecutive execution of the indicated interactive procedures allows one to ensure a rational compromise both between the system of requirements for external and internal indices of a product and between various groups of requirements in each system.

As a result of the simultaneous fulfillment of conditions (3.136) and (3.142) and as a corollary of conditions (3.131) and (3.132), the required Pareto set $P_{D,B}(\Phi)$ is obtained. The Pareto set characterized by a ternary $\langle D^*, B^*, \Phi \rangle$ ensures the systemic, mutual interval coordination of the definition domain D^* and a set of values B^* for each function of the set Φ and is described by the relationship [117, 120]:

$$P_{D,B}(\Phi) \triangleq \left\{ \langle D^*, B^*, \Phi \rangle \mid \left[\Phi : D^* \to B^*\right] \wedge \left[\Phi^{-1} : B^* \to D^*\right] \right\}.$$

Note an important feature of the products is that indices are elements of the Pareto set. Such products have a system-coordinated interconnection between a set of external indices and a set of internal indices. Therefore, at any choice of external-index value $y \in B^*$ for a set of internal indices the condition $x_1 \in D_1^*$ is fulfilled. The converse is also true: $\forall x_1 \in D_1^* \Rightarrow \exists y \in B^*$. At the same time, the corresponding conditions (3.129) and (3.130) are fulfilled for external-influence indices.

It is obvious that a set of such products is infinite. According to the Gödel principle of complementarity, for a choice of the only, rational product it is necessary to introduce an additional set of criteria and on that basis to make a final choice of the product.

Example. Some work within a given period of time is required, for example, digging a foundation pit. The work can be executed only with the help of a host of machinery: earth excavation is done by excavators, transportation is done by trucks or tractors with trailers, earth backfilling by leveling machines (bulldozers, soil compactors, etc.). The problem is to select indices of all these machines in such a way that will ensure their effective operation and the timely execution of the work.

The complex efficiency depends on several groups of parameters: foundation pit parameters (depth, length, width, and earth hardness), road parameters (length and quality), excavator parameters (the bucket capacity), truck parameters (the carrying capacity), bulldozer parameters (the capacity), etc. The machine complex should be such that it guarantees the timely execution of the work. The initial data are given in Table 3.3.

As a result of problem solving in the interactive dialog mode, the following coordinated Pareto set was obtained (Table 3.4).

Thus, the given example illustrates a practical possibility of finding a rational compromise between contradictory goals in the organization of some technological processes in the unified technology of construction, industrial, or other works while using different types of machinery and equipment.

Table 3.3. Initial data for example

Sampling number	Foundation pit depth, m	Foundation pit length, m	Foundation pit width, m	Ground hardness (hardness value)	Road length, km	Road quality, standard units	Excavator bucket capacity, m^3	Truck carrying capacity, tons	Bulldozer power, newtons	Cost	Timetable, h
1	8	8	3	1	3	0	3	2	3	400	18
2	1	4	2	2	5	1	5	3	1	100	14
3	5	5	3	1	3	0	2	4	1	500	12
4	10	8	4	1	5	1	4	5	2	300	20
5	2	7	3	1	5	0	1	2	2	200	10
6	2	6	2	2	3	0	3	4	3	100	11
7	6	3	1	1	3	0	5	3	1	300	17
8	7	5	2	1	5	1	2	2	2	400	19
9	6	4	3	2	1	0	2	5	2	500	13
10	8	4	3	1	4	1	3	2	3	250	10
11	2	3	3	2	1	1	1	3	1	270	14
12	5	5	2	1	2	0	2	2	1	370	17
13	9	8	1	1	5	0	5	5	3	550	19
14	3	7	4	1	3	0	4	4	2	150	15
15	4	6	3	1	2	1	3	3	1	370	13
16	5	5	2	2	1	0	2	2	2	420	18
17	7	2	1	2	1	0	1	4	2	170	11
18	3	9	2	2	5	0	3	3	1	120	10
19	4	7	5	1	4	0	5	2	1	280	17

Table 3.4. Coordinated Pareto set

Foundation pit depth, m	[1.709, 4.884]
Foundation pit length, m	[4.192, 6.595]
Foundation pit width, m	[1.688, 3.3]
Ground hardness (hardness value)	[1, 2]
Road length, km	[1.015, 5.11]
Road quality, standard units	[0, 1]
Excavator bucket capacity, m^3	[1.778, 3.898]
Truck carrying capacity, tons	[2.367, 4.046]
Bulldozer power, newtons	[1.589, 1.953]
Cost	[16.119, 559.599]
Timetable, h	[8.078, 20.374]

4 Disclosing Uncertainties in Problems of Interaction and Counteraction of Coalitions

In the previous chapter problems of disclosing various types of uncertainties were investigated: the uncertainty of goals and interests in practical activities of individual participants and information uncertainty during the interaction or counteraction of several participants not incorporated into certain groups. At the same time, in practice, complex organizational structures and forms of interaction and counteraction are realized. For example, in the world economy there exist different international associations, unions, and other forms of interconnections and interactions of manufacturers and consumers of various services, raw materials, and goods. In a market economy, simultaneous influences of uncertainty factors, risks, competition, interaction, or counteraction factors are typical for such participants. Therefore, research and development of methods for solving system problems of active interaction and counteraction in such associations and coalitions are of significant practical interest.

The purpose of this chapter is to develop the mathematical apparatus for formalizing and solving the system problems of disclosing uncertainties and optimization of goals in problems of active interaction and counteraction of several coalitions.

4.1 Mathematical Statement of Problem

First, we will briefly review the specificity and features of research objects. Generally the set of coalitions X can be represented as an association of two subsets

$$X = X_1 \bigcup X_2 \,. \tag{4.1}$$

Subset X_1 includes only those coalitions that cooperate among themselves as partners. We will call partners those participants of practical activities whose strategic goals coincide and whose interests may differ but are not antagonistic. Subset X_2 includes only those coalitions each of which operates in own interest and competes with other participants of a given type of practical activities. We will call competitors those participants of practical activities whose strategic goals are antagonistic, but actions for their achievement do not contradict established rules, for example, national or international law.

Along with the legislatively allowable types of competition on the national and world market, there is a form of counteraction by direct participants in which they act as opponents. In particular, an example is a type of unfair competition whose essence consists in infringement by one or several participants on the legal bases of competition for achievement of one's own goals. Such illegal actions in one's own interest cause direct damage to the other market participants. Injured participants may take countermeasures within the limits of the law.

Coalition counteractions are typical not only for national and regional markets but also for the different kinds of military conflicts, including local and world wars. In terms of the scale of operation and material damage and casualties, the Second World War is the most impressive example of coalition counteraction. Therefore research on coalition interaction and counteraction conditions represents a practical interest.

Let us turn to the mathematical description of the research objects. Subsets X_1 and X_2 are characterized by the quantitative and qualitative composition of coalitions. We will present the quantitative characteristic of these subsets as the association of a finite number of coalitions:

$$X_1 = \bigcup_{\kappa_1=1}^{K_1} X_{1\kappa_1}\,, \quad X_2 = \bigcup_{\kappa_2=1}^{K_2} X_{2\kappa_2}\,,$$

where κ_1 and κ_2 define a coalition location in association X_1 and X_2, respectively, by a certain attribute, for example, by the date of joining the coalition. The quantitative structure of each coalition is an important characteristic of its possibility in attaining the participants' common goals. Therefore, it is necessary to put coalitions in a specific order in X_1 and X_2 by this parameter. The coalition serial number r grows with a reduction in the coalition's quantitative structure. Then we have

$$\begin{gathered} X_{11} \succ X_{12} \succ \ldots \succ X_{1r} \succ X_{1r+1} \succ \ldots \succ X_{1K_1}\,; \\ X_{1r} \in X_1; \quad r \in K_{01}; \quad K_{01} = [1, K_1]\,; \\ X_{21} \succ X_{22} \succ \ldots \succ X_{2r} \succ X_{2r+1} \succ \ldots \succ X_{2K_2}\,; \\ X_{2r} \in X_2; \quad r \in K_{02}; \quad K_{02} = [1, K_2]\,. \end{gathered}$$

However, with respect to the level of success achieved by each participant, the qualitative structure of coalition participants, which is defined by many factors, is also of great importance. The major ones of these factors are the goals and objectives of each participant's association in the coalition. Obviously, each coalition participant's potentialities differ in many parameters, in particular, in financial, personnel, and other resources. As a consequence, the sizes of participants' contributions in the attainment of goals may also differ. Therefore, the position of each participant in the coalition should be based on a generalized quality indicator.

We will represent sets P_{1r} and P_{2r} of coalition participants $X_{1r} \in X_1$ and $X_{2r} \in X_2$ in the following form:

$$P_{1r} = \left\{P_{1i_r} | i_r = \overline{1, m_{1r}}\right\}, \quad P_{2r} = \left\{P_{2i_r} | i_r = \overline{1, m_{2r}}\right\}, \tag{4.2}$$

where i_r is a position of coalition r participant, his contribution to the achievement of the coalition's goals, determined by a quality indicator P_{ir}, and m_{ir} is the total number of coalition r participants. Indices 1 and 2 define a coalition belonging to a subset X_1 and X_2, respectively. Then the compositions of each coalition $X_{1r} \in X_1$ and $X_{2r} \in X_2$ are characterized by the conditions

$$X_{1r} = \left\{P_{1i_r} | P_{1i_r} \in P_{1r}; P_{1i_r} \succ P_{1(i_r+1)}; [i_r, i_{r+1}] \in I_{01r}\right\}; \quad I_{01r} = [1, m_{1r}];$$
$$X_{2r} = \left\{P_{2i_r} | P_{2i_r} \in P_{2r}; P_{2i_r} \succ P_{2(i_r+1)}; [i_r, i_{r+1}] \in I_{02r}\right\}; \quad I_{02r} = [1, m_{2r}].$$

The quantities M_{01} and M_{02} of the participants in coalitions X_1 and X_2 are defined by the relationships

$$M_{01} = \sum_{r=1}^{K_{01}} m_{1r} \quad \text{and} \quad M_{02} = \sum_{r=1}^{K_{02}} m_{2r}.$$

Later on, when discussing research on processes of only interactions or only counteractions of coalitions we will exclude indices 1 and 2 in the description of the aforementioned properties.

Let us turn to the mathematical description of coalition goals. We will start with the formalization of goals of coalition interaction of subset X_1. For coalitions $X_{11} \in X_1$ and $X_{12} \in X_1$ according to (4.2) participant composition is defined by the following formulas:

$$P_1 = \left\{P_{i_1} | i_1 = \overline{1, m_1}\right\},$$
$$P_2 = \left\{P_{i_2} | i_2 = \overline{1, m_2}\right\}.$$

Each participant of any coalition has his own goal vector function. The set of individual vector functions of all coalition participants with the number r is described as

$$\overline{f}_r^m = \{\overline{f}_{ri_r} | i_r = \overline{1, m_r}\}, \tag{4.3}$$

where $\overline{f}_{ri_r}$ is a vector function of participant i_r of coalition r. The index m shows that $\overline{f}_r^m$ is a set that consists of m_r elements of a type $\overline{f}_{ri_r}$, and the parameter i_r defines the position of an element in the given set. For coalitions whose positions are 1 and 2, the considered sets are defined by relationships

$$\overline{f}_1^m = \{\overline{f}_{1i_1} | i_1 = \overline{1, m_1}\},$$
$$\overline{f}_2^m = \{\overline{f}_{2i_2} | i_2 = \overline{1, m_2}\}.$$

In any coalition, the argument of each participant's vector function is defined by two vectors: the participant's individual vector of strategic decisions and a coalition vector of strategic decisions, which are common for all participants. Vector $\overline{x}_{ri_r}$ of the individual strategy of participant i_r of coalition r whose participant i_r values change at his own discretion values is presented as

$$\overline{x}_{ri_r} = \left\{ x_{ri_r}^{(j)} | j = \overline{1, n_{ri_r}} \right\} , \tag{4.4}$$

where j is the position of a component of an individual strategy vector of participant i_r of coalition r and $x_{ri_r}^{(j)}$ is a component whose position is j. For participant i_1 of the first coalition $(r = 1)$ and for participant i_2 of the second coalition $(r = 2)$ we have

$$\overline{x}_{1i_1} = \left\{ x_{1i_1}^{(j)} | j = \overline{1, n_{1i_1}} \right\}; \quad \overline{x}_{2i_2} = \left\{ x_{2i_2}^{(j)} | j = \overline{1, n_{2i_2}} \right\} .$$

The common strategy vector of coalition r is determined by the formula

$$\overline{\chi}_r = \left\{ \chi_{rl_r} | l_r = \overline{1, \rho_r} \right\} , \tag{4.5}$$

where $\overline{\chi}_r$ is the common strategy vector of coalition r, χ_{rl_r} is component l_r of vector $\overline{\chi}_r$, l_r is the component position of vector $\overline{\chi}_r$, and ρ_r is the total number of components of vector $\overline{\chi}_r$.

For the first and second coalition $(r = 1; 2)$ on the basis of (4.5) we find the common strategy vectors

$$\overline{\chi}_1 = \left\{ \chi_{1l_1} | l_1 = \overline{1, \rho_1} \right\} ,$$
$$\overline{\chi}_2 = \left\{ \chi_{2l_2} | l_2 = \overline{1, \rho_2} \right\} .$$

The goals of participant i_r of coalition r are characterized by the vector function

$$\overline{f}_{ri_r} \left(\overline{x}_{ri_r}, \overline{\chi}_r \right) = \left\{ f_{ri_r k_r} \left(\overline{x}_{ri_r}, \overline{\chi}_r \right) | k_r = \overline{1, L_{ri_r}} \right\} , \tag{4.6}$$

where the arguments are the individual strategy vector and the common coalition strategy vector. For participant i_1 of the first coalition and participants i_2 of the second coalition we have

$$\overline{f}_{1i_1} \left(\overline{x}_{1i_1}, \overline{\chi}_1 \right) = \left\{ f_{1i_1 k_1} \left(\overline{x}_{1i_1}, \overline{\chi}_1 \right) | k_1 = \overline{1, L_{1i_1}} \right\} ,$$
$$\overline{f}_{2i_2} \left(\overline{x}_{2i_2}, \overline{\chi}_2 \right) = \left\{ f_{2i_2 k_2} \left(\overline{x}_{2i_2}, \overline{\chi}_2 \right) | k_2 = \overline{1, L_{2i_2}} \right\} .$$

Each coalition forms the common vector function of the coalition participants' interaction goals as partners. For coalition r the vector function of the participants' interaction goals is written in the form

$$\overline{F}_r = \left\{ F_{r\gamma_r} \left(\overline{\varphi}_r, \overline{f}_r \right) | \gamma_r = \overline{1, \Gamma_r} \right\} , \tag{4.7}$$

where $F_{r\gamma_r}$ is a component γ_r of the vector function of coalition r's participant interaction goals, Γ_r is the common number of components of vector function F_r, $\overline{\varphi}_r$ is the vector function of coalition r's common goals, and

$\overline{f_r}$ is the vector function of coalition r's individual goals, determined by relationship (4.3).

For the first ($r = 1$) and second ($r = 2$) coalitions the vector function of participant interaction goals based on (4.7) may be presented in the form

$$\overline{F}_1 = \left\{ F_{1\gamma_1}\left(\overline{\varphi}_1, \overline{f}_1\right) | \gamma_1 = \overline{1, \Gamma_1} \right\} ,$$
$$\overline{F}_2 = \left\{ F_{2\gamma_2}\left(\overline{\varphi}_2, \overline{f}_2\right) | \gamma_2 = \overline{1, \Gamma_2} \right\} .$$

The vector function of each coalition's common goals depends both on common coalition strategies and on the individual strategies of all participants. For coalition r it is presented in the form

$$\overline{\varphi}_r\left(\overline{x}_r, \overline{\chi}_r\right) = \left\{ \varphi_{rs_r}\left(\overline{x}_r, \overline{\chi}_r\right) | s_r = \overline{1, s_{0r}} \right\} . \tag{4.8}$$

Here $\overline{x}_r$ is a vector of individual strategies of all coalition r participants, which is determined by the relationship

$$\overline{x}_r = \left\{ x_{ri_r} | i_r = \overline{1, m_r} \right\} . \tag{4.9}$$

For the first and second coalition ($r = 1,\ 2$) the vector functions of common goals are presented as

$$\overline{\varphi}_1\left(\overline{x}_1, \overline{\chi}_1\right) = \left\{ \varphi_{1s_1}\left(\overline{x}_1, \overline{\chi}_1\right) | s_1 = \overline{1, s_{01}} \right\} ;$$
$$\overline{\varphi}_2\left(\overline{x}_2, \overline{\chi}_2\right) = \left\{ \varphi_{2s_2}\left(\overline{x}_2, \overline{\chi}_2\right) | s_2 = \overline{1, s_{02}} \right\} .$$

Here $\overline{x}_1$ and $\overline{x}_2$ are the individual strategy vectors of all participants for coalitions 1 and 2 and $\overline{\chi}_1$ and $\overline{\chi}_2$ are the vectors of a common coalition strategies.

The vector function $\overline{f}_r$ of the individual interaction goals of all coalition r participants as a constituent of the vector function $\overline{F}_r$ of coalition r participant interaction goals can be written in the following form:

$$\overline{f}_r\left(\overline{x}_r, \overline{\chi}_r\right) = \left\langle f_{ri_r}\left(\overline{x}_{ri_r}, \overline{\chi}_r\right),\ i_r = \overline{1, m_r} \right\rangle . \tag{4.10}$$

Here $\left\langle f_{ri_r}\left(\overline{x}_{ri_r}, \overline{\chi}_r\right),\ i_r = \overline{1, m_r} \right\rangle$ denotes a sequence of functions as a unified integral object that consists of m_r elements of f_{ri_r} arranged in ascending order by serial number. The difference between $\overline{f}_r$ and $\overline{f}_r^m$ is that $\overline{f}_r^m$ determines a set in which the element is f_{ri_r} and which represents a set m_r of separate elements f_{ri_r}.

Thus $\overline{f}_r$ is a unified, integral object of m_r interconnected elements. On the basis of (4.10) it is possible to formulate the vector function of active interaction goals of all participants of the first and second coalitions having the respective serial numbers $r = 1$ and $r = 2$ in the form

$$\overline{f}_1\left(\overline{x}_1, \overline{\chi}_1\right) = \left\langle f_{1i_1}\left(\overline{x}_{1i_1}, \overline{\chi}_1\right),\ i_1 = \overline{1, m_1} \right\rangle ,$$
$$\overline{f}_2\left(\overline{x}_2, \overline{\chi}_2\right) = \left\langle f_{2i_2}\left(\overline{x}_{2i_2}, \overline{\chi}_2\right),\ i_2 = \overline{1, m_2} \right\rangle .$$

Note one point of the description: each vector function $\overline{f}_{ri_r}$ depends on vector $\overline{x}_{ri_r}$ of an individual strategy only of the i_r coalition participant, but for the total set of common strategies of coalition r there is a dependence on vector $\overline{\chi}_r$. In practice this means that the coalition's common decisions influence the interests of each coalition participant, and individual decisions $\overline{x}_{ri_r}$ of participant i_r allow him to adjust his own strategy for developing coalition interaction conditions. As a result of similar possibilities for all coalition participants we obtain that vector function $\overline{f}_r$ of individual interaction goals of all participants of coalition r will depend both on the total set of the common coalition strategies $\overline{\chi}_r$ and on the total set of all participants' individual strategies, i.e., $\overline{f}_r = \overline{f}_r\left(\overline{x}_r, \overline{\chi}_r\right)$.

In view of the above remark from formula (4.7) on the basis of (4.8)–(4.10) it follows that each component of the vector function $\overline{F}_r(\cdot)$ depends on $\overline{x}_r$ and $\overline{\chi}_r$. Therefore, we shall present

$$\overline{F}_r = \left\{F_{r\gamma_r}\left(\overline{x}_{0r}\right)|\gamma_r = \overline{1, \Gamma_r}\right\} ,$$

where $\overline{x}_{0r}$ is a vector of generalized solution of coalition r, which we shall define as a sequence

$$\overline{x}_{0r} = \langle \overline{x}_r, \overline{\chi}_r \rangle \ . \tag{4.11}$$

This sequence consists of consecutive components of vectors $\overline{x}_r$ and $\overline{\chi}_r$.

Now, we shall consider cases of coalition counteractions. We will proceed to a mathematical description of the coalition goals of subset X_2 by the example of two coalition counteraction. Let us present the **vector function of two coalition counteraction goals** as:

$$\overline{F}_{rr+1}\left(\overline{x}_{0r},\, \tilde{\overline{x}}_{0r+1}\right) = \overline{F}_{rr+1}\left\{\overline{F}_r\left(\overline{x}_{0r}, \tilde{\overline{x}}_{0r+1}\right),\, \tilde{\overline{F}}^{(r)}_{r+1}\left(\overline{x}_{0r},\, \tilde{\overline{x}}_{0r+1}\right)\right\} , \tag{4.12}$$

$$\overline{F}_{r+1r}\left(\overline{x}_{0r+1},\, \tilde{\overline{x}}_{0r}\right) = \overline{F}_{r+1r}\left\{\overline{F}_{r+1}\left(\overline{x}_{0r+1},\, \tilde{\overline{x}}_{0r}\right),\, \tilde{\overline{F}}^{(r+1)}_{r}\left(\overline{x}_{0r+1},\, \tilde{\overline{x}}_{0r}\right)\right\} . \tag{4.13}$$

Assuming for simplicity $r = 1$, we have

$$\overline{F}_{12}\left(\overline{x}_{01},\, \tilde{\overline{x}}_{02}\right) = \overline{F}_{12}\left\{\overline{F}_1\left(\overline{x}_{01},\, \tilde{\overline{x}}_{02}\right),\, \tilde{\overline{F}}^{(1)}_{2}\left(\overline{x}_{01},\, \tilde{\overline{x}}_{02}\right)\right\} ,$$

$$\overline{F}_{21}\left(\overline{x}_{02},\, \tilde{\overline{x}}_{01}\right) = \overline{F}_{21}\left\{\overline{F}_2\left(\overline{x}_{02},\, \tilde{\overline{x}}_{01}\right),\, \tilde{\overline{F}}^{(2)}_{1}\left(\overline{x}_{02},\, \tilde{\overline{x}}_{01}\right)\right\} .$$

Here it is accepted that goal vector functions of one coalition partner's interaction are unknown to the other partner, which corresponds to real counteraction conditions for the case of intercoalition competition or antagonism.

Then we assume that coalition partners' interaction functions are precisely known only within their own coalition, that is, for coalition $r = 1$ vector function $\overline{F}_1$ is precisely known, for coalition $r = 2$, vector function $\overline{F}_2$ is precisely known. Vector function $\tilde{\overline{F}}_2^{(1)}$ describes the interaction goals of the second coalition partners in the form of its interpretation by the first coalition in vector function $\overline{F}_{12}$ and is approximate in the sense that the first coalition has no complete information on the second coalition's goals and decisions and forms the given vector function (i.e., the form and number of this function's components, the form and number of the vector's components and arguments $\tilde{\overline{x}}_{02}$) in conditions of incompleteness and uncertainty of the initial information on the second coalition's goals and decisions. Similarly, vector function $\tilde{\overline{F}}_1^{(2)}$ describes the interaction goals of the first coalition partners in the form of its interpretation by the second coalition in vector function $\overline{F}_{21}$ and is approximate in the sense that the second coalition has no complete information on the first coalition's goals and decisions and forms the given vector function (i.e., the form and number of this function components, the form and number of the vector's components and arguments $\tilde{\overline{x}}_{01}$) in conditions of incompleteness, inaccuracy, and uncertainty of the initial information on the first coalition's goals and decisions. Due to the difference in the level of information availability upon formation of pair $\overline{F}_1$ and $\tilde{\overline{F}}_1^{(2)}$ and pair $\overline{F}_2$ and $\tilde{\overline{F}}_2^{(1)}$ for each pair it is valid that

$$\overline{F}_1 \neq \tilde{\overline{F}}_1^{(2)}, \quad \overline{F}_2 \neq \tilde{\overline{F}}_2^{(1)} .$$

Each vector function $\tilde{\overline{F}}_1^{(2)}$ and $\tilde{\overline{F}}_2^{(1)}$ depends both on its own generalized coalition strategy and on the counteractive coalition strategy. According to the earlier accepted assumption and informedness of one's own coalition's actions and incomplete knowledge of the opposing coalition's actions for the functions $\tilde{\overline{F}}_1^{(2)}$ and $\tilde{\overline{F}}_2^{(1)}$ we have

$$\tilde{\overline{F}}_1^{(2)} = \tilde{\overline{F}}_1^{(2)}\left(\tilde{\overline{x}}_{01}^{(2)}, \overline{x}_{02}\right), \quad \tilde{\overline{F}}_2^{(1)} = \tilde{\overline{F}}_2^{(1)}\left(\overline{x}_{01}, \tilde{\overline{x}}_{02}^{(1)}\right) .$$

Here, $\overline{x}_{02}$ and $\overline{x}_{01}$ are exact values of the generalized strategies of the second and first coalition, respectiveley, which are precisely known only to direct participants of a corresponding coalition; $\tilde{\overline{x}}_{01}^{(2)}$ is the approximate values of the first coalition's generalized strategy vector in the form of an approximate interpretation by the second coalition; $\tilde{\overline{x}}_{02}^{(1)}$ is the approximate values of the second coalition's generalized strategy vector in the form of its interpretation by the first coalition.

Let us consider the problem of disclosing uncertainties for the case of **coalition counteraction taking into account** the situational uncertainty. For coalitions $r = 1$ and $r = 2$ the natural, climatic, and other situational uncertainty factors are accounted for in a problem of coalition interactions by vector functions of the following form:

$$\overline{F}_{1\alpha} = \overline{F}_{1\alpha}(\overline{\alpha}_1)\,, \quad \overline{\alpha}_1 = [\alpha_{11}, \ldots, \alpha_{1k_1}, \ldots, \alpha_{1k_{01}}]\,, \tag{4.14}$$

$$\overline{F}_{2\alpha} = \overline{F}_{2\alpha}(\overline{\alpha}_2)\,, \quad \overline{\alpha}_2 = [\alpha_{21}, \ldots, \alpha_{2k_2}, \ldots, \alpha_{2k_{02}}]\,. \tag{4.15}$$

Each coalition independently forms both a system of indicators and methods of evaluating their influences on the results of the coalition's activities; therefore, functions and their arguments are different for different coalitions.

In the general case, in view of uncertainty factors, the **vector functions determining the goals of the two coalitions' counteraction** may be presented in the form

$$\overline{F}_{\alpha 12} = \overline{F}_{\alpha 12}\left\{\overline{F}_1(\overline{x}_{01}, \tilde{\overline{x}}_{02})\,,\ \tilde{\overline{F}}_2^{(1)}(\overline{x}_{01}, \tilde{\overline{x}}_{02})\,,\ \overline{F}_{1\alpha}(\alpha_1)\right\}\,, \tag{4.16}$$

$$\overline{F}_{\alpha 21} = \overline{F}_{\alpha 21}\left\{\overline{F}_2(\overline{x}_{02}, \tilde{\overline{x}}_{01})\,,\ \tilde{\overline{F}}_1^{(2)}(\overline{x}_{02}, \tilde{\overline{x}}_{01})\,,\ \overline{F}_{2\alpha}(\alpha_2)\right\}\,. \tag{4.17}$$

Vector functions $\overline{F}_{1\alpha}(\overline{\alpha}_1)$ and $\overline{F}_{2\alpha}(\overline{\alpha}_2)$ determine partners' interactions in coalitions $r = 1$ and $r = 2$, respectively, under conditions of situational uncertainty factor influence and are determined by the following relationships:

$$\overline{F}_{1\alpha}(\overline{\alpha}_1) = \overline{F}_{1\alpha}(\overline{\varphi}_{1\alpha}, \overline{f}_{1\alpha})\,, \tag{4.18}$$

$$\overline{F}_{2\alpha}(\overline{\alpha}_2) = \overline{F}_{2\alpha}(\overline{\varphi}_{2\alpha}, \overline{f}_{2\alpha})\,. \tag{4.19}$$

Here, the vector functions of common goals and participants' individual goals are presented as

$$\overline{\varphi}_{1\alpha} = \overline{\varphi}_{1\alpha}(\overline{x}_1, \overline{\chi}_1, \overline{\alpha}_1)\,, \quad \overline{\varphi}_{2\alpha} = \overline{\varphi}_{2\alpha}(\overline{x}_2, \overline{\chi}_2, \overline{\alpha}_2)\,,$$
$$\overline{f}_{1\alpha} = \overline{f}_{1\alpha}(\overline{x}_1, \overline{\chi}_1, \overline{\alpha}_1)\,, \quad \overline{f}_{2\alpha} = \overline{f}_{2\alpha}(\overline{x}_2, \overline{\chi}_2, \overline{\alpha}_2)\,.$$

On the basis of relationships (4.16) and (4.17) it is possible to describe both partners' interaction processes and two coalitions' counteractions. The given relationships can be easily generalized **for the case of several coalitions' interactions**. Let, as is assumed in (4.1), X_1 coalitions cooperate as partners. Then each coalition is described by mathematical models that are defined by relationship (4.7) and those related to it. As a result, we get the following set of vector functions determining partners' interaction goals in X_1 coalitions:

$$\overline{F}^0_{X_1} = \left\{F^0_{X_{1j}}\left(\overline{\varphi}_{X_{1j}}, \overline{f}_{X_{1j}}\right) | j = \overline{1, p_1}\right\}\,; \tag{4.20}$$

$$\overline{F}^\alpha_{X_1} = \left\{F^\alpha_{X_{1j}}\left(\overline{\varphi}_{X_{1j}}, \overline{f}_{X_{1j}}, \overline{f}^\alpha_{X_{1j}}\right) | j = \overline{1, p^\alpha_1}\right\}\,, \tag{4.21}$$

where relationship (4.20) describes the interaction process in the absence of situational uncertainty and relationship (4.21), obtained from (4.7) taking into account (4.18), characterizes the partners' interaction process in conditions of situational uncertainty.

Taking into account that X_1 coalitions are united in a supercoalition, every X_{1j} coalition $\left(j = \overline{1, p_1}\right)$ will be considered as an interacting participant in a supercoalition, taking into account that participants' interaction is described by a relationship of the form (4.7). Then, for a supercoalition we have

$$\overline{F}^0 = \overline{F}^0\left(\overline{\Phi}^0_{X_1}, \overline{F}^0_{X_1}\right); \quad \overline{F}^\alpha = \overline{F}^\alpha\left(\overline{\Phi}^\alpha_{X_1}, \overline{F}^\alpha_{X_1}\right),$$

where $\overline{F}^0_{X_1}$ and $\overline{F}^\alpha_{X_1}$ are determined by relationships (4.20) and (4.21) and $\overline{\Phi}^0_{X_1}$ and $\overline{\Phi}^\alpha_{X_1}$ are vector functions of a supercoalition's common goals in conditions of situational uncertainty absence and presence, accordingly.

Vector functions of a supercoalition's common goals are formed by analogy with relationships (4.8) and (4.9) and have the form

$$\overline{\Phi}^0_{X_1} = \left\{\Phi^0_{sX_1}\left(\overline{X}, \overline{\Xi}\right) | s_{X_1} = \overline{1, s_{0X_1}}\right\},$$
$$\overline{\Phi}^\alpha_{X_1} = \left\{\Phi^\alpha_{sX_1}\left(\overline{X}, \overline{\Xi}, \overline{A}\right) | s_{X_1} = \overline{1, s_{0X_1}}\right\}.$$

Here vector $\overline{X}$ defines all individual decisions of a supercoalition's participants and consists of components each of which determines decisions of the corresponding coalition exercising the rights of a participant in a supercoalition

$$\overline{X} = \left\{\overline{X}_{X_1 j_k} | k = \overline{1, p_{2j}},\ j = \overline{1, p_1}\right\},$$

where $\overline{X}_{X_1 j_k}$ is a vector of individual decisions of participant k of coalition $X_1 j_k$. Vector $\overline{\Xi}$ is a vector of a supercoalition's common decisions and is described as

$$\overline{\Xi} = \left\{\Xi_{l0} | l_0 = \overline{1, N_0}\right\},$$

where Ξ_{l0} is the l_0th component of the supercoalition's common decisions, accepted by all participants in coordination.

Vector $\overline{A}$ determines the influence of situational uncertainty factors on the supercoalition activities and is formed as

$$\overline{A} = \left\{\alpha_{X_1 j_k} | k = \overline{1, p_{2j}},\ j = \overline{1, p_1}\right\},$$

where $\alpha_{X_1 j_k}$ characterizes quantitative influence of the kth factor of situational uncertainty on the activity of the $X_1 j_k$ coalition.

The models describing the counteraction processes of X_2 coalitions are built similarly on the basis of (4.16) and (4.17) but will differ in the number of variables and goal vector functions, which will be equal to the number of counteracting coalitions in X_2.

4.2 General Strategy for Solving Problems of System Interactions or Counteractions of Coalitions

Approaches, principles, and methods of solving problems of coalition system interaction or counteraction are based on the use of the basic principles of system analysis. At the same time, at the first stage of the solution, the researcher should answer the following basic questions:

- What are the interests and goals of each of the coalition's participants and what contribution can he make to their attainment?
- What approaches are used by each participant for goal attainment?
- What methodes of counteracting the opponent may be used to attain the goals and what are their advantages and disadvantages?
- What objectively existing and subjectively formed restrictions and obstacles may influence the realization of the goals in practice?

Participants' interests are completely defined by their striving to achieve the following basic results:

1. An increase in the value of their goals' vector functions in the coalition's practical activities.
2. Minimization of the risk of possible damage in conditions of uncertainty in the competitors' actions.

To realize their interests the participants should observe certain rules of choice and of decision-making principles. From a set of possible approaches and principles of practical activities we point out a number of principles most important for decision making in the development and realization of coalitions' interaction and counteraction strategy. We present them in the form of the following principles of practical actions:

- Maximization of one's own goal functions up to certain limits, coordinated with partners, on the basis of a rational choice of one's own decisions.
- Striving for a rational compromise of the participants' actions and contributions in goal attainment while choosing and realizing the coalition's common decisions.
- Realization of the priority attitude in choosing and attaining the coalition's common goals.
- Reasonable caution in the conditions of situational uncertainty or information incompleteness regarding the counteracting party.
- Adherence to rational actions aimed at minimization of the risk level and risk degree of each participant of the coalition and the coalition as a whole.
- A rational choice of action strategy on the basis of preliminary formation of a Pareto set.

These principles have a number of practically important features. To a certain extent they are traditional for game theory, decision making, and operations research. They may be formalized by using fuzzy sets in certain conditions of information incompleteness and the presence of various uncertainties. At the same time, following general principles of caution, here a new approach and ways of its practical realization are suggested. The essence of the approach consists in replacing the guaranteed result principle, as the main principle of the theory of operations research, by the major principle of the theory of risk—the principle of risk minimization. At first sight it seems that between these approaches there is no basic difference, since the principle of guaranteed result can ensure minimization of risk under certain conditions. But such an inference is true if a number of principally important features and differences of these two approaches are not taken into consideration. The main difference is that the principle of guaranteed result does not take into account a probabilistic character of risk—only the extent of damage is taken into account, but the question of degree of probability of the most adverse situation occurring is not considered. Consequently, with this approach the degree of risk, which is understood as the probability of occurrence of an adverse, unpredictable risky situation, is not taken into account. And if the probability of such a situation is small, it is obvious that this approach results in great losses due to the missed opportunities. In rational estimation of risk degrees, as recommended by the theory of risk, the credibility of the obtained estimation is considerably higher, as a rule, and can ensure the principle of guaranteed result.

Moreover, in a number of economic applications the principle of guaranteed result may lead to absurd decisions. For example, in planning the manufacture of a new product that has neither prototypes nor analogs the worst situation will be a zero level of realization. In this situation it is senseless to solve the income optimization problem since it is impossible to find $\max_{x_1} \min_{x_2} f(x_1, x_2)$ at $\min_{x_2} f(x_1, x_2) = 0$ at any x_1, where x_1 is the product price, x_2 is the level of sale equal to zero, and $f(x_1, x_2)$ is the income level. There is only one answer here: it is inexpedient to manufacture the product, as the guaranteed result of income is equal to zero; therefore, it is guaranteed that the manufacturer's losses will be equal to the cost of the product instead of the profit. This example shows that the guaranteed result principle is practically inefficient in the conditions of the maximum situation uncertainty, i.e., in those situations where a theoretical estimation is especially required. It does not mean that the indicated principle is ineffective in practice. It may be effective if it is known a priori that an adverse situation probability is high and the guaranteed result is other than zero. And, finally, it is necessary to point out the following factor—what can ensure a "guaranteed result"? As is known, this result is determined by the maxmin function, i.e., $\max_{x_1} \min_{x_2} f(x_1, x_2)$. It means that in the worst case of $\min_{x_2} f(x_1, x_2)$, it is necessary to choose the best solution of $\max_{x_1} f(x_1, x_2)$. An issue naturally arises, which is that if the solution x_1 is not the best, then obviously $\max_{x_1} f(x_1, x_2)$ will not be ensured, and the obtained result will be worse than the guaranteed one. Consequently, the guaranteed result principle provides a certain result only for the best solution in the worst situation.

But it is known that in practice, a search for the best result may be either too labor-intensive or too difficult to realize. Therefore, it is frequently expedient to limit oneself to the rational solution, which is close enough to the optimum one. But the degree of its closeness to the optimum solution cannot be determined and guaranteed. Therefore, the guaranteed result principle for such a case does not allow one to obtain an estimation of the solution quality since there is nothing to compare the obtained result with because, until the best solution is found, the guaranteed result will not be obtained. Another practically important difference between the risk minimization principle and the guaranteed result principle consists in a choice of the obtained result count. The guaranteed result principle uses the point count, taking as the starting point the level of guaranteed result. And the risk minimization principle uses the interval count, choosing the worst and best results as its boundaries. The advantages and disadvantages of the point and interval count also are connected with the fact that the interval count allows one to determine the relative level of the obtained result in comparison with the worst and best results possible in the developing conditions. It makes it possible to assign beforehand the lower boundary of the maximum permissible damage, irrespective of the absolute values of the obtained result and estimation interval boundaries. At the same time a relative count shows to what extent the obtained result exceeds the worst one and does not reach the best one.

None of these possibilities is provided by the point count. One more difference of the considered principles is as follows. All methods of using the guaranteed result principle are focused on the condition—the initial information is complete, accurate, and credible. However, real problems are characterized by the incompleteness, inaccuracy, uncertainty, and contradictoriness of the initial information. And one of the most important ways of eliminating this drawback is the use of heuristic approaches, knowledge, and experience of experts. In particular, the risk minimization principle allows one to use them in solving the following problems:

1. Determination of the maximum permissible level of damage (in economics this is the level below which bankruptcy occurs; in military science, this is the damage level below which one or all participants are defeated).
2. Determination of a level of attainment of the possible limit result—the maximum and minimum possible results of predicted situations.

At the same time, other approaches, such as methods of fuzzy sets, situation forecasting, etc., are not excluded.

Further, we consider formalization of participants' strategies of interaction or counteraction and the development of methods of strategic problem solving.

4.3 Formalization of Coalition Counteraction Strategy

Let us formalize the basic characteristics of coalitions' counteraction strategy with the assigned restrictions, for the purpose of developing its rational realization approaches.

At the first stage we consider the case of two coalitions' counteraction and then generalize the result for the case of supercoalition counteractions.

1. The degree of attainment of the first coalition's goals will be characterized by the interval estimates

$$\bar{I}'_{12}\left(\bar{x}_{01}, \bar{\bar{x}}_{02}\right) = \frac{\bar{F}_{12}\left(\bar{x}_{01}, \bar{\bar{x}}_{02}\right) - \bar{F}_{12}^{-}}{\bar{F}_{12}^{+} - \bar{F}_{12}^{-}}, \tag{4.22}$$

$$\bar{I}''_{12}\left(\bar{x}_{01}, \bar{\bar{x}}_{02}\right) = \frac{\bar{F}_{12}^{+} - \bar{F}_{12}\left(\bar{x}_{01}, \bar{\bar{x}}_{02}\right)}{\bar{F}_{12}^{+} - \bar{F}_{12}^{-}}, \tag{4.23}$$

where $\bar{F}_{12}^{+}$ and $\bar{F}_{12}^{-}$ are, respectively, the maximum and minimum values of the goal function in a normal situation $\bar{F}_{12}\left(\bar{x}_{01}, \bar{\bar{x}}_{02}\right)$, and $\bar{I}'_{12}$ and $\bar{I}''_{12}$ determine a relative level of difference $\bar{F}_{12}\left(\bar{x}_{01}, \bar{\bar{x}}_{02}\right)$ between the minimum and maximum interval value, respectively.

With $\bar{F}_{12}\left(\bar{x}_{01}, \bar{\bar{x}}_{02}\right) = \bar{F}_{12}^{-}$ we have $\bar{I}'_{12} = 0$, which indicates the absence of a relative difference from the minimum interval value. Simultaneously $\bar{I}''_{12} = 1$, which indicates an extremely big relative difference $\bar{F}_{12}\left(\bar{x}_{01}, \bar{\bar{x}}_{02}\right)$ from the minimum value.

Similarly, with $\bar{F}_{12}\left(\bar{x}_{01}, \bar{\bar{x}}_{02}\right) = \bar{F}_{12}^{+}$ we have $\bar{I}'_{12} = 1$ and $\bar{I}''_{12} = 0$, which allows one to draw the opposite conclusion in comparison with the previous one—there is no difference $\bar{F}_{12}\left(\bar{x}_{01}, \bar{\bar{x}}_{02}\right)$ from the maximum interval value, and there is an extremely big difference from $\bar{F}_{12}^{-}$.

Note that the following relationships are true:

$$0 \le \bar{I}''_{12} \le 1, \quad 0 \le \bar{I}'_{12} \le 1, \tag{4.24}$$

$$\bar{I}'_{12} + \bar{I}''_{12} = 1. \tag{4.25}$$

Inequality (4.24) is true for the conditions corresponding to normal situations. From (4.25) it follows that it is enough to determine one interval estimate that may be considered as a relative goal function. Taking into account that $\bar{F}_{12}\left(\bar{x}_{01}, \bar{\bar{x}}_{02}\right)$ is the vector function, it is obvious that $\bar{I}'_{12}$ and $\bar{I}''_{12}$ will also be vector functions of the corresponding order.

The conducted analysis of the properties of functions $\bar{I}''_{12}$ and $\bar{I}'_{12}$ allows one to formalize the basic goal of the strategy of realizing the first coalition's interests by maximizing the corresponding vector functions in the following form:

$$\bar{F}_{12} \to \max; \quad \bar{I}'_{12} \to \max; \tag{4.26}$$

$$\bar{F}_{12} \to \max; \quad \bar{I}''_{12} \to \min. \tag{4.27}$$

From $\bar{F}_{12} \to \max$ it follows that $\bar{I}'_{12} \to \max$ or $\bar{I}''_{12} \to \min$.

It should be emphasized that relationships (4.26) and (4.27) are mathematically equivalent, and the difference is caused by the distinction of the physical meaning of $\bar{I}'_{12}$ and $\bar{I}''_{12}$. Further transformations are expedient to conduct after formalization of restrictions, conditions of a rational compromise, and risk minimization.

Similarly for the second coalition:

$$\bar{F}_{21} \to \max; \quad \bar{I}'_{21} \to \max ,$$
$$\bar{F}_{21} \to \max; \quad \bar{I}''_{21} \to \min ,$$

where

$$\bar{I}'_{21}\left(\bar{x}_{02}, \bar{\bar{x}}_{01}\right) = \frac{\bar{F}_{21}\left(\bar{x}_{02}, \bar{\bar{x}}_{01}\right) - \bar{F}_{21}^{-}}{\bar{F}_{21}^{+} - \bar{F}_{21}^{-}} , \tag{4.28}$$

$$\bar{I}''_{21}\left(\bar{x}_{02}, \bar{\bar{x}}_{01}\right) = \frac{\bar{F}_{21}^{+} - \bar{F}_{21}(\bar{x}_{02}, \bar{\bar{x}}_{01})}{\bar{F}_{21}^{+} - \bar{F}_{21}^{-}} . \tag{4.29}$$

2. Formalization of normal situations is traditional and consists in setting the variation range of goal vector function arguments.

The variation range of decisions for the first coalition participants will be presented in the following form:

$$\bar{x}_{11} \in \bar{D}_{11}, \ldots, \bar{x}_{1i} \in \bar{D}_{1i}, \ldots, \bar{x}_{1m_1} \in \bar{D}_{1m_1} ,$$

where

$$\bar{x}_{1i} = \left\{x_{1ij} | j = \overline{1, n_{1i}}\right\}; \quad \bar{x}_1 = \left\{\bar{x}_{1i} | i = \overline{1, m_1}\right\} ,$$
$$\bar{D}_{1i} = \left\{D_{1ij} | j = \overline{1, n_{1i}}\right\}; \quad \bar{D}_{1ij} = \left\{x_{1ij} | d_{1ij}^{-} \le x_{1ij} \le d_{1ij}^{+}\right\} .$$

Here $\bar{x}_1$ is a vector of all participants' own decisions for coalition 1 (the index 1 defines the coalition number), i is a coalition participant's position, and j is a coalition participant's decision position.

Similarly for the second coalition:

$$\bar{x}_2 = \left\{\bar{x}_{2i} | i = \overline{1, m_2}\right\}; \quad \bar{x}_{2i} = \left\{x_{2ij} | j = \overline{1, n_{2i}}\right\} ;$$
$$\bar{D}_{2i} = \left\{D_{2ij} | j = \overline{1, n_{2i}}\right\}; \quad \bar{D}_{2ij} = \left\{x_{2ij} | d_{2ij}^{-} \le x_{2ij} \le d_{2ij}^{+}\right\} .$$

The permissible variation range $\bar{\chi}_1$ of common coalition decisions (4.5) will be recorded as:

$$\bar{\chi}_1 \in \bar{D}_1, \quad \bar{\chi}_1 = \left\{\chi_{1i_1} | i_1 = \overline{1, p_1}\right\} ; \tag{4.30}$$

$$\bar{D}_1 = \left\{D_{1i_1} | i_1 = \overline{1, p_1}\right\}; \quad \bar{D}_{1i_1} = \left\{x_{1i_1} | d_{1i_1}^{-} \le x_{1i_1} \le d_{1i_1}^{+}\right\} . \tag{4.31}$$

Here χ_{1i_1} is a component of the vector of the common coalition decisions of coalition 1 and $\bar{D}_{1i_1}$ is a permissible interval for component i_1 of vector $\bar{x}_1$.

Similarly for the second coalition:

$$\bar{\chi}_2 \in \bar{D}_2, \quad \bar{\chi}_2 = \left\{\chi_{2i_2} | i_2 = \overline{1, p_2}\right\} ; \tag{4.32}$$

$$\bar{D}_2 = \left\{D_{2i_2} | i_2 = \overline{1, p_2}\right\}; \quad \bar{D}_{2i_2} = \left\{x_{2i_2} | d_{2i_2}^{-} \le x_{2i_2} \le d_{2i_2}^{+}\right\} . \tag{4.33}$$

3. Formalization of restrictions on situational uncertainty factors will be carried out both with the use of traditional approaches, characteristic of the operations research and decision-making theories, and on the basis of methodical techniques of risk theory.

Situational uncertainty factors (natural and others) with interaction between coalitions will be characterized with the help of a range of permissible vector parameter values (parameters $\bar{\alpha}_1$ for coalition 1 and $\bar{\alpha}_2$ for coalition 2), and by the probability of certain situations, each of which is characterized by a definite set of indices.

The approaches to formalizing the permissible variation range of vectors $\bar{\alpha}_1$ and $\bar{\alpha}_2$ are similar to those considered above. Taking into account relationships (4.14) and (4.15) and taking into consideration the approaches to forming relationships (4.30)–(4.33), we write the following:

$$\begin{aligned}
\bar{\alpha}_1 \in \bar{D}_{\alpha_1}, \quad \bar{\alpha}_1 &= \left\{\alpha_{1k_1} | k_1 = \overline{1, k_{01}}\right\} ; \\
\bar{D}_{\alpha_1} &= \left\{D_{\alpha_1 k_1} | k_1 = \overline{1, k_{01}}\right\} ; \\
D_{\alpha_1 k_1} &= \left\{\alpha_{1k_1} | \alpha_{1k_1}^- \le \alpha_{1k_1} \le \alpha_{1k_1}^+\right\} ; \\
\bar{\alpha}_2 \in \bar{D}_{\alpha_2}, \quad \bar{\alpha}_2 &= \left\{\alpha_{2k_2} | k_2 = \overline{1, k_{02}}\right\} ; \\
\bar{D}_{\alpha_2} &= \left\{D_{\alpha_2 k_2} | k_2 = \overline{1, k_{02}}\right\} ; \\
D_{\alpha_2 k_2} &= \left\{\alpha_{2k_2} | \alpha_{2k_2}^- \le \alpha_{2k_2} \le \alpha_{2k_2}^+\right\} .
\end{aligned}$$

We now proceed to the formalization of situation probabilistic properties using risk theory. Suppose that there is a certain set of situations in which coalition counteraction may take place. Each situation for coalitions 1 and 2 will be formally described accordingly in the following form:

$$\begin{aligned}
S_{L_1} &= \{\eta_{L_1}, \bar{\alpha}_{L_1}\}, \quad L_1 = \overline{1, L_{01}} ; \\
S_{L_2} &= \{\eta_{L_2}, \bar{\alpha}_{L_2}\}, \quad L_2 = \overline{1, L_{02}} .
\end{aligned}$$

Here L_1 and L_2 are the positions of the situations developing for coalitions 1 and 2, respectively; η_{L_1} and η_{L_2} are, respectively, the probability of occurrence of situations S_{L_1} and S_{L_2}; $\bar{\alpha}_{L_1}$ and $\bar{\alpha}_{L_2}$ are numerical values of the corresponding vectors $\bar{\alpha}_1$ and $\bar{\alpha}_2$, determined in the form:

$$\begin{aligned}
\bar{\alpha}_{L_1} &= \left\{\alpha_{1k_1}^{L_1} \in D_{\alpha_1 k_1} | k_1 = \overline{1, k_{01}}\right\} ; \\
\bar{\alpha}_{L_2} &= \left\{\alpha_{2k_2}^{L_2} \in D_{\alpha_2 k_2} | k_2 = \overline{1, k_{02}}\right\} .
\end{aligned}$$

Then, a set of situations corresponding to coalitions 1 and 2 can be described in the following form:

$$S_{01} = \left\{S_{L_1} | L_1 = \overline{1, L_{01}}\right\} ; \tag{4.34}$$

$$S_{02} = \left\{S_{L_2} | L_2 = \overline{1, L_{02}}\right\} . \tag{4.35}$$

Now proceed to the analysis and formalization of risk properties and parameters. We use two basic risk parameters—risk degree and risk level. Risk degree is defined as the probability of an event that will result in consequences undesirable for the corresponding coalition goals. Risk level is defined as the extent of potential damage for a coalition under the influence of risk factors.

Assume that the nature of risk is conditioned by the following groups of factors:

1. Risk factors of unpredictable situations with coalitions' counteraction.
2. Factors of force majeure risk.
3. Factors of information risk caused by inaccuracy, incompleteness, and noncredibility of the initial information about the intentions and goals of the counteracting parties.

Let us analyze the **first group of risk factors**. It is obvious that it is practically impossible to foresee all probable situations of counteraction, since their set is basically infinite. But the probability of many situations is low. Therefore, in practice a certain threshold of the probability of the most typical situation is established, and all others with probabilities lower than the threshold level are excluded from consideration.

Taking into account the above it is possible to present sets S_{01} and S_{02} in the following form:

$$S_{01} = \left\{ S_{1L_1} \Rightarrow \langle \eta_{1L_1}, \bar{\alpha}_{1L_1} \rangle \, | \eta_{1L_1} \geq \eta_{01}, \; L_1 = \overline{1, L_{01}} \right\} ; \tag{4.36}$$

$$S_{02} = \left\{ S_{2L_2} \Rightarrow \langle \eta_{2L_2}, \bar{\alpha}_{2L_2} \rangle \, | \eta_{2L_2} \geq \eta_{02}, \; L_2 = \overline{1, L_{02}} \right\} , \tag{4.37}$$

where η_{01} and η_{02} are the threshold probability values set by coalitions 1 and 2, respectively.

In the general case sets S_{01} and S_{02} include both predicted and unpredictable situations of coalitions' counteraction. Since a set of the considered situations is assumed to be finite, it forms a complete group of events. Supposing that predicted situations are independent, the probability of even one situation of the respective predicted sets for coalitions 1 and 2 will be defined by the relationship

$$\eta_{1ps} = 1 - \prod_{L_1=1}^{L_{1ps}} \left(1 - \tilde{\eta}_{1L_1}\right), \quad \eta_{2ps} = 1 - \prod_{L_2=1}^{L_{2ps}} \left(1 - \tilde{\eta}_{2L_2}\right) , \tag{4.38}$$

where L_{1ps} and L_{2ps} are the number of the predicted situations in S_{01} and S_{02}, respectively, and $\tilde{\eta}_{1L_1}$ and $\tilde{\eta}_{2L_2}$ are the probability of the predicted situations L_1 and L_2 from sets S_{01} and S_{02}, respectively.

In this case the probability η_{1ns} and η_{2ns} of unpredictable situations from S_{01} and S_{02} is determined by the relationship

$$\eta_{1ns} = 1 - \eta_{1ps}, \quad \eta_{2ns} = 1 - \eta_{2ps} . \tag{4.39}$$

The assumption that both sets S_{01} and S_{02} are finite and form a complete group of events is rigid enough and is not always fulfilled in practice.

Let us proceed to the analysis of the **second group of risk factors**, caused by the action of force majeure factors that lead to extreme and unfavorable events under the given conditions (flooding, storms, earthquakes, and others). In theory and practice these events may be called insuperable force factors or force majeure factors. There are in the world various national and international observing systems that allow one to identify the main characteristics of such events. In particular, there exists the possibility of timely warning about an approaching storm or flood. At the same time, at present there is no possibility of early warning about a number of other disasters, for example earthquakes. Therefore, methods of estimating the risk degree and risk level under the influence of such factors are being developed, in particular, expert methods. These methods are based on the use of long-term observation results and on expert knowledge and experience. On the one hand, it allows one to reveal a risk degree of a certain type of natural disaster for a definite period of time (a risk degree of hurricane occurrence within a day, a typhoon, etc.). On the other hand, the methods may be focused on the estimation of the risk degree, for example, for a certain region with the influence of one or several types of natural disasters. So, a risk degree of earthquake occurrence for various regions is unequal, and its magnitude depends on a region type (mountains, valleys, etc.) as well as the dynamics of geological processes in similar type regions. For example, the risk degree of earthquakes is different for the Carpathians, Pamirs, and Tibet.

Without going into details of the essence and features of these methods, we assume that a risk degree of force majeure factors is characterized by the probability of these events and designate it as η_{fm}.

Let us now proceed to an analysis of the third risk group caused by the incompleteness, inaccuracy, and noncredibility of information on the counteracting coalitions' goals and actions. This type of risk is caused also by the insufficient informedness of each counteracting party regarding the other party's goals and actions. Mathematically it is expressed in the fact that the vector function of the second coalition partner's interaction, determined as $\bar{F}_2$, is unknown for the first coalition's participants, and they interpret it as $\bar{\bar{F}}_2$. At the same time, coalition 1 participants know neither the number of goals (i.e., the number of components of vector function $\bar{F}_2$ is unknown) nor the corresponding form of each goal expression. That is, coalition 1 participants know neither the components of vector function $\bar{F}_2$ nor the components of functions $\bar{\varphi}_2$ and $\bar{f}_2$ (4.7). Such function arguments are unknown (i.e., unknown are both vectors of each participant's decisions in coalition 2 $\bar{x}_{21}, \ldots, \bar{x}_{2i_2}, \ldots, \bar{x}_{2m_2}$, and common coalition decision vectors $\bar{\chi}_{21}, \ldots, \bar{\chi}_{2i_2}, \ldots, \bar{\chi}_{2\rho_2}$ (4.4) and (4.5)). In coalition 1 all these functions and arguments are expressed in their own interpretation.

First, it is necessary to pay attention to the complex character of the difference between $\bar{F}_2(\bar{x}_{02}, \bar{\bar{x}}_{01})$ and $\bar{\bar{F}}_2^{(1)}(\bar{x}_{01}, \bar{\bar{x}}_{02})$, which is expressed in quantitative and qualitative distinction. The qualitative distinction is defined by the difference in structure of vector functions $\bar{F}_2(\bar{x}_{02}, \bar{\bar{x}}_{01})$ and $\bar{\bar{F}}_2^{(1)}(\bar{x}_{01}, \bar{\bar{x}}_{02})$, namely, by the difference in the form and number of these functions, components and also by the form and number of components of $\bar{\varphi}_2$ and $\bar{\bar{\varphi}}_2$, $\bar{f}_2$ and $\bar{\bar{f}}_2$. That is, in the condition $n'_{2i_2} \neq n_{2i_2}$, and $\rho'_2 \neq \rho_2$. The quantitative difference is defined by the distinction of numerical values of the aforementioned functions:

$$\left\{\bar{x}_{2i_2},\ i_2 = \overline{1, n_{2i_2}}\right\}, \quad \left\{\bar{\bar{x}}_{2i_2},\ i_2 = \overline{1, n'_{2i_2}}\right\},$$
$$\left\{\bar{\chi}_{2i_2},\ i_2 = \overline{1, \rho_2}\right\}, \quad \left\{\bar{\bar{\chi}}_{2i_2},\ i_2 = \overline{1, \rho'_2}\right\}.$$

From here arise a number of practically important problems:

- Estimating the degree of difference of vector functions $\bar{F}_2(\bar{x}_{02}, \bar{\bar{x}}_{01})$ and $\bar{\bar{F}}_2^{(1)}(\bar{x}_{01}, \bar{\bar{x}}_{02})$ in given conditions.
- Expressing quantitatively the risk degree as a measure of influence of the incompleteness, inaccuracy, and uncertainty of information on counteracting parties' goals and actions when the indicated deficit of information is expressed in such a many-sided difference of $\bar{F}_2(\bar{x}_{02}, \bar{\bar{x}}_{01})$ and $\bar{\bar{F}}_2^{(1)}(\bar{x}_{01}, \bar{\bar{x}}_{02})$.
- Defining the variation in risk degree while specifying some data on the counteracting party's goals and actions, that is, an increased level of informedness.

The complexity of solving these practical problems is determined by the fact that, owing to a complex difference in the indicated functions, none of the enumerated problems is reduced directly to known problems and cannot be solved by known methods. In fact, if $\bar{\varphi}_2$ and $\bar{\bar{\varphi}}_2$ only differed in the form of their components, then the degree of difference could be easily defined on the basis of the approximating function theory by component distinction with a subsequent use of any metric, Euclidean for example. Similarly, if these functions' arguments differed only in their numerical value, then determining the relative and absolute error would be an elementary problem.

It is necessary to draw attention to one more principal peculiarity of the difference between the considered practical problem and the known problems of the approximating function theory, operations research, and decision-making theories. In all these disciplines, functions $\bar{F}_2(\bar{x}_{02}, \bar{\bar{x}}_{01})$ and $\bar{\bar{F}}_2^{(1)}(\bar{x}_{01}, \bar{\bar{x}}_{02})$ are assumed to be known. And in a practical problem it is necessary to construct them on the basis of available information. That is, in addition to the considered problems, the problem of information provision arises—construction of a model in the form $\bar{\bar{F}}_2^{(1)}(\bar{x}_{01}, \bar{\bar{x}}_{02})$. Obviously, it may be assumed that, owing to the above arguments, a risk degree caused by insufficient informedness is a function of the quantity and quality of information. A risk degree of the influence of a given factor is conditioned by the fact that incompleteness, inaccuracy, and uncertainty of information leads to the occurrence of events that are characterized by undesirable consequences for the corresponding coalition's goals, in this case, for coalition 1. In this situation we designate the risk degree of the given type as η_{in}.

Further, it is necessary to consider the problem of determining the risk degree in the case of influence of all previously mentioned risk factors. In this analysis, first of all, it is necessary to take into account that the previously considered physical factors form events that are independent in the sense of probability. Second, we consider the influence of these factors for two conditions:

1. Occurrence of an undesirable event connected with an action of at least one of the three groups of risk factors listed above.
2. Occurrence of an event undesirable for coalition 1 connected with the simultaneous influence of the listed groups of risk factors.

By virtue of the independence of various groups of events, the probability of an undesirable event connected with the influence of even one of the listed groups of factors, owing to the independence of various groups of events, is defined by the relationship

$$\eta_{1\Sigma} = 1 - (1 - \eta_{1ns})\,(1 - \eta_{1fm})\,(1 - \eta_{1in})\ , \tag{4.40}$$

where η_{1ns}, η_{1fm}, and η_{1in} are respectively the risk degrees of counteraction of unpredictable situations, force majeure events, and information uncertainty for coalition 1.

For coalition 1 the probability of an undesirable risk event connected with simultaneous influence of all risk group factors, owing to the independence of various groups of events, is defined by the relationship

$$\eta_{1\Pi} = \eta_{1ns} \cdot \eta_{1fm} \cdot \eta_{1in}\ . \tag{4.41}$$

For coalition 2 relationships (4.40) and (4.41) are of a similar form.

In the general case the number of risk-factor groups for the rth coalition can be equal to N_f. In this case, relationships (4.40) and (4.41) will be transformed as follows:

$$\eta_{r\Sigma} = 1 - \prod_{k1=1}^{N_f} (1 - \eta_{k1})\ , \tag{4.42}$$

$$\eta_{r\Pi} = \prod_{k1=1}^{N_f} (\eta_{k1})\ . \tag{4.43}$$

The results of formalization of coalitions' counteraction strategies, obtained in this section, allow one to consider more complicated cases of coalitions' interaction and counteraction.

4.4 Formalization of Risks in Problems of Coalitions' System Interaction or Counteraction

Let us now proceed to the analysis of the degree of fulfillment of each coalition's interests considering risk factors. Therefore, we take into account that expressions (4.22), (4.23) and (4.28), (4.29) are true for normal situations, and the influence of risk factors leads to abnormal situations. The influence of abnormal situations may result in direct and indirect damage. Direct damage is a direct reduction in the level of goal achievement (for example, a reduction in production output or production distribution), and indirect damage is a deterioration of functioning conditions that results in the reduction in values of goal functions with respect to their values in normal situations (for example, an increase in the initial raw cost, transport charges, value added taxes, etc.). A level of direct damage will be defined in the form

$$\bar{J}_{12\eta_1}\left(\bar{x}_{01}, \bar{\bar{x}}_{02}, \bar{\alpha}_1, \bar{\eta}_1\right) = \frac{\bar{\Phi}_{12\eta_1}\left(\bar{x}_{01}, \bar{\bar{x}}_{02}, \bar{\alpha}_1, \bar{\eta}_1\right) - \bar{\Phi}^-_{12\eta_1}}{\bar{\Phi}^+_{12\eta_1} - \bar{\Phi}^-_{12\eta_1}}, \qquad (4.44)$$

where $\bar{\Phi}^+_{12\eta_1}$ and $\bar{\Phi}^-_{12\eta_1}$ are respectively the maximum and minimum values of damage resulting from the risk factors for coalition 1; $\bar{\Phi}_{12\eta_1}\left(\bar{x}_{01}, \bar{\bar{x}}_{02}, \bar{\alpha}_1, \bar{\eta}_1\right)$ is the damage for coalition 1 in the analyzed risk situation $\bar{\eta}_1 = \left(\eta_{ns}, \eta_{fm}, \eta_{in}\right)$.

Relationship (4.44) determines the risk level as a measure of possible direct damage from the influence of all risk-factor groups.

For the influence of various groups of risk factors—unpredictable counteraction situations, force majeure factors, factors of information uncertainty—the extent of damage will be defined by the following relationships:

$$\bar{J}_{12ns}\left(\bar{x}_{01}, \bar{\bar{x}}_{02}, \bar{\alpha}_1, \eta_{ns}\right) = \frac{\bar{\Phi}_{12\eta_{ns}}\left(\bar{x}_{01}, \bar{\bar{x}}_{02}, \bar{\alpha}_1, \eta_{ns}\right) - \bar{\Phi}^-_{12\eta_{ns}}}{\bar{\Phi}^+_{12\eta_{ns}} - \bar{\Phi}^-_{12\eta_{ns}}}; \qquad (4.45)$$

$$\bar{J}_{12fm}\left(\bar{x}_{01}, \bar{\bar{x}}_{02}, \bar{\alpha}_1, \eta_{fm}\right) = \frac{\bar{\Phi}_{12\eta_{fm}}\left(\bar{x}_{01}, \bar{\bar{x}}_{02}, \bar{\alpha}_1, \eta_{fm}\right) - \bar{\Phi}^-_{12\eta_{fm}}}{\bar{\Phi}^+_{12\eta_{fm}} - \bar{\Phi}^-_{12\eta_{fm}}}; \qquad (4.46)$$

$$\bar{J}_{12in}\left(\bar{x}_{01}, \bar{\bar{x}}_{02}, \bar{\alpha}_1, \eta_{in}\right) = \frac{\bar{\Phi}_{12\eta_{in}}\left(\bar{x}_{01}, \bar{\bar{x}}_{02}, \bar{\alpha}_1, \eta_{in}\right) - \bar{\Phi}^-_{12\eta_{in}}}{\bar{\Phi}^+_{12\eta_{in}} - \bar{\Phi}^-_{12\eta_{in}}}. \qquad (4.47)$$

Owing to the independence of the influence of various risk-factor groups a resulting risk estimation can be obtained on the basis of various approaches:

- Taking into account the factor influence of each risk group separately.
- Taking into account the availability of factor influence of at least one group.
- Taking into account the joint factor influence of all risk groups.

The most expedient approach is to make an estimation of influence of factors independently and, by virtue of various groups' influence independence, finding the resulting risk level of the joint influence of several risk-factor groups based on a summation of possible damage.

As the risk degree from the influence of various groups' factors is different, a generalized risk level from the influence of each group of factors will be considered in the following form:

$$\bar{U}_{1ns} = \bar{f}_{1ns}(\eta_{ns})\,\bar{J}_{12ns}\,, \tag{4.48}$$

$$\bar{U}_{1fm} = \bar{f}_{1fm}(\eta_{fm})\,\bar{J}_{12fm}\,, \tag{4.49}$$

$$\bar{U}_{1in} = \bar{f}_{1in}(\eta_{in})\,\bar{J}_{12in}\,, \tag{4.50}$$

where $\bar{U}_{1ns}$, $\bar{U}_{1fm}$, and $\bar{U}_{1in}$ are the risk levels for coalition 1 under the influence of factors connected with unpredictable situations, force majeure events, and information uncertainty, respectively; $\bar{J}_{12ns}$, $\bar{J}_{12fm}$, and $\bar{J}_{12in}$ denote the damage determined by relationships (4.45)–(4.47); $\bar{f}_{1ns}(\eta_{ns})$, $\bar{f}_{1fm}(\eta_{fm})$, and $\bar{f}_{1in}(\eta_{in})$ are vector functions that take into account the form of damage level dependences from the risk degree of unpredictable situations, force majeure events, and information uncertainty factors, respectively. In the simplest case it is assumed that:

$$\bar{f}_{1ns}(\eta_{ns}) = \eta_{ns}, \quad \bar{f}_{1fm}(\eta_{fm}) = \eta_{fm}, \quad \bar{f}_{1in}(\eta_{in}) = \eta_{in}\,. \tag{4.51}$$

Considering the influence of one risk-factor group, the goal functions for coalition 1 will be presented as:

$$\hat{\bar{F}}_{12\eta_{ns}}(\bar{x}_{01}, \bar{\bar{x}}_{02}, \bar{\alpha}_1, \eta_{ns}) = \hat{\bar{f}}_1(\eta_{ns})\,\bar{I}'_{12}(\bar{x}_{01}, \bar{\bar{x}}_{02}) \\ - \bar{f}_{1ns}(\eta_{ns})\,\bar{J}_{12ns}(\bar{x}_{01}, \bar{\bar{x}}_{02}, \bar{\alpha}_1, \eta_{ns})\,, \tag{4.52}$$

where $\hat{\bar{F}}_{12\eta_{ns}}(\bar{x}_{01}, \bar{\bar{x}}_{02}, \bar{\alpha}_1, \eta_{ns})$ is a goal vector function of coalition 1 in the conditions of influence of unpredictable situations of counteraction; $\bar{I}'_{12}(\bar{x}_{01}, \bar{\bar{x}}_{02})$ is a goal vector function of coalition 1 in the absence of risk factors, determined by relationship (4.22); $\hat{\bar{f}}_1(\eta_{ns})$ is a vector function that takes into account the risk degrees of indirect influence of unpredictable situations of counteraction on the level of fulfilling the interests of coalition 1; $\bar{f}_{1ns}(\eta_{ns})$ is a vector function that takes into account the form of dependence of direct damage from a risk degree of unpredictable situations; and $\bar{J}_{12ns}(\bar{x}_{01}, \bar{\bar{x}}_{02}, \bar{\alpha}_1, \eta_{ns})$ is the level of direct damage.

In (4.52) the first summand accounts for indirect damage, the second for direct damage from predictable risk factors.

With the simultaneous influence of risk factors of the three mentioned groups, owing to the independence of their influence, goal functions for coalition 1 will be defined by the following relationship:

$$\hat{\bar{F}}_{\Sigma 12}(\bar{x}_{01}, \bar{\bar{x}}_{02}, \bar{\alpha}_1, \bar{\eta}) = \hat{\bar{f}}_1(\eta_{ns})\,\hat{\bar{f}}_1(\eta_{fm})\,\hat{\bar{f}}_1(\eta_{in})\,\bar{I}'_{12}(\bar{x}_{01}, \bar{\bar{x}}_{02}) \\ - \left(\bar{U}_{1ns} + \bar{U}_{1fm} + \bar{U}_{1in}\right)\,. \tag{4.53}$$

Taking into account that in the simplest case

$$\hat{\hat{f}}_1(\eta_{ns}) = 1 - \eta_{ns}, \quad \hat{\hat{f}}_1(\eta_{fm}) = 1 - \eta_{fm}, \quad \hat{\hat{f}}_1(\eta_{in}) = 1 - \eta_{in}\,. \tag{4.54}$$

Taking into account (4.48)–(4.50), relationship (4.53) is transformed into

$$\begin{aligned}\hat{\hat{F}}_{\Sigma 12}(\bar{x}_{01}, \bar{\bar{x}}_{02}, \bar{\alpha}_1, \bar{\eta}) &= (1-\eta_{ns})(1-\eta_{fm})(1-\eta_{in})\, I'_{12}(\bar{x}_{01}, \bar{\bar{x}}_{02}) \\ &\quad - \left(\eta_{ns}\bar{J}_{12ns} + \eta_{fm}\bar{J}_{12fm} + \eta_{in}\bar{J}_{12in}\right)\,. \end{aligned} \tag{4.55}$$

In the general case with simultaneous influence of N'_f risk-factor groups the following conditions are fulfilled:

$$\begin{aligned}\bar{F}_{\Sigma 12}(\bar{x}_{01}, \bar{\bar{x}}_{02}, \bar{\alpha}_1, \eta_\Sigma) &= \hat{\hat{f}}_1(\eta_\Sigma)\, \bar{I}'_{12}(\bar{x}_{01}, \bar{\bar{x}}_{02}) \\ &\quad - \sum_{k=1}^{N'_f} \bar{f}_{1k}(\eta_k)\, \bar{J}_{12k}(\bar{x}_{01}, \bar{\bar{x}}_{02}, \bar{\alpha}_1, \eta_k)\,, \end{aligned} \tag{4.56}$$

where $\hat{\hat{f}}_1(\eta_\Sigma)$ is a vector function that takes into account a risk degree of indirect influence of all factors of all risk groups on the level of attainment coalition 1's interests; $\bar{f}_{1k}(\eta_k)$ is a vector function that represents the dependence type of direct damage level of the influence of risk factors of group kth to a level of attainment of interests for coalition 1; $\bar{J}_{12k}(\bar{x}_{01}, \bar{\bar{x}}_{02}, \bar{\alpha}_1, \eta_k)$ denotes the level of direct damage from the influence of the kth risk-factor group.

In the simplest case, accounting for relationships (4.51) and (4.54), relationship (4.56) will be transformed as follows:

$$\begin{aligned}\bar{F}_{\Sigma 12}(\bar{x}_{01}, \bar{\bar{x}}_{02}, \bar{\alpha}_1, \eta_\Sigma) &= \bar{I}'_{12}(\bar{x}_{01}, \bar{\bar{x}}_{02}) \prod_{k=1}^{N'_f}(1-\eta_k) - \\ &\quad - \sum_{k=1}^{N'_f}\left(\eta_k \bar{J}_{12k}(\bar{x}_{01}, \bar{\bar{x}}_{02}, \bar{\alpha}_1, \eta_k)\right)\,. \end{aligned} \tag{4.57}$$

Relationships (4.56) and (4.57) are more general as compared with those traditionally accepted in operations research theory and define the goal-function dependences both on the decisions taken by counteracting participants and on the parameters of risk factors. Relationship (4.57) will be transformed into (4.22) in the absence of the risk-factor influence ($\eta_k = 0$, $k = \overline{1, N'_f}$).

Note in the general case the dependences of the risk level and of the degree to which one's interests are fulfilled on the risk degree are nonlinear. Mathematically it is convenient to present this dependence in (4.56) as exponential functions of the following form:

$$\bar{f}_{1k}(\eta_k) = 1 - \exp(b_k \eta_k)\,.$$

For $\eta_k = 0$, for all $k = \overline{1, N_f'}$ the initial conditions are to be fulfilled:

$$\hat{\bar{f}}_1(\eta_\Sigma) = 1, \quad \bar{f}_{1k}(\eta_k) = 0,$$

which corresponds to the absence of risk-factor influence.

In particular, for the conditions to be fulfilled the function $\hat{\bar{f}}_1(\eta_\Sigma)$ can be presented as:

$$\hat{\bar{f}}_1(\eta_\Sigma) = \exp\left(\frac{\sum_{k=1}^{N_f'} a_k(1-\eta_k)}{\sum_{k=1}^{N_f'} a_k} - 1\right).$$

According to (4.26) the strategy of coalition 1 based on relationship (4.56) and considering risk factors is determined by maximization of goal vector functions:

$$\hat{\bar{F}}_{\Sigma 12}(\bar{x}_{01}, \tilde{\bar{x}}_{02}, \bar{\alpha}_1, \eta_\Sigma) \to \max. \tag{4.58}$$

The given condition can be fulfilled in various ways:

1. By optimization of the decisions of both coalitions' participants $\bar{x}_1 = \{\bar{x}_{1i_1} | i_1 = \overline{1, m_1}\}$ and the common coalition decisions of the following form: $\bar{\chi}_1 = \{\bar{\chi}_{1l_1} | l_1 = \overline{1, \rho_1}\}$ with the use of the condition

$$\bar{x}_{01}^0 = \operatorname{Arg}\max \bar{I}_{12}'(\bar{x}_{01}). \tag{4.59}$$

2. By minimization of the risk degree $\left\{\eta_k | k = \overline{1, N_k'}\right\}$ based on an increase in the decision maker's level of informedness regarding risk factors:

$$\bar{\eta}_k^0 = \operatorname{Arg}\min_{\eta_k} \bar{J}_{12k}. \tag{4.60}$$

3. By system coordination and maximization of one's own goal vector function $\bar{I}_{12}'$ and minimization of risk level $\bar{J}_{12k}$ owing to the optimization of coalition common decisions and the coalition participants' coordinated decisions with a certain assigned risk degree.
4. By simultaneous system-coordinated maximization of $\bar{I}_{12}'$ and minimization of $\bar{J}_{12k}$ owing to the realization of two conditions: first, optimization of coalition collective decisions, and second, an increase in the level of informedness of each participant of the coalition regarding risk factors.

Owing to the similarity of the approaches to constructing goal functions for coalitions 1 and 2, we arrive at the conclusion that all earlier argumentation is true for the strategy of coalition 2. From here, comparing relationships

(4.12) and (4.13), it is possible to obtain relationship similar to (4.56) in the following form:

$$\hat{\bar{F}}_{\Sigma 21}\left(\bar{x}_{02}, \tilde{\bar{x}}_{01}, \bar{\alpha}_2, \eta_\Sigma\right) = \bar{f}_2\left(\eta_\Sigma\right) \bar{I}'_{12}\left(\bar{x}_{02}, \tilde{\bar{x}}_{01}\right) - \\ - \sum_{k=1}^{N'_f} \bar{f}_{2k}\left(\eta_k\right) \bar{J}_{21k}\left(\bar{x}_{02}, \tilde{\bar{x}}_{01}, \bar{\alpha}_2, \eta_k\right) . \quad (4.61)$$

Let us indicate one principal peculiarity of the optimization problem for which the basic optimality criterion is described by relationship (4.60). This problem is conditional, as compared to the problem where the basic criterion is defined by relationship (4.59). The conditional character of this problem is connected with the fact that here the optimization is carried out owing to an increase in the level of completeness, credibility, and accuracy of the information used by the decision maker regarding factors of the corresponding risk group. This result is achieved by searching for a certain value of the resulting vector $\bar{\eta}_1$ whose components determine a risk degree of a certain group of factors. Hence, the optimization problem on criterion (4.60) should be considered as a maximization problem for a level of the decision maker's informedness. In addition, optimization problem (4.59) should be considered a problem of defining the coalition participants' own decisions.

It should also be pointed out that formula (4.56) allows one to determine the extent to which coalition 1's goals may be attained in normal and abnormal situations taking into account risk factors. In this way it is possible to optimize the decision made in real conditions of people's practical activities, where the influence of risk factors cannot be eliminated in principle. In fact, all practical activities are carried out under the conditions of various risk factors, for example, under the conditions of risks of external natural influences, natural disasters, competition, information incompleteness, etc. These risks are principally unremovable due to their nature. Therefore, accounting for the risk level and risk degree in solving applied problems of both interaction and counteraction of participants in certain activities is a real and important problem.

Relationships (4.56) and (4.61) are the basis for the research of a two-coalition counteraction process in real conditions of influence of risk factors of various natures. One should consider the influence of each of these factors using two parameters: risk degree and risk level. The first parameter characterizes the probability of undesirable influence of each risk factor. The second parameter characterizes the extent of potential damage caused by undesirable influence of the revealed risk factor.

4.5 Methods and an Example of Coalition Interaction and Counteraction Problem Solving

We now turn to an analysis of the capabilities of various methods for solving coalition counteraction problems taking into account risk factors. The choice of method in many respects is determined by the problem peculiarities. The main peculiarity of the considered class of problems is determined by the conditions under which a strategy of each coalition's actions is realized. These conditions result directly from the action strategy of each coalition, which consists in raising the level of their intentions realization by increasing the values of goal vector functions and lowering the risk degree and risk level connected with the influence of uncertainty factors regarding the opposing party's intentions and actions. From here it follows that practical problems of disclosing the uncertainty of coalitions' counteractions differ from typical extreme problems in the presence of principally new properties and features. Among these features the specificity of goal function formalization should be singled out. The goal functions of counteracting parties are interconnected, which directly follows from counteraction factors. But it is particularly important that they are formed in conditions of conceptual uncertainty. This is conditioned by the fact that, due to the contradiction of goals and interests, each coalition is not interested in disclosing its intentions. So, as a rule, there is no credible information in counteracting coalitions. Therefore, planned goals and actions are unknown, and, consequently, neither goal functions nor their arguments are known. Each coalition has to interpret the probable actions of the counteracting party according to its own estimations and prognoses. On this basis it has to form the counteracting party goal functions and their arguments. Some other peculiarities of coalition counteraction problems complicating the solution for each participant are as follows:

- Presence of contradictions between the coalition's common goals and coalition participants' individual goals.
- Presence of contradictions between individual interests of each coalition's participants.
- Interdependence of individual goals of each coalition's participants and the coalition common goals and coalition common decisions.
- Presence and possible influence of a great volume of unpredictable natural, climatic, ecological, technological, and other risk factors, which may essentially change the conditions and results pertaining to the achievement of goals.
- Direct influence of a time factor on effectiveness of each participant's action strategy and the coalition as a whole.
- The necessity of reaching system-coordinated decisions and actions of the coalition participants in the presence of a sufficiently large set of vector functions and their arguments.

Taking into consideration the above-indicated features, we try to formulate the basic requirements for the problem-solving method. It should ensure:

- The possibility of timely realization of purposeful actions of each coalition participant within the framework of the coalition-coordinated strategy.
- Quantitative estimation of the degree of attainment of potentially predicted best results of activities and the level of difference in the predicted minimally permissible result from the potentially possible worst result.
- Estimation of errors of the obtained solution for action strategies, conditioned by incompleteness, uncertainty, and contradictoriness of the initial information about the counteracting coalitions' goals, decisions, and actions.
- The possibility of specifying the decisions by raising the level of informedness regarding the strategy and intentions of the counteracting coalition.
- Estimation of the risk degree and risk level in conditions of possible influence of force majeure factors, factors of situational uncertainty regarding the goals and actions of the counteracting coalition, and other risk factors.
- Estimation of consequences of the influence of unpredictable risk factors on the level of attainment of coalition goals.

Using as the basis relationship (4.57) and condition (4.58), let us turn to the stated approach to problem solving. The decomposition of the initial problem will be made in accordance with the action strategy of each partner and the coalition as a whole, namely, with their striving to raising the level of goal achievement and minimize the risk in uncertainty conditions. Firstall, we must take into account the features of the goal function (4.57). We must also take into account the additivity of this function and the physical meaning of the first summand, which determines the level of goal achievement, as well as the physical meaning of the second summand determining the extent of damage. With these assumptions, the problem of goal function maximization will be presented in the following form:

$$\hat{\bar{F}}_{\Sigma 12}\left(\bar{x}_{01}, \bar{\bar{x}}_{02}, \bar{\alpha}_1, \eta_\Sigma\right) \to \max\,, \tag{4.62}$$

which is fulfilled under the following conditions:

$$\hat{\bar{f}}_1\left(\eta_\Sigma\right) I'_{12}\left(\bar{x}_{01}, \bar{\bar{x}}_{02}\right) \to \max\,, \tag{4.63}$$

$$\sum_{k=1}^{N'_f} \bar{f}_{1k}\left(\eta_k\right) \bar{J}_{12k}\left(\bar{x}_{01}, \bar{\bar{x}}_{02}, \bar{\alpha}_1, \eta_k\right) \to \min\,. \tag{4.64}$$

Here relationship (4.63) defines the condition of attainment of the vector function maximum level under the influence of risk factors. In this case, the risk-factor influence is determined by the function $\bar{f}_1(\eta_\Sigma)$, but it does not take into consideration the extent of damage from the influence of factors. Relationship (4.64) determines the condition of the minimal common risk level with all risk-factor interactions. Here, the risk degree, caused by the influence of each group of factors and the corresponding extent of damage, caused by that influence, are considered. From relationships (4.63) and (4.64) it is obvious that condition (4.62) is fulfilled upon simultaneous fulfillment of the following conditions:

$$\hat{\bar{f}}_1(\eta_\Sigma) \to 1; \quad \bar{I}'_{12}(\bar{x}_{01}, \tilde{\bar{x}}_{02}) \to 1; \tag{4.65}$$

$$\bar{f}_{1k}(\eta_k) \to 0; \quad \bar{J}_{12k}(\bar{x}_{01}, \tilde{\bar{x}}_{02}, \bar{\alpha}_1, \eta_k) \to 0; \quad k = \overline{1, N'_f}. \tag{4.66}$$

These conditions characterize the necessity of maximizing each factor of (4.63) and of efficiently minimizing each factor of (4.64).

The limits determined by relationship (4.65) follow from the physical meaning of factors in the absence of influence by unpredictable situations. Such an assumption is true under the condition that the probability of unpredictable-factor influence is extremely small. The second relationship in (4.65) characterizes the condition of attainment of potentially possible, maximum level of goal functions, which directly follows from (4.22).

Now attention should be paid to the fact that functions $\hat{\bar{f}}_1(\eta_\Sigma) \to 1$ and $\bar{f}_{1k}(\eta_k) \to 0$ are interdependent, which follows from the physical meaning of the risk degree as a probabilistic measure. From relationship (4.54) and from the condition independence of various risk factor influences it follows that the boundaries of the limiting values of a risk degree under the influence of a set of risk factors are in the interval [0,1]. In fact, according to (4.57), in the simplest case the following dependences are fulfilled:

$$\hat{\bar{f}}_1(\eta_\Sigma) = \prod_{k=1}^{N'_f} (1 - \eta_k), \quad \bar{f}_{1k}(\eta_k) = \eta_k; \quad k = \overline{1, N'_f}.$$

We now turn to the fact that the risk degree from the influence of various risk factors is determined by the level of the decision maker's informedness regarding the probability of undesirable events. Conditionally we record it in the form $\eta = \eta(\bar{U})$, where we understand the vector of parameters determining the level of the decision maker's informedness as $\bar{U}$; their physical meaning, approaches, and methods of definition will be considered in the course of carrying out information analysis.

From relationships (4.65) and (4.66), taking into account both this dependence and interconnections of the functions $\hat{\bar{f}}_1(\eta_\Sigma)$ and $\bar{f}_{1k}(\eta_k)$, $k = \overline{1, N'_f}$ the following condition for determining a rational informedness level $\bar{U}^*$ is obtained:

$$\hat{\bar{f}}_1(\eta_\Sigma(\bar{U})) \to 1; \tag{4.67}$$

$$\bar{f}_{1k}(\eta_k(\bar{U})) \to 0, \quad k = \overline{1, N'_f}.$$

Thus, in order for a degree of undesirable influence to be minimal, it is necessary to ensure a minimal risk probability due to informedness maximization.

The given problem solving will be reduced to the Chebyshev approximation problem, presenting the considered condition in the form of a system of equations $\left(N_f^{'}+1\right)$ of the form

$$\bar{f}_k\left(\bar{U}\right)=0; \quad k=\overline{1,N_{0f}^{'}}. \tag{4.68}$$

Here it is designated

$$\bar{f}_k\left(\bar{U}\right)=\bar{f}_{1k}\left(\eta_k\left(\bar{U}\right)\right) \quad \text{at} \quad k=\overline{1,N_f^{'}};$$
$$\bar{f}_k\left(\bar{U}\right)=\hat{\bar{f}}_1\left(\eta_\Sigma\left(\bar{U}\right)\right)-1 \quad \text{at} \quad k=N_{0f}^{'}=\overline{1,N_f^{'}}+1.$$

Taking the value as a Chebyshev approximation measure

$$\Delta=\max_k\left|\bar{f}_k\left(\bar{U}\right)\right|,$$

we consider the approximation problem for system (4.68) as a problem of finding a value $\bar{U}^*$ at which the value Δ will be minimum in the assigned interval of values $\bar{U}^-\leq\bar{U}^*\leq\bar{U}^+$. In that case for Δ^0 the following condition is fulfilled:

$$\Delta^0=\Delta\left(\bar{U}^*\right)=\min_{\bar{U}}\Delta, \tag{4.69}$$

and the argument value, which ensures the fulfillment of the above condition, is defined by the relationship

$$\bar{U}^*=\arg\min_{\bar{U}}\max_k\left|\bar{f}_k\left(\bar{U}\right)\right|. \tag{4.70}$$

The approximation problem of system (4.68) on criterion (4.69) is formalized with an equal significance of all groups of risk factors. In the case of a different significance of risk factors, the following value should be taken as the Chebyshev approximation measure:

$$\Delta_v=\max_k v_k\left|\bar{f}_k\left(\bar{U}\right)\right|.$$

Then relationships (4.69) and (4.70) can be transformed into

$$\Delta_v^0=\Delta_v\left(\bar{U}^*\right)=\min_{\bar{U}}\Delta_v;$$
$$\bar{U}^*=\arg\min_{\bar{U}}\max_k\left|v_k\bar{f}_k\left(\bar{U}\right)\right|. \tag{4.71}$$

Here $1>v_k>0$ and is normalized by the condition

$$\sum_{k=1}^{N_f^{'}}v_k=1.$$

Solving system (4.68) on criterion (4.71) determines a rational value $\bar{U}^*$ under the condition that the risk degree of various groups of factors is different and is characterized by the weighting coefficients v_k; $k = \overline{1, \left(N_f' + 1\right)}$.

Note that in disclosing the uncertainty of system (4.68) on criterion (4.69) or (4.71), we obtain the rational solution values, but not the optimal solution values. In principle, it is impossible to reduce the risk degree to zero, since, on the one hand, any observations have a limited degree of accuracy, and, on the other hand, by raising the informedness level, for example, due to an increase in the measurement accuracy, the expenditures for obtaining the corresponding information grow. And under certain conditions these expenditures may exceed the level of possible damage from the consequences of the influence of risk factors.

Therefore, a practical problem of searching for a rational compromise arises—the determination of such quantitative and qualitative characteristics of information resources at which a certain rational balance of possible expenditures upon obtaining the corresponding information about factors of risk and possible damage from their influence is ensured.

It is natural that in the international norms of trade the conditions for each party concluding an agreement are specially stipulated for the case of the influence of force majeure factors.

It should be pointed out that the conditions determining a rational level of informedness in the form (4.68) may be directly considered as a multigoal optimization problem. In this case, there arises the necessity of disclosing the uncertainty of many goals. The methodology of a similar problem was considered above.

Now we turn to an analysis of conditions for attaining the limiting values of goals and risk levels. Let us define the interval limits. For goals such a limit should determine the values of the maximum possible success, and for a risk level, the minimum possible damage. For this case from relationships (4.65) and (4.66) it is possible to obtain the limits for the above indicated levels. Therefore, the level of goal attainment can be determined by the formulas

$$\begin{aligned} \bar{I}'_{12}\left(\bar{x}_{01}, \bar{\bar{x}}_{02}\right) &\to 1\,; \\ \bar{J}_{12k}\left(\bar{x}_{01}, \bar{\bar{x}}_{02}, \bar{\alpha}_1, \eta_k\right) &\to 0; \quad k = \overline{1, N_f'}\,. \end{aligned} \tag{4.72}$$

Note that coalition 1, for which the considered problem is being solved, could on its own control only $\bar{x}_{01}$. Hence, it is possible to determine the form and quality of this vector's components and to choose the numerical values of each component based on certain considerations. Under the conditions of the coalition counteraction, as was pointed out for the components of vector $\bar{\bar{x}}_{02}$, coalition 1 does not have accurate information about either the quantity of the vector components or their numerical values. Moreover, coalition 1 does not have accurate information about a goal vector function of coalition 2. Therefore, coalition 1 on its own forms an analytical view of the vector function of coalition 2's goals and chooses numerical values of all components of vector $\bar{\bar{x}}_{02}$. In these conditions the coalitions' striving simultaneously to maximize interests and minimize risk dictates as a rational solution a choice of values $\bar{x}^*_{01}$ and $\bar{\bar{x}}^*_{02}$ at which simultaneously the difference of a level of the achieved interests from potentially possible result and the difference of the inflicted damage from the minimum possible damage will be minimum. Thus, we arrive at the statement of the Chebyshev approximation problem.

For the system of equations

$$\bar{G}_{1k}\left(\bar{x}_{01}\right) = 0, \quad k = \overline{1, N_f^{'}}, \tag{4.73}$$

to find a value $\bar{x}^*_{01}$ at which

$$\Delta_G = \max_{k \in \overline{1, N_f^{'}}} \mu_k \left| \bar{G}_{1k}\left(\bar{x}^*_{01}\right) \right| , \tag{4.74}$$

taken for system (4.73) as a Chebyshev approximation measure, will be minimum

$$\Delta_G^0 = \Delta_G\left(\bar{x}^*_{01}\right) = \min_{\bar{x}_{01} \in D_{01}} \Delta_G\left(\bar{x}_{01}\right) . \tag{4.75}$$

In (4.74) it is accepted that values μ_k, $1 > \mu_k > 0$, and $k = \overline{1, N_f^{'}}$ are weighting coefficients that take into account the importance degree both of different types of damage and the goal vector function. They are coordinated by the relationship

$$\sum_{k=1}^{N_f^{'}} \mu_k = 1 .$$

In conformity with the strategy of risk minimization the functions $\bar{G}_{1k}\left(\bar{x}_{01}\right)$ are determined by the relationship

$$\bar{G}_{1k}\left(\bar{x}_{01}\right) = \max_{\bar{\bar{x}}_{02} \in \bar{D}_{02}} \bar{J}_{12k}\left(\bar{x}_{01}, \bar{\bar{x}}_{02}, \bar{\alpha}_1, \bar{\eta}_k\right); \quad k = \overline{1, N_f^{'}}; \tag{4.76}$$

$$\bar{G}_{1k}\left(\bar{x}_{01}\right) = \max_{\bar{\bar{x}}_{02} \in \bar{D}_{02}} \left| I_{12}^{'}\left(\bar{x}_{01}, \bar{\bar{x}}_{02}\right) - 1 \right| ;$$
$$k = N_{0f}, \quad N_{0f} = N_f^{'} + 1 . \tag{4.77}$$

In relationships (4.76) and (4.77) it is taken into account that under the coalitions' counteraction each of them chooses its own decision vector in such a way that the counteracting party has the minimum level of attainment of its interests and the maximum level of damage. According to (4.72) the minimum level of attaining interests is expressed by the maximum deviation from the potentially achievable level, which is equal to 1.

In formalizing conditions (4.72) in the form (4.73)–(4.75) it is taken into account that coalition 1 can optimize only its own solution $\bar{x}_{01}$, and vector

$\bar{\bar{x}}_{02}$ may be the type of vector for which the least favorable condition for coalition 1 is created, namely, the value of interest achievement level will be minimum on $\bar{\bar{x}}_{02}$, and the extent of damage will be maximum on $\bar{\bar{x}}_{02}$. And for these conditions values $\bar{\bar{x}}_{02}$ are selected in such a way as to ensure the best result, i.e., to provide a minimal deviation from the potentially achievable conditions determined by (4.72).

Note that problem (4.73)–(4.75) principally differs from the classical formulation of the Chebyshev approximation problem. The difference of the incompatible system of nonlinear equations is that the enclosed extreme problem (4.76) and (4.77) is within problem (4.73)–(4.75). Therefore, we turn to the analysis of problem (4.76)–(4.77). Problem (4.76)–(4.77) is a typical multigoal optimization problem and, like formula (4.72), may be presented as

$$\left| I_{12}^{'}\left(\bar{x}_{01}, \bar{\bar{x}}_{02}\right) - 1 \right| \to \max_{\bar{\bar{x}}_{02} \in \bar{D}_{02}} ; \quad \bar{J}_{12k}\left(\bar{x}_{01}, \bar{\bar{x}}_{02}, \bar{\alpha}_{1}, \eta_{k}\right) \to \max_{\bar{\bar{x}}_{02} \in \bar{D}_{02}} . \tag{4.78}$$

It should be noted that at first sight it seems that problem (4.78) is opposite in meaning to the problem of determining the best guaranteed result, as here the maximum of the corresponding functions is determined, and according to the guaranteed result principle some function minimum is defined. However, this is not so. Here as well as in the problem of determining the best guaranteed result, relationship (4.78) determines the worst conditions for coalition 1 that coalition 2 may create. The difference is conditioned by the physical meaning of functions included in (4.78). The problem of disclosing many goals' uncertainty is reduced to a problem of the form (4.78). The solving of these problems was considered earlier, and we will not consider it in detail here.

Thus, solving problem (4.72) is reduced to a sequence of problems:

1. Defining the functions $\bar{G}_{1k}\left(\bar{x}_{01}\right)$ in the form (4.76) and (4.77) on the basis of solving problem (4.78).
2. The solution to the Chebyshev approximation problem for system (4.73) based on criterion (4.74) subject to (4.76).

The solutions to the given problems may be presented in the form

$$\bar{\bar{x}}_{02} = \operatorname{Arg} \max_{\bar{\bar{x}}_{02} \in \bar{D}_{02}} \left\{ \bar{G}_{2k}\left(\bar{x}_{01}, \bar{\bar{x}}_{02}, \bar{\alpha}_{1}, \bar{\eta}_{k}\right), \quad k = \overline{1, N_{f}^{'}} \right\} ,$$

$$\bar{G}_{2k}\left(\bar{x}_{01}, \bar{\bar{x}}_{02}, \bar{\alpha}_{1}, \bar{\eta}_{k}\right) = \begin{cases} \bar{J}_{12k}\left(\bar{x}_{01}, \bar{\bar{x}}_{02}, \bar{\alpha}_{1}, \eta_{k}\right), & k = \overline{1, N_{f}^{'}} ; \\ \left| I_{12}^{'}\left(\bar{x}_{01}, \bar{\bar{x}}_{02}\right) - 1 \right|, & k = N_{0f}, \quad N_{0f} = N_{f}^{'} + 1 , \end{cases}$$

where

$$\bar{x}_{01}^{*} = \operatorname{Arg} \min_{\bar{\bar{x}}_{02} \in \bar{D}_{02}} \max_{k} \mu_{k} \left| \bar{G}_{2k}\left(\bar{x}_{01}, \bar{\bar{x}}_{02}, \bar{\alpha}_{1}, \bar{\eta}_{k}\right) \right| ,$$

$$\bar{G}_{1k}\left(\bar{x}_{01}\right) = \max_{\bar{\bar{x}}_{02} \in \bar{D}_{02}} \left\{ \bar{G}_{2k}\left(\bar{x}_{01}, \bar{\bar{x}}_{02}, \bar{\alpha}_{1}, \bar{\eta}_{k}\right) \right\} .$$

Hence, solving the initial problem (4.62) subject to (4.63) and (4.64) can be reduced to the Chebyshev approximation problem for system (4.68) based on criterion (4.69) and to a Chebyshev approximation problem for system (4.73) based on criterion (4.74) in which it is necessary to solve problem (4.78).

To sum up, note that in problem (4.67) it is assumed that the corresponding functions are determined only by the level of informedness. In real conditions of coalition counteraction the degree and level of the influence of risk factors depends not only on the informedness level, but also on the decisions taken by the coalitions. So, for the general case, for coalition 1 it is necessary to assume that the functions in (4.67) are dependent both on $\bar{U}$ and on $\bar{x}_{01}, \bar{\bar{x}}_{02}$. However, it is possible to show that this general case can be reduced to condition (4.67) on the basis of the approaches used in problem (4.72) to a sequence of problems: the problem of determining $\bar{G}_k(\bar{x}_{01})$ in the form (4.76) and (4.77) and the Chebyshev approximation problem for system (4.73) based on criterion (4.75) under condition (4.74).

Formalization and solving the problem for coalition 2 are carried out similarly.

Considering the problem of a two-coalition counteraction, we have left open the questions of making common coalition decisions and decisions of each coalition's individuals participants, i.e., we have not considered the coalitions interaction cases. It should be noted that the participants' interaction in the coalition in which their number is more than 2 differs in some features, properties, and capabilities from other interactions. Thus, the coalition that unites resources and capabilities in an integral object makes it possible, under certain conditions, to obtain a common result in goal attainment that is greater than the sum of results of participants acting in isolation. Here, the so-called superadditive law of a complex system comes into play. Physically, the action of this law may be explained by a wider possibility to maneuver forces and resources, a reduction in nonproductive expenses, and other factors. For example, before uniting in a coalition (for example, an industrial corporation) participants could have sales regions randomly distributed by distance from the producers. After uniting in a coalition, the sales regions could be redistributed in such a way as to possibly reduce the distances between all or most producers, thereby reducing the corporation's transport expenditures and increase profit. A more rational labor redistribution would also be possible by having some producers closer to the raw markets, and a number of other opportunities.

These factors lead to the emergence of global associations of producers in various spheres. In particular, it is characteristic of the information and computerization field.

At the same time, for the coalition to realize its advantages it must meet certain requirements, in particular, exchange information among its participants. But here there are a number of contradictions. For example, the exchange of information increases, on the one hand, the coalition's possibilities and, as a consequence, each participant's possibilities. But on the other hand, such an exchange may be undesirable to one or some participants if it leads to disclosing their trade secrets. Many peculiarities and disadvantages result from the difference in strategies of the coalition's participants, and, as a result, some participants may obtain more advantages, while others may enjoy fewer advantages, and still others may lose. That is why it makes sense for a participant to join a coalition under certain conditions: in particular, if he can improve his own results.

Let us consider an example of coalitions' counteraction taking into account risk factors.

Example. Solving the problem of the coalition counteraction. Let us consider a problem of competition between two corporations. We shall state and solve this problem as a choice of a rational strategy for success in conditions of electronic commerce given certain risk factors. The problem is stated and solved from the point of view of corporation 1 in conditions of a priori unknown responses of corporation 2.

Contensive problem statement. Assume that the products of corporations 1 and 2 are similar and belong to a class of high-technology products. The corporations operate in the market in the duopoly mode, i.e., they are not bound by any agreements on prices and types of products or on the division of market spheres of influence. It is known that corporation 2 is the basic producer of product 1, and it is corporation 1's main competitor; therefore, in the analysis of its position in the market the possibilities and actions of other competitors are not taken into account. The primary goal of corporation 1 is the attainment of long-term commercial success on the basis of launching a new product (henceforth product 2). Corporation 1's strategy is aimed at a quick launch of product 2 and a subsequent replacement of product 1. Corporation 2's strategy is aimed at attracting new customers as a result of improved service and advertising. The two corporations' actions and counteractions become clear at the production stage and at the launch of product 2 by corporation 1 and are adjusted in accordance to the initial response to their products.

Each corporation **must** define a rational strategy of production and actions they will take in the developing conditions of uncertainty and risk at the stage of production and launch of the new product 2 by corporation 1.

Conditions and results of the two corporations' actions. We take into account that for the purposes of attaining its basic goal corporation 1 must above all consider the properties, conditions, and factors that determine the profit outlook and the effectiveness of production and the launch of product 2. Therefore, we first carry out an analysis of the initial position of each corporation regarding production and the marketing of product 1 based on the following indicators.

The total volume of product 1 production $V_{\Sigma 1}$ is distributed in the proportion

$$V_{11} = \beta_{11} V_{\Sigma 1}; \quad V_{21} = \beta_{21} V_{\Sigma 1}; \quad \beta_{21} > \beta_{11}; \quad V_{\Sigma 1} = V_{11} + V_{21}, \tag{4.79}$$

where V_{11} and V_{21} are volumes of product 1 production by corporations 1 and 2, respectively. Henceforth the first index corresponds to the corporation number and the second index to the product number.

Consumer demand for product 1 of the corporations is characterized by two groups:

$$N_{11} = \gamma_{11} N_{\Sigma 1}; \quad N_{21} = \gamma_{21} N_{\Sigma 1}; \quad \gamma_{21} > \gamma_{11}; \quad N_{\Sigma 1} = N_{11} + N_{21}, \tag{4.80}$$

where N_{11} and N_{21} are indicators of consumer demand for product 1 of corporations 1 and 2, respectively, and $N_{\Sigma 1}$ is the total demand.

Product 1 of the two corporations is similar in function but differs in its quality, design, and cost:

$$C_{11} = \chi_{11} C_{21}; \quad \chi_{11} > 1, \tag{4.81}$$

where C_{11} and C_{21} are the cost of a unit of product 1 of corporations 1 and 2, respectively. Product of corporation 1 has higher quality indicators and a higher cost; however, it is inferior in its design.

It is known that the real gross income of the corporations is determined by the cost and sales volume of product 1 of each corporation:

$$B_{11} = C_{11} \gamma_{11} V_{\Sigma 1}; \quad B_{21} = C_{21} \gamma_{21} V_{\Sigma 1}, \tag{4.82}$$

and the predicted gross income of the corporations is estimated on the basis of the forecast of the cost and production volume of product 1:

$$\tilde{B}_{11} = C_{11} V_{11}; \quad \tilde{B}_{21} = C_{21} V_{21}. \tag{4.83}$$

This analysis shows that before product 2 of corporation 1 is launched on the market, the results of the two corporations' activities are characterized by the following indicators:

$$\beta_{11} = 0.27; \quad \beta_{21} = 0.73; \quad \gamma_{11} = 0.24; \quad \gamma_{21} = 0.76; \quad C_{11} = 1.08 C_{21}. \tag{4.84}$$

According to (4.79)–(4.83) for corporations 1 and 2 the functional dependences of their incomes from the sales of product 1 are determined as follows:

$$\begin{aligned}
\tilde{B}_{11} &= C_{11} V_{11} = 1.08 C_{21} \cdot 0.27 V_{\Sigma 1} = 0.2916 C_{21} \cdot V_{\Sigma 1}; \\
\tilde{B}_{21} &= C_{21} V_{21} = C_{21} \cdot 0.73 V_{\Sigma 1} = 0.73 C_{21} \cdot V_{\Sigma 1}; \\
\tilde{B}_{11} &= 0.399 \tilde{B}_{21}, \quad \tilde{B}_{21} = 2.503 \tilde{B}_{11}; \qquad (4.85) \\
B_{11} &= 1.08 C_{21} \cdot 0.24 V_{\Sigma 1} = 0.2592 C_{21} \cdot V_{\Sigma 1}; \\
B_{21} &= C_{21} \cdot 0.76 V_{\Sigma 1} = 0.76 C_{21} \cdot V_{\Sigma 1}; \\
B_{11} &= 0.341 B_{21}, \quad B_{21} = 2.932 B_{11}. \qquad (4.86)
\end{aligned}$$

From relationships (4.85) and (4.86) it follows that the incomes B_{21}, $\tilde{B}_{21}$ of corporation 2 from product 1 sales exceeds by almost three times the incomes B_{11}, $\tilde{B}_{11}$ of corporation 1, and relationships (4.79), (4.80), and (4.84) prove that the demand γ_{11} for product 1 of corporation 1 is almost 13% lower than the production volume β_{11}. These circumstances are the primary factor that determines an advantage of corporation 2's product 1 over corporation 1's product 1. From here, as a consequence, the necessity and expediency arise for corporation 1 to develop qualitatively new high-technology products.

To attain the set goal, corporation 1 solved the problem of creating a new product, product 2, which belonged to a class of high-technology products and which was supposed to be considerably superior to the prototypes available on the market. Therefore, corporation 1 had to organize rationally the following activities: advertising and popularization of a new product; determination of a reasonable volume of production determination and presentation of a reasonable volume of typical samples of production in the market. To arrange these activities corporation 1 had to determine the new product basic indicators upon introduction on the market and when initial demand growth patterns became clear. In conditions of uncertainty the basic indicators are formed as interval estimations. Corporation 1, when launching product 2, establishes the predicted interval estimations of the product for the following basic indicators:

- Production volume $V_{12} = \beta_{12} V_{\Sigma 12}$
- Demand volume $N_{12} = \gamma_{12} N_{\Sigma 12}$
- Market value $C_{12} = \chi_{12} C_{\Sigma 21}$
- Gross income $\tilde{B}_{12} = k_{12} C_{12} V_{\Sigma 12}$

Upon launching product 2, corporation 1 made the following interval estimations of indicators:

$$\begin{aligned} \beta_{12} &= [0.06; 0.12]\,; \quad \gamma_{12} = [0.05; 0.11]\,; \\ \chi_{12} &= [1.08; 1.12]\,; \quad k_{12} = [0.09; 0.15]\,. \end{aligned} \tag{4.87}$$

For the initial demand growth patterns for product 2 the following interval estimations of indicators were predicted:

$$\begin{aligned} \beta_{12} &= [0.12; 0.18]\,; \quad \gamma_{12} = [0.15; 0.21]\,; \\ \chi_{12} &= [1.10; 1.18]\,; \quad k_{12} = [0.15; 0.25]\,. \end{aligned} \tag{4.88}$$

Similar estimations were made by corporation 2 at the stage of initial demand growth for product 2; these will be presented as the solution to the problem is worked out.

On the basis of the accepted values of interval estimations of indicators (4.87) and (4.88), corporation 1 forms interval estimations for goal functions and the boundaries of the expected results. Similar estimations are formed by corporation 2. According to (4.22) the degree to which the corporations reached their goals is characterized by the following interval estimations:

$$I'_{12}(x_1, \tilde{x}_2) = \frac{F_{12}(x_1, \tilde{x}_2) - F_{12}^-}{F_{12}^+ - F_{12}^-},$$
$$I'_{21}(x_2, \tilde{x}_1) = \frac{F_{21}(x_2, \tilde{x}_1) - F_{21}^-}{F_{21}^+ - F_{21}^-}. \tag{4.89}$$

Here, F_{12}^+, F_{12}^- and F_{21}^+, F_{21}^- are the boundaries of the expected results for corporations 1 and 2, respectively, which are predicted by them independently in the given market conditions and without taking into account possible influences of risk factors. Vectors x_1 and $\tilde{x}_2$ of corporation 1 and vectors x_2 and $\tilde{x}_1$ of corporation 2 consist of the indicators characterizing the basic results and prospects of their practical actions. They are formed on the basis of relationships (4.79)–(4.83), where x_1 and x_2 define the interval boundaries of the predicted incomes for corporations 1 and 2 and $\tilde{x}_2$ and $\tilde{x}_1$ define the interval boundaries of the predicted influences of the competitors' actions on the incomes of corporations 1 and 2, respectively.

By analogy with (4.82) and (4.83) it is expedient to define all vector components of the predicted and real gross incomes of corporations 1 and 2 upon introducing of product 2 and initial demand growth for product 2, taking into account relationships (4.87) and (4.88) on the basis of relationships

$$\tilde{B}_{ij} = \tilde{k}_{ij}\tilde{x}_{ij} = \tilde{k}_{ij}C_0V_0, \quad \tilde{x}_{ij} \in [\tilde{x}_1; \tilde{x}_2];$$
$$B_{ij} = k_{ij}x_{ij} = k_{ij}C_0V_0, \quad x_{ij} \in [x_1; x_2].$$

Values C_0 and V_0 are known for each corporation. They are evaluated on the basis of the analysis of the average cost C_0 of a product unit and the predicted volume demand V_0 for products 1 and 2. Therefore, the interval estimations define only the boundary values of indicators k_{ij} for components of the specified vectors that are presented in Tables 4.1 and 4.2.

One more practically important characteristic of conditions of the two corporations' activities is an estimation of the impact of risk factors that exert a considerable influence on the effectiveness of practical activities. Among the most important factors are those of counteraction situations, information uncertainty, and force majeure factors. The analysis and definition of possible influences of the indicated risk factors will be carried out as the solution to the considered problem is worked out.

Mathematical statement of the problem. The numerical values of the main indicators of activities of corporations 1 and 2, before product 2 of corporation 1 is introduced, **are known**. These values are given in relationships (4.84), (4.85), and (4.86). Intervals of predicted indicators of gross incomes B_{11} and B_{12} for products 1 and 2 are established by corporation 1; also, a forecast of variation intervals of the indicators of gross incomes $\tilde{B}_{21}$

and $\tilde{B}_{22}$ of corporation 2 is made. This variation is caused by a partial reorientation of demand for corporation 1's product 2 and by a reduction in demand for corporation 2's product 1. The intervals of the predicted indicators of corporation 1's gross incomes are determined on the basis of Table 4.1. The intervals of the predicted indicators of gross income changes B_{21} and B_{22} are established by corporation 2 as a result of advertising and an improvement in product 1's quality. In addition, a forecast is made for intervals of possible change of gross income indicators $\tilde{B}_{11}$ and $\tilde{B}_{12}$ for corporation 1 due to a partial reorientation of demand for corporation 2's product 1 to corporation 1's product 2, and due to a reduction in demand for corporation 1's product 1. The intervals of the predicted indicators of gross incomes are determined on the basis of Table 4.2.

On the basis of expert estimations, possible risk situations and the basic limiting values of risk indicators are determined for each corporation. Each corporation establishes the functional dependences of mutual influence of its competitor's actions and its own actions based on the incomes from their practical activities in the established market conditions. The corporations set the initial and predicted values of the main indicators based on duopoly conditions, according to which the two corporations have no information on the intentions and actions of its competitor.

Table 4.1. k_{ij} values for corporation 1

k_{ij} for x_1				$\tilde{k}_{ij}$ for $\tilde{x}_2$			
k_{11}^-	k_{11}^+	k_{12}^-	k_{12}^+	$\tilde{k}_{21}^-$	$\tilde{k}_{21}^+$	$\tilde{k}_{22}^-$	$\tilde{k}_{22}^+$
0.2352	0.3472	0.0925	0.1330	0.6560	0.8510	0.6110	0.7123

Table 4.2. k_{ij} values for corporation 2

k_{ij} for x_2				$\tilde{k}_{ij}$ for $\tilde{x}_1$			
k_{21}^-	k_{21}^+	k_{22}^-	k_{22}^+	$\tilde{k}_{11}^-$	$\tilde{k}_{11}^+$	$\tilde{k}_{12}^-$	$\tilde{k}_{12}^+$
0.6745	0.8611	0.6324	0.7623	0.1712	0.2625	0.0678	0.0944

Each corporation plans its strategy and predicts its competitor's actions based on its own assumptions, forecasts, and estimations.

Each corporation **must**

1. Determine the rational functional dependences of the mutual influence of its competitor's actions and the two corporations' own actions based on the results of their activities in the established market conditions.

2. Find out and estimate conditions and risk factors that have the most and the least negative influence upon attainment of long-term commercial success by corporation 1.
3. Predict and estimate corporation 2's possible counteractions, which are factors of negative influence on the attainment of goals by corporation 1 in the conditions of risk-factor influence.

The solution. 1. Corporations 1 and 2 are working out the indicators of their activities and predict their competitor's actions. The results are obtained in the form of analytical dependences of goal functions $F_{12}(x_1, \tilde{x}_2)$ and $F_{22}(x_2, \tilde{x}_1)$ and the initial boundary values F_{12}^-, F_{12}^+ and F_{21}^-, F_{21}^+, which have the following form:

$$
\begin{aligned}
F_{12}(x_1, \tilde{x}_2) &= T_{11}k_{11}x_{11} + T_{12}k_{12}x_{12} + \tilde{T}_{21}\tilde{k}_{21}\tilde{x}_{21}^- - \tilde{T}_{22}\tilde{k}_{21}\tilde{x}_{22}^-\,, \qquad (4.90)\\
F_{12}^+ &= k_{11}^+x_{11} + k_{12}^+x_{12} + \tilde{k}_{21}^+\tilde{x}_{21} - \tilde{k}_{22}^+\tilde{x}_{22}\,;\\
F_{12}^- &= k_{11}^-x_{11} + k_{12}^-x_{12} + \tilde{k}_{21}^-\tilde{x}_{21} - \tilde{k}_{22}^-\tilde{x}_{22}\,,\\
F_{21}(x_2, \tilde{x}_1) &= T_{21}k_{21}x_{21} + T_{22}k_{22}x_{22} - \tilde{T}_{11}\tilde{k}_{11}^-\tilde{x}_{11} - \tilde{T}_{12}\tilde{k}_{12}^-\tilde{x}_{12}\,, \qquad (4.91)\\
F_{21}^+ &= k_{21}^+x_{21} + k_{22}^+x_{22} - \tilde{k}_{11}^+\tilde{x}_{11} - \tilde{k}_{12}^+\tilde{x}_{12}\,;\\
F_{21}^- &= k_{21}^-x_{21} + k_{22}^-x_{22} - \tilde{k}_{11}^-\tilde{x}_{11} - \tilde{k}_{12}^-\tilde{x}_{12}\,.
\end{aligned}
$$

Here T_{11} and T_{12} determine the duration of periods during which the gross incomes B_{11} and B_{12} are formed due to sales of products 1 and 2 by corporation 1; $\tilde{T}_{21}$ and $\tilde{T}_{22}$ predict the duration of periods during which emerge changes in the gross incomes B_{21} of the two corporations due to the reorientation of consumer demand from corporation 2's product 1 to corporation 1's product 2 and the new gross income B_{22} of corporation 2 due to reorientation of demand for corporation 1's product 1 to corporations 2's product 2.

For corporation 2 the indicators T_{21} and T_{22} determine the duration of the period during which the incomes B_{21} and B_{22} are established. Income B_{21} is formed due to the typical sales mode of product 1, and income B_{22} determines an increase in sales of product 1 due to the expansion of advertising. $\tilde{T}_{11}$ and $\tilde{T}_{12}$ determine, as predicted by corporation 2, the duration of the period of negative influence of corporation 1 due to reorientation of demand for corporation 2's product 1 of to corporation 1's product 2 and due to a drop in demand for corporation 2's product 1 because it does not meet the demands of consumers.

The goal function $F_{12}(x_1, \tilde{x}_2)$ according to (4.90) determines an aggregate planned gross income that includes the income $k_{11}x_{11}$ for the period T_{11} and the income $k_{12}x_{12}$ for the period T_{12}, and also the predicted gross income $\tilde{k}_{21}x_{21}$ for the period $\tilde{T}_{21}$ and the predicted loss $\tilde{k}_{21}\tilde{x}_{21}$ for the period $\tilde{T}_{22}$. The goal function $F_{21}(x_2, \tilde{x}_1)$ according to (4.91) determines an aggregate planned income of corporation 2 that includes the income $k_{21}x_{21}$ for the period T_{21} and $k_{22}x_{22}$ for the period T_{22}, and also the predicted losses $\tilde{k}_{11}\tilde{x}_{11}$ and $\tilde{k}_{12}\tilde{x}_{12}$ for periods $\tilde{T}_{11}$ and $\tilde{T}_{12}$, respectively.

For function $F_{12}(x_1, \tilde{x}_2)$ the numerical values of periods in months are determined by corporation 1 as follows: $T_{11} = 12$, $T_{12} = 5$; $\tilde{T}_{21} = 9$; $\tilde{T}_{22} = 3$, and the numerical values of indicators are: $k_{11} = 0.2912$; $k_{12} = 0.1129$; $\tilde{k}_{21} = 0.6571$; $\tilde{k}_{22} = 0.6117$. Similarly for the function $F_{21}(x_2, \tilde{x}_1)$ corporation 2 determines the numerical values of periods $T_{21} = 9$; $T_{22} = 5$; $\tilde{T}_{11} = 9$; $\tilde{T}_{12} = 1,5$ and the numerical values of indicators: $k_{21} = 0.7679$; $k_{22} = 0.11285$; $\tilde{k}_{11} = 0.21625$; $\tilde{k}_{12} = 0.1622$.

As a result, the basic indicators of goal functions and of boundary values are characterized by the following numerical values

$$\begin{aligned}
F_{12}(x_1, \tilde{x}_2) &= 3.4981x_{11} + 0.5642x_{12} + 5.914\tilde{x}_{21} - 1.835\tilde{x}_{22}\,,\\
F_{12}^{-} &= 0.2352x_{11} + 0.0925x_{12} + 0.656\tilde{x}_{21} - 0.611\tilde{x}_{22}\,,\\
F_{12}^{+} &= 0.3472x_{11} + 0.1332x_{12} + 0.8513\tilde{x}_{21} - 0.712\tilde{x}_{22}\,,\\
F_{21} &= 6.912x_{21} + 2.0924x_{22} - 1.9512\tilde{x}_{11} - 0.2436\tilde{x}_{12}\,,\\
F_{21}^{-} &= 0.6745x_{21} + 0.6324x_{22} - 0.1712\tilde{x}_{11} - 0.0678\tilde{x}_{12}\,,\\
F_{21}^{+} &= 0.8613x_{21} + 0.7623x_{22} - 0.2613\tilde{x}_{11} - 0.0944\tilde{x}_{12}\,.
\end{aligned}$$

2. On the basis of (4.89) and the obtained dependencies of goal functions $F_{12}(x_1, \tilde{x}_2)$ and $F_{21}(x_2, \tilde{x}_1)$ and the initial boundary values F_{12}^{-}, F_{12}^{+} and F_{21}^{-}, F_{21}^{+}, the interval estimations $I'_{12}(x_1, \tilde{x}_2)$ and $I'_{21}(x_2, \tilde{x}_1)$ of the incomes from practical activities of corporations 1 and 2 are determined. Also were obtained the functional dependences $I'_{12}(x_1, \tilde{x}_2)$ and $I'_{21}(x_2, \tilde{x}_1)$, which additionally account for the mutual influences of the corporations under the conditions of simultaneous market launch of the initial product 1 by corporations 1 and 2 and a new product 2 by corporation 1.

$$\begin{aligned}
I'_{12}(x_1, \tilde{x}_2) &= \frac{3.2629x_{11} + 0,4717x_{12} + 5.258\tilde{x}_{21} - 1.224\tilde{x}_{22}}{0.1121x_{11} + 0.0407x_{12} + 0.19533\tilde{x}_{21} - 0.1012\tilde{x}_{22}}\,;\\
x_1 &= \langle x_{11}, x_{12}\rangle\,, \quad \tilde{x}_2 = \langle \tilde{x}_{12}, \tilde{x}_{22}\rangle\,;\\
I'_{21}(x_2, \tilde{x}_1) &= \frac{6.2375x_{21} + 1,4612x_{22} - 1.7798\tilde{x}_{11} - 0.1758\tilde{x}_{12}}{0.1867x_{21} + 0.1299x_{22} - 0.0913\tilde{x}_{11} - 0.02661\tilde{x}_{12}}\,;\\
\tilde{x}_1 &= \langle \tilde{x}_{11}, \tilde{x}_{12}\rangle\,, \quad x_2 = \langle x_{12}, x_{22}\rangle\,.
\end{aligned}$$

$I'_{12}(x_1, \tilde{x}_2)$ and $I'_{21}(x_2, \tilde{x}_1)$ determine the relative difference of contribution of periods T_{11}, T_{12}, $\tilde{T}_{21}$, $\tilde{T}_{22}$ and T_{21}, T_{22}, $\tilde{T}_{11}$, $\tilde{T}_{12}$ to the aggregate gross income of corporations 1 and 2, respectively. The indicators x_{11} and x_{12} of corporation 1 and x_{21} and x_{22} of corporation 2 determine the gross income of the respective components of practical activities; the indicators $\tilde{x}_{21}$, $\tilde{x}_{22}$ and $\tilde{x}_{11}$, $\tilde{x}_{12}$ determine the level of counteraction of the competing corporation. In particular, the indicator $\tilde{x}_{11}$ characterizes, as predicted by corporation 2, the level of damage due to the market launch of product 2 by corporation 1, and the indicator $\tilde{x}_{21}$ characterizes the predicted growth in income of corporation 1 from sales of product 2 due to a reduction in demand for corporation 2's product 1.

3. On the basis of (4.57) the functional dependences of the mutual influence $F_{\Sigma 21}(x_2, \tilde{x}_1, \eta_{ns}, \eta_{in}, \eta_{fm})$ of actions of corporation 2 and corporation 1's own actions $F_{\Sigma 12}(x_1, \tilde{x}_2, \eta_{ns}, \eta_{in}, \eta_{fm})$ should be determined based on the results of corporation 1's activities in contingency situations under the impact of various groups of risk factors. To reveal the impact of risk factors of abnormal situations and to estimate their contribution to the aggregate gross income from the practical activities of corporations 1 and 2, respectively, matrices R_1, R_2 are formed in accordance with expert estimations:

$$R_1 = \left| \begin{array}{cccc} S_1 & S_2 & S_3 & S_4 \\ 0.11 & 0.01 & 0.3 & 0.4 \\ 0.23 & 0.05 & 0.1 & 0.3 \\ 0.32 & 0.15 & 0.15 & 0.25 \\ \tilde{x}_2^1 & \tilde{x}_2^2 & \tilde{x}_2^3 & \tilde{x}_2^4 \end{array} \right| \begin{array}{c} \eta_{ns} \\ \eta_{fm} \\ \eta_{in} \end{array},$$

$$R_2 = \left| \begin{array}{ccccc} S_1 & S_2 & S_3 & S_4 & S_5 \\ 0.15 & 0.21 & 0.16 & 0.25 & 0.32 \\ 0.17 & 0.22 & 0.08 & 0.35 & 0.38 \\ 0.27 & 0.35 & 0.11 & 0.30 & 0.42 \\ \tilde{x}_1^1 & \tilde{x}_1^2 & \tilde{x}_1^3 & \tilde{x}_1^4 & \tilde{x}_1^5 \end{array} \right| \begin{array}{c} \eta_{ns} \\ \eta_{fm} \\ \eta_{in} \end{array}.$$

Values $J_{12ns}; \ldots; J_{21in}$ of the risk level are given in the form:

$$\begin{aligned} J_{12ns}(x_1; \tilde{x}_2) &= \alpha'_{12} I_{12}(x_1; \tilde{x}_2); \quad \alpha'_{12} \in [0.05; 0.15], \\ J_{12fm}(x_1; \tilde{x}_2) &= \alpha''_{12} I_{12}(x_1; \tilde{x}_2); \quad \alpha''_{12} \in [0.05; 0.25], \\ J_{12in}(x_1; \tilde{x}_2) &= \alpha'''_{12} I_{12}(x_1; \tilde{x}_2); \quad \alpha'''_{12} \in [0.05; 0.35]; \end{aligned}$$

$$\begin{aligned} J_{21ns}(\tilde{x}_1; x_2) &= \alpha'_{21} I_{21}(\tilde{x}_1; x_2); \quad \alpha'_{21} \in [0.05; 0.15], \\ J_{21fm}(\tilde{x}_1; x_2) &= \alpha''_{21} I_{21}(\tilde{x}_1; x_2); \quad \alpha''_{21} \in [0.05; 0.25], \\ J_{21in}(\tilde{x}_1; x_2) &= \alpha'''_{21} I_{21}(\tilde{x}_1; x_2); \quad \alpha'''_{21} \in [0.05; 0.3]; \end{aligned}$$

$$\begin{aligned} \tilde{x}_1 &= \langle \tilde{x}_{11}; \tilde{x}_{12} \rangle, \quad \tilde{x}_{11} = \langle \tilde{x}_{11}^-; \tilde{x}_{11}^+ \rangle, \quad \tilde{x}_{12} = \langle \tilde{x}_{12}^-; \tilde{x}_{12}^+ \rangle; \\ \tilde{x}_2 &= \langle \tilde{x}_{21}; \tilde{x}_{22} \rangle, \quad \tilde{x}_{22} = \langle \tilde{x}_{22}^-; \tilde{x}_{22}^+ \rangle, \quad \tilde{x}_{21} = \langle \tilde{x}_{21}^-; \tilde{x}_{21}^+ \rangle. \end{aligned}$$

Calculations are made for the following variants of the initial data:

$$\begin{aligned} x_2^1 &= \langle \tilde{x}_{21} \in [10; 25]; \tilde{x}_{22} \in [5; 10] \rangle, \\ x_2^2 &= \langle \tilde{x}_{21} \in [35; 65]; \tilde{x}_{22} \in [10; 25] \rangle, \\ x_2^3 &= \langle \tilde{x}_{21} \in [45; 75]; \tilde{x}_{22} \in [15; 25] \rangle, \\ x_2^4 &= \langle \tilde{x}_{21} \in [60; 95]; \tilde{x}_{22} \in [25; 30] \rangle; \end{aligned}$$

$$\tilde{x}_1^1 = \langle \tilde{x}_{11} \in [5;10] \, ; \tilde{x}_{12} \in [12;25] \rangle \; ,$$
$$\tilde{x}_1^2 = \langle \tilde{x}_{11} \in [12;24] \, ; \tilde{x}_{12} \in [17;35] \rangle \; ,$$
$$x_1^3 = \langle \tilde{x}_{11} \in [15;30] \, ; \tilde{x}_{12} \in [37;52] \rangle \; ,$$
$$x_1^4 = \langle \tilde{x}_{11} \in [25;40] \, ; \tilde{x}_{12} \in [57;75] \rangle \; ,$$
$$x_1^5 = \langle \tilde{x}_{11} \in [40;60] \, ; \tilde{x}_{12} \in [55;85] \rangle \; .$$

4. Some results of calculating the mutual influence of one's competitor's actions and the two corporations' own actions based on the results of their activities for situation S_1 are shown in graphic form in Figs. 4.1 and 4.2, on the basis of functional dependences $I'_{12}(x_1, \tilde{x}_2)$ and $I'_{21}(x_2, \tilde{x}_1)$ for normal conditions and $F_{\Sigma 12}(x_1, \tilde{x}_2, \eta_{ns}, \eta_{in}, \eta_{fm})$ and $F_{\Sigma 21}(x_2, \tilde{x}_1, \eta_{ns}, \eta_{in}, \eta_{fm})$, taking into account risk factors.

The results given in Fig. 4.1, based on functional dependences $I'_{12}(x_1, \tilde{x}_2)$ and $I'_{21}(x_2, \tilde{x}_1)$, show that the values of $I'_{12}(x_1, \tilde{x}_2)$ exceed the values of $I'_{21}(x_2, \tilde{x}_1)$ under counteraction conditions, without taking into account risk factors. The results given in Fig. 4.2, based on the functional dependences $F_{\Sigma 12}(x_1, \tilde{x}_2, \eta_{ns}, \eta_{in}, \eta_{fm})$ and $F_{\Sigma 21}(x_2, \tilde{x}_1, \eta_{ns}, \eta_{in}, \eta_{fm})$, show that the values of $F_{\Sigma 12}(x_1, \tilde{x}_2, \eta_{ns}, \eta_{in}, \eta_{fm})$ essentially exceed the values $F_{\Sigma 21}(x_2, \tilde{x}_1, \eta_{ns}, \eta_{in}, \eta_{fm})$ under counteraction conditions, considering risk factors. This fact proves the expediency of the decision taken by corporation 1 to shift to the stage of initial demand growth for product 2.

The results of estimating the risk situation's impact on the aggregate gross income of corporations 1 and 2 prove that the decisions taken by corporation 1 and actions aimed at developing innovative products turned out to be efficient and made it possibe to gain an advantage over corporation 2 not only in terms of product quality but also in total market demand.

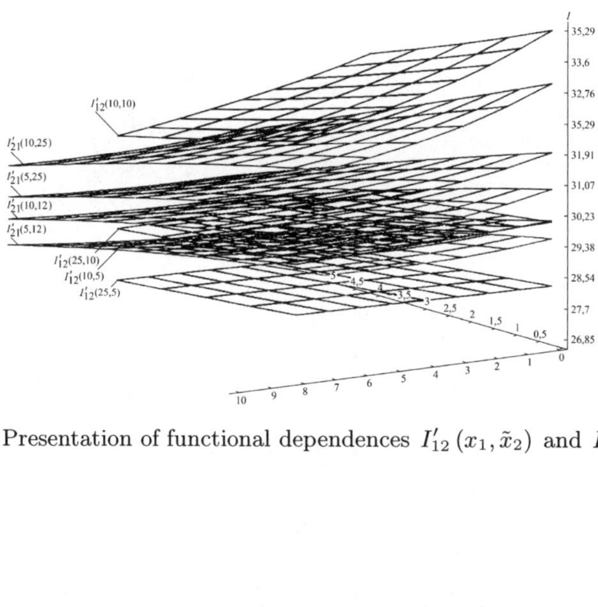

Fig. 4.1. Presentation of functional dependences $I'_{12}(x_1, \tilde{x}_2)$ and $I'_{21}(x_2, \tilde{x}_1)$

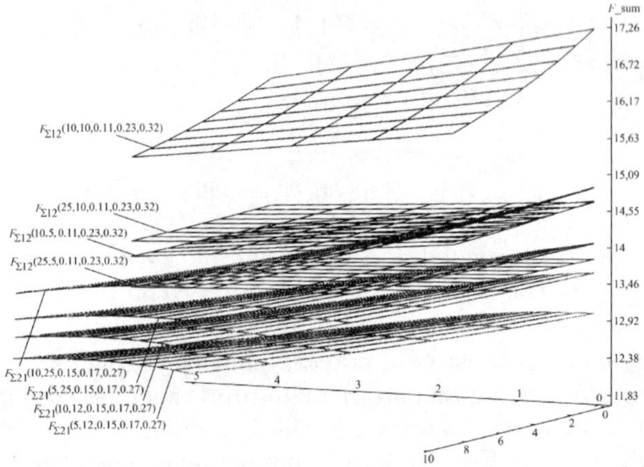

Fig. 4.2. Presentation of functional dependences $F_{\Sigma 12}$ and $F_{\Sigma 21}$

5 Information Analysis of System Problems

The qualitative characteristics of information that is put to use are of great importance in formalizing and solving real problems of system analysis.

Different disciplines, such as informatics, information theory, radiolocation theory, identification theory, and others, study information comprehensively. Characteristics of information quantity evaluation have been introduced. Shannon's [20, 172], Kolmogorov's [59–62], and Kulbak's [72] approaches are well known. Other characteristics and criteria of information property evaluation are also available [41, 58, 60–63, 163, 177]. The main feature of these approaches is that they are used for evaluating quantitative characteristics only. But such characteristics of information as completeness, credibility, and others are very rarely evaluated separately. At the same time, validity, credibility, and efficiency of applied system analysis problem solving depends directly not only on quantitative, but also on qualitative characteristics of information [89]. From the above it follows that there appear fundamentally new requirements for the procedure of information analysis, in general, and evaluation of the initial information.

Let us state the main goals and objectives of information analysis as one of important tools of formalizing and solving system problems. The goals consist in ensuring a practically necessary and technologically possible level of information credibility and validity of applied system problem solving. The objective is to create methodological and mathematical tools for achieving these goals.

In this chapter we shall try to formulate a unified approach from the position of decision makers to system analysis of quantitative and qualitative characteristics of information and evaluation of their influence on completeness, credibility, and timeliness of system problem solving.

5.1 Analysis of Quantitative and Qualitative Characteristics of Information

The methodological complexity of forming such an important field of knowledge as information analysis is prompted by the numerous interpretations of the notion of "information", on the one hand, and by the absence of generally accepted definition, on the other.

Well-known definitions of information describe only separate sides of this notion, which is in fact many-sided, as to its properties and areas of application. These notions are characterized by an ambiguity of interpretation in their practical applications; thus, they provide no means of obtaining accurate estimates of information quantity [61]. There are definitions that describe information as a fundamental notion because of its specific role and negation of its similarity to other fundamental notions, such as matter or energy. In particular, Wiener in his fundamental work in cybernetics postulates: "Information is information, neither matter nor energy". The meaning of this interpretation is explained by him in the following definition: "Information is designation of content, which we draw from the outer world during processes of accommodation to it and bringing our thinking to conformity with it" [189].

Regarding the term "information" we must point out that in information theory there is a mathematically justified definition of quantity of information, but to date, there is no unique definition of information. We think it necessary to introduce the following definition of information. **"Information" is an ordered sequence of contentwise mutually coordinated and structurally interconnected words, pictures, diagrams, tables, and/or other means of written, oral, visual, or technical presentation of states, actions, arrangements, and other properties and/or processes of any object under investigation**.

Let us analyze briefly the present data pertaining to information theory and find out their practical potentialities for forming mathematical tools of information analysis and conditions of complex system functioning.

Shannon's works [170, 171] used to be considered the first in the development of information theory, though they were oriented toward research of communication systems as a concrete field of activities. Such an approach is conditioned by the fact that in those works presented the first definition of information quantity. The definition is given by Shannon on the basis of the notion of information entropy and is described by the following formula:

$$H = -\sum_{i=1}^{n} p_i \log p_i \,. \tag{5.1}$$

The formula determines the entropy of a complete group of random events or random conditions. It accounts for the fact that entropy by its content is an inverse value to the quantity of information. H is a measure of uncertainty of a set that consists of n random events with probabilities $p_1, \ldots, p_n$. It follows from (5.1) that $H = 0$ on the condition that out of a whole set of events only one will take place, and a situation where two or more events take place simultaneously is categorically impossible. This condition is fulfilled in successive letter-by-letter message transmissions.

Expansion of information theory capabilities and its area of application and its further emergence as an independent science are closely connected with the development of cybernetics, the foundation of which was laid by the fundamental work of Wiener [189]. The qualitatively new problems posed by cybernetics made new demands on the formation of initial information. "Logical independence of information from any probabilistic assumptions" [62] became an especially essential requirement. It generated an important practical task of developing new principles and approaches to forming and evaluating information quantity. A degree of strictness in the research of information properties and features used in the first works of Shannon was found to be insufficient for solving such problems [62]. The necessity arose to reach a qualitatively new level of research, namely, a level of strict mathematical justification of the basic properties of information. This was accomplished in [46–48, 61, 62, 85]. Among the main results of the above-mentioned works, special significance is attached to the ideas on the determination of the quantity of information suggested by Kolmogorov on the basis of which combinatorial and algorithmic approaches were developed. These approaches created a number of new potentialities in information theory. First, they ensured the logical independence of information properties from probabilistic assumptions. Second, there appeared the possibility of determining the quantity of information, which is a characteristic of an individual object realized in the form of a separate word or a continuous message in the form of a diagram. A great contribution made by Kolmogorov, his disciples, and coauthors was that they brought information theory to the level of a strict mathematical discipline, laid the basis for a new scientific direction—algorithmic information theory—and considerably extended its area of application [62].

At the same time, even after the appearance of a strict definition of information quantity and development of the discipline known as "information theory", some of its parts are insufficiently studied.

First is the notion of "quantity of information", but there is no sufficiently strict notion of "information". Second, the introduced notions have some specific features and restrictions. Thus the quantity of information can be defined by Shannon's formula only on the condition that an object under investigation is an element of a definite discrete set with an a priori given distribution of probabilities. Such a variant is realized in a communication system: the transmission of messages consisting of separate signs, for example, letters. In the case of switching to continuous messages, such an evaluation becomes impossible: all natural analogs of entropy become equal to infinity [60, 62]. The combinatorial and algorithmic approaches of Kolmogorov exclude the probability restriction of Shannon and considerably extend the range of objects for which the quantity of information may be determined. The quantity of information of an object under investigation, for example, a book being published, may be determined on the basis of its comparison with another object, in particular, a book already published. We must mention a common feature of Shannon's and Kolmogorov's approaches: none of them takes into account the heterogeneous nature of information, which is typical for many system problems.

At the same time, in practice, the necessity of evaluating the quantity of information often arises under conditions by which the aforesaid restrictions and assumptions do not work. In particular, in evaluating risk factors it is impossible to determine the composition of a abnormal situation set and the probability of each event.

Seemingly, the rapid development and wide application of personal computers and highly developed information systems brought the problem close to its solution—the quantity of information is easily determined by the number of binary digits in bytes. But even in this case, a problem cannot be solved unambiguously—in different text editors the volume of information becomes different for the same text. In addition, the same message can be transmitted by different words with similar meanings (synonyms), which have different computer word lengths, and phrases may differ in style. It should be noted that works on information theory can be divided into three groups [89]:

- Those works in which quantitative properties of information are qualitatively characterized: for example, information may be maximal, minimal, average, and so on, and in these cases in essence the notion of information is of no interest.
- Those works in which the notion of information is used without its contensive explanation, for example, "transmission of information", "signal-based information transmission", "information confined in a symbol set", and so on, as if we were dealing with a notion well known to everyone.
- Those works in which attempts are made to explain the meaning of the notion of information by quite general interpretations, for example, using such terms as "knowledge", "content", and so on. Some such definitions are given above.

It is necessary to point out that by using well-known approaches to defining the quantity of information, we can solve problems from the positions of analysis of general properties of an object, but the position of a decision maker is not taken into consideration. **That is why the necessity arises of defining the quantity and quality of information, not just for describing properties and features of an object being studied but also from the position of forming and reaching goals set by a decision maker on the basis of his vision of necessary properties and features of an object and ways and means for their realization.**

For that a decision maker should have a certain level of informedness about the object. With such an approach to information analysis some questions arise:

- What is understood as a decision maker's informedness?

- Which indicators are expedient to be used for evaluating a decision maker's level of informedness?
- How can we quantitatively evaluate a decision maker's level of informedness about certain possible situations, in particular, the informedness about risk factors?

By analogy with the definition of information quantity, let us introduce the following definitions:

A **decision maker's informedness** is understood as a change in the level of knowledge uncertainty with regard to a situation or an object being analyzed, as a result of getting information.

The **level of a decision maker's informedness** is an indicator of the level of knowledge about an object being analyzed or investigated. To characterize quantitatively the level of a decision maker's informedness we use the value of changing the level of knowledge uncertainty as a result of getting information.

It should be mentioned, that when information is obtained, the level of uncertainty of a situation may decrease if the information is accurate, but it also may increase if the information is deliberately garbled or noncredible (i.e., not confirmed by experiments, documents, calculations or in any other way).

In addition, **the uncertainty of a situation can be evaluated proceeding from the goals of system analysis**, in particular, by evaluating the degree and level of risk. For each concrete situation uncertainty is characterized by the following types of knowledge uncertainty:

- Uncertainty connected with the possibility of a given situation.
- Uncertainty characterized by the quality of the available information and information being received.
- Uncertainty connected with the degree of a situation's influence on the risk level.

From the above it follows that an increase in the level of a decision maker's informedness does not always result in disclosing the uncertainty of a given situation, as was assumed while forming relationship (5.1).

Therefore, it is expedient to determine the level of a decision maker's informedness, taking into consideration all the indicated factors. First, let us determine the inverse value—the level of uninformedness.

Let us analyze some methods and the essence of disclosing **the uncertainty of a situation's occurrence**. Let us assume that the level of uninformedness is the uncertainty of knowledge about the occurrence of an alternative outside the predicted set of situations. The value of knowledge uncertainty can be evaluated by different methods. Let the set of possible situations M_s be discrete and let each element S_i from M_s be characterized by a definite probability ρ_i for $i = \overline{1, m_s}$. Then, the value of uninformedness

H_s will be defined as the level of knowledge uncertainty about M_s. Therefore, we have a condition similar to that for formula (5.1). That is why the condition of uncertainty can be determined by the entropy

$$H_s = -\sum_{i=1}^{m_s} \rho_i \log \rho_i \,. \tag{5.2}$$

Note that for equiprobable events $\rho_i = 1/m_s$ and $H_s = \log m_s$.

Now, let us analyze the **second factor**, namely, the influence of information quality on the level of a decision maker's informedness. Note that the evaluation of information quality is the least explored element in both informatics and in other disciplines connected in some way with information—optimal control theory, decision-making theory, and others.

At present there is no accepted system of indicators used for evaluating information qualitative characteristics. That is why it is not expedient to dwell for too long on the analysis of different approaches to formalizing qualitative characteristics of information, since they are inapplicable for solving the majority of practical system analysis problems.

Let us note only the most essential qualitative properties of information, which are fundamentally important in solving problems of system analysis, in particular, problems of estimation of the degree and level of risk in normal, abnormal, and critical situations. Among these properties are uncertainty, inaccuracy, incompleteness, fuzziness, untimeliness, noncredibility, and contradictoriness.

Uncertainty: a property that indicates the presence of several alternative descriptions of a situation.

Inaccuracy: a property that indicates the presence of a certain interval of tolerances or an error in measurements or calculation of quantitative parameters and (or) qualitative characteristics of a situation description.

Incompleteness: a property that indicates the presence of information gaps in the description of a situation (something is missing, something is insufficiently described, and so on).

Fuzziness: a property that characterizes the indistinctness of a situation description when the presence or absence of a certain property or its exact quantitative characteristic cannot be precisely indicated (for example, such notions as good weather and favorable conditions cannot be quantitatively described accurately—they are subjective and indistinct).

Untimeliness: a property that characterizes the time correlation between the moment when an event occurs and the moment when information about this event is obtained. If a decision maker does not have enough time to form and make a decision based on the information obtained, this information is untimely.

Incredibility: a property that indicates the presence of quantitative data or qualitative characteristics that do not correspond to the actual situation.

Contradictoriness: a property that indicates the presence of quantitative or qualitative characteristics whose meaning or essence contradicts other data.

The above definitions are to be taken into consideration when establishing the indicators of a decision maker's informedness.

5.2 Formalization of Characteristics and Indicators of a Decision Maker's Informedness

On considering the qualitative properties of information from the system analysis position, let us proceed to the formalization of qualitative and quantitative indicators of a decision maker's informedness [126, 136].

Obviously, it is important for a decision maker that the received information possess to a minimum extent the above presented properties, such as incompleteness, incredibility, untimeliness, and others. That is why the initial notions in defining a decision maker's informedness are the properties of information opposite to those stated above.

From a decision maker's position the most important properties are completeness, credibility, and timeliness of a decision maker's informedness.

Completeness of informedness: a property that characterizes the conformity of the quantity of information received by a decision maker to the quantity of information required for decision making.

Timeliness of informedness: a property that determines the conformity of a decision maker's time resource for forming and making decisions to the time margin from the moment of receiving information to the moment of the solution realization.

Credibility of informedness: a property that characterizes the conformity of the information received by a decision maker to the actual state of the situation.

Qualitatively we will characterize completeness of informedness by an indicator of the informedness completeness I_C:

$$I_C = \frac{C - C^-}{C^+ - C^-}, \qquad (5.3)$$

where C^+ and C^- are respectively maximum appropriate and minimum permissible quantities of information necessary for decision making in definite conditions and C is the quantity of information received by a decision maker in the current situation.

The value I_C determines the level of informedness completeness from the point of view of how the relative quantity of the received information exceeds the minimum permissible amount of information for decision making, i.e., this indicator qualitatively characterizes the level of a decision maker's informedness completeness proceeding from the relative minimum permissible quantity of informedness C^-. It should be noted that $I_C = 0$, with

$C = C^-$, where $C^- > 0$. It follows that the computing origin of the informedness completeness level is assumed to be the value corresponding to a certain minimum-permissible body of information in real conditions. Here, the notion of real conditions means conditions of interaction (or counteraction) of partners in possible or predicted sets of situations. The value C characterizes the body of information for a concrete situation that arises in the process of interaction (or counteraction) of a situation from the predicted set of situations. At the same time, each kth concrete situation has its corresponding value C_k. And the values C^- and C^+ are common for all situations from the investigated set. It is significant that the value C^+ does not coincide with the value C_M^+, which determines the ultimate complete body of information about the set of situations.

We must mention here the notion of **maximum appropriate body of information** for decision making. At first sight it seems that increasing the body of information always leads to increasing the solution justification. But in practice such an opinion often does not come true because the solution justification increases upon obtaining not just any kind of information but only information that is useful from the position of decision making. At the same time, certain quantitative restrictions on the maximum appropriate body of information for decision making exist because with an increasing volume of information the time required for information processing also increases. It may turn out that increasing the volume of information above some level may cause unacceptable increasing of the time required for its processing. This will also be true if the information is useful for decision making. In fact, if the information duplicates or just slightly supplements the available data, then it should be considered redundant. But, along with information useful for decision making, there may be information that contains data that are unnecessary for decision making in the current situation.

Information that does not contain data useful for decision making is "parasitic". In addition, in a coalition, the information of each participant may be deliberately distorted. Information where data are deliberately distorted is called misinformation.

A comment should be made regarding the value C^-. Here, the value C^- is understood as the lower boundary of information in the sense that a decision can be made on the condition $C \geq C^- + \varepsilon$, where ε is a sufficiently small value since when $C = C^-$, $I_C = 0$.

Schematically, the relationship between the values that characterize the notion of informedness completeness I_C can be presented in a structure chart (Fig. 5.1).

In Fig. 5.1 the following symbols are used:

C_m^-: level of absolute absence of information about factors of a given set of situations;

C^-: level of minimum permissible body of information for decision making in certain conditions;

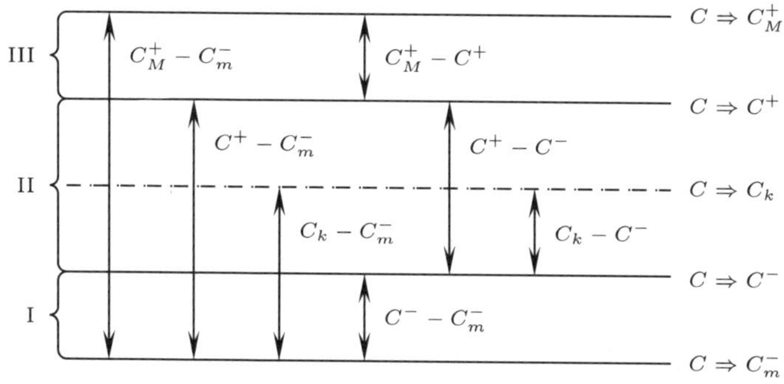

Fig. 5.1. Relationship between the values characterizing informedness completeness I_C

C^+: level of maximum appropriate body of information for decision making in certain conditions;

C_M^+: level of maximum complete information about factors of a given set of situations;

C_k: level that corresponds to the kth situation from the given set.

Here, the following regions also are introduced:

Region I corresponds to a level of information in the interval $[C_m^-, C^-]$ that defines a region of insufficient informedness completeness.

Region II corresponds to a level of information in the interval $[C^-, C^+]$ that corresponds to a region of rational completeness of a decision maker's informedness.

Region III corresponds to a level of information in the interval $\left[C^+, C_M^+\right]$ that corresponds to a region of the superfluous completeness of a decision maker's informedness.

In general, the level of completeness of a decision maker's informedness depends on which of these three regions a level of information C_k (for the kth concrete analyzed situation) corresponds to and the probability of each situation's "hitting" the given set of situations.

The value $(C^- - C_m^-)$ is determined by the level of information deficiency for decision making, and the value $\left(C_M^+ - C^+\right)$ characterizes the level of redundancy of information for decision making. As is evident from Fig. 5.1, an indicator of informedness completeness I_C should be used for characterizing region II. Therefore, we shall introduce additional indicators for regions I and III:

– An indicator of redundant informedness completeness I_C^+:

$$I_C^+ = \frac{C_M^+ - C^+}{C^+ - C^-}\,.$$

– An indicator of deficiency informedness I_C^-:

$$I_C^- = \frac{C^- - C_m^-}{C^+ - C^-}\,.$$

It is evident that $I_C^- = C^- / \left(C^+ - C^-\right)$ when $C_m^- = 0$.

It should be noted that in the presented definitions the word "completeness" cannot be omitted, and the terms "indicator of informedness redundancy" and "indicator of deficiency informedness" cannot be used because the level of redundant and deficiency informedness depends not only on the level of informedness completeness but also on other properties—credibility, timeliness, and so on.

Let us now analyze the property of informedness timeliness I_T. This property follows from the notion of informedness' untimeliness, which is in opposition to it. Let us analyze this property using the structure chart in Fig. 5.2:

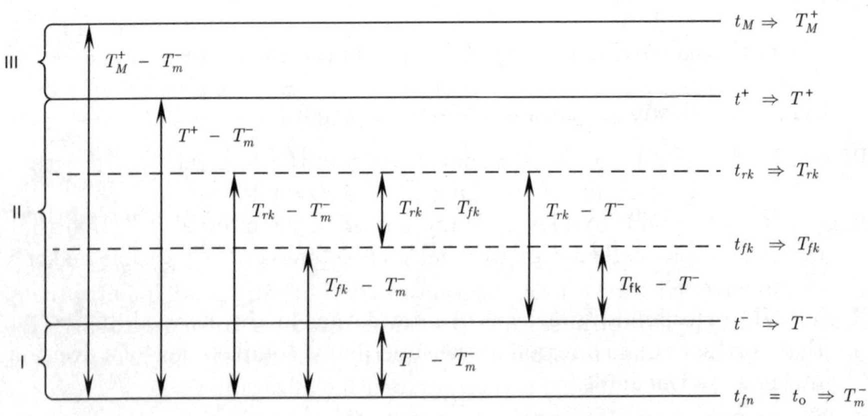

Fig. 5.2. Relationship between values characterizing the notion of informedness timeliness I_T

In this diagram the following symbols are used:

t_0: the moment when information is received;
t^-: the moment when the process of generating a decision has ended on the condition that the duration of this procedure is T^-;
t^+: the moment when the process of generating a decision has ended on the condition that the duration of this procedure is T^+;
t_{rk}: the moment a decision is realized for the kth situation;
t_{fn}: the moment a decision for the kth situation is just starting to be generated;
t_{fk}: the moment a decision for the kth situation has been generated;

T^-: the minimum possible length of time for generating a decision for a given set of situations;

T^+: the maximum possible length of time for generating a decision for a given set of situations;

T_{rk}: the length of time beginning from when a decision is generated to when it is realized for a definite kth situation;

T_M^+: the maximum possible length of time from when a decision is generated to when it is realized, for a given set of situations i.e., T_M^+ is the maximum of T_{rk} on the set of situations;

T_m^-: the minimum possible length of time from when a decision is generated to when it is realized, for a given set of situations, i.e., T_m^- is the minimum of T_{rk} on the set of situations;

T_{fk}: length of time from when a decision for kth situation is just starting to be generated to when it has been formed.

It is assumed that the moment a decision begins to be generated coincides with the moment information $t_{fn} = t_0$ is received.

Let us point out three regions:

Region I corresponds to $t \in [T_m^-, T^-]$ and is the region of untimeliness of a decision maker's informedness because $t_{rk} < T^-$ and, hence, of a decision is impossible to generate even for the minimum possible time.

Region II corresponds to $t \in [T^-, T^+]$ and is a region of risky timeliness, in the sense that there is some risk that the information may be untimely.

Region III corresponds to $t \in [T^+, T_M^+]$ and is a region of guaranteed timeliness because for all t_{rk} the condition $t_{rk} > T^+$ is true.

Let us proceed to the formalization of the indicator that quantitatively characterizes the **timeliness of a decision maker's informedness**. This indicator presents the correlation between the time margin for decision generation and the amount of time margin for its realization.

The **time margin for decision generation** for the kth situation from a given set of situations is the period of time from when information t_0 is received to when the final solution t_{fk} for the kth situation is generated.

The **time margin for decision realization** for the kth situation from a given set of situations is the period of time from the moment the information t_0 is received to the moment the decision is realized t_{rk} for the kth situation.

The quantitative values of the above indicated margines shall be defined in relative values by normalizing them with respect to the maximum interval allowed for working out a solution. First, let us obtain the relationship for region II. In this case the time margin for generating decision for the kth situation can be presented as:

$$R_{fk} = \frac{T_{fk} - T^-}{T^+ - T^-}, \tag{5.4}$$

where $T_{fk} = t_{fk} - t_{fn} = t_{fk} - t_0$.

Similarly, the time margin for decision realization for the kth situation is characterized by the relationship:

$$R_{rk} = \frac{T^+ - T_{rk}}{T^+ - T^-}, \tag{5.5}$$

where $T_{rk} = t_{rk} - t_0$.

Note in particular one feature of normalization used in relationships (5.4) and (5.5). In (5.4) we consider that time spent on decision generation should be kept to a minimum, i.e., it is expedient that the margin for decision generation must be as small as possible. The minimum possible time for decision generation in region II is equal to T^-; hence, the numerator in formula (5.4) shows how much T_{fk} differs from the minimum possible value of t in the interval $[T^-, T^+]$ and determines the difference between the time spent on decision generation for the kth situation and the minimum possible value of t.

Normalization in relationship (5.5) is performed based on other considerations. For a decision maker it is advisable that the moment of decision realization come as late as possible, because in this case much more time for generating decision will be available. That is why it is advisable that T_{rk} be as close as possible to the maximum allowed value of t in the considered interval $[T^-, T^+]$. The numerator in (5.5) determines the absolute difference between T_{rk} and T^+, and in general R_{rk} determines the difference between T^+ and T_{rk}.

Formalization of an informedness timeliness indicator. This indicator can be presented based on the following physical preconditions: informedness is timely if the moment of decision realization t_{rk} does not come before the decision has been generated, i.e., no earlier than the moment t_{fk}. Then, this indicator can be presented through the duration of the corresponding periods:

$$I'_T = \frac{T_{rk} - T_{fk}}{T^+ - T^-}. \tag{5.6}$$

Let us present I'_T through the introduced time margines R_{fk} and R_{rk} using the following transformations:

$$I'_T = \frac{T_{rk} - T^+ + T^+ - T^- + T^- - T_{fk}}{T^+ - T^-}.$$

It follows that $I'_T = 1 - R_{fk} - R_{rk}$.

It is easy to notice that I'_T changes in the interval $[-1, 1]$. In fact, when $T_{fk} = T^+$, $R_{fk} = 1$, $T_{rk} = T^-$, and $R_{rk} = 1$, then $I'_T = -1$. When $T_{fk} = T^-$, $R_{fk} = 0$, $T_{rk} = T^+$, and $R_{rk} = 0$, then $I'_T = 1$.

The given interval is not convenient for characterizing the informedness timeliness because the insufficient level of informedness is characterized by the negative value. It is expedient that this indicator should change in the interval $[0,1]$. By substituting the variables $I'_T = 2I_T - 1$ we pass to the indicator I_T that is characterized by the following relationship:

$$I_T = 1 - \frac{R_{fk} + R_{rk}}{2} \,. \tag{5.7}$$

In this case I_T changes in the interval $[0,1]$. In fact, when $R_{fk} = R_{rk} = 1$, we have $I_T = 0$, and when $R_{fk} = R_{rk} = 0$, $I_T = 1$. Thus, the degree of timeliness of a decision maker's informedness is characterized by the indicator I_T, which is determined by relationship (5.7).

Note that $I_T = 1$, in the case $T_{fk} = T^-$ and $T_{rk} = T^+$. This means that decision is being generated in the minimum possible time frame (i.e., with minimum time margin), decision realization should be carried out in the maximum possible time frame for the given region, i.e., $T_{rk} = T^+$.

For region I, where for each kth situation the moment of decision realization comes earlier than the decision generation, because $T_{rk} < T^-$, we can introduce an indicator of **informedness untimeliness**. This indicator will be determined by the formula:

$$I_T^- = \frac{T^- - T_m^-}{T^+ - T^-} \,. \tag{5.8}$$

For region III, where $T_{rk} > T^+$, i.e., it is guaranteed that the moment of decision realization will come later than its generation with the maximum possible duration, we shall introduce an indicator of the **level of development of informedness timeliness**. Quantitatively this indicator will be determined by the formula:

$$I_T^+ = \frac{T_M^+ - T^+}{T^+ - T^-} \,. \tag{5.9}$$

Contensively the value I_T^+ in relative units means the level of informedness timeliness in region III exceeds the best result in region II. Indeed, the best result in region II is the value $I_T = 1$. Let us transform relationship (5.9):

$$I_T^+ = \frac{T_M^+ - T^- - T^+ + T^-}{T^+ - T^-} = \frac{T_M^+ - T^-}{T^+ - T^-} - 1 \,. \tag{5.10}$$

Since $T_M^+ > T^+$, then $\left(T_M^+ - T^-\right) / \left(T^+ - T^-\right) > 1$, and the difference shows how I_T^+ exceeds the maximum value $I_T = 1$. Now, after having introduced the indicators for all regions, the properties of region II should be analyzed in more detail. This necessity is conditioned by one feature of this region, i.e., under certain conditions for an arbitrary kth situation from the given set of situations one of the following alternatives can be realized:

1. Informedness timeliness is guaranteed with some time reserve.
2. Informedness timeliness is ensured.

3. Informedness timeliness is not ensured.

Let us write down a set of situations in the form

$$S_0 = \{S_k \mid k = \overline{1, N_s}\} . \tag{5.11}$$

Now, let us analyze the conditions that result in the above indicated alternatives. To solve this problem, first of all, we should choose from the given set of alternatives those situations that by the property of informedness timeliness belong to regions I, II, and III. For that we shall classify the set of situations S_0 into subsets in which only one of the above stated conditions of informedness timeliness is true. For that we shall pick out from set S_T, where index "T" means that the classification is made by the property of informedness timeliness, the subsets with the following properties:

$$\begin{aligned} S_{T1} &= \{S_{k_1} \in S_0 \mid T_m^- \le T_{rk_1} < T^-,\ k_1 = \overline{1, N_{T1}}\} , \\ S_{T2} &= \{S_{k_2} \in S_0 \mid T^- \le T_{rk_2} \le T^+,\ k_2 = \overline{1, N_{T2}}\} , \\ S_{T3} &= \{S_{k_3} \in S_0 \mid T^+ < T_{rk_3} \le T_M^+,\ k_3 = \overline{1, N_{T3}}\} , \\ & N_{T1} + N_{T2} + N_{T3} = N_{Ts} . \end{aligned} \tag{5.12}$$

Here S_{T1}, S_{T2}, and S_{T3} are sets of situations for which the level of informedness timeliness is determined by regions I, II, and III, respectively. The properties of sets S_{T1} and S_{T3} were examined above. Therefore, S_{T1} determines the situations for which informedness timeliness is not ensured, and S_{T3} determines the situations for which informedness timeliness is guaranteed with some time reserve. It is significant that the level of informedness timeliness for the given subsets is determined, as was noted above, in the following way: $I_T = 0$ for S_{T1}, $0 \le I_T \le 1$ for S_{T2}, and $I_T \ge 1$ for S_{T3}.

Recall that $I_T > 1$ means that there is a time allotment for decision generation, because $T_{rk} \ge T^+$ (i.e., the time from when information is received to when the decision realization is longer than the maximum time allotted for generation the decision).

Now, let us proceed to an analysis of set S_{T2} whose informedness timeliness properties are characterized by region II. Let us pick out in S_{T2} those subsets that differ in terms of the duration of their decision generation:

$$\begin{aligned} S_{T21} &= \{S_{k_{21}} \in S_{T2} \mid T_{fk_{21}} > T_{rk_{21}},\ k_{21} = \overline{1, N_{T21}}\} , \\ S_{T22} &= \{S_{k_{22}} \in S_{T2} \mid T_{fk_{22}} = T_{rk_{22}},\ k_{22} = \overline{1, N_{T22}}\} , \\ S_{T23} &= \{S_{k_{23}} \in S_{T2} \mid T_{fk_{23}} < T_{rk_{23}},\ k_{23} = \overline{1, N_{T23}}\} . \end{aligned} \tag{5.13}$$

Here S_{T21} is a set of situations for which the duration of decision generation exceeds the duration of its realization period. The realization period is an interval from the moment of receipt of information to the moment of decision realization. Set S_{T22} is a set of situations where the period of decision generation is equal to the period of its realization, i.e., the moment of decision realization coincides with the moment when a final decision is generated. Set S_{T23} is a set of situations for which the period of solution generation is less than the period of decision realization, i.e., the moment of decision generation for any situation from the given set ends earlier than the moment of its realization. It follows from the properties of regions I–III that set S_{T21} coincides in its properties with set S_{T1} in the sense that the informedness timeliness is not ensured for $S_k \in S_{T21}$ and $S_k \in S_{T1}$. Set S_{T22} is a set of situations for which informedness timeliness is ensured, but without a time allotment. Set S_{T23} is a set of situations where informedness timeliness is ensured with some allotment of time.

Set S_0 can be divided into two subsets based on features, determined by relationships (5.12) and (5.13): S_T^+ and S_T^-, where S_T^+ is a subset for whose elements informedness timeliness is ensured, and S_T^- is a subset for whose elements informedness timeliness cannot be ensured. Taking into consideration the properties of the above introduced subsets, we can state:

$$S_T^+ = S_{T22} \cup S_{T23} \cup S_{T3}, \quad S_T^- = S_{T1} \cup S_{T21}\,. \tag{5.14}$$

It follows that for practical estimation of an informedness timeliness indicator, in the conditions of the given set of situations, first, it is necessary to solve the problem of recognition, namely, to define for each situation which of the above introduced sets it belongs to.

Let us proceed to the formalization of the third property of the decision maker's informedness—**credibility** I_D. In this case, the credibility of the decision maker's informedness is a property that characterizes the degree of conformity of the information received by the decision maker to a real situation. In this case, we shall examine informedness credibility I_D as an integral property and understand it in the widest sense, i.e., as a property that stands in opposition to informedness noncredibility, assuming that the latter is conditioned by a number of factors—incompleteness, inaccuracy, contradictoriness, fuzziness, and uncertainty of the initial information. For the analysis of the mentioned property we shall use the scheme shown in Fig. 5.3, where different levels of informedness credibility are presented.

The following symbols are introduced:

- D_m^-: a level corresponding to the absence of any knowledge about the analyzed set of situations;
- D_M^+: a level corresponding to the absolute, complete, precise, noncontradictory knowledge about the analyzed set of situations;
- D^+: the maximum, practically expedient level of informedness credibility for a given set of situations;
- D^-: the minimum, practically expedient level of informedness credibility for a given set of situations;
- D_k: the value of informedness credibility about the kth situation for given set of situations.

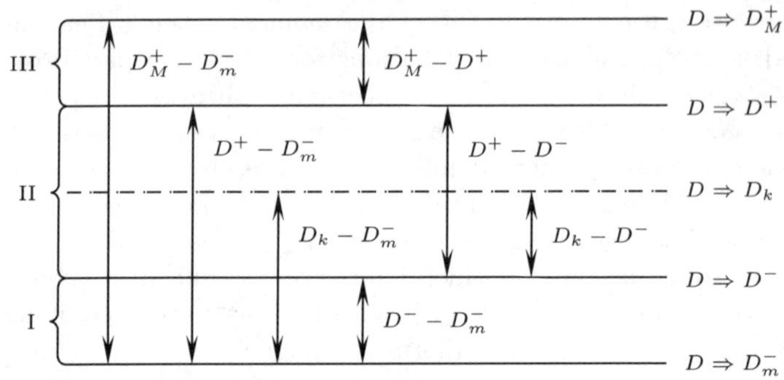

Fig. 5.3. Presentation of different levels of informedness credibility I_D

The levels D_m^-, D_M^+ are the ultimate, potentially possible boundary levels of informedness credibility. And here, D_m^- is practically inexpedient and D_M^+ is practically unattainable due to tool limitations in terms of accuracy, resolution, and other characteristics.

By analogy with the above accepted method let us single out three regions:

Region I corresponds to the interval $D \in [D_m^-, D^-]$ and defines a region of unacceptable level of informedness credibility.

Region II corresponds to the interval $D \in [D^-, D^+]$ and defines a region of practically expedient informedness credibility.

Region III corresponds to $D \in [D^+, D_M^+]$ and defines a region of potential informedness credibility.

Let us introduce the indicators of informedness credibility for each region. For region I we shall introduce an indicator of the level of informedness credibility deficit:

$$I_D^- = \frac{D^- - D_m^-}{D^+ - D^-} . \tag{5.15}$$

For region II we shall introduce an indicator of the level of credibility in the form:

$$I_D = \frac{D_k - D^-}{D^+ - D^-} . \tag{5.16}$$

For region III we shall introduce an indicator of the level of credibility allotment in the form:

$$I_D^+ = \frac{D_M^+ - D^+}{D^+ - D^-} . \tag{5.17}$$

For the purpose of completing the formalization let us classify a set of situations by the indicators of informedness completeness and credibility. Since these indicators are classified identically by their properties, we shall consider them simultaneously.

We shall transform set S_0, determined by relationship (5.11), in such a way that the introduced subsets of situations will have properties determined by regions I, II, and III, respectively, by the indicators of informedness completeness and credibility. In this case, the classification by the given properties will be similar to the classification that is determined by relationship (5.12).

Let us pick out subsets in S_0 by informedness completeness:

$$\begin{aligned} S_{C1} &= \left\{S_{k_1} \in S_0 \mid C_m^- \leq C_{k_1} < C^-,\ k_1 = \overline{1, N_{n1}}\right\} , \\ S_{C2} &= \left\{S_{k_2} \in S_0 \mid C^- \leq C_{k_2} \leq C^+,\ k_2 = \overline{1, N_{n2}}\right\} , \\ S_{C3} &= \left\{S_{k_3} \in S_0 \mid C^+ < C_{k_3} \leq C_M^+,\ k_3 = \overline{1, N_{n3}}\right\} . \end{aligned} \tag{5.18}$$

Similarly, let us pick out subsets from S_0 by informedness credibility:

$$\begin{aligned} S_{D1} &= \left\{S_{k_1} \in S_0 \mid D_m^- \leq D_{k_1} < D^-,\ k_1 = \overline{1, N_{n1}}\right\} , \\ S_{D2} &= \left\{S_{k_2} \in S_0 \mid D^- \leq D_{k_2} \leq D^+,\ k_2 = \overline{1, N_{n2}}\right\} , \\ S_{D3} &= \left\{S_{k_3} \in S_0 \mid D^+ < D_{k_3} \leq D_M^+,\ k_3 = \overline{1, N_{n3}}\right\} . \end{aligned} \tag{5.19}$$

It should be pointed out that, unlike the classification by timeliness, in this case, there is no necessity of introducing additional classification for sets S_{C2} and S_{D2}. This is explained by the fact that for the property of informedness timeliness the unique relationships between the moment of completion of decision generation and the moment of its realization is typical. In fact, informedness is timely if a decision formation is made before the necessity for its realization arises, and the opposite property corresponds to the opposite condition. Undoubtedly, there is no such strict relationship for the properties being examined. The indicators of completeness and credibility in the general case are functions of time in the sense that in the period from when the initial information is received and when the decision generation process begins to the moment of its completion some additional information may appear that will change in some way (will increase if information is credible or decrease if it is misinformation) the indicators of informedness completeness and credibility. If for the situation being analyzed the conditions determining its belonging to region II change, then, as a result, the indicators of informedness in the sense of completeness and (or) credibility will change for the better or worse.

In the most general case the characteristics C_k and D_k should be considered as functions of time, i.e., we assume $C_k(t)$ and $D_k(t)$ and examine changes in these values at any moment t from the moment $t_{fk} = t_0 = 0$ to the moment of solution realization $t_{rk} = T_{rk}$. At the same time, it is necessary to take into consideration that those two moments—first, the moment when the decision generation process is complete $t = t_{fk} = T_{fk}$, and second, the moment when decision realization of $t = t_{rk} = T_{rk}$, are important or, more precisely, critical. In fact, at the moment t_{fk} the quality of the decision being generated is determined, and at the moment t_{rk} the quality of the decision being realized is determined, i.e., the quality of the decision at the moment of its realization. In the general case, the quality of the decision generated and realized may not coincide at the indicated moments, because, when $t_{rk} > t_{fk}$ in the period from the completion of the decision generation to the moment of its realization, some information may appear that will change the estimation of the solution quality. And, finally, it should be pointed out that besides region II it is necessary to introduce a classification for sets S_{C1} and S_{D1} because in the periods $t \in [0, t_{fk}]$ and $t \in [0, t_{rk}]$ examined above, to determine S_k it is possible to switch to region II by the indicators of informedness completeness and credibility. Similarly, for region III it is possible that, based on informedness completeness and credibility, certain situations, will transition to region II. Therefore, for these informedness properties the given classification at a certain moment may change all three subsets.

Thus the necessity arises to introduce a classification according to completeness and credibility for the final result at the moment of completion of decision generation and at the moment of its realization. Taking into consideration that while forming the classifications (5.18) and (5.19) no time limitations were introduced, and considering that it is possible to change a decision only once it has been made (i.e., when $t \in [0, t_{fk}]$), we assume that the classifications in the form of (5.18) and (5.19) determine the classes of situations by the properties of informedness completeness and credibility when a decision has been generated, i.e., when $t = t_{fk}$. Let relationships (5.18) and (5.19) define the classification of set S_0 by the indicators of a decision maker's informedness completeness and credibility when a decision has been generated.

We shall refer to the classification at the moment of decision realization $t = t_{rk}$ as the classification of set S_0 by the completeness and credibility of justification of the decision being realized. This means that set S_0 is classified based on the decision maker's informedness completeness and credibility at the moment of decision realization, i.e., taking into consideration the information that arrived in the period between the completion of the decision generation process and the moment of its realization.

This type of classification by completeness and credibility of justification of the solution being realized may be presented in the following form:

$$
\begin{aligned}
S^r_{C1} &= \left\{S_{k_1} \in S_0 \mid C^-_m \le C_{k_1}(t) < C^-,\ k_1 = \overline{1, N^r_{C1}},\ t = T_{rk}\right\} , \\
S^r_{C2} &= \left\{S_{k_2} \in S_0 \mid C^- \le C_{k_2}(t) \le C^+,\ k_2 = \overline{1, N^r_{C2}},\ t = T_{rk}\right\} , \\
S^r_{C3} &= \left\{S_{k_3} \in S_0 \mid C^+ < C_{k_3}(t) \le C^+_M,\ k_3 = \overline{1, N^r_{C3}},\ t = T_{rk}\right\} , \\
& N^r_{C1} + N^r_{C2} + N^r_{C3} = N_{S1} ,
\end{aligned}
\tag{5.20}
$$

$$
\begin{aligned}
S^r_{D1} &= \left\{S_{k_1} \in S_0 \mid D^-_m \le D_{k_1}(t) < D^-,\ k_1 = \overline{1, N^r_{D1}},\ t = T_{rk}\right\} , \\
S^r_{D2} &= \left\{S_{k_2} \in S_0 \mid D^- \le D_{k_2}(t) \le D^+,\ k_2 = \overline{1, N^r_{D2}},\ t = T_{rk}\right\} , \\
S^r_{D3} &= \left\{S_{k_3} \in S_0 \mid D^+ < D_{k_3}(t) \le D^+_M,\ k_3 = \overline{1, N^r_{D3}},\ t = T_{rk}\right\} , \\
& N^r_{D1} + N^r_{D2} + N^r_{D3} = N_{S2} .
\end{aligned}
\tag{5.21}
$$

Summing up, let us note that the classifications in the form (5.18) and (5.19) should be considered the main ones on which we can determine the quality of the decision generated, and classifications (5.20) and (5.21) are control ones, in the sense that they determine the quality of the decision justification at the moment of its realization.

5.3 Classification and Recognition of Situations Through Integral and Partial Indicators

Let us now turn to the final stage of formalization of informedness properties. The classification of situations from the given set S_0 by the properties of completeness, timeliness, and credibility of a decision maker's informedness was carried out in the previous section. As a result, their qualitative interconnection with the levels of completeness, timeliness, and credibility of the corresponding decision was established. However, it is quite obvious that the decision that was made will have a required quality and efficiency level if while it was being made the corresponding levels of completeness, credibility, and timeliness of informedness in the developing situation were simultaneously ensured for the purpose of risk level minimization.

In this case it is necessary:

1. To form an integral indicator of informedness that will take into account the influence of the degree and level of each of the introduced indicators I_C, I_T, I_D on the degree of achievement of the decision maker's goals.
2. To suggest classifications of the given set of situations S_0 using the unified system of interconnected indicators I_C, I_T, I_D or the unified integral indicator of informedness.
3. To develop methods and procedures of recognition of belonging of the concrete situation S_k from the given set of situations S_0 to a certain class of objects from the introduced classification.
4. To develop ways and methods of estimating the risk degree and level for various types of situations.
5. To suggest, under conditions of origin of the given set of situations, ways and procedures of reducing the risk degree and level on the basis of a rational strategy of a decision maker's informedness during the decision-making process.

Let us consider a solution of the first problem and give a common formulation and mathematical statement of the problem of estimating the level of a decision maker's informedness about risk factors [125].

In forming an integral indicator of informedness it is necessary to take into account certain properties and features of the introduced parameters I_C, I_T, I_D. The following major properties belong to them:

1. The level of informedness grows continuously with an increase of the indicators I_C, I_T, I_D as a group or any of them individually.

2. The common level of informedness increases with an increase in its completeness, credibility, and timeliness according to the nonlinear law: the growth of the informedness level gradually slows down as the indicators of completeness, credibility, and timeliness approach gradually their limiting values.
3. If the value of one of the indicators decreases below some threshold value, then the informedness level cannot be compensated by an increase in other indicators.
4. With a zero value of any indicator I_C, I_T, I_D the common informedness level is also zero.

The first property is obvious—with a growth in the completeness, timeliness, and credibility of information the values I_C, I_T, I_D grow, along with, naturally, a decision maker's level of informedness.

The second property follows from human psychometric features: to simultaneously perceive, remember, and analyze a certain volume of new information. Therefore, as a volume of the given information approaches the threshold of perception, the growth in a decision maker's informedness level slows down. This property is very essential for human life and activities and is studied in detail by a number of disciplines: above all by psychology, including such subdisciplines as psychophysics, engineering psychology, and the psychology of labor. The practical importance of this property is proved by such phenomena as the presence of the loudness pain barrier of human beings, a change in perception of the sound volume according to the logarithmic law depending on an increase in sound pressure, etc. In practice, humans constantly face different manifestations of the indicated property. One such manifestation is known from statistics: given a statistical sample whose credibility is increasing by one, the same relative number sharply grows when the credibility approaches one. For example, when the credibility of the information increases by 10% (from 0.9 to 0.99), the required sample is almost 50 times larger than in the case of an increase from 0.8 to 0.88.

The third and fourth properties require more detailed explanations. Let us consider the third property. It should be noted that $I_C = 0$ or $I_T = 0$ or $I_D = 0$ takes place when, respectively, the level of completeness, timeliness, and credibility about a concrete situation is lower than or equal to a certain minimum value, which is determined respectively by the sizes C_m^-, T_m^-, and D_m^-. This means that under this condition the quantity of information does not satisfy the given values with respect to completeness, timeliness, or credibility. Under existing conditions it is impossible to make a decision of the required quality level. Let us explain this using some examples. If the information given to a decision maker is absolutely credible but untimely, then it is impossible to make a decision and the level of the general informedness is assumed to be equal to zero. For example, if the complete and credible information about a military armed attack on a certain target is received after the attack, then it is senseless to set about solving the problem of defending this target. Similarly, if the information is received in a timely manner and is complete, but it is noncredible, then a decision maker's informedness level will also be equal to zero. For example, if the complete information about an attack on a target is received in time, but the location of the target is incorrect (for example, a similar object will be attacked in another city), then obviously the decision to tighten security on the first target will be useless. From the above it follows that it is impossible to compensate the impropriety of information either by the level of its completeness or by the level of its credibility.

Similarly, it is impossible to compensate any other property if its indicators are below the allowable threshold values of C^-, T^-, or D^-.

These properties determine the requirement for the integral informedness indicator I, as a function of partial indicators I_C, I_T, and I_D, characterizing respectively the completeness, timeliness, and credibility of a decision maker's informedness. But it is obvious that the level of informedness depends not only on qualitative, but also on quantitative characteristics of information. Considering earlier the defined quantity of information, we obtained the relationship for the level of uninformedness in the following form:

$$H_s = -\sum_{i=1}^{m_s} \rho_i \log \rho_i ,$$

where m_s is the total number of elements of set M_s, which unites all theoretically possible situations, and ρ_i is the probability of occurrence of the situation $\rho_i \in M_s$.

Obviously, in practice the probability of a number of situations from set M_s will be so low that it is possible not to take them into consideration at all. For example, in theory, it is possible that it will snow in Ukraine in the middle of June, but the probability of this event is so low that when solving practical problems for this season, this situation may be excluded from the analysis. As a result, we obtain set S_0 of the given situations, where the number of elements is $N_s \leq m_s$. Set S_0 is formed on the basis of expert estimations, statistical analysis, and other a priori knowledge about situations. All qualitative characteristics of a decision maker's informedness are determined exactly for set S_0.

It should be noted that the inverse value to the uninformedness level is the informedness level I_S, determined by the quantitative characteristics of the information. The level of informedness is introduced here proceeding from the following considerations. This level measures the possibility of analyzing set S_0. The measure of possibility is used in treating possibility theory [54].

If the available information about set S_0 makes it possible to analyze any situation $S_i \in S_0$, then we assume that the level of informedness is $I_S = 1$, i.e., $I_S = 1$ is an indicator of the possibility of analyzing any situation $S_i \in S_0$. Obviously, in this case the entropy level (the level of uninformedness of set S_0) is equal to zero. If there is no any information about set S_0, then the possibility of analyzing any situation from set S_0 is equal to zero, i.e., $I_S = 0$, which means it is not possible to analyze any situation from S_0. The entropy of set S_0 is the highest possible in that case.

Let us denote the maximum value of the entropy of set S_0 by H_s^+:

$$H_s^+ = \max H_s .$$

From these considerations it follows that $I_S = 1$ when $H_s = 0$ and $I_S = 0$ when $H_s = H_s^+$ and $0 \le H_s \le H_s^+$. Let us introduce I_S in the following way:

$$I_S = 1 - \frac{H_s}{H_s^+} = 1 - \left(- \sum_{i=1}^{N_s} \rho_i \log \rho_i \right) \Big/ H_s^+ . \tag{5.22}$$

If abnormal and critical situations are equiprobable and there is no a priori information, then $H_s^+ = \log N_s$.

Let us proceed to a formal description of the properties of an integral indicator of a decision maker's informedness. It is obvious that in the general case the informedness an integral indicator I depends on all indicators introduced above: I_S, I_C, I_T, I_D, I_C^-, I_C^+, I_T^-, I_T^+, I_D^-, I_D^+. However, in an analysis it is expedient to consider separately informedness indicator regions I, II, and III. First, let us consider region II for all partial indicators. The informedness an integral indicator I is presented as a function of partial indicators in the following form:

$$I = f_I \left(\overline{I_c} \right) , \tag{5.23}$$

where $\overline{I_c}$ is a vector of partial indicators, which may be presented by means of the following equivalent expressions:

$$\overline{I_c} = \langle I_S, I_C, I_T, I_D \rangle , \tag{5.24}$$

$$\overline{I_c} = \left\{ I_q \mid q \in Q_c,\ Q_c = \overline{1, q_0} \right\} , \tag{5.25}$$

then q_0 is should be equal to 4 ($q_0 = 4$), but in the general case, it is necessary to assume $q = q_\Sigma = 10$ for all regions in relationship (5.25) (i.e., to take into account all the above listed partial indicators, including I_C^-, I_C^+ etc.).

Let us now consider the peculiar features of function (5.23), taking into account the above formulated properties. The **first feature** follows from the threshold property of partial indicators—it is impossible to compensate the deficit of one indicator at the expense of others if its value is below the threshold level. In accordance with the imposed boundaries, such threshold levels are C^-, T^-, and D^-. From here it follows that

$$f_I \left(\overline{I_c} \right) = 0 \tag{5.26}$$

on the condition that there exists such $q \in Q_c$ for which the $I_q = 0$ condition is fulfilled.

From (5.26) it follows that function (5.23) should be a multiplicative function of partial indicators:

$$f_I\left(\overline{I_c}\right) = \prod_{q=1}^{q_0} f_q\left(I_q\right). \tag{5.27}$$

Or, taking into account (5.24) as a product of fuzzy relationships, the given relationship is presented in the form:

$$f_I\left(\overline{I_c}\right) = f_S\left(I_S\right) \circ f_C\left(I_C\right) \circ f_T\left(I_T\right) \circ f_D\left(I_D\right). \tag{5.28}$$

Here, it should be noted that for any q it is true that:

$$f_q\left(I_q\right) = 0, \quad \text{at} \quad I_q = 0. \tag{5.29}$$

The **second feature** is that the restrictions $0 < I_C \leq 1$, $0 < I_T \leq 1$, and $0 < I_D \leq 1$ are true for the indicators I_C, I_T, and I_D in region II. It allows one to introduce norming for I, assuming $f_I\left(\overline{I_c}\right) = 1$, for any

$$q \in Q_c \to I_q = 1. \tag{5.30}$$

The **third feature** of function (5.23) is determined by the above considered nonlinear dependence of the informedness level on each of the partial indicators, which is preconditioned by human psychometric characteristics. In accordance with the given feature each relationship that is a part of the multiplicative function (5.27) or (5.28) should be described by a function of the logarithmic type

$$f_q\left(I_q\right) = \log\left(1 + \sum_{l=0}^{m_q} b_{ql} I_q^l\right), \tag{5.31}$$

or in the elementary case

$$f_q\left(I_q\right) = a_q \lg\left(1 + b_q I_q\right), \quad f_q\left(I_q\right) = a_q \ln\left(1 + b_q I_q\right). \tag{5.32}$$

In conclusion, we shall consider peculiarities of the formalized description of the informedness an integral indicator I, which follows from the pattern of change in time of the partial indictors I_S, I_C, I_T, I_D. Here, it is necessary to point out that changes in time of the indicator I_T principally differ from those of other indicators. In fact, in the process of making a decision, which starts at $t = t_{fk} = 0$ and is over at $t = t_{rk}$, some additional information is received that results in an increase in the completeness, credibility, and quantity of information.

From here it follows that, in the general case, the indicators I_S, I_C and I_D are increasing functions of time. However, for the indicator I_T the dependence turns out to be different, which is determined by a tight interconnection between the moment when the decision generation process is complete and the moment of its realization. The timeliness indicator is proved to be a decreasing function of time because the moment of realization is fixed and determined by external factors. For example, the moment of realization is determined by the concrete and announced date of a new product release, the seasonal changes on highways, etc. It follows from a well-known fact that, with an increase in information volume, the time required for its processing also increases, which leads to a reduction of the time allotted before the moment of decision realization. The considered properties can be presented schematically by Fig. 5.4.

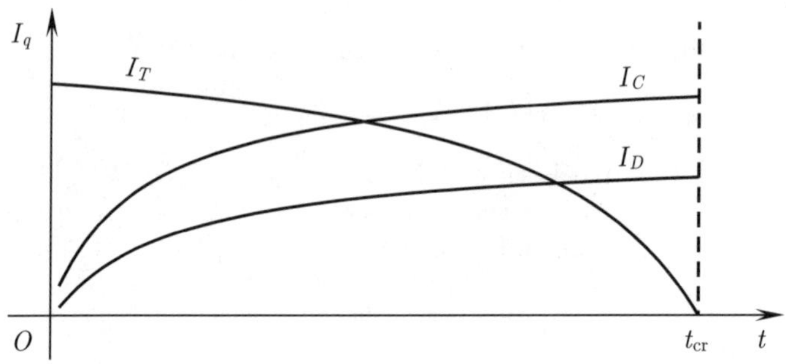

Fig. 5.4. Behavior of informedness qualitative characteristics in time

From the above premises it follows that there is a contradiction between the completeness and credibility levels of informedness on the one hand and the level of timeliness on the other. Therefore, it is practically important to find a rational compromise between the levels I_C, I_D, and I_T in the decision-making process, taking into account the requirement to reduce the time spent on that process.

To solve the above problem it is necessary to account for the time dependence of I_S, I_C, I_T, I_D, i.e., we assume that:

$$I_S = I_S(t), \quad I_C = I_C(t), \quad I_T = I_T(t), \quad I_D = I_D(t)\,.$$

The concrete functional dependence of the indicators in the form (5.32) is determined for a concrete set of the situations S_0. It is most convenient to present this dependence as a polynomial. First, it allows one to determine polynomial coefficients with the help of the known interpolation techniques. Secondly, any analog function in accordance with the Weierstrasse theorem can be approximated by a polynomial with any prescribed accuracy. Therefore, from here we obtain:

$$\begin{aligned} I_S(t) &= \sum_{k=0}^{N_s} a_{Sk}t^k, \quad I_C(t) = \sum_{k=0}^{N_C} a_{Ck}t^k\,, \\ I_T(t) &= \sum_{k=0}^{N_T} a_{Tk}t^k, \quad I_D(t) = \sum_{k=0}^{N_D} a_{Dk}t^k\,. \end{aligned} \tag{5.33}$$

Thus, the informedness indicator I for region II is determined by relationships (5.23), (5.27), (5.31), and (5.33). It is expedient to use the indicators I_C^-, I_C^+, I_T^-, I_T^+, I_D^-, I_D^+ for the formation of restrictions in optimization problems.

Let us proceed to the analysis of the classification task of a given set S_0 of situations by a system of indicators. The purpose of this task consists in finding out what class of informedness indicators a concrete analyzed situation belongs to. For this it is necessary to execute two steps: (1) introduce the classification of set S_0 of situations by informedness indicators and (2) formalize the recognition procedure, which helps to determine what class a concrete situation belongs to.

Thus let us turn to classification of set S_0 by the totality of informedness indicators. In the previous section we considered the procedure of the classification separately by each of the indicators I_C, I_T and I_D. Now, we shall classify it by their totality.

Here, two approaches are possible:

1. Classify set S_0 by an integral informedness indicator I, which, as shown above, unites all partial indicators in a single whole.
2. Classify set S_0 directly by the totality of partial indicators, not reducing them to a partial indicator.

Each of the approaches has its advantages and disadvantages; therefore, we shall briefly consider them both, which will allow us to analyze their properties and features and to obtain comparative estimations.

It is expedient to classify set S_0 by an integral informedness indicator on the basis of the same method that was used for the classification by separate informedness partial indicators I_C, I_D, etc. Hence, we assume that the possible values of indicator I are placed in three regions, as presented in Fig. 5.5.

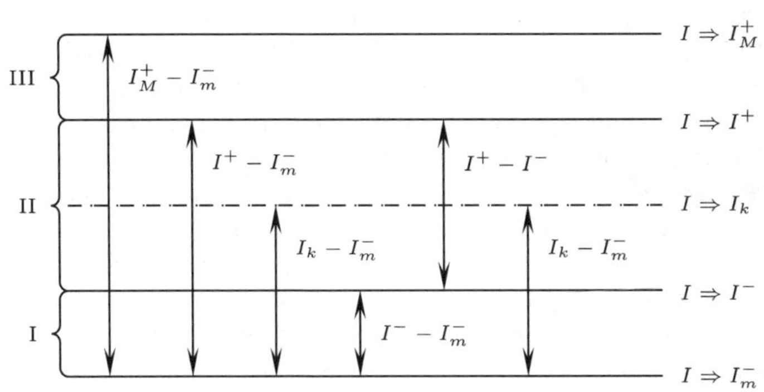

Fig. 5.5. Set S_0 classification by an integral informedness indicator: I_k is an integral informedness indicator of the kth situation and I_m^-, I^-, I^+ and I_M^+ are region boundaries

Suppose that the limiting values of the regions are determined by relationships (5.26)–(5.30) and correspond to the following values of partial informedness indicators:

$$I^+ \Rightarrow I_C^+,\ I_T^+,\ I_D^+\,,$$
$$I_m^- \Rightarrow I_C = 0,\ I_T = 0,\ I_D = 0\,.$$

The boundaries I^+ and I^-, which determine region II, are found on practical considerations on the basis of expert estimations and should satisfy the following conditions:

$$1 \le I^+ < I_M^+, \quad I_m^- \le I^- < 1\,.$$

Since the value of informedness level $I^- = 0$ has no practical meaning, and the level $I^+ = I_M^+$ is practically unachievable, we suppose that

$$I^+ \in [0.8; 0.95], \quad I^- \in [0.05; 0.15]\,. \tag{5.34}$$

Then, the initial set S_0 can be divided into three subsets:

$$S_{I1} = \left\{S_{k_1} \in S_0 \mid I_m^- \le I_{k_1} < I^-\right\}\,,$$
$$S_{I2} = \left\{S_{k_2} \in S_0 \mid I^- \le I_{k_2} \le I^+\right\}\,,$$
$$S_{I3} = \left\{S_{k_3} \in S_0 \mid I^+ < I_{k_3} \le I_M^+\right\}\,. \tag{5.35}$$

At the prescribed limiting values the procedure of situation recognition and their attribution to a concrete subset becomes considerably easier.

But it is necessary to pay attention to one principal feature of the introduced classification—the boundaries of region II are ambiguously interconnected with the limits of the partial indicators I_C, I_T, and I_D. It is caused by the fact that the numerical value of I depends not only on numerical values of the appropriate partial indicators but also on the form of $f_I\left(\overline{I}_c\right)$ in relationships (5.27) and (5.28). Therefore, if the form of the function $f_I\left(\overline{I}_c\right)$ has been changed, the values of I^+ and I^- obtained from condition (5.34) will be different for different values of the informedness indicators $\overline{I}_c$. For example, increasing or decreasing the degree of polynomial in formula (5.31). This circumstance complicates the comparison procedure of various decisions.

The second peculiarity of the given classification is the impossibility of unambiguously determining whether set S_{I2} is a union of sets $S_{C2} \cup S_{T2} \cup S_{D2}$ only or a union of other sets, such as $S_{C3} \cup S_{T2} \cup S_{D2}$ or $S_{C2} \cup S_{T2} \cup S_{D3}$, etc., i.e., the subset of the first partial indicator corresponds to region II, that of the second to region III, of the third to either region II or III. Here the drawbacks of one partial indicator are partly compensated by stronger properties of another.

Now let us consider classification by a partial indicator system. When classifying, it is necessary to pay attention to the following feature. To carry out classification by several indicators we need to answer the following questions:

1. What class or subset should an analyzed situation be referred to if by one indicator it belongs to one class, by another to another class, by a third indiciator to a third class?
2. Is it possible to be limited only by three classes (subsets), or is it necessary to introduce one or more additional classes (subsets) related to various combinations corresponding to different indicators?

The answers to the given questions will be formulated while carrying out the formalization procedure. Obviously, the easiest way is to introduce subsets by the totality of indicators for the simplest special case, when the elements included in them correspond to the identical regions I, II or III by all indicators. Taking into account these conditions, we introduce the following subsets:

$$\begin{aligned} S_{\Sigma 1} &= \left\{ S_{k_1} \in S_0 \mid I_{qm}^- \leq I_q < I_q^-,\ \forall\, q \in Q_c \right\}, \\ S_{\Sigma 2} &= \left\{ S_{k_2} \in S_0 \mid I_q^- \leq I_q \leq I_q^+,\ \forall\, q \in Q_c \right\}, \\ S_{\Sigma 3} &= \left\{ S_{k_3} \in S_0 \mid I_q^+ < I_q \leq I_{qM}^+,\ \forall\, q \in Q_c \right\}. \end{aligned} \tag{5.36}$$

Here, I_q is determined by relationships (5.24) and (5.25). Note that an integral informedness indicator will be equal to zero ($I = 0$), in conformity with (5.26), if there exists $q \in Q_c$ for which $I_q = 0$. But at $I_q = 0$ and in view of the classification by each of the partial indicators, the level of the corresponding indicator of element $S_k \in S_0$ will be less than the minimally required levels I_C^-, I_T^-, or I_D^-. This element belongs to the subset S_{C1} and S_{T1} or S_{D1}. Therefore, it is necessary to take into account that in cases when for some situation the condition $I_q = 0$ is fulfilled for any $q \in Q_c$, then the given situation belongs to the subset $S_{\Sigma 1}$. This ensures that the considered situation S_{k_1} belongs to $S_{\Sigma 1}$. Thus, the element $S_{k_1} \in S_{\Sigma 1}$ if the condition $I_q < I_q^-$ is fulfilled at least for one q, taking into account that I_q is a physically nonnegative value ($I_q \geq 0$). Then, set $S_{\Sigma 1}$ is described as:

$$S_{\Sigma 1} = \left\{ S_{k_1} \in S_0 \mid q \in Q_c \Rightarrow I_{qm}^- \leq I_q < I_q^- \right\}. \tag{5.37}$$

Further, let us consider the cases where a condition corresponding to $S_{\Sigma 2}$ is satisfied by one indicator, and $S_{\Sigma 3}$ by other indicators. Here, it is important to decide whether we should separate or unite sets $S_{\Sigma 2}$ and $S_{\Sigma 3}$, since the condition of information sufficiency is fulfilled both for elements of set $S_{\Sigma 2}$ and of set $S_{\Sigma 3}$. But when we unite $S_{\Sigma 2}$ and $S_{\Sigma 3}$, the information on the structure of informedness partial indicators is lost. Therefore, it seems expedient to introduce a double-sided classification: at the first level to reject those situations that belong to set $S_{\Sigma 1}$, and at the second level to single out situations that belong to sets $S_{\Sigma 2}$ and $S_{\Sigma 3}$, and also to set $S_{\Sigma 21}$, that is:

$$\begin{aligned} S_{\Sigma 21} = \{ S_{k_{21}} \in S_0 \mid & q' \in Q_c \Rightarrow I_{q'} > I_q^+, \\ & q'' \in Q_c \setminus \{ q' \} \Rightarrow I_{q''} \in [I_q^-, I_q^+] \}. \end{aligned} \tag{5.38}$$

Thus, set $S_{\Sigma 21}$ consists of the elements $S_{k_{21}} \in S_0$. For these elements one of the partial indicators lies in the interval corresponding to region III for this indicator, and the others lie in the interval corresponding to region II. Let us determine set $S_{\Sigma 31}$, which consists of the elements $S_{k_{31}} \in S_0$. For these elements one of the partial indicators lies in the interval corresponding to region II for this indicator, and the others lie in the intervals corresponding to region III of each indicator:

$$S_{\Sigma 31} = \Big\{ S_{k_{31}} \in S_0 \,|\, q' \in Q_c \Rightarrow I_{q'} \in [I_q^-, I_q^+], \\ q'' \in Q_c \setminus \{q'\} \Rightarrow I_{q''} \in [I_q^+, I_{qM}^+] \Big\} . \tag{5.39}$$

This method of classification by the totality of indicators makes it possible to avoid the above enumerated weaknesses of the classification by an integral informedness indicator. However, this approach can be more laborious unless we confine ourselves only to a rejection of situations belonging to set $S_{\Sigma 1}$. If it is possible to reject $S_{k_1} \in S_{\Sigma 1}$ only, then the given approach could be less laborious, in comparison with an approach based on classification by an integral index, since the former does not require calculation of the functions of the form of (5.31) and (5.27).

Now, the process of situation recognition should be formalized and related to concrete classes. Formalization will be similar to that by an integral indicator and by the totality of partial indicators.

5.4 Recognition of Situations in Conditions of Information Fuzziness

The necessity of recognizing properties and features of situations is typical for various fields of practical activities. The following three types of problems are the most important in this class: a definition of situations (1) in conditions of complete certainty of the initial information, (2) in conditions of incompleteness and fuzziness of information, and (3) in the process of changing a decision maker's informedness. Let us formulate these problems.

Recognition of situations in conditions of complete certainty of the initial information. Let us split set S_0 into a finite number of (classes) subsets Ω_r, $r = \overline{1, R_0}$, $\bigcup_{r=1}^{R_0} \Omega_r = S_0$. Each subset is characterized by the information about properties or features of elements defined by quality and quantity. Situation S_k is characterized by a certain information I_k whose qualitative characteristics I_q can be defined from the given set of characteristics $q \in Q_c$. Each value I_q may take a fixed value from the given set of permissible values $I_q \in \left[I_{qm}^-, I_{qM}^+\right]$. It is necessary to define the relation of situation S_k to a certain class (subset) Ω_r based on the given information I_k about this situation.

This statement corresponds to strict boundaries of each subset $\Omega_r \in S_0$ and strict setting of information about each feature (or characteristic) I_q contained in the information I_k about situation S_k being analyzed. Solving this problem allows us to determine what class, by the totality of the informedness attributes (I_C, I_T, I_D, I_S), the current situation S_k belongs to if the classification S_0 (i.e., splitting of S_0 into classes or subsets by the integral informedness attribute or by the totality of particular indicators) is known. Leaving open the question about the expedient methods of such problem solving, we shall simply note that it has more theoretical than practical meaning. It is conditioned by the following factors:

- In practice, information about a concrete situation, for example, in problems of interaction or counteraction of subjects, is characterized by incompleteness, inaccuracy, and contradictoriness of information about different external factors [weather conditions, market situation changes (if a counteraction is competition), about plans of the opponent, and so on].
- Information about a concrete situation is continuously being updated, and the updated information may both specify (i.e., raise the level of the completeness and credibility) and deny the previous information (i.e., decrease the level of the completeness and credibility of information because of the contradictoriness of the data already available and received).

That is why in practice it is necessary to solve a more complicated problem of recognition, where both changes of a situation and incompleteness, inaccuracy, and contradictoriness of the information should be considered. The indicated features are typical for each concrete situation because its precise prediction is fundamentally impossible. For example, it is impossible to describe all the parameters of competition in the market of software products in the current year because from experience we know that each competitor keeps his classified business information about new research a secret and reveals it only when the product is ready for sale so that competitors do not have time to make a knock-off. That is why the problem of situation recognition should always be formulated taking into consideration the incompleteness and inaccuracy of the initial information.

Recognition of situations in conditions of incompleteness and inaccuracy of information. Let the given set S_0 of situations $S_k \in S_0$ be split into a finite set of subsets (classes) Ω_r, $r = \overline{1, R_0}$, for which it is true that $\bigcup_{r=1}^{R_0} \Omega_r = S_0$. At the same time, we shall take into consideration that splitting is not completely defined; only some incomplete, fuzzy information I_r about classes Ω_r, $r = \overline{1, R_0}$ is assigned. The definition I_r of the class Ω_r may be a set of membership functions $\mu_{\Omega_r}\left(\overline{I}_{\Omega_r}\right)$, where $\overline{I}_{\Omega_r}$ is a vector of the partial indicators of the informedness about the elements of the class Ω_r, $\overline{I}_{\Omega_r} = \{I_{rq} \mid q \in Q_c\}$. Each object $S_k \in S_0$ being examined is described by the fuzzy information $\overline{I}_k$, which represents a set of membership functions $\mu_k\left(\overline{I}_k\right)$, where $\overline{I}_k = \{I_{kq} \mid q \in Q_c\}$ is a vector of partial indicators of the informedness about the object S_k. It is necessary based on the given information I_k about the situation $S_k \in S_0$ to define which class Ω_r the current situation belongs to.

In the given statement the problem of situation recognition is more general, as compared to that examined above, because, as is known from the theory of fuzzy sets and fuzzy correlations, exact (common) sets are particular cases of fuzzy ones when the meaning of the membership function is equal to 1. At the same time, this statement does not consider the character of changing the quality and quantity of the information in the process of decision making. We shall provide a statement of the situation recognition problem taking into consideration the decision-making process and under changing informedness about the situation S_k.

Recognition of situations while a decision maker's informedness status is changing. For the given set of situations S_0, presented in the form of a split into a finite set of classes Ω_r, $r = \overline{1, R_0}$, it is necessary, based on the available, but incomplete, information I_k about situation S_k, to define which class $\Omega_r \in S_0$ S_k belongs to at any given moment $t_{rk} \geq t_0 \geq t_{fk}$ of the decision-making process if the partial indicators of informedness change in the time $I_q = I_q(t) \; \forall \, q \in Q_c$.

Now, let us examine the methods of solving the above stated problems. However, let us first explain certain features of these methods.

Such problems correspond to the class of image recognition problems. In contemporary image recognition theory this term is used in quite broad sense—it is understood to mean some structured approximated description (sketch) of an object or event being examined, and here a partial definability of the description is a fundamental feature of the image. The main purpose of these descriptions consists in their use in the process of setting the conformity between objects, i.e., in proving their identity, similarity, likeness, and so on, which is realized by comparison.

Let us define the term "description" more concretely, as applied to the stated problems. Assume that each class of objects in the general case is described or characterized:

1. By a certain set of attributes. In the case being examined, they may be presented by partial indicators of informedness in the form of (5.25) or by the integral informedness indicator I, which is defined by relationship (5.23).
2. By the concrete numerical values of each attribute. For example, by the integral informedness indicator, in conformity with relationship (5.35), three intervals for each attribute are introduced, each corresponding to a certain set of situations S_{I1}, S_{I2}, S_{I3}. Similarly, the classifications by the partial indicators of informedness are brought in. Next, each situation S_k is characterized by the description I_k, which is a totality of attributes and their values, for example, in the form of the membership function for the fuzzy sets or in the form of the function that takes 0 or 1.

Let us consider the methods of solving the stated problems for a number of cases. First, we shall assume that there are only two subsets (classes): Ω_2, a class of situations for which an integral indicator of informedness I takes on the value, which is not less than the minimum allowed level; and Ω_1, a class of situations for which an integral indicator of informedness is less than the minimum allowed level.

These classes, in accordance with the introduced system of notation (5.1)–(5.35), correspond to: Ω_2—the union of sets S_{I2} and S_{I3}, i.e., $\Omega_2 = S_{I2} \cup S_{I3}$, and Ω_1—the class S_{I1}. This problem belongs to the class of the easiest classification problems and is solved by checking the condition that follows from relationship (5.35).

If $I_k\left(S_k\right) \geq I^-$, then $S_k \in \Omega_2$, else $S_k \in \Omega_1$. As a result, as follows from (5.35), the total set S_0 will be split into two subsets:

$$\Omega_1 = \left\{S_{k_1} \in S_0 \mid I_k < I^-\right\} ,$$
$$\Omega_2 = \left\{S_{k_2} \in S_0 \mid I_k \geq I^-\right\} .$$

and any situation being analyzed will be referred to Ω_1 or Ω_2 quite easily.

The solution is a bit more complicated if the analyzed situation S_k is described by the system of partial indicators

$$I_k\left(S_k\right) = \left\{I_{kq}\left(S_k\right) \mid q \in Q_c\right\} .$$

In this case the classification in two classes is not enough for a decision maker because his purpose is not classification but making some decision in any situation that arises, even in a situation where the informedness level by one attribute or a number of attributes is insufficient. In this case, the situations that have been referred to the class of insufficient informedness Ω_1 by different attributes will not be equivalent. Indeed, if because of impropriety of information arrival S_k is referred to the class Ω_1 only by the attribute I_T, then a decision maker should analyze the following three alternatives: whether it is possible to make a decision within the given time or it is necessary to make an immediate decision, which was preliminarily predicted based on the analysis of possible situations or it is absolutely impossible to realize the decision for the given situation. Another problem arises if the situation is referred to the class Ω_1 by the incompleteness indicator, but it satisfies the timeliness and credibility indicators. In this case, a variant of the decision's unrealizability should not be examined; instead, the alternatives should be analyzed: we should make decisions about an increase in informedness completeness or a decision in conditions of information incompleteness.

The above reasoning shows that the classification S_0 on Ω_1 and Ω_2 is quite complicated, and in addition as was noted above, cassification by the integral indicator I has its own additional shortcomings. That is why in solving recognition problems by the system of partial indicator attributes the necessity arises of specifying the classification (5.39) and introduce new classes that will specify the basis of the attributes by which an examined object was referred to class Ω_1 or Ω_2. For this, it is expedient to make the procedure of classification multilayer. It will allow one, on the one hand, to increase the accuracy of the procedure and, as a result, provide a decision maker with more accurate information about the concrete situation, and, on the other hand, will allow simplify and accelerate the procedure of recognition.

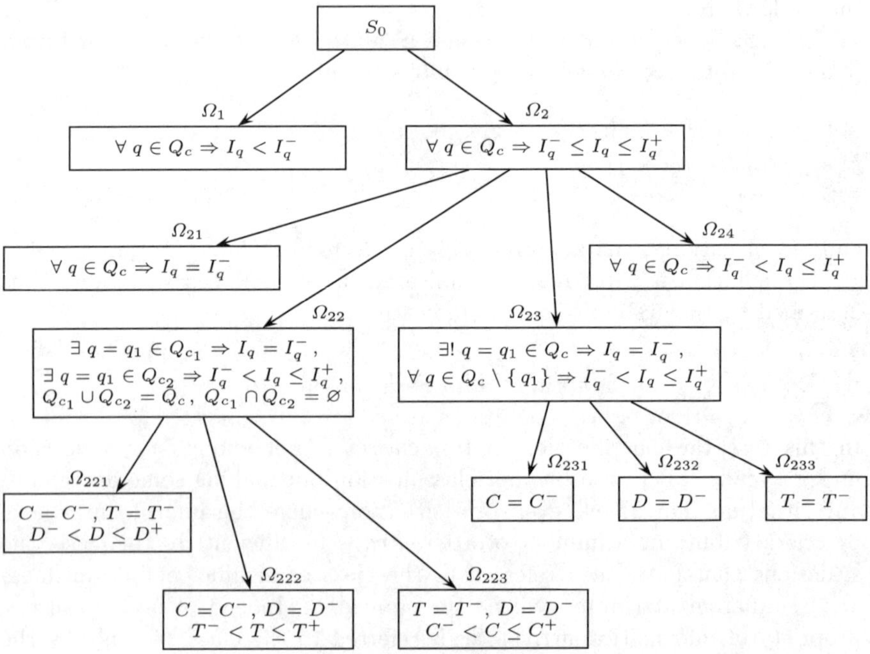

Fig. 5.6. Procedure of recognition by partial indicators of informedness

The procedure of recognition, consisting in referring the examined object S_k to a concrete class of each hierarchy level, is reduced to a consistent checking of the conditions that define properties of the corresponding classes. The sequence of steps in the procedure of recognition is illustrated in Fig. 5.6.

Let us move on to solving the next two problems of recognition, for conditions of incompleteness and information fuzziness. It should be pointed out that the problem of fuzzy information and the problem for the conditions of incompleteness, uncertainty, inaccuracy, and contradictoriness of the information differ in an essential way. At the same time, mathematically they may be reduced to similar procedures. Let us first examine the problem with fuzzy information. While solving this problem the following procedures should be formalized:

1. Classification of set S_0 with fuzzy information about the partial informedness indicators.
2. Setting the fuzzy relationship that will allow one to compare unambiguously the fuzzy description of the class Ω_r and fuzzy description of the examined situation S_k by each indicator from the system of the particular informedness indicators.
3. Composition of the fuzzy relationships that will allow one to compare unambiguously S_k and classes Ω_r by the system of the partial indicators.

Now let us move on to the formalization of these procedures. The set of the relationships both fuzzy and common is quite varied and very powerful. But, of all the variety in those relationships, we are interested only in the binary relationships (since they compare object S_k with one of the classes Ω_r) and certain operations on relationships (since the comparison is made by the totality of indicators). Evidently, of numerous possible types of classification the most convenient type is one that will coincide with the introduced multilayer classification at the value of the membership function corresponding to the characteristic function of a common set. Further, in the procedure of classification of set S_0 by the fuzzy assigned information, it is necessary to determine the choice of the level of class comparison, assuming that set S_0 will be presented in the form of a fuzzy-set union.

For realization of these conditions let us introduce a system of the following fuzzy sets. Their properties will be described verbally:

$$
\begin{aligned}
A_{qm}^- &= \{I_q \mid \text{value } I_q \text{ is very close to } I_{qm}^-,\ \forall\, q \in Q_c\}\ , \\
A_q^- &= \{I_q \mid \text{value } I_q \text{ is very close to } I_q^-,\ \forall\, q \in Q_c\}\ , \\
A_q^+ &= \{I_q \mid \text{value } I_q \text{ is very close to } I_q^+,\ \forall\, q \in Q_c\}\ , \\
A_{qM}^+ &= \left\{I_q \mid \text{value } I_q \text{ is very close to } I_{qM}^+,\ \forall\, q \in Q_c\right\}\ .
\end{aligned}
\tag{5.40}
$$

These relationships determine fuzzy boundaries that correspond to strict boundaries in the form $A_{qm}^- \Rightarrow I_{qm}^-$, $A_q^- \Rightarrow I_q^-$, $A_q^+ \Rightarrow I_q^+$, $A_{qM}^+ \Rightarrow I_{qM}^+$.

It is convenient to present each given fuzzy set in the form of a table, where meanings of the membership function for different $q \in Q_c$ may be different, but the structure of the sets stays the same. As an obligatory condition, it is essential to mention the following one. Since $I_{qm}^- = 0\ \forall\, q \in Q_c$, each set A_{qm}^- should be set in the absolute values I_q. For example, for $q = 1$, assuming $I_q = I_1$, we shall present the given set in the form of Table 5.1.

Then, for the membership function $\mu_{A_{m1}^-}(I_1)$ of set A_{m1}^- we can assume $I_1 \in V_1$, where $V_1 = [0; 0.12]$ and, by supposing that $\mu_{A_{m1}^-}(I_1)$ is determined from Table 5.1, we may present the given set in the standard form:

$$
A_{m1}^- = \left\{I_1 \mid I_1 \in V_1,\ \mu_{A_{m1}^-}(I_1) \ge 0\right\}\ .
$$

Table 5.1. Value $I_q = I_1$ in absolute numbers

I_1	0	0.01	0.02	0.04	0.06	0.08	0.1	0.12
$\mu_{A^-_{m1}}(I_1)$	1	0.99	0.9	0.8	0.7	0.5	0.25	0

For other sets the values I_q are convenient to set in the relative values, for example, using Table 5.2, where γ_q^- determines the relative value of I_q for set A_q^-:

$$\gamma_q^- = \frac{I_q}{I_q^-}.$$

Table 5.2. Value I_q in relative numbers

γ_q^-	0.9	0.93	0.95	0.98	1	1.02	1.05	1.07	1.1
$\mu_{A_q^-}(\gamma_q^-)$	0	0.5	0.7	0.9	1	0.9	0.7	0.5	0

Then, for set $A_q^-(I_q)$ we can assume $I_q = \gamma_q^- I_q^-$, where $\gamma_q^- \in V_q^-$, $V_q^- = [0.9, 1.1]$. Supposing that $\mu_{A_q^-}(\gamma_q^-)$ is determined from Table 5.2, set A_q^- may be presented in the standard form in the following way:

$$A_q^- = \left\{ I_q \mid I_q = \gamma_q^- I_q^-, \mu_{A_q^-}(\gamma_q^-) \geq 0 \right\}.$$

Similarly, we can formalize all other fuzzy sets presented by relationships (5.40). For set A_{qM}^+ it is necessary to determine γ_{qM}^+ on the asymmetric interval $[d^+, 1]$, where d^+ is the value of γ_{qM}^+ when $\mu_{A_{qM}^+}(d^+) = 0$ and $d^+ < 1$. It should be noted that the formation of the examined fuzzy sets is a nonformalizable problem, in the sense that Tables 5.1 and 5.2 are created by experts or they are generated on the basis of statistical processing of a certain array of experimental data. These tables cannot be obtained analytically or on the basis of some calculations.

Taking into consideration the well-known properties of fuzzy sets of the form (5.40) and assigning the tables for them for all $q \in Q_c$, we can build fuzzy sets that describe the classification S_0 for the classes Ω_r, $r = \overline{1, R_0}$. Let us present sets (5.40) graphically by the following scheme (Fig. 5.7). This scheme represents the following objects: region I: $I \in [I_{qm}^-, I_q^-]$; region II: $I \in [I_q^-, I_q^+]$; region III: $I \in \left[I_q^+, I_{qM}^+\right]$.

Let us denote the fuzzy sets that correspond to the mentioned regions by the symbols A_{I}, A_{II} and A_{III}. These fuzzy sets are determined by the fuzzy boundaries A_{qm}^-, A_q^-, A_q^+ and A_{qM}^+, whose membership functions are presented in the scheme. It is necessary to get a mathematical description of

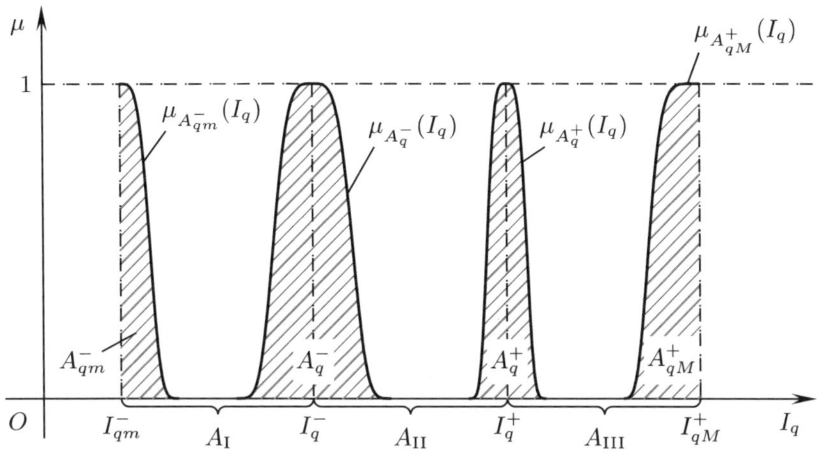

Fig. 5.7. Graphic representation of fuzzy sets described verbally by relationships (5.40)

the mentioned fuzzy sets $A_{\mathrm{I}}, A_{\mathrm{II}}$, and A_{III}. Taking into consideration (5.40), each set A_{I}, A_{II}, and A_{III} is determined by the respective set of the number of axis points that satisfy the following conditions:

for region I: $I^-_{qm} < I_q < I^-_q$;
for region II: $I^-_q < I_q < I^+_q$;
for region III: $I^+_q < I_q < I^+_{qM}$,

and have the corresponding membership functions.

Consequently, in this case it is necessary to get descriptions for the intervals (or periods) that characterize regions I, II, and III based on the fuzzy assigned bounds.

At the same time, the following restrictions should be observed:

1. The variables I_q for each q and for each region I, II, and III cannot be equal to the boundary values of the corresponding intervals from I^-_{qm} to I^+_{qM}. Consequently, the fuzzy sets we look for should have values of the corresponding membership functions equal to zero when I_q is equal to boundary values I^-_{qm}, I^-_q, I^+_q, and I^+_{qM}.
2. It is necessary to take into consideration that boundaries of the numerical sets are determined not by the fixed values (points) but by fuzzy sets A^-_{qm}, A^-_q, A^+_q, and A^+_{qM}, and, therefore, beyond them the corresponding membership functions of the sought-for sets will be equal to 1. It follows that on the boundaries the fuzzy-set membership function values are determined as the membership functions of the complement to fuzzy sets A^-_{qm}, A^-_q, A^+_q, and A^+_{qM}. That is why it is necessary to build complements to each boundary fuzzy set A^-_{qm}, A^-_q, A^+_q, and A^+_{qM} within the limits of the corresponding regions I, II, and III.

For region I we shall build complements $A_{qm}^{-'}$ and $A_q^{-'}$ of fuzzy sets A_{qm}^{-} and A_q^{-}, respectively. Fuzzy set $A_{qm}^{-'}$, taking into account the membership function in the form

$$\mu_{A_{qm}^{-'}}(I_q) = 1 - \mu_{A_{qm}^{-}}(I_q) ,$$

will be characterized by the relationship

$$A_{qm}^{-'} = \left\{ I_q \mid I_q \in V_1, \, \mu_{A_{qm}^{-'}}(I_q) = 1 - \mu_{A_{qm}^{-}}(I_q) \right\} . \qquad (5.41)$$

Here, I_q is determined on the interval of region I, i.e., $I_q \in V_1$, where $V_1 = \left[I_{qm}^{-}, I_q^{-} \right)$. Fuzzy set $A_q^{-'}$, taking into account the membership function in the form

$$\mu_{A_q^{-'}}(I_q) = 1 - \mu_{A_q^{-}}(I_q) ,$$

is described by the relationship

$$A_q^{-'} = \left\{ I_q \mid I_q \in V_1, \, \mu_{A_q^{-'}}(I_q) = 1 - \mu_{A_q^{-}}(I_q) \right\} . \qquad (5.42)$$

The sought-for set A_{I} will be determined by sets $A_{qm}^{-'}$ and $A_q^{-'}$.

Let us find the intersection of fuzzy sets $A_{qm}^{-'}$ and $A_q^{-'}$ and denote it by A_{I}. Determine the membership function A_{I} in the form

$$\mu_{A_{\mathrm{I}}}(I_q) = \mu_{A_{qm}^{-'}}(I_q) \cap \mu_{A_q^{-'}}(I_q) = \min\left(\mu_{A_{qm}^{-'}}(I_q), \mu_{A_q^{-'}}(I_q) \right) . \qquad (5.43)$$

Analyze relationship (5.43) on the following intervals.

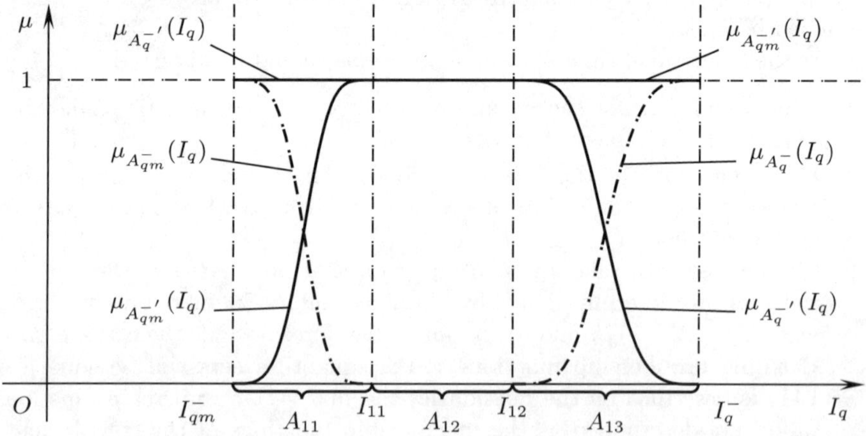

Fig. 5.8. Schematic representation of fuzzy set A_{I}, $I_{qm}^{-} < I_q < I_q^{-}$

1. In the interval $\left[I_{qm}^{-}, I_{11}\right)$, we have $\mu_{A_{qm}^{-\prime}}\left(I_q\right) < \mu_{A_q^{-\prime}}\left(I_q\right)$ and, consequently, $\mu_{A_{qm}^{-\prime}}\left(I_q\right) < 1$. Thus, function $\mu_{A_{11}}\left(I_q\right)$, in accordance with (5.43), satisfies the condition

$$\mu_{A_{11}}\left(I_q\right) = \mu_{A_{qm}^{-\prime}}\left(I_q\right) \quad \text{when} \quad I_q \in \left[I_{qm}^{-}, I_{11}\right).$$

2. In the interval $\left[I_{11}, I_{12}\right]$ we have $\mu_{A_q^{-\prime}}\left(I_q\right) = \mu_{A_{qm}^{-\prime}}\left(I_q\right) = 1$ and, consequently, $\mu_{A_{12}}\left(I_q\right) = \mu_{A_q^{-\prime}}\left(I_q\right) = \mu_{A_{qm}^{-\prime}}\left(I_q\right) = 1$.
3. In the interval $\left(I_{12}, I_q^{-}\right]$ we have $\mu_{A_q^{-\prime}}\left(I_q\right) < \mu_{A_{qm}^{-\prime}}\left(I_q\right)$, restriction $\mu_{A_q^{-\prime}}\left(I_q\right) < 1$ is true, and, consequently, $\mu_{A_{13}}\left(I_q\right) = \mu_{A_q^{-\prime}}\left(I_q\right) < 1$.

Thus, for set A_{I}, which is the intersection of fuzzy sets $A_q^{-\prime}$ and $A_{qm}^{-\prime}$, we have the relationship:

$$\begin{aligned} A_{\mathrm{I}} = \Big\{ I_q \mid & \Big(\mu_{A_{11}}\left(I_q\right) = \mu_{A_{qm}^{-\prime}}\left(I_q\right) < 1 \,\forall\, I_q \in \left[I_{qm}^{-}, I_{11}\right)\Big), \\ & \Big(\mu_{A_{12}}\left(I_q\right) = \mu_{A_q^{-\prime}}\left(I_q\right) = \mu_{A_{qm}^{-\prime}}\left(I_q\right) = 1 \,\forall\, I_q \in \left[I_{11}, I_{12}\right]\Big), \\ & \Big(\mu_{A_{13}}\left(I_q\right) = \mu_{A_q^{-\prime}}\left(I_q\right) < 1 \,\forall\, I_q \in \left(I_{12}, I_q^{-}\right]\Big)\Big\}. \end{aligned} \tag{5.44}$$

Consequently, set A_{I} is determined in three intervals, using different membership functions whose values in the boundary intervals are less than 1, and in the middle interval $\left[I_{11}, I_{12}\right]$ the membership function is equal to 1.

Schematically the initial sets $A_q^{-\prime}$ and $A_{qm}^{-\prime}$ and their intersection in the form of set A_{I} are shown in Fig. 5.8.

It should be noted that the membership function $\mu_{A_{\mathrm{I}}}\left(I_q\right)$ keeps its form proceeding from the definition of the intersection in the form of an algebraic product.

Similar to the A_{I} we can find fuzzy sets A_{II} and A_{III}. Set A_{II} is described in the form

$$A_{\mathrm{II}} = \left\{I_q \mid I_q \in V_{\mathrm{II}}^q, \mu_{A_{\mathrm{II}}}\left(I_q\right) = 1\right\}, \tag{5.45}$$

where $V_{\mathrm{II}}^q = \left[I_q^{-}, I_q^{+}\right]$ and $\mu_{A_{\mathrm{II}}}\left(I_q\right)$ are determined as the membership function of sets $A_q^{-\prime}$ and $A_q^{+\prime}$, which are the complements to the sets

$$A_q^{-} = \left\{I_q \mid I_q \in V_{\mathrm{II}}^q, \mu_{A_q^{-}}\left(I_q\right) \geq 0\right\},$$
$$A_q^{+} = \left\{I_q \mid I_q \in V_{\mathrm{II}}^q, \mu_{A_q^{+}}\left(I_q\right) \geq 0\right\},$$

respectively.

Here, A_q^{-} and A_q^{+} are fuzzy sets that describe the boundaries I_q^{-} and I_q^{+}. Set A_{III} is described by the relationship:

$$A_{\mathrm{III}} = \left\{I_q \mid I_q \in V_{\mathrm{III}}^q, \mu_{A_{\mathrm{III}}}\left(I_q\right) \geq 0\right\}, \tag{5.46}$$

where $V_{\mathrm{III}}^q = \left(I_q^{+}, I_{qM}^{+}\right]$.

Note that sets A_{I}, A_{II}, and A_{III} determine the classification by the partial indicator I_q; we shall denote them by $A_{q\mathrm{I}}$, $A_{q\mathrm{II}}$, and $A_{q\mathrm{III}}$. The fuzzy sets Ω_r, which describe the classification S_0 by the partial indicators of informedness, are similarly built.

Now, let us move to an analysis of the procedure of **situation recognition** in conditions of fuzzy information. The essence of this procedure lies in a comparison of the fuzzy sets that describe classes and the sets that describe the corresponding situations. Comparing these sets allows us to correlate unambiguously, in the general case, the class Ω_r and the analyzed situation S_k by each of the partial indicators of informedness I_q. In accordance with the accepted statement of the problem, the situation S_k, as an object of examination, will be described by set $C_q(S_k)$ of the indicators I_q:

$$C_q(S_k) = \{I_q \mid I_q \in W_q, \mu_{S_k}(I_q) \geq 0\} , \tag{5.47}$$

where $W_q = \left[I_{qm}^-, I_{qM}^+\right]$. Recognition of S_k will be considered, in particular, as a problem of determining which set $A_{q\mathrm{I}}$, $A_{q\mathrm{II}}$, $A_{q\mathrm{III}}$ set $C_q(S_k)$ belongs to. For this it is necessary to find the intersection of the following fuzzy sets:

$$C_q(S_k) \cap A_{q\mathrm{I}}, \quad C_q(S_k) \cap A_{q\mathrm{II}}, \quad C_q(S_k) \cap A_{q\mathrm{III}} .$$

Note that while assigning $C_q(S_k)$ in the form (5.47), it may turn out that the object S_k simultaneously belongs to two or even three classes. That is why, while recognizing S_k, it is expedient to pass from fuzzy set $C_q(S_k)$ to an exact set formed on the basis of a certain level α of the given fuzzy set:

$$C_q^\alpha(S_k) = \{I_q \mid I_q \in W_q, \mu_{S_k}(I_q) \geq \alpha\} . \tag{5.48}$$

Then, the problem of recognition will consist in defining the intersections:

$$C_q^\alpha(S_k) \cap A_{q\mathrm{I}}, \quad C_q^\alpha(S_k) \cap A_{q\mathrm{II}}, \quad C_q^\alpha(S_k) \cap A_{q\mathrm{III}} .$$

The solution to the recognition problem for S_k will be single-valued if the condition is satisfied that only one of the intersections creates a nonempty set and others, an empty set. For example, for

$$\mu_{C_q^\alpha(S_k)} \cap \mu_{A_{q\mathrm{I}}} = 0, \quad \mu_{C_q^\alpha(S_k)} \cap \mu_{A_{q\mathrm{II}}} \neq 0, \quad \mu_{C_q^\alpha(S_k)} \cap \mu_{A_{q\mathrm{III}}} = 0 ,$$

we have

$$C_q^\alpha(S_k) \cap A_{q\mathrm{I}} = \varnothing, \quad C_q^\alpha(S_k) \cap A_{q\mathrm{II}} \neq \varnothing, \quad C_q^\alpha(S_k) \cap A_{q\mathrm{III}} = \varnothing . \tag{5.49}$$

The recognition problem for situation S_k in the conditions of fuzzy information is schematically represented in Fig. 5.9.

It should be pointed out that after passing to the level $\alpha = \alpha_2$ the fuzzy set $C_q(S_k)$ is reduced to a common set $C_q^\alpha(S_k)$, which consists of those elements of set $C_q(S_k)$ for which a membership function exceeds the fixed

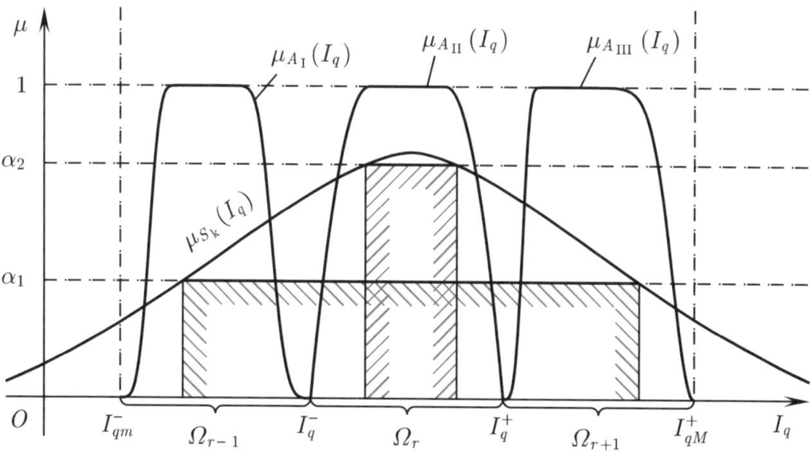

Fig. 5.9. Schematic representation of recognition problem for S_k situation

level α. That is why the problem of recognition for S_k when the information is fuzzy is reduced to a standard problem of recognition with unambiguous and accurate information. The value α is a degree of credibility of the fact that S_k is described by set $C_q^\alpha(S_k)$. The further procedure recognizing S_k is fulfilled according to the multilayer classification scheme (Fig. 5.6) examined above and consists in checking condition (5.48) on each hierarchical level for each of the indicators.

At the same time, the given procedure of recognition has some specific features that should be mentioned.

The choice of the level α is a nonformalizable problem and is made directly by a decision maker. Here, there is one important feature—with increasing α the probability that S_k will simultaneously belong to several classes at one or several levels of the classification decreases. But at the same time, the probability of occurrence of the situation where some objects will stay beyond any class increases. A priori, we cannot exclude the possibility that there will be situations S_k for which in the conditions $1 > \mu_{S_k} \geq 0$ it may turn out that $\mu_{S_k}(I_q) < \alpha$. This follows from the fact that it is not necessary for $\mu_{S_k}(I_q)$ to reach the maximum value equal to 1 for all $S_k \in S_0$, because physically the value $\mu_{S_k}(I_q)$ can be interpreted as a result of group expert evaluation. While the value α is decreasing, not only is set $C_q^\alpha(S_k)$ extended but also the probability increases that in (5.49) there will be no empty intersections at all. That is why determining the rational level of α is a separate problem the solution to which for every concrete practical application is expedient to place on an adaptive intellectual system of decision support.

The second feature of recognition of S_k belonging to a concrete class Ω_r in case of fuzzy initial information is a possible ambiguity of recognition at any value of α. This is connected with the fact that set $C_q^\alpha(S_k)$ is an interval, not a point, as in the case of precisely assigned information, when I_q is equal to a definite number. In this case it is impossible to give simple recommendations to a decision maker. Consequently, making various alternative recommendations and estimating their efficiency in the given conditions also is a separate problem, and it is expedient to entrust it to an adaptive intellectual system of decision support. The problem consists in choosing a rational alternative out of a set of possible ones, based on their multifactor comparative estimation in the sense of the assigned preference relation. Note that intellectual systems of decision support of the stated problems should be hybrid expert systems and ensure both forming and processing of expert information, so the analysis and processing of the statistic or other, a priori, information and the forming of the recommendations should be carried out taking into consideration new knowledge and using the knowledge database previously formed.

Not mentioning other features of intellectual systems of decision support, we shall pay attention to the property of ambiguity recognition of S_k belonging to the introduced classes Ω_r. In the general case, a number of variants of ambiguity is possible:

Variant 1. For any values α_1 and α_2 the object S_k belongs to several classes (Fig. 5.10).

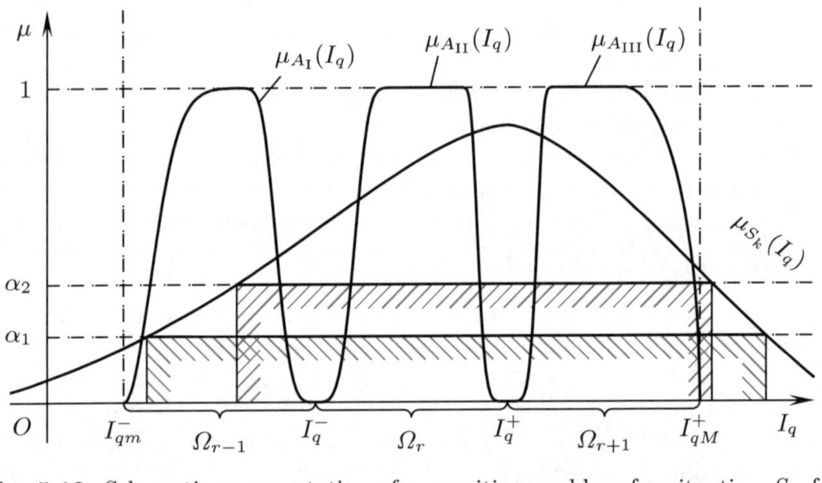

Fig. 5.10. Schematic representation of recognition problem for situation S_k for different levels α

Variant 2. While increasing α from α_1 to α_2 for $\alpha_2 \to 1$, the number of classes to which S_k belongs will decrease, but the number of classes Ω_r will be not less than 2 (Fig. 5.11).

Variant 3. When α_2 increases to 1, the number of classes in Ω_r decreases to 1 (Fig. 5.9).

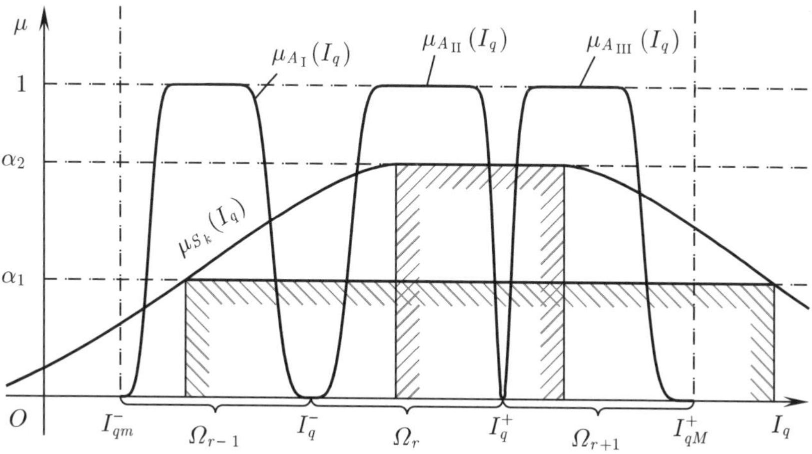

Fig. 5.11. Schematic representation of recognition problem for situation S_k for case $\alpha_2 \to 1$ (number of cases Ω_r is no less than two)

From the above examples it follows that fundamentally unavoidable ambiguity (variants 1 and 2), conditioned by the ambiguity of the fuzzy S_k estimation (of membership function), is possible. In this case the degree of the membership in a concrete class Ω_r can be determined by comparing the membership function of the class and the membership function of an object for the assigned level α, using the procedure of comparing two intervals. Such procedures are developed in possibility theory [54].

5.5 Determining the Permissible Time for Generating and Realizing Decisions to Prevent Catastrophic Situations: Example

The practical need for investigation of system problems of recognition and prevention of abnormal, critical, and catastrophic situations in complex systems of various purposes directly follows from the tendencies and features of modern technology. Modern technology is characterized by a transition to sophisticated machinery and complex engineering systems that have high unit capacity and high efficiency along with a dramatic increase in power rating and thermomechanical loading of structural components [33].

In this section we shall give a mathematical statement of system problems of recognition and prevention of critical and catastrophic situations and offer an algorithm of their solution by the example of a power-plant turbogenerator.

Mathematical statement of problem. In the process of functioning of power-plant turbogenerators under the influence of the set

$\Phi = \{\Phi_j \mid j = \overline{1,m}\}$ of uncontrollable risk factors Φ_j, the normal situation S_i may turn into a critical, extreme, or catastrophic one. Such a turnabout in a situation can occur during a period of time of unknown duration. This period depends on the amount, properties, and exposure time of the factors $\Phi_j \in \Phi$.

One **must** determine the permissible time T_0 for generating and realizing a decision, for which the probability of transition of situation S_i into critical, extreme, or catastrophic will not exceed the assigned value $\eta = \eta_{\mathrm{adm}}$.

The quantity of risk factors and situations will be assigned by means of Table 5.3, where the "+" sign means that under the action of the corresponding factor the normal situation turns into a critical, extreme, or catastrophic one, and the sign "0" means that the risk factor does not influence the situation. Notice that the method and algorithm of the problem solution is applicable for the finite values i and j.

Table 5.3. Risk factors Φ_j influencing the transition of power-plant turbogenerator functioning into a critical, extreme, or catastrophic situation

S_i	Decrease in frequency to 49.7 Hz	Increase in frequency above 50.1 Hz	Erroneous actions of operations staff	Failure of antidamage automatic equipment	Spontaneous forces	Decrease in frequency below 49 Hz	Emergency switching of high power capacity
Variation in capacity of power system	+	+	+	+	0	+	+
Using nuclear power plant units as auxiliaries	0	0	+	+	0	0	+
Asynchronous mode of power system functioning	0	0	+	+	+	+	+
Division of power systems into parts	0	0	+	+	+	+	0

Problem solution. The probability of transition of a situation S_i into a critical, extreme, or catastrophic one under the influence of the factor $\Phi_j \in \Phi$, $j \in [1, 7]$ depends on the time variation of completeness I_C^{ij}, credibility I_D^{ij}, and timeliness I_T^{ij} of a decision maker's informedness. The probability of such an event is determined by the relationship:

$$\eta_{ij} = 1 - \lg\left(1 + a_{ij} I_{ij}(t)\right), \quad I_{ij}(t) = I_C^{ij}(t) I_T^{ij}(t) I_D^{ij}(t).$$

To make a decision it is necessary to find a rational compromise between the levels of I_C^{ij}, I_T^{ij}, and I_D^{ij} for the purpose of reducing the time needed for its formation and realization.

The indicators of completeness I_C^{ij} and validity I_D^{ij} of informedness increase over time and are determined by the following conditions:

$$I_C^{ij} = \begin{cases} \hat{I}_C^{ij}\left(1+\alpha_{ij}t\right), & \text{if } \hat{0} < I_C^{ij}\left(1+\alpha_{ij}t\right) < 1\,, \\ 1, \quad \text{if } \hat{I}_C^{ij}\left(1+\alpha_{ij}t\right) \geq 1\,, \end{cases}$$

$$I_D^{ij} = \begin{cases} \hat{I}_D^{ij}\left(1+\gamma_{ij}t\right), & \text{if } \hat{0} < I_D^{ij}\left(1+\gamma_{ij}t\right) < 1\,, \\ 1, \quad \text{if } \hat{I}_D^{ij}\left(1+\gamma_{ij}t\right) \geq 1\,. \end{cases}$$

Simultaneously with an increase in the time of the influence of the factors $\Phi_j \in \Phi$ the level of the indicator of informedness timeliness I_T^{ij} decreases in compliance with its properties characterized by the relationship:

$$I_T^{ij} = \begin{cases} \hat{I}_T^{ij}\left(1-\beta_{ij}t^2\right), & \text{if } \hat{0} < \hat{\beta}_{ij}t^2 < 1\,, \\ 0, \quad \text{if } \beta_{ij}t^2 \geq 1\,. \end{cases}$$

And, as a consequence, this reduces the duration of the time period for generating, making, and realizing a decision by a decision maker for the purpose of preventing the transition of an investigated situation into a critical, extreme, or catastrophic one.

The factors α_{ij}, β_{ij}, and γ_{ij} characterize the dynamics of changes in informedness indicators. They are determined by the following dependences:

$$\alpha_{ij} = \begin{cases} e^{\hat{a}_{ij}}\hat{I}_C^{ij}\cdot 0.5, & \text{if } 0 < \hat{a}_{ij} \leq 1\,, \\ 0, \quad \text{if } \hat{a}_{ij} > 1\,, \end{cases}$$

$$\beta_{ij} = \begin{cases} \left(\hat{a}_{ij}+\gamma_{ij}\right)\hat{I}_T^{ij}\cdot 10^{-5}, & \text{if } 0 < \hat{a}_{ij} \leq 1\,,\ 0 < \hat{\gamma}_{ij} \leq 1\,, \\ 0, \quad \text{if } \hat{a}_{ij} > 1\,, \end{cases}$$

$$\gamma_{ij} = \begin{cases} e^{\hat{I}_D^{ij}}\hat{a}_{ij}\cdot 0.05, & \text{if } 0 < \hat{a}_{ij} \leq 1\,, \\ 0, \quad \text{if } \hat{a}_{ij} > 1\,. \end{cases}$$

Values of $\hat{I}_C^{ij}$, $\hat{I}_T^{ij}$, and $\hat{I}_D^{ij}$ are the preliminary estimates of the corresponding indicators determined by experts at the moment of detection of an abnormal mode of functioning of a turbogenerator, and factors $\hat{a}_{ij}$ characterize the level of influence of each of factors $\Phi_j \in \Phi$, $j \in [1,7]$ on properties of the situations S_i, $i \in [1,4]$. Values of the indicators $\hat{I}_C^{ij}$, $\hat{I}_T^{ij}$, and $\hat{I}_D^{ij}$ and factors $\hat{a}_{ij}$ are presented in Table 5.4.

In order to determine the duration of the permissible period $T_0 = [T_1, T_2]$, where T_1 and T_2 are the bottom and top limits of a range, respectively, it is necessary to find the solution of the inequality:

$$0 \leq 1 - \lg\left(1 + \hat{a}_{ij}I_C^{ij}I_D^{ij}I_T^{ij}\left(1+\alpha_{ij}t\right)\left(1+\gamma_{ij}t\right)\left(1-\beta_{ij}t^2\right)\right) \leq \eta_{\text{p}}\,. \quad (5.50)$$

Table 5.4. Values of coefficients $\hat{a}_{ij}$ and indicators $\hat{I}_C^{ij}$, $\hat{I}_D^{ij}$, and $\hat{I}_T^{ij}$

	S_i, Φ_j	1	2	3	4	5	6	7
$\hat{a}_{ij}$								
	1	0.5	0.6	0.65	0.5	—	0.7	0.6
	2	—	—	0.6	0.7	—	—	0.4
	3	—	—	0.7	0.7	0.4	0.55	0.65
	4	—	—	0.75	0.6	0.4	0.5	—
$\hat{I}_C^{ij}$								
	1	0.6	0.7	0.4	0.8	—	0.7	0.6
	2	—	—	0.5	0.6	—	—	0.5
	3	—	—	0.4	0.4	0.4	0.8	0.6
	4	—	—	0.6	0.3	0.35	0.6	—
$\hat{I}_D^{ij}$								
	1	0.7	0.8	0.4	0.7	—	0.7	0.7
	2	—	—	0.3	0.8	—	—	0.8
	3	—	—	0.3	0.8	0.4	0.6	0.6
	4	—	—	0.5	0.7	0.3	0.7	—
$\hat{I}_T^{ij}$								
	1	0.8	0.8	0.6	0.8	—	0.8	0.9
	2	—	—	0.7	0.9	—	—	0.6
	3	—	—	0.5	0.8	0.5	0.7	0.75
	4	—	—	0.8	0.75	0.55	0.8	—

By the initial data set in Table 5.4 we determine factors α_{ij}, β_{ij}, and γ_{ij} in the form of the following matrices:

$$\alpha_{ij} = \begin{pmatrix} 0.494 & 0.637 & 0.383 & 0.66 & 0 & 0.705 & 0.547 \\ 0 & 0 & 0.456 & 0.604 & 0 & 0 & 0.373 \\ 0 & 0 & 0.403 & 0.403 & 0.298 & 0.693 & 0.575 \\ 0 & 0 & 0.635 & 0.273 & 0.261 & 0.494 & 0 \end{pmatrix},$$

$$\gamma_{ij} = \begin{pmatrix} 0.05 & 0.067 & 0.0495 & 0.050 & 0 & 0.070 & 0.06 \\ 0 & 0 & 0.040 & 0.078 & 0 & 0 & 0.045 \\ 0 & 0 & 0.047 & 0.078 & 0.03 & 0.05 & 0.062 \\ 0 & 0 & 0.062 & 0.060 & 0.027 & 0.05 & 0 \end{pmatrix},$$

$$\beta_{ij} = 10^4 \begin{pmatrix} 0.389 & 0.563 & 0.259 & 0.568 & 0 & 0.62 & 0.546 \\ 0 & 0 & 0.347 & 0.613 & 0 & 0 & 0.254 \\ 0 & 0 & 0.225 & 0.385 & 0.164 & 0.52 & 0.478 \\ 0 & 0 & 0.558 & 0.249 & 0.158 & 0.435 & 0 \end{pmatrix}.$$

From inequality (5.50) we find permissible periods T_0 at various values $\eta_p = 0.5, 0.6, 0.7, 0.8, 0.9$. The results of the solution for $\eta_p = 0.5$ are presented in Table 5.5.

Table 5.5. Permissible period T_0 of decision making

S_i	Φ_1	Φ_2	Φ_3	Φ_4	Φ_5	Φ_6	Φ_7
S_1	$[0, 30.2]$	$[0, 33.4]$	$[0, 36.2]$	$[0, 32.2]$	—	$[0, 27.2]$	$[0, 29.1]$
S_2	—	—	$[0, 39.4]$	$[0, 36.3]$	—	—	$[0, 41.3]$
S_3	—	—	$[0, 45.5]$	$[0, 35.4]$	$[0, 43.9]$	$[0, 28.2]$	$[0, 32.1]$
S_4	—	—	$[0, 31.3]$	$[0, 35.3]$	$[0, 30.1]$	$[0, 33.3]$	—

Thus, for a situation S_1 the permissible time to forming, making, and realizing the decision should not exceed $T_2 = 27.2$; for situation S_2—$T_2 = 36.3$; for situation S_3—$T_2 = 28.2$; for situation S_4—$T_2 = 30.1$.

Remark. The choice of the unit of T_0 for various engineering systems depends on the dynamic properties of an investigated process: a microsecond, a second, a minute, or an hour.

6 Structural-Functional Analysis of Complex Hierarchical Systems

Let us consider the main problems that compose the general problem of the system analysis of complex multilevel hierarchical systems (CMHSs).

The structural-functional analysis (SFA) of specified class systems relates to a number of the most complicated but, in practical terms, extremely important problems of modern science and technology. The importance and role of these problems constantly grows, which is determined by industrial needs for a wide introduction and development of high technologies. Problems of the SFA by level of complexity belong to a class of organized complexity problems. The given class of problems occupies the main part of the complexity spectrum. The boundaries of the complexity spectrum are two, in some sense, opposite classes of problems—organized simplicity and unorganized complexity. This means that, except for the two boundaries, the main part of this spectrum is not supported methodologically, in the sense that it is impossible to solve the corresponding problems either by analytical or by statistical methods. The overwhelming majority of practical problems in various branches of science and technology, economics, social life, etc. belong to this part of the spectrum. The peculiarity of the indicated problems is that they describe objects that possess fundamentally new properties, such as organization, hierarchy, cosubordination, adaptability, stability, and controllability. Due to these properties the indicated class of problems is known as ***problems of organized complexity***.

The objects for which these problems are characteristic have the following distinctive features.

First, for these problems it is impossible in principle to ignore most influential factors; second, for these objects it is difficult to get meaningful statistical estimations; third, these objects are so complex that most applied problems in their analysis are impossible to formalize (for example, choice of an object's structure, choice of criteria of preference of variants, etc.); fourth, such an analysis of objects must be carried out under real conditions of multifactorial risks characterized by the uncertainty, incompleteness, inaccuracy, and contradictoriness of the initial information; fifth, a considerable part of practically important properties of objects cannot be precisely described (for example, aesthetics, convenience of use, etc.).

Therefore, the methodical approaches to solving system problems of the given category of complexity are not sufficiently developed. The development of these approaches is frequently based on heuristic methods and methods of creating the intellectual means of support of decision making on the basis of systematization, generalizations, and accumulation of developers' knowledge and experience.

In this chapter the principles and approaches to solving the specified class of problems are formed, and practical procedures of their algorithmization and numerical solving are proposed.

6.1 Basic Properties and Features of Complex Hierarchical Systems

An important feature of decision making in complex hierarchical systems is granting the decision maker a certain freedom of action at different hierarchical levels in forming and choosing decisions. In this connection a complex nonformalizable problem arises consisting of a rational distribution of efforts and authorities on decision making between the decision makers at different hierarchical levels of the system. Such an approach is caused by the necessity to realize the potential capabilities of the complex hierarchical structure in saving various resources and organizing its rational behavior.

However, solving the indicated problem gets complicated in many ways due to the following peculiarities of complex hierarchical structures.

1. At higher levels of a hierarchy there is deeper understanding of the purpose, functions, and possibilities of an object and various aspects of its behavior, but at the same time the level of incompleteness, uncertainty, inaccuracy, and contradictoriness of the initial information is higher.

2. At a consecutive transition to a lower level of the hierarchy, the level of uncertainty decreases, and the possibility of a more detailed, specific description of the structure and design of functional elements (FEs) increases, and the possibility to concretize various works and problems also increases.

These features generate a contradiction: at a higher level of hierarchy the understanding of goals and problems of an object is essentially higher; however, at the same time, the level of understanding of concrete ways of their achievement, and concrete possibilities of finding alternative design solutions for concrete FEs of lower levels are essentially lower. At the lowest level the situation is the opposite: there is almost complete understanding of concrete technical, technological, and other possibilities of alternative design solutions of FEs, but at the same time there is almost a complete absence of understanding of the role and place of a concrete FE in achieving conceptual goals of a complex hierarchical system as a whole.

Resolving this contradiction is one of the major problems of the general strategy of ***structural functional analysis*** (SFA). Due to the above indicated peculiarities the contradiction is resolved by:

1. Forming a hierarchy of problems of SFA consecutively from top to bottom based on the hierarchical structure of decision making.
2. Forming a solution of the initial problem of SFA consecutively from top to bottom based on the hierarchical structure.
3. The consecutive coordination of goals of SFA and the obtained results of the solution by interactive realization of items (1) and (2).

One more contradiction is caused by the specificity of decision making. As is known, in any real decision-making procedure there are two extremely simple, but principally important, peculiarities.

1. In real situations, the time for making a decision and its implementation is limited, and untimely actions may have irreparable consequences.

2. Incompleteness of the initial information and the absence of sufficient knowledge about the reasons and consequences of the influence of uncontrollable factors hamper the understanding and evaluation of the developing situation.

These features lead to the following contradiction: on the one hand, to ensure a timely solving of such problems, it is desirable to reduce the time of formation and justification of the solution, and on the other hand, to increase the level of justification and credibility of the solution, it is desirable to increase the time of its formation for the purpose of obtaining more complete information and more detailed analysis of the developing situation. The practical importance of resolving this contradiction is obvious from the features considered above. Attention should also be given to the threshold mechanism of the time factor action—an absolutely proven and most efficient solution becomes useless if it is untimely.

The techniques for resolving this contradiction depend on many factors among which are the following:

- Level of required efficiency of decision making.
- Extent of possible damage from an unjustified solution.
- Level of a priori informedness of the decision maker about the object of SFA and available possibilities for getting information in the course of solving the problem.

In addition, it is important to take into account the following peculiarities: (1) the stage at which the life cycle of an object SFA is carried out—at the stage of its development or at the stage of its operation; (2) conditions in which the operation of an object SFA is carried out—in normal mode or in an abnormal, critical, or extreme situation.

At the same time, it is possible to suggest some general methods that are applicable in practically any variant of SFA:

1. A rational choice of the number of levels of decision making depending on the time limit.

2. Formation of a rational hierarchy of problems by degree of importance at each level of decision making.
3. Ensuring rational coordination of activities of all levels of making and taking decisions.

The presented properties and features of real, complex hierarchical systems characterize the degree and aspects of their influence on the strategy of solving an SFA problem.

On the one hand, structural properties of cosubordination, interdependence, and determinacy in general are differed for technical, organizational, and organizational–technical systems. But on the other hand, for all these systems there exists the generality in the indicated properties. We shall mention only ***general, functional properties of different types and kinds of systems***.

Observability: expressed by the possibility of determining the system state in the past or at the present moment by results of observations over certain parameters and characteristics of a system.

Controllability: expresses the capability of a system as a whole and its FEs to react adequately to environmental influences. For example, FEs of the top hierarchical level control FEs of the bottom hierarchical level.

Sensitivity: characterizes the severity of system reactions to changes in influencing factors and its own parameters.

Stability: determines the capability of a system to maintain the qualitative characteristics of its behavior in the set limits under the influence of uncontrollable factors.

Coordinateness: characterizes the capability of FEs of one hierarchical level to coordinate their actions with FEs of other levels on the basis of attaining the general goals of a complex system as a whole.

Adaptability: determines the capability of a system to change the structure and parameters of its FEs for the purpose of adapting to changing external conditions.

Complexity: characterizes the degree of saturation of some process or system with FEs, and also the number and variety of connections between them. It is possible to distinguish the following types of complexity:

- ***Structural***: determines the level of complexity of structural interconnections between FEs in a complex hierarchical system.
- ***Functional***: determines the level of complexity of functions realized by a system in general and its FEs.
- ***Technological***: determines the complexity of technologies laid at the foundation of a system and technologies that control processes of functioning.
- ***Constructive***: determines the complexity of creation and realization of a system and its FEs.

– ***Information***: characterizes volumes and complexity of information flows, the variety of ways to transform information into complex hierarchical systems.

Efficiency: productivity of attainment by a system of the goals of some process or system.

Technological efficiency: characterizes the effectiveness in achieving technological goals.

Technical-economic efficiency: characterizes the degree of commensurability of the achieved result with the expenses, for example, the relationship of technological efficiency to the expenses of resources spent on its achievement.

Reliability: determines a system's ability to function without breakdowns within a set time interval while preserving all its qualitative characteristics and modes.

Survivability: determines a system's ability to maintain a certain level of serviceability with breakdowns occurring in its separate FEs.

Further we shall consider the *general constructive-technological properties of the given class of systems.*

Material consumption: characterizes the levels of consumption of various materials for realization of a system's constructive elements.

Metal intensity: characterizes the level of consumption of various metals, including precious metals, for realization of all FEs and system structures. This property is additional to the previous one and distinguishes metals from general material consumption for constructive elements of a system.

Power intensity: characterizes the level of consumption of various kinds of energy or power resources for maintenance of a system in functioning condition.

Capital intensity: characterizes the level of general financial expenses for development, manufacture, and operation of a system.

Laboriousness: characterizes the level of labor expenses for the development, manufacture, and operation of a system.

Let us introduce also some restrictions and assumptions. We shall assume, first of all, that of all the possible stages of system's life cycle only the development cycle (designing and experimental adjustment) is considered. Problems of system analysis for other stages of the life cycle, in particular, for production stages and operation, are not only similar, but in many respects are special cases of the problem of an object system analysis at the development stage. Therefore, as an ***assumption*** we shall take the condition that a problem of system analysis of a complex hierarchical system at the development stage may be considered as a generalized problem of system analysis during the entire life cycle of a product.

As a ***restriction*** we assume that the structure of a complex hierarchical system is strictly defined. For example, the hierarchical levels for some technical systems are strictly determined by tests. In particular, for radio-electronic systems the following hierarchical structure is accepted: ***object*** $\Rightarrow$ ***functional system*** $\Rightarrow$ ***functional device*** $\Rightarrow$ ***functional block*** $\Rightarrow$ ***functional module***.

Note that the basic peculiarity of all design stages of complex hierarchical systems is a combination of various unforeseen nonformalizable and rather labor-intensive computing problems. A principally unavoidable feature of requirements for complex hierarchical systems is their contradictoriness. For example, the requirement of high reliability contradicts the requirement of low cost, the requirement of a reduction in metal intensity contradicts the requirement of an increase in reliability, etc.

Incompleteness, uncertainty, inaccuracy, contradictoriness, and fuzziness of the initial information at the initial stage of development of a complex hierarchical system determines the necessity of solving some important problems interconnected by the unique goal of meeting the requirements specifications within the limits of the allocated resources and the assigned time limit.

First of all, such problems include:

- Estimation of the principal possibility of meeting the assigned requirements.
- Specification of conditions of functioning and operation of the designed system.
- Choice and justification of design goals.
- Choice and justification of the principles of the system's functional capabilities.
- Formation and justification of an allowable set of alternative variants of the product hierarchical structure.
- Choice and justification of the system's rational structure, composition, and interconnections of FEs of each hierarchical level.
- Choice and justification of the system functions in general and functions of all elements for each hierarchical level.
- Rational decomposition of system requirements in the form of requirements for FEs for all hierarchical levels.
- Formation and justification of alternative variants of design solutions of FEs for all hierarchical levels.
- Choice and justification of a group of criteria for estimating the quality of design solutions for all hierarchical levels.
- Choice and justification of rational design solutions of FEs for all hierarchical levels.
- Choice and justification of estimation criteria of the system (structural, functional, constructive-technological) properties.
- Estimations of the degree of conformity of the realized structural, functional, and constructive-technological properties of the system to the assigned requirements.

– Estimation of complexity and technical-economic effectiveness of a system.

It seems expedient to unite several of the enumerated problems in three important groups:

1. Problems of SFA of complex hierarchical systems.
2. Problems of analysis of functioning control of complex hierarchical systems.
3. Problems of technical-economical analysis of complex hierarchical systems.

The solution to each of these groups of problems is reached in conditions of multifactorial risks, determined by the incompleteness, uncertainty, and contradictoriness of the initial information.

This overview of the basic properties and features of complex hierarchical systems allows us to give the following contensive formulation of ***the general problem of system analysis of CMHSs.***

The data on a complex multilevel hierarchical system, determining its applications area, the basic functions and general characteristics, and also the basic requirements for functional, technological, constructive, and operation parameters of quality of its functioning ***are known***. It is a priori known that these data are functionally incomplete, contradictory, and inaccurate.

It is required to define the general structure of a system, to distribute rationally requirements among FEs of all hierarchical levels, to choose and justify a group of criteria for estimation of quality and design solutions of a system as a whole and its FEs, to optimize the design solutions of FEs regarding the accepted group of criteria, and to fulfill the assigned requirements to a system.

The hierarchical structure of a system determines the necessity of introducing the multilevel procedure of forming and taking a decision. It is reached by the consecutive solution of the three groups of problems indicated above. The process of solving is interactive. For example, after obtaining unacceptable results while solving the third problem, one must repeat the solving procedure, returning to the first one. An aggregation of results of solutions of all three problems gives the required solution of the general problem of system analysis of CMHSs.

6.2 Formalization of Structural-Functional Analysis Problem

Formalization of the problem of SFA follows from the general problem of the system analysis of complex hierarchical systems.

First of all, we shall give the contensive statement of the problem of SFA. For the most general case the given problem may be presented in the following form.

The data presented in formalized form that determine the purpose, general characteristics, and properties of a system, and also the basic requirements for its technical, constructive, technological, operational, and economic parameters ***are known***.

It is required to determine the structure of a complex hierarchical system, to justify the requirements of each FE for all hierarchical levels, and to choose and justify functions of each FE for all hierarchical levels proceeding from the condition of meeting the assigned requirements.

Let us consider the formalization of the given problem. With this goal in mind we shall carry out the structuring of a complex hierarchical system. Here it is appropriate to cite the statement of two well-known experts on system theory. Gauguin and Varel in one of their works pointed out: "The world generally is not divided for us into systems, subsystems, environment, etc. We ourselves subdivide it in such a way, proceeding from the different reasons usually reduced to one general reason—for convenience".

This phrase precisely defines the main peculiarity of system structuring—ensuring the convenience of its description from the position of the specific researcher. From here it follows that the given process is principally nonformalizable and subjective. Therefore, we shall consider only the general ideas and methods of structuring and give the formalized description from the position of the formulated problem.

Let us turn to the structuring and formalized description of the properties of a system and its requirements.

We shall present the set of properties of the analyzed system in the form of a structure of classes ordered by importance:

$$B_0 = \left\{B_v | v = \overline{1, m_0}\right\} . \tag{6.1}$$

Here B_0 is a set of properties of the system, B_v is the vth class and unites a certain category of the properties possessing common manifestations, for example, a class of structural properties, a class of functional properties, a class of constructive-technological properties, and also other classes—a class of ergonomic properties, a class of aesthetic properties, etc.

Each class B_v is defined by a set of properties:

$$B_v = \left\{b_{vi} | i = \overline{1, m_v}\right\} . \tag{6.2}$$

For example, in the class of functional properties it is possible to take as b_{vi} controllability, stability, adaptability, and other properties.

Each ith property b_{vi} of the class B_v is characterized by a set of indicators:

$$Y_{vi} = \left\{y_{vik} | k = \overline{1, k_{vi}}\right\} , \tag{6.3}$$

where y_{vik} is the kth indicator of the ith property of the vth class B_v.

The requirements for the properties of the system could be generally presented as follows.

Requirements for the properties of the vth class are determined by the set A_v in the form

$$A_v = \left\{ Z_{vi} | i = \overline{1, m_v} \right\} , \tag{6.4}$$

where Z_{vi} is a set of requirements for the ith property of class B_v, determined by the relationship:

$$Z_{vi} = \left\{ z_{vik} | k = \overline{1, k_{vi}} \right\} , \tag{6.5}$$

where z_{vik} is the requirements for the kth indicator of the ith property of class B_v.

The requirements for indicators are usually assigned taking into account the interval of allowable values or required value in one of the following forms:

$$z_{vik}^- \le z_{vik} \le z_{vik}^+, \quad z_{vik} = z_{vik}^0 \pm \Delta z_{vik}, \quad z_{vik} = z_{vik}^0 \left(1 \pm \delta z_{vik}\right) ,$$

where Δz is an absolute value of allowance and δz is a relative (in %) value of the allowance.

Conditions of system functioning are determined by a set of operation conditions S_e that characterize some set of external influencing factors by the nomenclature and allowances in the form

$$S_e = \left\{ w_{je} | w_{je}^- \le w_{je} \le w_{je}^+, \ j_e = \overline{1, Q_e} \right\} , \tag{6.6}$$

where w_{je} is an indicator of one of the factors of operation conditions, for example, an indicator of temperature, humidity, vibration, or other external factors; w_{je}^- and w_{je}^+ are the allowable minimal and maximal values of the considered indicator, respectively. For example, for some class of engineering objects the temperature interval may be set in the limits $[-40\,°\mathrm{C},\ +20\,°\mathrm{C}]$. Situations and risk factors in normal, abnormal, and critical situations are defined in the course of system development or are briefly described in the requirements specification.

Let us turn to a ***formalized description of the system structure and functions***. Assume that the system structure corresponds to the following hierarchical structure: ***system*** $\Rightarrow$ ***functional system*** $\Rightarrow$ ***functional device*** $\Rightarrow$ ***functional block*** $\Rightarrow$ ***functional module***.

Suppose that each qth hierarchical level consists of a set of FEs V_q described as

$$V_q = \left\{ V_{qp} | q = \overline{1, Q}, \ p = \overline{1, P_0} \right\} , \tag{6.7}$$

where V_{qp} is the pth FE of the qth hierarchical level and Q is the total number of hierarchical levels.

Each FE is characterized by a vector of indicators which for elements $V_{qp} \in V_q$ is determined by the relationship

$$X_{qp} = \left\{ x_{qpj} | j = \overline{1, n_{xp}} \right\} , \tag{6.8}$$

where x_{qpj} is the jth indicator of the pth FE of the qth hierarchical level Q.

Each FE performs some set of functions, for example, certain functions of signal transformation, functions of information processing, etc.

A set of functions of an element $V_{qp} \in V_q$ is described as follows:

$$\Phi_{qp} = \left\{ f_{qpk} | k = \overline{1, n_{qp}} \right\} , \tag{6.9}$$

where f_{qpk} is the kth function of the pth FE of set V_q.

Each function f_{qpk} depends on a vector of indicators $\overline{X}_{qp}$:

$$f_{qpk} = f_{qpk} \left(\overline{X}_{qp} \right) ,$$

where the composition and form of the function is to be defined in the course of the system analysis.

Each FE influences the general properties of the system. The degree of its influence is generally described as

$$F : X \to Y , \tag{6.10}$$

where X is a set of indicators of the system's FEs, Y is a set of indicators determining the system properties, and F is a FE that transforms set X into set Y.

Formation and realization of transformation (6.10) is one of the main goals of the SFA of a complex hierarchical system by a group of requirements assigned in the form (6.4)–(6.6). The given goal is achieved upon a simultaneous informal choice of the system structure as a whole and FEs of all hierarchical levels.

The given formalized description of conditions and system structure allows us to move on to a mathematical statement of the problem.

In the development of complex hierarchical systems, the problem of the SFA is formed at the stage of outline design, where also the ways of solving it are indicated. Further specification of both the problem statement and its solution is carried out at the stage of technical design.

The considered problem may be solved by application of the simulation modeling method for the object properties for the purpose of assessing the possibility of meeting the assigned requirements and disclosing the uncertainty of goals and conditions. Then, a rational compromise of contradictory system requirements is sought and a rational distribution of FE parameter requirements at all hierarchical levels and their structural interconnections is fulfilled to form a unified integral object.

Formalized problem description. Let requirements for system properties being designed be defined as:

$$A_0 = \left\{ A_v | v = \overline{1, m_0} \right\} , \tag{6.11}$$

which do not completely describe its characteristics and parameters as expressed by (6.1)–(6.5). The basic conditions determining a system's ability to function in the form (6.6)–(6.9), as well as its purpose and application area, are specified.

It is required to develop the hierarchical structure of the system, taking into account (6.11) to prove the requirements for FEs of all levels, to suggest and justify mathematical models that establish the interconnections of FE parameters and system property parameters in the form of (6.10), to develop a set of criteria by which to verify that the assigned requirements have been fulfilled, to choose and justify design solutions that ensure the minimal difference between the obtained set of properties B_0 and set A_0 determined by condition (6.11) in the sense of the accepted criteria and assigned assumptions.

Briefly the given formalized problem may be presented as the following sequence of problems.

1. Problem of requirements decomposition. A set of requirements A_0 ***is assigned***. ***It is required*** to construct a sequence of transformations:

$$F_A : A_0 \to \bigcup_{q=1}^{Q} A_q, \quad F_{Ap} : A_q \to \bigcup_{p=1}^{P_0} A_{qp}, \quad F_{qp} : A_{qp} \to \bigcup_{k=1}^{K_{pk}} z_{qpk}, \tag{6.12}$$

where A_0, A_q, and A_{qp} are, respectively, a set of system requirements, a set of requirements for the FEs of the qth hierarchical level, and a set of requirements for indicators of the pth FE of the qth hierarchical level. The functionals F_A, F_{Ap}, and F_{qp} and sets A_q, A_{qp}, and z_{qpk} are determined from the condition assigned by relationships (6.15) and (6.16).

2. Problem of multilevel structure formation. ***It is required*** under the conditions determined by relationship (6.12) to construct a sequence of transformations:

$$f_q : V_0 \to \bigcup_{q=1}^{Q} V_q, \quad f_q^- : \bigcup_{q=1}^{Q} V_q \to V_0, \quad f_{qp} : V_q \to \bigcup_{p=1}^{P_0} V_{qp}, \tag{6.13}$$

where V_0 is a set of FEs of a system as a whole, V_q is a set of FEs of the qth hierarchical level, and V_{qp} is the pth FE of the qth hierarchical level. Functionals f_q and f_{qp} and sets V_0, V_q, and V_{qp} can be determined from conditions (6.15) and (6.16).

3. Problem of a rational choice of FEs indicators. For the structure determined by relationship (6.13), under conditions (6.12), (6.15), and (6.16), it is required to determine direct and reverse fuzzy relations:

$$\begin{aligned} &A_0 R_0 Y_0, \quad Y_0 R_0^{-1} A_0, \quad A_q R_q Y_q, \quad Y_q R_q^{-1} A_q, \\ &A_{qp} R_{qp} Y_{qp}, \quad Y_{qp} R_{qp}^{-1} A_{qp}, \quad Z_{qpk} R_{qpk} X_{qpk}, \quad X_{qpk} R_{qpk}^{-1} Z_{qpk}, \end{aligned} \tag{6.14}$$

where fuzzy relations R_0 and R_0^{-1} define the interconnection of requirements for properties of the system as a whole and the total set of values of the realized parameters of FEs for all hierarchical levels. Fuzzy relations R_q and R_q^{-1}

define the relationships between the requirements for parameters and their values for the qth hierarchical level. Fuzzy relations R_{qp} and R_{qp}^{-1} define the relationships between the requirements for parameters of the pth FE of the qth hierarchical level and numerical values of its parameters. Relations R_{qpk} and R_{qpk}^{-1} define the relationship between the requirements and numerical values of the kth parameter of the pth FE of the qth level. Fuzzy relations $R_0, R_0^{-1}, R_q, R_q^{-1}, R_{qp}, R_{qp}^{-1}, R_{qpk}$, and R_{qpk}^{-1} and sets Y_0, Y_q, Y_{qp}, and X_{qpk} are defined from condition (6.15).

4. Problem of a multipurpose estimation of system functional quality indicators. According to the requirements specification, a system of estimating the solution quality of the total set of problems 1–3 must be constructed. The estimation of the solution quality should be carried out on the basis of using the functional that allows to associate any pair of elements of the set with the value, allowing one to define the distance these elements appear from each other according to some fundamental ordering.

This problem solving consists of two stages.

Stage 1. Construction of the FE

$$\delta_n : W_A \times W_y \rightarrow R_n \,, \tag{6.15}$$

where R_n is the n-dimensional numerical space; W_A is a set unambiguously determined by the functional transformation of set A, set S_e, and a set of risk factors Ω_0; W_y is a set unambiguously determined by the functional transformation of sets Y and S_e and risk factors Ω_0.

Stage 2. Finding solutions to problems 1–3 for which the value

$$R^* = \max_n R_n \left(W_y\right) ,$$

taken as a measure of the object quality, was minimally possible:

$$R^* = R^0, \quad R^0 = \min_Y R^*(Y) \,. \tag{6.16}$$

It is necessary to indicate the number of principal features of the SFA problem.

1. Direct and reverse interconnections of the problem. In reality, to carry out the decomposition of requirements, it is necessary to know the hierarchical structure of the system. But to create such a structure, one should know the requirements of the FEs of all levels.

Further, problems (6.13) and (6.14) are to be solved under condition (6.16). But condition (6.16) can be ensured only with the known solution to problem (6.15). At the same time, problem (6.15) can be solved using the known solutions (6.13) and (6.14). Such repeated interconnection is a characteristic feature of the system analysis problems, which distinguishes these problems from the problems of axiomatic scientific directions and theories (theory of operations research, theory of decision making, etc.).

2. Problems are structurally incompletely determinate. In reality, in problem (6.12) only set A_0 is known. Other sets and functionals are to be obtained in the process of solving the considered problem. Thus, in problem (6.13) all values and all functionals are unknown, i.e., $V_0, V_q, V_{qp}, f_q, f_{qp}, f_q^{-1}$, and f_{qp}^{-1} are unknown.

3. Problems are partially or completely nonformalizable. In particular, it is principally impossible to formalize the procedure for forming a set of alternative design solutions at each hierarchical level. It is possible to formalize a procedure for choosing the best alternative in a certain sense from the assigned set.

The indicated peculiarities prove that there are basic difficulties in solving problems of the given class. Thus, it is necessary to take into account that relationships (6.13)–(6.16) give only a mathematical description of the problem, and on their basis it is impossible directly to organize the computing process.

6.3 General Strategy for Solving the Problem of Structural-Functional Analysis

The earlier considered properties and features of the problem of SFA allow us to consider a strategy for solving it. We understand strategy as the general sequence of principles, approaches, and methods whose implementation will allow us to formalize a concrete practical problem and help achieve the general goal of SFA.

The general goal of SFA is determined by the general statement of the problem of analysis and consists in obtaining a solution with the assigned degree of credibility and justification of the above stated partial problems of SFA upon rational fulfillment of the set requirements, limitations, and conditions. To achieve this goal, the strategy should ensure a rational choice of [118]:

- Hierarchical structure of the system.
- Distribution of requirements and functions among different hierarchical levels.
- Elements and structure of functional interconnection of FEs of different hierarchical levels.
- Elements and structure of each hierarchical level.
- Functions and parameters of FEs on each hierarchical level.

The considered problems retain such properties and features as nonformalizability and multilevel direct and reverse parametric interrelation. The last feature requires explanation. Its essence consists in the following. The direct interrelation indicates the dependence of solving a problem of a lower hierarchical level on the solution accepted at a higher hierarchical level. For example, requirements for FEs of hierarchical levels are determined by the general requirements for an object as a whole. The reverse interrelation indicates the dependence of properties, capabilities, and other characteristics of FEs of a higher hierarchical level on the corresponding solutions accepted at a lower hierarchical level. For example, all dimensions and material consumption of an object as a whole are determined by decisions taken while selecting FEs of lower levels. Thus, the direct and reverse parametric interrelation may determine the functional and parametrical interdependence of FEs of different hierarchical levels and, as a consequence, the corresponding type of dependence of the problems being solved.

Let us consider the analysis of one more feature of SFA problems. In all the statements given above the problem of a rational choice, and not optimization, is set. The difference between these statements is fundamental. The solution of an optimization problem must be unique, but the solution of a rational choice problem is one of the solutions in a Pareto set. The necessity of the accepted statement is determined by the basic nonformalizability of the considered problems of SFA. For example, principally nonformalizable are problems of forming a set of possible hierarchical structures of an object, forming a set of criteria for comparative estimation of structures, forming a set of FEs of a certain hierarchical level, and also a number of other SFA problems.

The major property of such problems is the ambiguity of the solution. It is caused by the ambiguity of forming sets of possible alternatives and sets of possible criteria for comparing alternatives by different decision makers. Decisions of decision makers are determined by subjective factors—experience, intuition, knowledge, preference, etc. From here it is obvious that even identical sets of alternatives may be ranked by a decision maker in the form of different sequences.

The indicated features and properties, and also those that not considered here, determine the necessity of using the following methods and approaches in the implementation of a strategy for solving a problem of SFA.

1. *Orientation toward a rational shared use of abilities of human experts and capabilities of modern intellectual information systems and technologies* (the ability to store large data sets; the ability to parallele computation processes; the ability to perform certain intellectual functions, in particular, accumulating and using knowledge, etc.). Such an approach allows one to use the strengths of the person and computer facilities and jointly compensate to some extent for the drawbacks of both, if they work separately. In addition, this technique allows one to use the modern and potential capabilities of personal computers.

2. *Orientation toward multiloop use of iterative procedure for solving a sequence of SFA problems.* Such an approach allows one to reduce expenses of time and other resources by excluding from the further analysis less perspective alternatives of solutions at each level in a sequence of SFA problems.

3. *Application of an interactive mode in the implementation of a strategy for solving all SFA problems.* First, such an approach provides a rational use of computing systems for solving nonformalizable problems by giving experts auxiliary information (initial data, various scales for estimation of certain properties, etc.); second, such an approach would enhance a decision maker's ability to change some data in the process of solving a general SFA problem, with an orientation toward the obtained solutions of partial problems of the system analysis (for example, to change the order of criteria preference, admissible intervals of change of parameters, etc.); third, such an approach would enable one to obtain the required justification and credibility of solutions only for promising alternatives by carrying out a preliminary analysis of alternatives on the basis of more rough estimates.

4. *A rational use of capabilities of facilities and methods of intellectual systems of support of solutions in the form of informational-analytical and expert-advising systems.* This approach allows effectively accumulate, store and use data and knowledge of a concrete subject domain. First of all, there should be referred data and knowledge about development, tests, technical diagnostics and operation of analogues and prototypes of an object being developed, data about materials and completing products, and also other reference data.

The considered techniques and approaches may be applied in the process of development of new types of systems of various purposes, for example, new engineering products. Here, under new engineering products we shall understand the engineering products of any purpose, developed on the basis of new ideas, technologies, materials, and also the products adapted for new conditions of operation and application. These approaches and methods are applicable also for solving problems of the analysis of operating complex systems in various conditions of their operation and application, including in estimation of properties and capabilities of such systems in critical and extreme situations.

Not dwelling on many peculiarities of these problems, we consider only some aspects of algorithmic support. First of all, it is necessary to devote some attention to the distinction between the notions "strategy" and "algorithm" in problem solving. ***Strategy*** covers a wider arsenal of approaches, methods, and examples, including, among others, both a statement and formalization of a problem. ***Algorithm*** includes only a certain sequence of actions in solving an already stated problem. As a consequence, the same strategy may be realized by different algorithms.

In addition, a strategy, unlike an algorithm, covers not only technical, but also organizational aspects of problem solving. The practical importance of rational organization of the procedure of solving problems of SFA is determined by the necessity to coordinate all actions of developers in regard to time, goals, and expected results. Experience shows that practical SFA problems may be solved by many institutions and companies.

The essence of planning and its importance in practical activities was very precisely and figuratively expressed by one of the founders of systemology (the science of complex engineering systems), the American professor Russell Ackoff: "Planning is designing of the desirable future and effective ways of its achievement".

Such an understanding of planning precisely and completely reflects the essence of SFA and its goals. In reality, SFA involves designing a model and the structural-functional bases of a desirable future object, the necessary properties of which are defined by the requirements specification. In the course of SFA the possibility of meeting the set requirements is estimated, possible variants of realization of a system are analyzed, and the expediency of a certain variant of its structure and its general functional exterior is justified.

Let us consider an analysis of the above indicated methods and approaches to realizing the strategy of solving a SFA problem. First of all, we shall examine the structure of the generalized SFA algorithm (Fig. 6.1). This algorithm reflects the general sequence of the basic procedures of SFA strategy. It is necessary to be attentive to a number of peculiarities of these procedures.

1. The algorithm does not set as its goal to obtain all parameters of all FEs of a system being created. Its goal is to show that under the assigned requirements it is fundamentally possible or impossible to realize a system with the corresponding properties using existing technologies of industrial production or by introducing approved new technologies.

2. The algorithm sets as its goal to reveal a rational structural functional exterior of the future system and to obtain data that confirm the expediency or inexpediency of its realization meeting the assigned requirements. Therefore, the algorithm provides an analysis of the reasons for possible flaws of a system and possible ways to eliminate them (procedure 21–29 on the scheme).

3. The algorithm's structure is oriented toward a rational use of the above indicated methods and approaches to strategy realization. In particular, the presence of a multiloop procedure of iterative search of solutions for basic SFA problems is directly reflected in the block diagram of the algorithm. The expediency of organization of the algorithm's realization mode follows from the requirement that a decision maker should take decisions at many stages of problem solving. In addition, the interactive mode makes it possible to control continuously the performance of the algorithm's basic procedures and is a certain guarantee of avoiding infinite cycles.

Note the presence in the algorithm of two kinds of level numberings: in procedures 5–8 numbering is made from the top level to the bottom one, and $i = 1$ corresponds to the object as a whole and $i = I$ to the lowest level (in particular, to a level of modules); in procedures 12–18 the numbering is made from the lowest level to the top one, and $q = 1$ corresponds to the lowest level (in particular, to a level of modules) and $q = Q$ to the highest level and to the object as a whole.

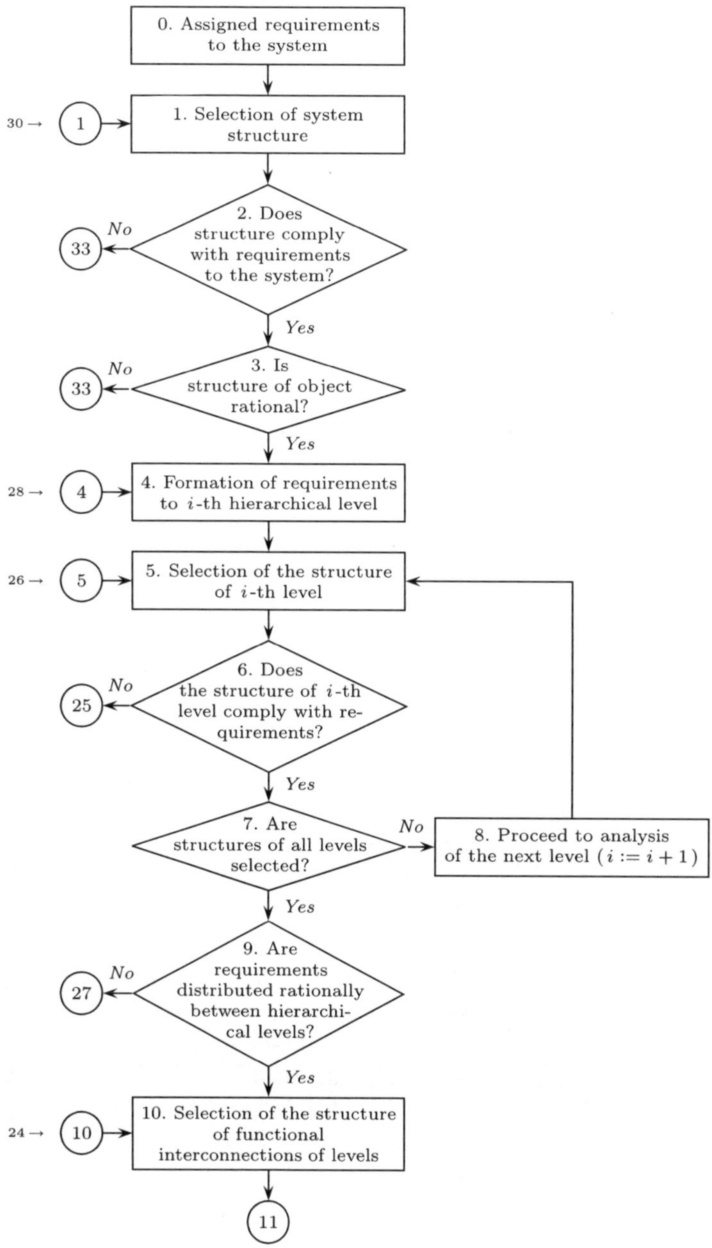

Fig. 6.1. Block diagram of generalized algorithm of SFA *(to be continued on following pages)*

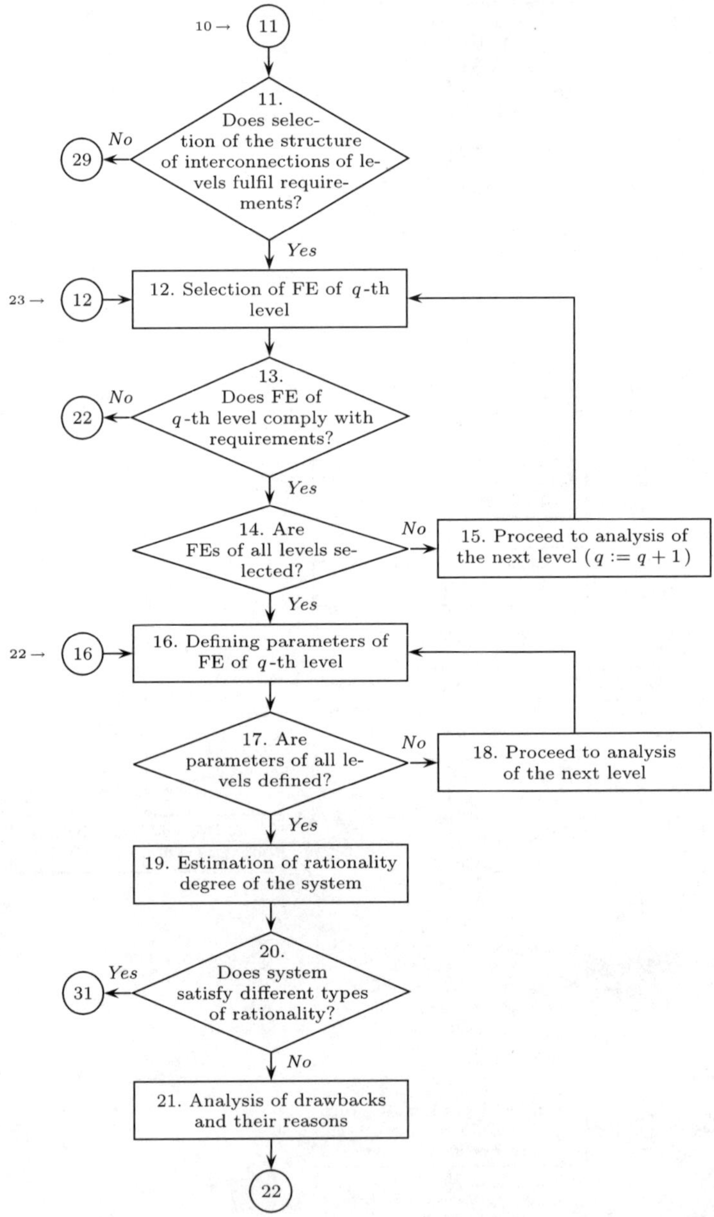

Fig. 6.1. Block diagram of generalized algorithm of SFA *(continuation)*

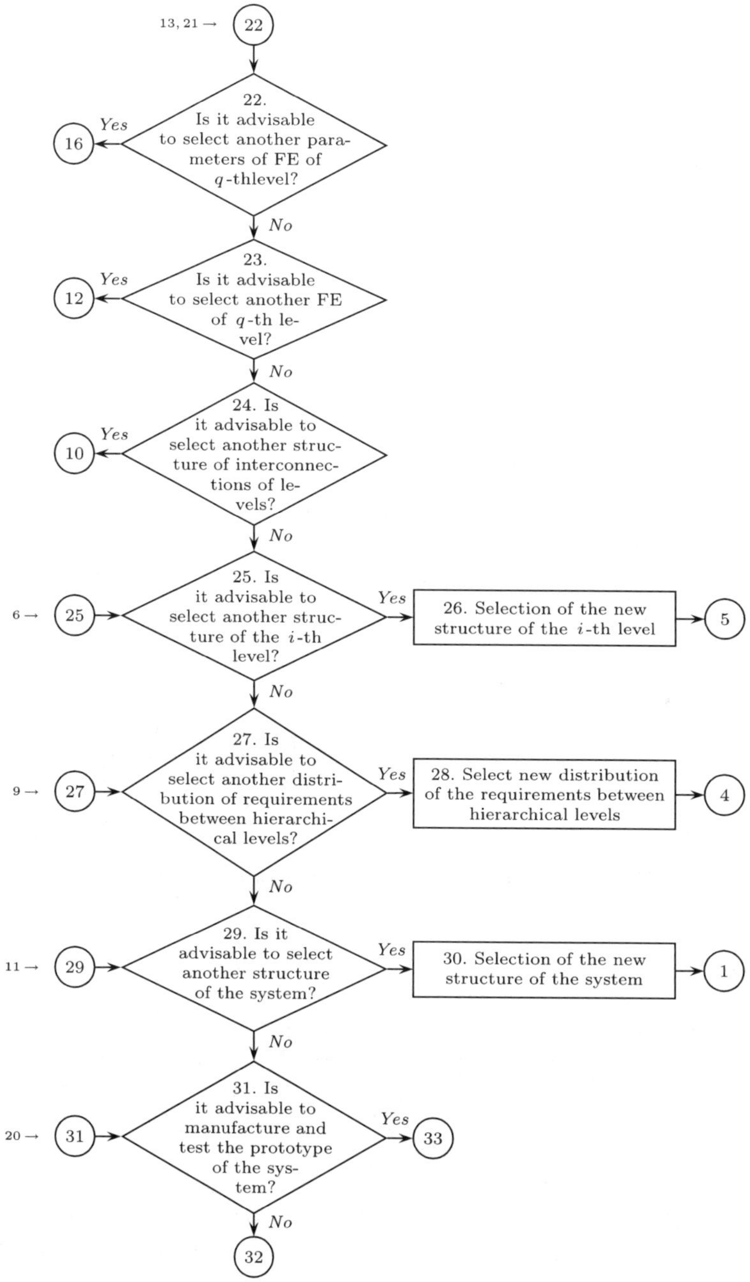

Fig. 6.1. Block diagram of generalized algorithm of SFA *(continuation)*

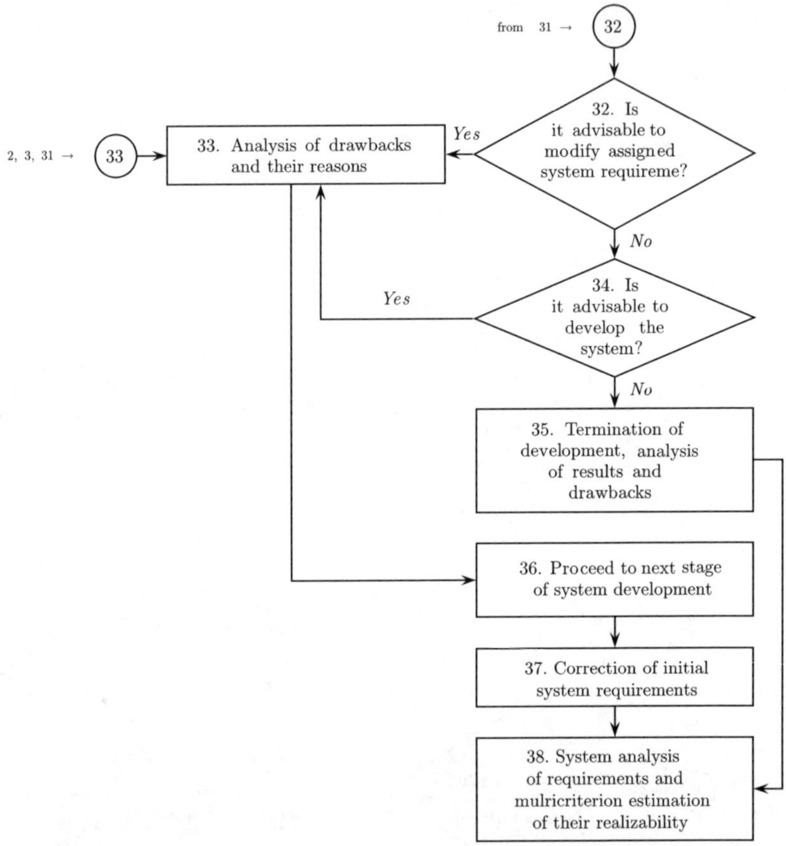

Fig. 6.1. Block diagram of generalized algorithm of SFA *(ending)*

The necessity of such numbering is determined by the fact that the requirements for hierarchical levels are set from top to bottom, and designing (in particular calculation) is done from bottom to top. Here, it is assumed that $I = Q$; also, these values determine the total number of hierarchical levels in a system.

The practical expediency of using the indicated methods and approaches will be considered using an example of procedures in choosing FEs of different levels and determining their parameters. These procedures are typical to some extent and evidently show the peculiarities and complexities of the basic SFA procedures. Therefore, we present a more detailed algorithm for choosing FEs and determining their parameters for one qth level.

Let us introduce the following designations:

q: number of a level of hierarchical structure of the system $q = \overline{1, Q}$;
Q: total number of a system's hierarchical levels;

j: number of types of a FE $j = \overline{1, J_q}$;
J_q: total number of types of FEs at the qth hierarchical level;
l: number of an alternative variant of a FE $l = \overline{1, L_j}$;
L_j: total number of alternative variants of FEs of jth type;
B_{qjl}: lth alternative variant of FE of jth type of qth hierarchical level;
S_{qj}: set of alternative variants of jth type at qth hierarchical level:

$$S_{qj} = \left\{B_{qjl} | l = \overline{1, L_j}\right\} ;$$

K_{qjk}: kth criterion of estimation of a FE of jth type at qth hierarchical level $k = \overline{1, M_j}$;
M_j: total number of estimation criteria of a FE of jth type;
K_{qj}: a set of estimation criteria of a FE of jth type at qth hierarchical level $K_{qj} = \left\{K_{qjk} | k = \overline{1, M_j}\right\}$;
$\bar{x}_{qj}$: a vector of internal parameters of a FE of jth type at qth hierarchical level:

$$\bar{x}_{qj} = \left\{x_{qjr} | r = \overline{1, R_{qj}}\right\} ;$$

x_{qjr}: rth internal parameter of a FE of jth type at qth hierarchical level;
R_{qj}: number of internal parameters of a FE of the jth type;
$\bar{y}_{qj}$: vector of external indicators of quality of a FE of jth type at qth hierarchical level $\bar{y}_{qj} = \left\{y_{qjp} | p = \overline{1, P_{qj}}\right\}$;
y_{qjp}: pth external indicators of quality of a FE of jth type of the qth hierarchical level;
P_{qj}: number of external indicators of quality of a FE of jth type;
$\bar{X}_q$: vector of internal parameters of qth hierarchical level of a system $\bar{X}_q = \left\{\bar{x}_{qj} | j = \overline{1, I_q}\right\}$;
$\bar{Y}_q$: vector of external indicators of quality at qth hierarchical level of a system $\bar{Y}_q = \left\{\bar{y}_{qj} | j = \overline{1, I_q}\right\}$;
D_q^x: set of tolerances of internal parameters at qth hierarchical level of a system defined in the form:

$$\begin{aligned} D_q^x &= \left\{\bigcup D_{qj}^x | j = \overline{1, I_q}\right\} , \\ D_{qj}^x &= \left\{\bigcup D_{qjr}^x | r = \overline{1, R_j}\right\} , \\ D_{qjr}^x &= \left\{x_{qjr} | x_{qjr}^- \le x_{qjr} \le x_{qjr}^+\right\} ; \end{aligned} \tag{6.17}$$

D_q^y: a set of tolerances of external indicators of quality of qth hierarchical level of a system defined in the form:

$$\begin{aligned} D_q^y &= \left\{\bigcup D_{qj}^y | j = \overline{1, I_q}\right\} , \\ D_{qj}^y &= \left\{\bigcup D_{qjp}^y | p = \overline{1, P_j}\right\} , \\ D_{qjp}^y &= \left\{y_{qjp} | y_{qjp}^- \le y_{qjp} \le y_{qjp}^+\right\} ; \end{aligned} \tag{6.18}$$

$\bar{F}_q(\bar{X}_q)$: vector function that determines the dependence of a vector of the quality external indicators $\bar{Y}_q$ and a vector of internal parameters $\bar{X}_q$ at qth hierarchical level:

$$\bar{Y}_q = \bar{F}_q(\bar{X}_q) ; \tag{6.19}$$

$\bar{\Phi}_q(\bar{Y}_{q-1})$: vector function that determines interconnection of indicators of quality of considered qth hierarchical level and a lower $(q-1)$ hierarchical level of a system:

$$\bar{Y}_q = \bar{\Phi}_q(\bar{Y}_{q-1}) ; \tag{6.20}$$

$\bar{f}_q(\bar{F}_q(\bar{X}_q), \bar{\Phi}_q(\bar{Y}_{q-1}))$: vector function that determines the general dependence of an external parameter vector $\bar{Y}_q$ in a complex both on $\bar{X}_q$ and on $\bar{Y}_{q-1}$:

$$\bar{Y}_q = \bar{f}_q(\bar{F}_q(\bar{X}_q), \bar{\Phi}_q(\bar{Y}_{q-1})) . \tag{6.21}$$

In particular, $\bar{f}_q$ may be an additive or multiplicate function of components $\bar{F}_q(\cdot)$ and $\bar{\Phi}_q(\cdot)$. For example, relationship (6.21) can be presented in the form:

$$\bar{Y}_q = a_q \bar{F}_q(\bar{X}_q) \bar{\Phi}_q(\bar{Y}_{q-1}) , \tag{6.22}$$

or

$$\bar{Y}_q = [\bar{F}_q(\bar{X}_q)]^{a_F} [\bar{\Phi}_q(\bar{Y}_{q-1})]^{a_\Phi} , \tag{6.23}$$

$$\bar{Y}_q = a_F \bar{F}_q(\bar{X}_q) + b_\Phi \bar{\Phi}_q(\bar{Y}_{q-1}) . \tag{6.24}$$

Now it is necessary to explain an essential aspect of the considered problem of choosing FEs of the qth hierarchical level and determining the parameters of these FEs. In the beginning we shall consider the simplest variant of this problem for the lowest level, for which $q = 1$. In this case, there are no lower levels and, consequently, it is necessary to assume $\bar{Y}_0 = 0$ and $\bar{\Phi}_1(\bar{Y}_0) = 0$.

Then, expression (6.21) can be presented in the form:

$$\bar{Y}_1 = \bar{F}_1(\bar{X}_1) , \tag{6.25}$$

which directly follows from the physical meaning (there is no dependence $\bar{Y}_1$ on the quality indicators of a lower level due to its absence) and relationships (6.21). Thus, relationships (6.22) and (6.23) [when $\Phi_1(Y_0) = 0$] lose their physical meaning.

Under these conditions the problem of choosing FEs and the corresponding parameters consists in finding FEs for a level $q = 1$ (and it is equivalent to finding $\bar{Y}_1$, $\bar{X}_1$, and $\bar{F}_1(\cdot)$) such that equality (6.25) and the following condition should be simultaneously fulfilled:

$$\bar{Y}_1 \in D_1^y, \quad \bar{X}_1 \in D_1^x , \tag{6.26}$$

where D_1^y and D_1^x are determined by the assigned requirements (6.17) and (6.18) and physical restrictions. The complexity of the problem is connected with the fact that the choice of FEs does not unambiguously determine function $\bar{F}_1(\cdot)$. Depending on the structural interconnection, including various types of direct and reverse connections between FEs, the functional connection $\bar{X}_1$ and $\bar{Y}_1$, i.e., a form $\bar{F}_1(\cdot)$, (6.25) changes. In other words, the complexity of the problem is that a certain functional dependence between external indicators of the quality and internal parameters of FEs unambiguously corresponds to the assigned structure of FEs. But the reverse conformity is not fulfilled—a countable set of possible structures may correspond to any functional dependence $\bar{F}_q(x)$. A vivid example is that various broadcast radio receivers of one class have identical indicators of quality (sensitivity, sound quality, etc.), specified by accepted standards, but essentially differ in FEs of which they are made. From here it follows that the choice of the corresponding hierarchical level structure is a creative, nonformalizable problem and its solution depends on individual features of a decision maker.

Let us consider a more complicated version of the given problem that corresponds to $q > 1$. In this case, as is seen from (6.21), the problem becomes considerably more complicated since external indicators of quality $\bar{Y}_q$ depend both on internal parameters $\bar{X}_q$ and external indicators of quality of the $(q-1)$th lower hierarchical level. In its turn, the vector $\bar{Y}_{q-1}$ depends both on $\bar{X}_{q-1}$ and on $\bar{Y}_{q-2}$, etc.

If we sequentially determine $\bar{Y}_q$ for all values $q \in [1, Q]$, where $q = Q$ corresponds to the system as a whole, we get the following sequence:

$$
\begin{aligned}
\bar{Y}_1 &= \bar{F}_1\left(\bar{X}_1\right), \\
\bar{Y}_2 &= \bar{f}_2\left(\bar{F}_2\left(\bar{X}_2\right), \bar{\Phi}_2\left(\bar{Y}_1\right)\right), \\
\bar{Y}_3 &= \bar{f}_3\left(\bar{F}_3\left(\bar{X}_3\right), \bar{\Phi}_3\left(\bar{Y}_2\right)\right), \\
&\cdots\cdots\cdots\cdots\cdots\cdots \\
\bar{Y}_{Q-2} &= \bar{f}_{Q-2}\left(\bar{F}_{Q-2}\left(\bar{X}_{Q-2}\right), \bar{\Phi}_{Q-2}\left(\bar{Y}_{Q-3}\right)\right), \\
\bar{Y}_{Q-1} &= \bar{f}_{Q-1}\left(\bar{F}_{Q-1}\left(\bar{X}_{Q-1}\right), \bar{\Phi}_{Q-1}\left(\bar{Y}_{Q-2}\right)\right), \\
\bar{Y}_Q &= \bar{f}_Q\left(\bar{F}_Q\left(\bar{X}_Q\right), \bar{\Phi}_Q\left(\bar{Y}_{Q-1}\right)\right).
\end{aligned}
\tag{6.27}
$$

In the convoluted form for a system as a whole $\left(q = \overline{1, Q}\right)$ we have

$$
\begin{aligned}
\bar{Y}_Q = \bar{f}_Q\,&\left[\left(\bar{F}_Q\left(\bar{X}_Q\right),\ \bar{\Phi}_Q\left(\bar{f}_{Q-1}\left(\bar{F}_{Q-1}\left(\bar{X}_{Q-1}\right),\right.\right.\right.\right. \\
&\bar{F}_{Q-1}\left(\bar{f}_{Q-2}\left(\bar{F}_{Q-2}\left(\bar{X}_{Q-2}\right), \ldots\right]\right..
\end{aligned}
$$

Recall that in practice the requirements specification defines only requirements for a system as a whole, and only for external indicators of quality. Hence, in the requirements specification, using the terminology accepted here, only set D_Q^Y is defined. All other sets $D_{Q-1}^Y, \ldots, D_1^Y$, i.e., requirements in the form of tolerances to external indicators of quality of all other hierarchical levels should be formed during the design stage and is specifically in the process of SFA of a system being developed.

Similarly, sets $D_Q^X, \ldots, D_1^X$, i.e., restrictions on internal parameters, should also be defined in the course of SFA.

But to solve this problem it is necessary to know, ***first***, the hierarchical structure of a system, ***second***, all $\bar{F}_q(\cdot), \bar{\Phi}_q(\cdot)$ and $\bar{f}_q(\cdot)\, \forall q \in [1, Q]$. And to get these data it is necessary to choose sequentially all FEs at all hierarchical levels.

Thus, we come to the contradiction: to assign justified requirements to all FEs at all hierarchical levels it is necessary to define all FEs at all levels and their functional interconnections. But to define FEs it is necessary to know the requirements for all FEs of all hierarchical levels.

Attention should also be drawn to the principal difference between a practical problem of a rational choice of parameters of FEs from the traditional mathematical statement of the multipurpose optimization problem. As is known, traditionally it is assumed that functions $\bar{F}_q(\cdot)$, $\bar{\Phi}_q(\cdot)$, $\bar{f}_q(\cdot)$ are known, and the problem of determining those values of arguments of these functions at which the minimal or maximal value of functions is reached with the set restrictions on arguments is solved, or the Pareto area is sought.

In this case, there is another statement—it is assumed that restrictions are assigned on $\bar{Y}_q(\cdot)$ and $\bar{X}_q(\cdot)$, and one must find such functions that would provide simultaneous fulfillment of restrictions both on arguments of functions, and on values of these functions, i.e., the following condition must be satisfied:

$$\bar{Y}_q \in D_q^Y; \quad \bar{X}_q \in D_q^X \forall q \in [1, Q]$$

by choosing $\bar{F}_q(\cdot)$, $\bar{\Phi}_q(\cdot)$, $\bar{f}_q(\cdot)$.

The complexity of the problem is not only that effective methods of its solution in the direct statement do not exist, but also that one must find simultaneously all the functions in sequence (6.27). The necessity to fulfill the last condition follows directly from the requirements specification—only the requirements for quality indicators of a system as a whole, i.e., only D_Q^Y, are known. From here it follows that it is necessary to solve simultaneously, in a complex way, purposefully and in coordination, two problems: (1) the problem of choosing functions $\bar{F}_q(\cdot)$, $\bar{\Phi}_q(\cdot)$, and $\bar{f}_q(\cdot)$ and (2) the problem of choosing restrictions $D_q^Y \forall q \in [1, Q-1]$ and $D_q^X \forall q \in [1, Q]$.

In practice, to resolve the specified contradiction and obtain a method of solving the given two problems, intuition, and experience of a human expert are used by applying the trial-and-error method. On the basis of expert estimations, a certain hierarchical structure of a system is chosen, requirements for FEs of each hierarchical level, and restrictions on internal parameters are set, i.e., $D_q^Y \forall q \in [1, Q-1]$ and $D_q^X \forall q \in [1, Q]$.

Then, for the lowest level ($q = 1$) certain FEs are chosen and certain functional interconnections of FEs and a structure of the given level are set. It allows one to construct a certain mathematical model of the given level and in such a way as to establish functional dependence $\bar{Y}_1$ and $\bar{X}_1$ in the form (6.25). Further fulfillment of conditions (6.26) is checked. If the conditions are fulfilled, there is a transition to choosing FEs of the following level $(q = 2)$, otherwise new FEs or new types of interconnections are chosen. The procedure is repeated for fulfillment of the required conditions for the highest level $(q = Q)$, i.e., a system as a whole. The problem solving for all levels defines the corresponding Pareto set for $\bar{Y}_q(\cdot)$ and $\bar{X}_q(\cdot)$ in the form D_Q^Y.

By introducing in accordance with the Gödel principle of complementarity of some criteria, it is possible to solve the following problem: to determine whether the FEs of all levels have been chosen rationally, and whether D_q^Y; $D_q^X \forall q \in [1, Q]$ have been rationally determined. This problem is reduced to the earlier considered general problem of disclosing uncertainty of goals. The mathematical statement of this problem is reduced to solving the following system of equations:

$$\begin{aligned}
&\bar{F}_1\left(\bar{X}_1\right) - \bar{Y}_1^* &&= 0\,,\\
&\bar{f}_2\left(\bar{F}_2\left(\bar{X}_2\right), \bar{\Phi}_2\left(\bar{Y}_1^*\right)\right) - \bar{Y}_2^* &&= 0\,,\\
&\cdots\cdots\cdots\cdots\cdots\cdots\cdots\cdots\cdots\cdots\cdots\cdots\cdots\cdots\cdots\cdots\\
&\bar{f}_Q\left(\bar{F}_Q\left(\bar{X}_Q\right), \bar{\Phi}_Q\left(\bar{Y}_{Q-1}^*\right)\right) - \bar{Y}_Q^* &&= 0\,,
\end{aligned} \tag{6.28}$$

where $\bar{Y}_1^*, \bar{Y}_2^*, \ldots, \bar{Y}_Q^*$ are desirable values of the quality indicators of FEs of the corresponding hierarchical levels that are set by experts or on the basis of simulation modeling of the qualitative characteristics of a system on the basis of (6.27) and satisfying the conditions:

$$\bar{Y}_q^* \in D_q^Y \; \forall q = \overline{1, Q-1}\,. \tag{6.29}$$

The areas of tolerances $D_q^Y \forall q \in [1, Q-1]$ for $\bar{Y}_q^*$ may be corrected in the course of solving the given problem, i.e., the requirements to FEs of various hierarchical levels, except for the level—a system as a whole $(q = Q)$, may be corrected from the condition (6.29) determining the achievement of a rational quality of a system as a whole. As an additional condition the following restrictions are introduced

$$\bar{X}_q \in D_q^X \; \forall q = \overline{1, Q}\,.$$

Solving system (6.28) is reduced at the first stage to the Chebyshev problem of approximation, consisting in finding those values

$$\bar{X}_q^* \in D_q^X \; \forall q = \overline{1, Q}$$

for which a value

$$\Delta_q = v_q \left|Y_q^* - \bar{f}_q\left(\cdot\right)\right|\,,$$

taken as a measure of the Chebyshev approximation, was minimally possible:

$$\Delta_q^0 = \min_{\bar{X}_q \in D_q^X} \Delta_q ,$$

$$\bar{X}_q^* = \mathrm{Arg} \min_{X_q \in D_q^X} \Delta_q .$$

Here v_q is a coefficient of importance of the quality indicators of the qth hierarchical level, satisfying the condition $0 < v_q \leq 1$. The coefficient v_q is decreased in the process of reducing the degree of importance.

At the second stage as a criterion of the Chebyshev problem it is necessary to take the value

$$\Delta_Q = \left| \bar{Y}_Q^* - \bar{f}_Q(\cdot) \right|$$

and to solve system (6.28) from the condition

$$\Delta_Q^0 = \min_{\bar{X}_q \in \hat{D}_q^X} \min_{\bar{Y}_q \in \hat{D}_q^Y} \Delta_Q . \tag{6.30}$$

In relationship (6.30) it is necessary that for $\bar{Y}_q$ the value q satisfy the condition $q \in [1, Q-1]$ and for $\bar{X}_q$, $q \in [1, Q]$. Here $\hat{D}_q^X$ and $\hat{D}_q^Y$ define the areas of tolerances in the δ-vicinity of values $\bar{X}_q^*$ and $\bar{Y}_q^*$, respectively, and are described by the relationships of the form (6.17) and (6.18) in which it is assumed that

$$\hat{D}_{qjr}^x = \left\{ x_{qjr}^* | x_{qjr}^* (1 - \delta_{qjr}) \leq x_{qjr}^* \leq x_{qjr}^* (1 + \delta_{qjr}) \right\} ,$$

$$\hat{D}_{qjp}^y = \left\{ y_{qjp}^* | y_{qjp}^* (1 - \delta_{qjp}) \leq y_{qjp}^* \leq y_{qjp}^* (1 + \delta_{qjp}) \right\} .$$

The general block diagram of the algorithm for choosing FEs and determining their parameters is given in Fig. 6.2.

Note that the problem of the SFA of an operating system is in some sense a partial case of the considered problem of SFA of a system being developed. In reality, for all hierarchical levels of an operating system, all functional dependences of a system's quality indicators on the internal parameters of FEs are defined. In this case, the problem of SFA is reduced to checking the rationality of the accepted solutions, their practical realizability during testing, and experimental adjustment of a system.

The problem of SFA of an operating system may consist in estimating its ability to function in critical and extreme situations. Such a problem is especially acute for human-caused and ecologically dangerous events. Mathematically these problems are reduced to system (6.28) in which $\bar{F}_q(\cdot)$, $\bar{\Phi}_q(\cdot)$, $\bar{f}_q(\cdot)$ may be determined directly based on the measurements of quality indicators and parameters of a system.

The solution of a SFA problem forms the basis for two other types of system analysis problems of the second conceptual functional space of properties of a complex hierarchical system.

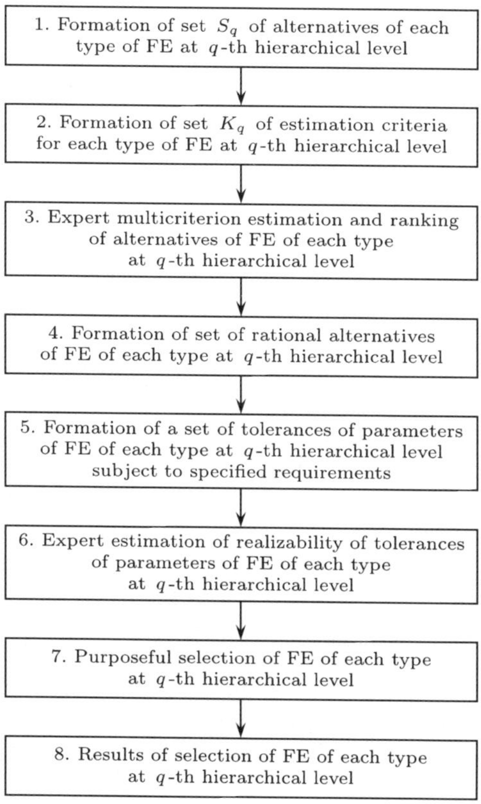

Fig. 6.2. Block diagram of algorithm for choosing FEs and determining their parameters

6.4 System Optimization of Complex Constructive Elements of Modern Equipment

The common trend of increasing the requirements for functionality, strength, reliability, profitability, safety, and other characteristics of modern equipment leads to a continuous increase in the complexity of equipment and conditions under which it operates. It sets a number of new system problems, such as designing, experimental adjustment, testing, and technical diagnostics [33, 55, 180, 194]. These problems include various optimization problems [169] such as the problem of searching a rational compromise with contradictory requirements for strength, reliability, adaptability for production, and technical and economic efficiency of structures, taking into account the normal operating conditions and risk factors of possible abnormal situations [29, 78, 98, 167, 168].

The practical necessity to continuously raise the efficiency and safety of equipment determines the necessity to account more fully and reliably for the risk factors of abnormal operating situations at the stage of designing complex objects. Here a new approach is suggested for solving the multicriterion problems of structural and parametrical optimizations of nonhomogeneous anisotropic constructive elements of complex engineering objects. It is based on a rational choice of the hierarchical structure of a complex object in the sense of a rational distribution of requirements for functional and constructive elements of each hierarchical level and a rational compromise of contradictory requirements for strength, reliability, and adaptability for production, and technical and economic efficiency of the object, taking into account the risk factors of possible abnormal situations. To solve the given problems new system methods of risk-factor estimation under abnormal and critical situations are used [134, 135].

The system structural optimization of complex objects. At the design stage of complex engineering objects, the most complicated task is formalization of a system problem and representation of the object in the form of a multilevel hierarchical system to organize the consecutive-parallel development of FEs of each hierarchical level in a rational way. In practice, the given problem frequently has the following features [118, 133]:

1. The choice of the structure for an object under development is a nonformalizable problem. The structure is devised based on the subjective understanding of the developer of certain types of interconnections between FEs of the given object, which are represented in the form of a multilevel hierarchical system.

2. The choice of the number of hierarchical levels and the number of FEs at each level and the formation and justification of criteria for choosing the FEs at each level are nonformalizable problems. Such choices are based on the experience, intuition, and knowledge of a decision maker.

3. There are many possible alternative solutions at each hierarchical level. Therefore, a direct enumeration of all possible variants of the structure leads to a problem of transcomputing complexity, where the solution is fundamentally impossible [54].

4. The initial information has a considerable level of uncertainty and incompleteness. In particular, only the set requirements for properties and characteristics of the designed object as a whole are known, and requirements for FEs of each hierarchical level should be formed in the course of its development.

To exclude a direct search of variants, the method of consecutive analysis is successfully being used and developed [19, 34, 169, 187]. At the same time, the system problem of formation of a rational hierarchical structure, including a rational choice of the number of hierarchical levels, rational formation of requirements for FEs of all levels, and a rational choice of FEs of each hierarchical level, has no effective solving methods.

Therefore, the development and improvement of procedures and methods that allow one to essentially reduce the volume of calculations and find the rational structure of an object, excluding a direct enumeration of all possible alternative variants of the structure, is a practically important problem.

The problem of structural optimization of complex objects is solved on the basis of the method of the purposeful choice of a rational hierarchical structure under the set requirements Q_0 for the object as a whole [118, 132]:

$$Q_0 = \left\{K_r^0 | K_r^- \leq K_r^0 \leq K_r^+; \quad r = \overline{1, R_0}\right\} , \tag{6.31}$$

where K_r^0 is the rth indicator of object quality.

The initial problem will be presented in the form of a sequence of the following problems:

1. The choice of a rational number of hierarchical levels.
2. Forming rational requirements for FEs of the hierarchical structure under the set requirements for the object.
3. The purposeful choice of FE hierarchical structure under the accepted requirements.

Let us consider these problems.

Problem 1. Choice of a rational number of hierarchical levels. Let us give a formalized description of a hierarchical structure using set theoretical concepts of the general theory of systems [96]. Considering that the number of hierarchical levels m is finite, we shall represent a model of the designed object in the form of a Cartesian product:

$$S_0 = S_1 \times S_2 \times \ldots \times S_i \times \ldots \times S_m . \tag{6.32}$$

Here S_0 is a hierarchical level that corresponds to the object as a whole and S_i is the ith hierarchical level, described in the form:

$$S_i = \langle M_i, f_i, \varphi_i, X_i, Y_i \rangle , \quad M_i = \left\{\Phi_{ij} | j \in \hat{N}_i, \hat{N}_i = \overline{1, N_i}\right\} , \tag{6.33}$$

$$\varphi_i : X_i \to Y_i ; \quad f_i : Y_i \to Y_{i-1} , \tag{6.34}$$

where M_i is a set of FEs of the ith level, Φ_{ij} is a FE of the jth type on the ith level, φ_i and f_i are functionals that define the interrelation of the corresponding parameters, X_i and Y_i are a set of internal and external FE parameters of the ith level, respectively, defined by the relationships

$$Y_i = \left\{Y_{ij} | j \in \hat{N}_i ; \hat{N}_i = \overline{1, N_i}\right\}, \quad X_i = \left\{X_{ij} | j \in \hat{N}_i ; \hat{N}_i = \overline{1, N_i}\right\} ,$$

where N_i is the number of types of FE at the ith hierarchical level.

On the basis of the given relationships we shall state a problem of finding the number of rational levels of the structure $\hat{m}$.

Let the goal function $g(m_v, \omega, u_v) = G(m_v, \omega, u_v, P(m_v, \omega))$ be defined on the basis of the function of transformation $P : Q_0 \times M_v \to S_{0v}$ of the estimated function $G : Q_0 \times \Omega \times S_{0v} \to C_v$ and the function of expense restriction $\tau : Q_0 \times \Omega \to C$. We must find such $\hat{m} \in [m^-, m^+]$ that for all $\omega \in \Omega$ upon realization of the set requirements the expenses do not exceed the permissible values $g(\hat{m}, \omega, \hat{u}) \leq \tau(\omega)$. Here, S_{ov} is a structure (6.32) consisting of m_v levels and M_v is a set of all FEs of the structure S_{0v}, $M_v = \{M_i | i = \overline{1, m_v}\}$, u_v is the control vector, Ω is a set of uncertain factors that influence the quality of the object and expenses, C_v is a set of expenses for the life cycle of the object with the chosen structure S_{0v}, $C_v = \{c_v | v = \overline{1, \chi_v}\}$, and C is a set of permissible expenses for the life cycle of the object.

Problem 2. Forming requirements for the FEs of a hierarchical structure. To form requirements for the FEs of a structure S_{0v} we shall establish the interrelation of external parameters of different levels [129]. From (6.34) it follows

$$Y_0 = f_0(Y_1), \ \ldots, \ Y_i = f_i(Y_{i+1}), \ \ldots, \ Y_{m-1} = f_{m-1}(Y_m) \, . \qquad (6.35)$$

For all $i = \overline{1, \hat{m}}$, assuming $Y_i = \{Y_{ir_i} | r_i = \overline{1, R_i}\}$ and considering that, according to (6.32), $Y_0 = Q_0$ from (6.33), (6.35), and (6.31) for the definition of requirements for a level $i = 1$ we obtain the following systems of equations:

$$K_r^- = f_{0r}\left(Y_{1r_1}^-\right); \quad K_r^+ = f_{0r}\left(Y_{1r_1}^+\right); \quad r = \overline{1, R_0}; \quad r_1 = \overline{1, R_1} \, .$$

The results of solving the given systems $\hat{Y}_1^- = \left\{\hat{Y}_{1r_1}^- | r_1 = \overline{1, R_1}\right\}$ and $\hat{Y}_1^+ = \left\{\hat{Y}_{1r_1}^+ | r_1 = \overline{1, R_1}\right\}$ are the initial data for the definition of $\hat{Y}_2^-$ and $\hat{Y}_2^+$. In the general case for $i \in [1, m]$ we have the following systems of equations:

$$\begin{aligned} \hat{Y}_{ir_i}^- &= f_{ir_i}\left(Y_{(i+1)r_{i+1}}^-\right); \quad r_i = \overline{1, R_i}; \\ \hat{Y}_{ir_i}^+ &= f_{ir_i}\left(Y_{(i+1)r_{i+1}}^+\right); \quad r_{i+1} = \overline{1, R_{i+1}} \, . \end{aligned} \qquad (6.36)$$

Hence, the definition of the requirements for the FEs of a structure is reduced to forming a sequence of equation systems (6.36). On the basis of the values $\hat{Y}_i^-$ and $\hat{Y}_i^+$, obtained from (6.36), the requirements for external parameters k_{ijq} of each type of FE Φ_{ij} are similarly defined in the form

$$\hat{K}_{ij} = \{k_{ijq} | k_{ijq}^- \leq k_{ijq} \leq k_{ijq}^+; q = \overline{1, \hat{q}_{ij}}\} \, , \qquad (6.37)$$

which are the initial data for the problem of choosing FEs S_0.

Let us formulate the problem of a structure choice [131, 132]. Let the structure $\hat{S}_0$ of the designed object consist of $\hat{m}$ hierarchical levels S_i, $i = \overline{1, \hat{m}}$. Each level S_i consists of types n_i of FEs Φ_{ij}, $j = \overline{1, n_i}$. Each FE Φ_{ij} is characterized by parameters k_{ijq}, $q = \overline{1, \hat{q}_{ij}}$. Alternative variants of a FE Φ_{ij} are defined by set $M\Phi_{ij}$. One must choose one FE of each jth type on each ith hierarchical level from condition (6.37) and construct a Pareto Π_s set of rational structures $\hat{S}_{0v}$ of the object.

For all $i = \overline{1,\hat{m}}$ and $j = \overline{1,n_i}$ sets, $M\Phi_{ij}$, which consist of alternative variants of FEs Φ_{ij}, are assumed to be known. Each set $M\Phi_{ij}$ consists of subsets $M\Phi_{ij}^-$ and $M\Phi_{ij}^+$, $M\Phi_{ij}^+ \bigcup M\Phi_{ij}^- = M\Phi_{ij}$; $M\Phi_{ij}^+ \bigcap M\Phi_{ij}^- = \oslash$; $M\Phi_{ij}^+ = \left\{\Phi_{ij\beta}^+ \Leftrightarrow k_{ij\beta} | k_{ij\beta} \in \hat{K}_{ij}; \beta = \overline{1,n_{ij}^+}\right\}$; $M\Phi_{ij}^- = \left\{\Phi_{ij\beta}^- \Leftrightarrow k_{ij\beta} | k_{ij\beta} \notin \hat{K}_{ij}\,; \beta = \overline{1,n_{ij}^-}\right\}$.

Elements of sets $M\Phi_{ij}^-$ and $M\Phi_{ij}^+$ are placed randomly in set $M\Phi_{ij}$. Numerical values are known ($n_{ij}^- > 0$ and $n_{ij}^+ > 0$), but it is not known what set $M\Phi_{ij}^-$ or $M\Phi_{ij}^+$, a concrete element Φ_{ij} of set $M\Phi_{ij}$, belongs to.

Problem 3. Purposeful choice of FEs and hierarchical structure. The idea of the purposeful choice method for a rational hierarchical structure of an object consists in the following [76,118,132,141]. The choice of elements from set $M\Phi_{ij}$ is carried out consistently. If at the next attempt α it turns out that $\Phi_{ij\alpha} \in M\Phi_{ij}^-$, then $\Phi_{ij\alpha}$ is excluded from $M\Phi_{ij}$. The subsequent choice is carried out from the obtained set $\hat{M}_{ij}$. The choice of an element of the jth type on the ith hierarchical level stops when at the next attempt γ the chosen element $\Phi_{ij\gamma}$ belongs to the set $M\Phi_{ij}^+$. Performance of the given procedure for all $i = \overline{1,\hat{m}}$, $j = \overline{1,n_i}$ allows one to obtain one structure of the object all of whose elements satisfy condition (6.37).

To ensure the possibility of a rational choice of the complex system structure, the final set of structures is formed. In this case a search for the second and subsequent structures of the system is carried out on the basis of a choice of elements from sets $\hat{M}_{ij}$, $i = \overline{1,\hat{m}}$, and $j = \overline{1,n_i}$, which are obtained as a result of choosing the previous structure. A set of such structures satisfies condition (6.37) and is a Pareto set.

The given method is realized in the form of a computation algorithm. The computation complexity of the algorithms used for choosing the system structure can be defined by the number of attempts that are made to choose a FE $\Phi_{ij}^+ \in M\Phi_{ij}^+$ for each $i = \overline{1,\hat{m}}$ and $j = \overline{1,n_i}$. Let us define the number of attempts of a choice using the suggested method. These attempts are to be made to guarantee choosing a FE of the jth type at the ith hierarchical level from condition (6.37).

We obtain the relationship for the probability of the guaranteed choice of a FE $\Phi_{ij} \in M\Phi_{ij}^+$ in the case of $j = \text{const}$ and $i = \text{const}$. Taking into account that in a sequence of attempts of choosing FEs from $M\Phi_{ij}$, the attempts k and $(k+1)$ are independent. Then, the probability of a choice $\Phi_{ij} \in M\Phi_{ij}^+$ at the first attempt $(k = 1)$ is defined by the relationship

$$P_1\left(\Phi_{ij} \in M\Phi_{ij}^+\right) = 1 - \left(1 - n_{ij}^+/n_{ij}\right)^{n_{ij}} . \tag{6.38}$$

Here it is considered that the probability of choosing a certain variant of a FE from $M\Phi_{ij}^+$ is equal to n_{ij}^+/n_{ij}. Therefore, the choice of any FE from

$M\Phi^+_{ij}$ will be determined by (6.38). If the first attempt is unsuccessful, the chosen element is excluded from $M\Phi_{ij}$ and the total number of elements of set $M\Phi_{ij}$ decreases by 1. Therefore, at the second attempt the probability of choosing any FE from $M\Phi_{ij}$ is defined by the relationship

$$P_2\left(\Phi_{ij} \in M\Phi^+_{ij}\right) = 1 - \left[1 - n^+_{ij}/\left(n_{ij} - 1\right)\right]^{(n_{ij}-1)} .$$

For the kth attempt it is:

$$P_k\left(\Phi_{ij} \in M\Phi^+_{ij}\right) = 1 - \left(1 - n^+_{ij}/\left[n_{ij} - (k-1)\right]\right)^{[n_{ij}-(k-1)]} .$$

For $k = n^-_{ij} + 1$ it is:

$$P_k\left(\Phi_{ij} \in M\Phi^+_{ij}\right) = 1 - \left(1 - n^+_{ij}/n^+_{ij}\right)^{n^+_{ij}} = 1 . \quad (6.39)$$

Here it is considered that $n_{ij} = n^-_{ij} + n^+_{ij}$. Hence, from (6.39) we have that the choice $\Phi_{ij} \in M\Phi^+_{ij}$ at $j =$ const and $i =$ const is guaranteed if the number of attempts of a choice is defined by the relationship

$$k^+_{ij} = n^-_{ij} + 1 . \quad (6.40)$$

Thus, when choosing the jth FE at the ith hierarchical level from condition (6.37), in order to form the set $M^+_{ij} = \left\{\Phi^+_{ij} \in M\Phi^+_{ij} | i \in [1, m] ; \; j = \overline{1, n_i}\right\}$ it is enough to enumerate k^+_{ij} at the FEs from this set $M\Phi_{ij}$.

The general number of FEs at the ith level is defined by the relationship

$$N_i = \sum_{j=1}^{n_i} n_{ij} .$$

Taking into consideration that for different $j = 1, \ldots, n_i$ the number of choices k^+_i is defined independently, we get $k^+_i = k^+_{i1} + \ldots + k^+_{in_i}$ or, taking into account (6.40):

$$k^+_i = \sum_{j=1}^{n_i} \left(n^-_{ij} + 1\right) .$$

Here, k^+_i is the number of attempts necessary for choosing all types of FEs of the ith level from a condition (6.37). We take into consideration that the choice of FEs at each hierarchical level is done independently. In this case the general number of attempts k^+ for a choice of all FEs for all levels of a structure is defined by the formula

$$k^+ = \sum_{i=1}^{m} k^+_i .$$

After solving a set of problems of structural optimization one must move on to the problem of parametrical optimization of a complex system. Let us consider this problem.

System parametrical optimization. Parametrical optimization is performed to achieve the following interconnected system goals: (1) ensuring a rational compromise of contradictory requirements for strength, reliability, adaptability to production, and technical and economic efficiency of the structure under the normal mode of functioning; and (2) ensuring the required level of safety on the basis of minimization of the degree and level of risk of potentially possible abnormal situations [50, 134, 135, 137]. Let us consider this problem by example of a complex engineering structure.

General problem statement. For a structure consisting of nonhomogeneous anisotropic material, operating under complex conditions of nonuniform force and temperature effects, it is required to:

- Find a rational compromise between contradictory requirements for strength, reliability, and technical, technological, and economic parameters of the structure.
- Determine the optimum values of parameters of the structure and materials meeting the conditions of a rational compromise of requirements for strength, reliability, adaptability to production, and technical and economic efficiency in normal modes of operation and under permissible restrictions of risks of abnormal situations.

To solve the given problem it is necessary to:

- Construct mathematical models that describe the dependence of indicators of strength $\bar{y}_s$, reliability $\bar{y}_r$, manufacturability $\bar{y}_t$, and technical and economic efficiency $\bar{y}_{te}$ on geometric parameters and physical-mechanical parameters of the structure and materials, respectively, in the following form:

$$\bar{y}_s = \bar{f}_s(\bar{x}), \quad \bar{y}_r = \bar{f}_r(\bar{x}), \quad \bar{y}_t = \bar{f}_t(\bar{x}), \quad \bar{y}_{te} = \bar{f}_{te}(\bar{x}). \tag{6.41}$$

- Define the values of the specified indicators

$$\bar{y}_s = \bar{b}_s^*, \quad \bar{y}_r = \bar{b}_r^*, \quad \bar{y}_t = \bar{b}_t^*, \quad \bar{y}_{te} = \bar{b}_{te}^*, \tag{6.42}$$

  that would ensure a rational compromise between contradictory requirements for strength, reliability, and technological, technical, and economic characteristics of the structure.
- Introduce indicators of risk of abnormal situations and define the permissible restrictions.
- Construct mathematical models that describe the dependence of indicators of abnormal-situation risk on geometrical and physical-mechanical indicators of the structure and of risk-factor parameters.
- Find the values $\bar{x}^0$ of a vector $\bar{x} = \{x_j | j = \overline{1, n}\}$ that allow one to meet condition (6.42) in normal modes of operation and at permissible restrictions of parameters of risk of abnormal situations.

Approaches to constructing mathematical models (6.41) are considered in [43, 67, 108, 109, 111, 113, 116, 123, 124]. Therefore, we shall move to substantiate the restrictions of normal modes of operation and restrictions of abnormal-situation risk that are to be considered while choosing a rational compromise [118]. In practice, values b_s^*, b_r^*, b_t^*, and b_{te}^* in condition (6.42) should be defined within the limits of certain intervals, which is caused by the action of physical, technological, operational, and other restrictions.

For normal modes of functioning of a structure such intervals are set, as a rule, in the form

$$b_i \in B_i, \quad B_i = \left[b_i^-, b_i^+\right] . \tag{6.43}$$

Here b_i are certain components of vectors $\bar{b}_s, \bar{b}_r, \bar{b}_t$, and $\bar{b}_{te}$ that make up a vector $\bar{b} = \left\{\bar{b}_i | i = \overline{1, m}\right\}$ of required indicators of the quality of a structure, and m is the total number of indicators. Each interval B_i defines the permissible limits of change of a corresponding indicator under the influence of the uncontrollable factors of the manufacture and operation of a structure. The major goals B_i of a choice are, first, preservation of a certain level of working capacity of a structure when any changes in uncontrollable factors take place; second, introduction of the condition that random changes in each uncontrollable factor occur in the permissible limits set in advance. In particular, the international and national standards on each kind of equipment define the permissible intervals of variation of environmental parameters—temperature, humidity, etc. Therefore, the normal conditions of the structure's operating are determined not only by a priori known intervals of permissible changes in the quality indicators of a structure but also by uncontrollable factors. As a result, any normal mode of functioning in the general understanding can be characterized as the determinate mode, because it is generally defined by the known data.

A fundamentally different approach is necessary for analyzing the structure in ***abnormal situations***. It is caused by a number of reasons. First, in abnormal situations there may occur not only random deviations of quality indicators and uncontrollable factors beyond the limits of the permissible intervals, but also random occurrence of specific structural changes in structural elements. For example, in aviation gas turbine engines, there may occur the following structural changes: plastic deformation, thermal fatigue crack, destruction of special coatings, fatigue cracks from internal defects (foundry cracks, gas bubbles, defects of heat treatment), etc. Second, abnormal situations may arise owing to the influence of secondary factors of operation. For example, in gas turbine engines the occurrence of shock loadings due to colliding with various extraneous objects the occurrence of a nonuniform field of temperatures owing to infringement of burning processes, etc. The given factors may result not only in an inadmissible decrease in quality indicators of the structure, but also in the destruction of constructive elements with catastrophic consequences. Hence, the mode of functioning of the structure in ***abnormal situations*** is fundamentally indeterminate. From here arises the necessity to analyze an abnormal situation from positions of the risk theory.

Risk of an abnormal situation will be considered on the basis of two indicators: ***degree of risk*** ρ_{os}, which is the probability of an abnormal situation appearance, and ***level of risk*** W_{os}, which is the extent of possible damage resulting from an abnormal situation, normalized to an economic indicator, for example, profit. To account for abnormal-situation factors in a problem of system optimization of complex structures, in particular, constructive elements of aviation and rocket equipment, it is useful to coordinate conditions (6.41) and (6.43) for rational compromises of requirements for normal situations with the permissible value of risk of abnormal situations. To this end we shall generate an interval of permissible risk. ***An interval of permissible risk*** is, in our interpretation, an interval w_{α_0} that defines the permissible variation of a vector $\bar{b}$ from the condition that the degree and level of risk of any situation S_v from a set S_{ps} of predicted abnormal situations will not exceed an a priori assigned level ρ_{α_0}, W_{α_0}. We shall represent the given interval in the form

$$\begin{aligned} w_{\alpha_0} &= \left\{\bar{b}|\rho^+_{os}\left(\bar{b},\bar{\zeta}\right) \le \rho_{\alpha_0},\ W^+_{os}\left(\bar{b},\bar{\zeta}\right) \le W_{\alpha_0},\ \bar{b} \in B_{ps},\ \bar{\zeta} \in \Xi_{ps}\right\}\,, \\ S_{ps} &= \left\{S_v \Rightarrow \left\langle \bar{b}_v, \bar{x}_v, \bar{\zeta}_v\right\rangle |\bar{b}_v \in B_{ps};\ \bar{x}_v \in X_{ps};\ \bar{\zeta}_v \in \Xi_{ps};\ v = \overline{1, N_{ps}}\right\}\,. \end{aligned} \tag{6.44}$$

Here $\rho^+_{os}(S_v)$ and $W^+_{os}(S_v)$ are respectively maximum the permissible degree and level of risk of any situation; $S_v \in S_{ps}$, B_{ps}, X_{ps}, and Ξ_{ps} are respectively the permissible interval of variation of vectors $\bar{b}$, $\bar{x}$, and $\bar{\zeta}$ for a set of predicted situations S_{ps}. These intervals are defined by the following relationships:

$$B_{ps} = \left\{B_{pi}|i = \overline{1,m}\right\}\,,\quad B_{pi} = \left\{b_i|b^-_{pi} \le b_i \le b^+_{pi}\right\}\,, \tag{6.45}$$

$$X_{ps} = \left\{X_{pj}|j = \overline{1,n}\right\}\,,\quad X_{pj} = \left\{\bar{x}_j|x^-_{pj} \le \bar{x}_j \le \bar{x}^+_{pj}\right\}\,, \tag{6.46}$$

$$\Xi_{ps} = \left\{\Xi_{pq}|q = \overline{1,Q}\right\}\,,\quad \Xi_{pq} = \left\{\zeta_q|\zeta^-_{pq} \le \zeta_q \le \zeta^+_{pq}\right\}\,. \tag{6.47}$$

Thus, relationships (6.43) and (6.44) in view of restrictions (6.46)–(6.47) allow one to describe the permissible intervals of variation of quality indicators of the structure at the assigned tolerances for normal modes of operation and with known restrictions for the degree and level of risk of predicted abnormal situations.

Problem-solving algorithm. The algorithm is based on a complex use of modern capabilities of system analysis theory and design methods [194], possibilities of means and methods of expert procedures [107], and an artificial intellect, in particular, possibilities of fuzzy situation algorithms of information processing [94], and also the method of generalized consecutive linearization and adaptive algorithms of Chebyshev's approximation of an incompatible system of nonlinear equations [77,150]. The algorithm defines the structure and interconnection of procedures of system modeling and parametrical optimization of the structure on the basis of the following system of equations:

$$f_i(\bar{x}) - b_i = 0, \quad \bar{x} = \{x_j | j = \overline{1,n}\}, \quad i = \overline{1,m}, \quad m > n.$$

The block diagram of the given algorithm is presented in Fig. 6.3.

The algorithm for solving the problem of parametrical optimization is realized in an interactive mode and includes the following generalized procedures.

1. The choice of a structure variant.
2. Construction of mathematical models (6.41) for the accepted structure variant.
3. The choice of a rational compromise of requirements (6.42).
4. System modeling of abnormal situations.
5. Solving a system optimization problem of the structure for the chosen requirements taking into account restrictions (6.43) and (6.44).
6. Estimation of conformity of the structure quality to a desirable level. Making decisions on the transition to procedure 7 or 9.
7. Specification of requirements to the structure. Making decisions on the transition to procedure 3 or 8.
8. Search for a new variant of the structure. Transition to procedure 2.
9. Termination of a dialog. Printing of problem solution results.

The necessity of operating in an interactive mode is determined by the fact that realization of procedures 1–3 and 6–8 requires expert estimation of certain parameters and making decisions regarding the choice of a certain alternative.

The most difficult procedure is 4. It is realized in the form of the following sequence of partial procedures.

4.1 Expert forecasting of a set of abnormal situations S_{ps}.
4.2 Construction of mathematical models of risk indicators $\rho_{os}(\bar{b}, \bar{\zeta})$ and $W_{os}(\bar{b}, \bar{\zeta})$ for a set S_{ps}.
4.3 Expert definition of the permissible level of risk ρ_{α_0}, W_{α_0}.
4.4 Formation of an interval of the permissible level of risk (6.44).
4.5 Estimation of risk indicators for an interval (6.43).
4.6 Making decisions on the expediency of realizing the accepted structure by results of procedure 4.5. Transition to procedure 5, 7, or 3.

The purpose of procedure 4 is to ensure the required accuracy, credibility, and justification of estimating the degree and level of risk of abnormal situations by system optimization of the structure. The most difficult step in the procedure is the construction of mathematical models of the risk indicators. These models should reflect the processes of fatigue, thermal, and other kinds

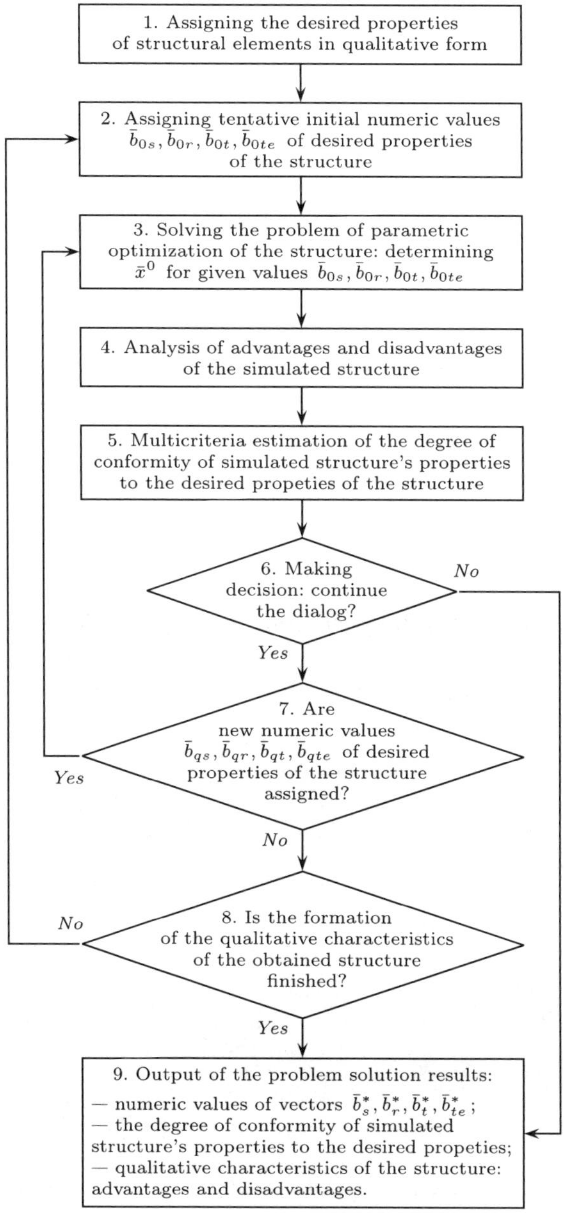

Fig. 6.3. Algorithm for solving a parametrical optimization problem

of aging mechanisms and the destruction of the structure under the influence of uncontrollable and secondary factors of operation.

The mathematical models of risk indicators are represented in the form

$$\rho_{os}(S_v) = F_\rho\left(\tilde{\tilde{\varsigma}}_v, \tilde{\tilde{\chi}}_v, \tilde{\tilde{\zeta}}_v\right); \quad W_{os}(S_v) = F_W\left(\tilde{\tilde{\varsigma}}_v, \tilde{\tilde{\chi}}_v, \tilde{\tilde{\zeta}}_v\right). \tag{6.48}$$

Here $\tilde{\tilde{\varsigma}}_v, \tilde{\tilde{\chi}}_v$, and $\tilde{\tilde{\zeta}}_v$ are the vectors of fuzzy indicators reflecting the degree of membership of situation S_v in a set S_{ps} by a level of deviation respectively $\bar{b}$, $\bar{x}$, and $\bar{\zeta}$ in intervals B_{ps}, X_{ps}, and Ξ_{ps} from the boundary conditions. Models (6.48) are constructed on the basis of a generalized functional-simulation model:

$$\tilde{y} = \tilde{\Phi}\left(\tilde{x}, \tilde{\zeta}\right),$$

where fuzzy input $\tilde{x}$, output $\tilde{y}$, and external uncontrollable indicators $\tilde{\zeta}$ are described by the relationships

$$\begin{aligned} \tilde{y} &= \left\{\tilde{y}_i | i = \overline{1,m}\right\}, \quad \tilde{x} = \left\{\tilde{x}_j | j = \overline{1,n}\right\}; \quad \tilde{\zeta} = \left\{\tilde{\zeta}_q | q = \overline{1,Q}\right\}, \\ \tilde{y}_i &\in \tilde{Y}_i; \quad \tilde{Y}_i = \left\{\mu_{Y_i}(y_i) | y_i \in Y_i;\ \mu_{Y_i}(y_i) \in [0,1]\right\}, \\ \tilde{x}_j &\in \tilde{X}_j; \quad \tilde{X}_j = \left\{\mu_{X_j}(x_j) | x_j \in X_j;\ \mu_{X_j}(x_j) \in [0,1]\right\}, \\ \tilde{\zeta}_q &\in \tilde{\Xi}_q; \quad \tilde{\Xi}_q = \left\{\mu_{\Xi_q}(\zeta_q) | \zeta_q \in \Xi_q;\ \mu_{\Xi_q}(\zeta_q) \in [0,1]\right\}. \end{aligned}$$

The initial information for construction of models are the results of tests of various materials, data of experimental adjustment and tests of constructive elements, the results of technical diagnostics of structures in operation, and also a priori data, restrictions, and laws of the theory of mechanical, physical, and other processes.

The problem of model construction and the multicriterion optimization problem of a structure are reduced to a sequence of Chebyshev problems of approximation described by incompatible systems of nonlinear equations. Such an approach to solving a multicriterion optimization problem of the structure for a normal mode of functioning is considered in detail in [130].

Let us consider an example of solving a structural optimization problem taking into account a purposeful choice of FEs.

6.5 Choice of a Rational Structure of a Personal Computer

It is required to choose the rational structure of a personal computer (PC). The hierarchical structure of a PC is presented in Fig. 6.4.

The requirements Q_0 for the rational structure of a PC as a whole object $Q_0 = \{K_r^0 \mid K_r^- \le K_r^0 \le K_r^+,\ r = \overline{1, R_0}\}$ are given in Fig. 6.5.

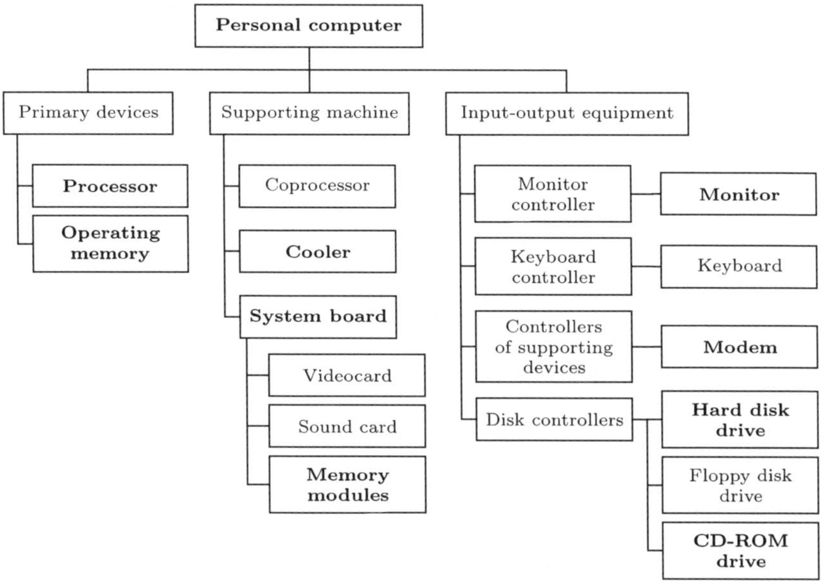

Fig. 6.4. Hierarchical structure of a PC

For a solution to this problem alternatives of parameters $k_{ij\beta}$ of FEs $\Phi_{ij\beta}$ of various types (processor, operating memory, system board, memory modules, cooler, hard disk, CD-ROM, monitor, modem) are proposed in Figs. 6.6–6.8.

Let us formulate a problem of choice of FEs of a hierarchical structure. Let the structure $\hat{S}_0$ of the design of a PC consist of $\hat{m} = 3$ hierarchical levels S_i, $i = \overline{1, \hat{m}}$. The first hierarchical level S_1 consists of $n_1 = 2$ types of FE Φ_{1j}, $j = \overline{1, n_1}$; the second S_2 hierarchical level consists of $n_2 = 3$ types of FEs Φ_{ij}, $j = \overline{1, n_2}$; and the third hierarchical level S_3 consists of $n_3 = 4$ types of FEs Φ_{ij}, $j = \overline{1, n_3}$. Each FE Φ_{ij} is characterized by parameters k_{ijq}, $q = \overline{1, \hat{q}_{ij}}$. Alternative variants of FEs Φ_{ij} are defined by a set $M\Phi_{ij}$. One must choose one FE of each jth type at each ith hierarchical level to construct a Pareto set $\prod$ of rational structures $\hat{S}_{0v}$ of the object.

Application of the algorithm of random search taking into account the number of alternatives at each level allows one to define the general number of possible alternatives of structures of the object [54]. Using a random search method is required to perform $K = 3^2 \cdot 3^3 \cdot 3^4 = 19683$ attempts.

Using the proposed method of the purposeful choice of FEs [120] the number of necessary attempts for a choice of the structure is $K = 15$. The rational structure of the personal computer, obtained on the basis of the method of the purposeful choice of FEs, is given in Fig. 6.5. Thus, it envisages the introduction of new alternatives for available FEs, and also updating methods for selecting variants and their priorities. The offered method allows

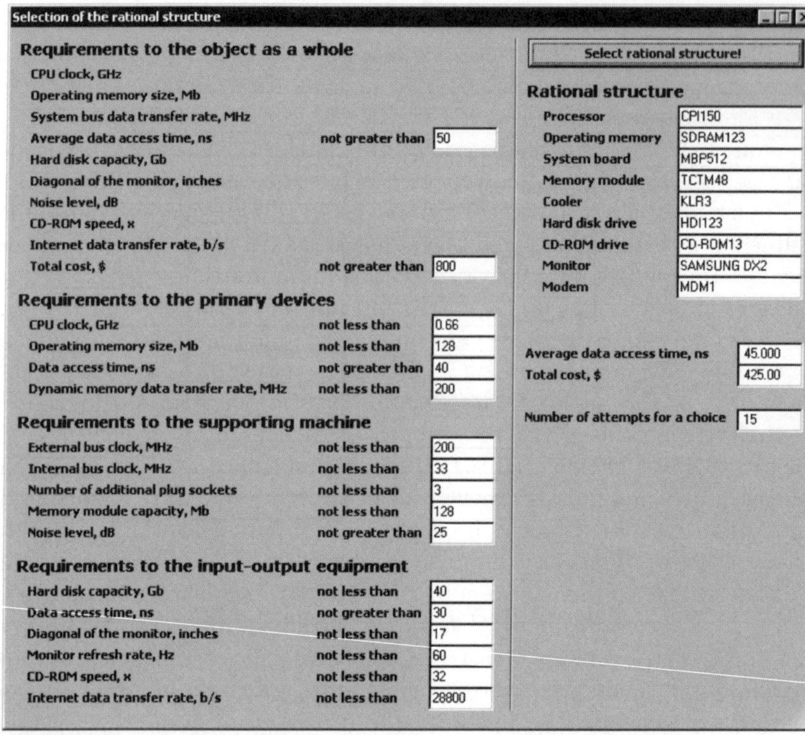

Fig. 6.5. Requirements and results of choice of rational structure of a PC as a whole object

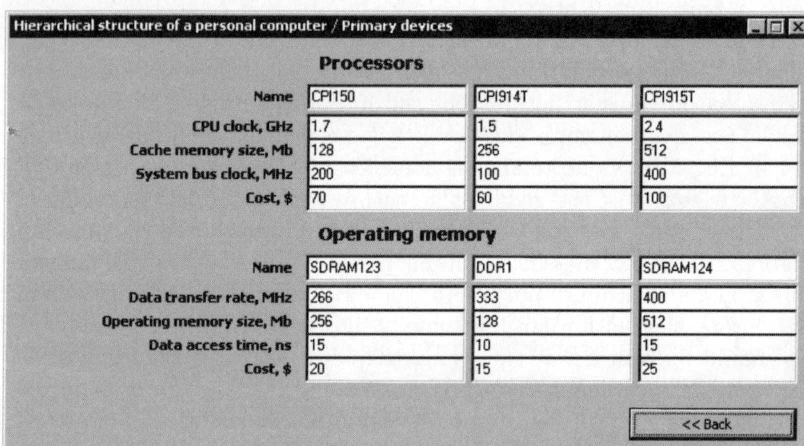

Fig. 6.6. Alternatives of FE parameters: processor and cache memory

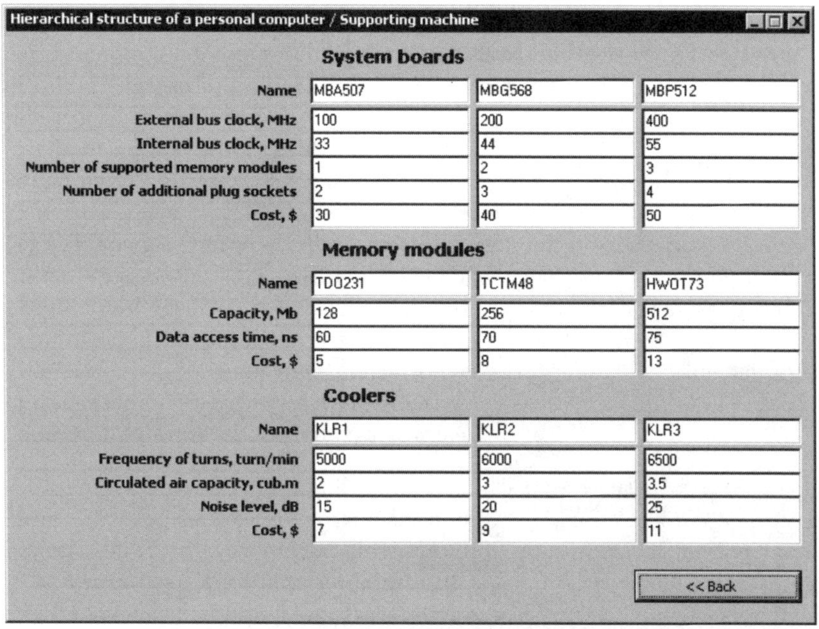

Fig. 6.7. Alternatives of FE parameters: system board, memory module, and cooler

one to essentially reduce the sample volume to guarantee that the required structure will be found.

To ensure the possibility of a rational choice of the structure of a complex system, a finite set of structures is formed. In this case, the second and subsequent variants of system structures are searched on the basis of a choice of elements from sets $\hat{M}_{ij}$, $i = \overline{1, \hat{m}}$, $j = \overline{1, n_i}$, which are obtained upon choosing the previous structure. The set of such structures is a Pareto set.

Thus, the given example shows that the proposed method of a purposeful choice of FEs allows one to reduce considerably the sample volume for the guaranteed finding of the required structure with a probability equal to 1; further, it gives the possibility of constructing Pareto sets of the permissible structures that meet the set requirements for all FEs and for the object as a whole. The algorithm of purposeful choice allows one to solve the problem of data processing with a large volume of information.

It should be noted that the complexity of complex hierarchical systems is characterized not merely by an increase in dimensions, but multifunctionality, multicriterionness, and hierarchy of its structure; the presence of many possible alternative solutions at each hierarchical level; a considerable level of uncertainty and incompleteness of the initial information; and the presence of subsystems of various purposes and of different physical natures. In spite of the fact that a complex system consists of a set of separate subsystems with their local control facilities, their functioning should be coordinated to

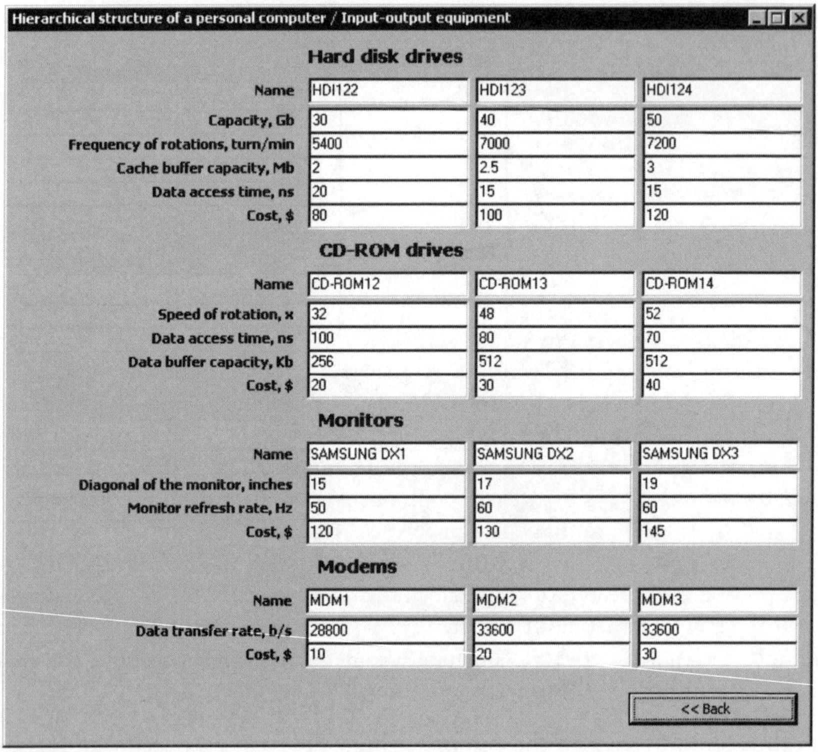

Fig. 6.8. Alternatives of FE parameters: hard disk, CD-ROM, monitor, and modem

achieve the common goal that is defined for the whole system. All this leads to the necessity of application of the system approach while designing and investigating complex hierarchical systems, taking into account structural, functional, and information components. Application of the system approach preconditions the guaranteed considerations of all possible alternative variants of the system structure and acceptance of such justified decision that would to the utmost extent meet the requirements set for the object.

7 Problems and Methods of System Analysis of Multifactor Risks

For complex systems of any nature an actual problem is a credible and timely foresight, forecast, and prevention of critical, extreme, and other undesirable situations and influences that may result in an abnormal functioning mode, accident, or catastrophe or considerably affect the serviceability, survivability, safety, efficiency, and other properties of such objects. The possibility of such situations and results of such influences, conditions, and factors are determined by random and chaotic processes that are characterized as risks by mechanisms of their influence. Risks are generated by various internal and external factors, and, therefore, the mechanisms of their influence on a complex engineering system are multifactor.

In this chapter we shall consider a general problem of multifactor risk analysis, analyze conditions, reasons for their occurrence, and peculiarities of detection of critical, extreme, and other risk situations, study sources and reasons for critical and catastrophic risks, methods of timely prevention of their turning into an accident or a catastrophe, propose principles for solving system problems that directly refer to the analysis of accidents and catastrophes of complex engineering systems.

7.1 General Problem of System Analysis of Multifactor Risks

We shall formulate the main problem of system analysis of multifactor risks in the generalized form [133,134]. The set M_0 of risk factors ρ_q is ***known*** from the data of testing a complex system of arbitrary nature and other a priori information:

$$M_0 = \left\{\rho_q | q = \overline{1, n_0}\right\} . \tag{7.1}$$

Each risk factor $\rho_q \in M_0$ is characterized by a set L_q of attributes l_{qj}:

$$L_q = \left\{l_{qj} | q \in N_0; j = \overline{1, n_q}\right\} , \quad N_0 = [1, n_0] . \tag{7.2}$$

Each attribute $l_{qj} \in L_q$ is determined by an information vector

$$I_{qj} = \left\{x_{qj} | x_{qj} = \left\langle x_{qjp} | p = \overline{1, n_{qj}}\right\rangle ; \; x_{qjp} \in H_{qjp}; \; q \in N_0; \; j \in N_q\right\} , \tag{7.3}$$

$$H_{qjp} = \left\{x_{qjp} | x_{qjp}^- \le x_{qjp} \le x_{qjp}^+\right\} , \quad N_q = [1, n_q] . \tag{7.4}$$

Based on I_{qj} sets, an information vector is formed for each risk factor ρ_q:

$$I_q = \left\{I_{qj} | q \in N_0;\ j = \overline{1, n_q}\right\} , \tag{7.5}$$
$$I_q = \left\{x_{qj} | x_{qj} = \left\langle x_{qjp} | p = \overline{1, n_{qj}}\right\rangle ;\ x_{qjp} \in H_{qjp};\ q \in N_0;\ j = \overline{1, n_q}\right\} . \tag{7.6}$$

The set M_0 corresponds to a definite, a priori predicted set S_0 of risk situations.

In the functioning of a complex system new risk factors affect it and are revealed, and the properties and indicators of a priori known risk factors $\rho_q \in M_0$ are changed. This results in quantitative and qualitative changes in the set of risk factors that determine the necessity to form a sequence of embedded sets of the form

$$M_0 \subset M_1 \subset \ldots \subset M_\tau \subset \ldots, \quad S_0 \subset S_1 \subset \ldots \subset S_\tau \subset \ldots , \tag{7.7}$$

where M_τ and S_τ are a set of risk factors and a set of risk situations at the moment $T_\tau \in T^{\pm}$, respectively, and $T^{\pm}$ is an assigned or predicted period of functioning of a complex system. Sets M_τ and S_τ are determined as

$$M_\tau = \left\{\rho_q^\tau | q = \overline{1, n_\tau}\right\}, \quad S_\tau = \left\{S_k^\tau | k = \overline{1, K_\tau}\right\} .$$

Each situation $S_k^\tau \in S_\tau$ is characterized by a set $M_k^\tau \in M_\tau$ of risk factors ρ_{qk}^τ

$$M_k^\tau = \left\{\rho_{q_k}^\tau | q_k = \overline{1, n_k^\tau}\right\} . \tag{7.8}$$

Each factor $\rho_{q_k}^\tau \in M_k^\tau$ is characterized by set $L_{q_k}^\tau$ of attributes $l_{q_k j_k}^\tau$:

$$L_{q_k}^\tau = \left\{l_{q_k j_k}^\tau | q_k \in N_k^\tau; j_k = \overline{1, n_{q_k}^\tau}\right\}, \quad N_k^\tau = [1, n_k^\tau] . \tag{7.9}$$

Each attribute $l_{q_k j_k}^\tau \in L_{q_k}^\tau$ is revealed based on the information obtained and processed by a diagnostic system. Information at the moment of measurement T_τ is characterized by its incompleteness, uncertainty, and inaccuracy. Information vector $I_{q_k j_k}^\tau$ for each attribute $l_{q_k j_k}$ is formed using the diagnostic information in the form of a fuzzy set:

$$\begin{aligned} I_{q_k j_k}^\tau &= \left\{\tilde{x}_{q_k j_k}^\tau | \tilde{x}_{q_k j_k}^\tau = \left\langle \tilde{x}_{q_k j_k p_k}^\tau | p_k = \overline{1, n_{q_k j_k}}\right\rangle ; \right. \\ &\qquad \left. q_k \in N_{q_k}^\tau;\ j_k \in N_{q_k}^\tau \right\}, \quad N_{q_k}^\tau = \lfloor 1, n_{q_k}^\tau \rfloor , \end{aligned} \tag{7.10}$$

$$\begin{aligned} \tilde{x}_{q_k j_k p_k}^\tau &= \left\langle x_{q_k j_k p_k}^\tau, \mu_{H_{q_k j_k p_k}}^\tau \left(x_{q_k j_k p_k}^\tau\right)\right\rangle ; \\ x_{q_k j_k p_k}^\tau &\in H_{q_k j_k p_k}^\tau;\quad \mu_{H_{q_k j_k p_k}}^\tau \in [0, 1] , \end{aligned} \tag{7.11}$$

$$H_{q_k j_k p_k}^\tau = \left\langle x_{q_k j_k p_k}^\tau | x_{q_k j_k p_k}^- \le x_{q_k j_k p_k}^\tau \le x_{q_k j_k p_k}^+ \right\rangle . \tag{7.12}$$

From the information base of sets $I_{q_k j_k}^\tau$; $q_k = \overline{1, n_k^\tau}$; and $j_k = \overline{1, n_{q_k}^\tau}$ an information vector is formed for each factor $\rho_{q_k}^\tau$; $q_k = \overline{1, n_k^\tau}$ in the following form:

$$I_{q_k}^{\tau} = \left\{ I_{q_k j_k}^{\tau} | q_k \in N_k^{\tau};\ j_k = \overline{1, n_{q_k}^{\tau}} \right\} , \qquad (7.13)$$

$$I_{q_k j_k}^{\tau} = \left\{ \tilde{x}_{q_k j_k p_k}^{\tau} | q_k \in N_k^{\tau}; j_k = \overline{1, n_{q_k}^{\tau}};\ p_k = \overline{1, n_{q_k j_k}^{\tau}} \right\} . \qquad (7.14)$$

Similarly, for each situation $S_k^{\tau} \in S_{\tau}$ an information vector is formed

$$I_k^{\tau} = \left\{ I_{q_k}^{\tau} | q_k = \overline{1, n_k^{\tau}};\ k = \overline{1, K^{\tau}} \right\} , \qquad (7.15)$$

$$I_{q_k}^{\tau} = \left\{ \tilde{x}_{q_k j_k p_k}^{\tau} | q_k = \overline{1, n_k^{\tau}};\ j_k = \overline{1, n_{q_k}^{\tau}};\ p_k = \overline{1, n_{q_k j_k}^{\tau}} \right\} . \qquad (7.16)$$

Information about risk factors can be supplemented or made more exact in the process of decision forming oriented at the prevention of undesired influence of risk factors under investigation. Obtaining additional information is ensured by the time, energy, and other resources spent. The time for forming a decision is limited, since the set M_{τ} contains a subset $\hat{M}_k^{\tau}$ of factors with a threshold action mechanism. When one or several factors $\rho_{q_k}^{\tau} \in \hat{M}_k^{\tau}$ reach the threshold level at some a priori unknown critical moment T_{cr}, undesired consequences, like accidents, catastrophes, or ecological disasters, produce their effect.

In the process of controlling a complex system functioning on a true scale of the set moments of time T_{τ} or with a certain interval of moments $\tilde{T}_{\tau} \in T_{\tau}^{\pm}$, $T_{\tau}^{\pm} = \left\{ \tilde{T}_{\tau} | T_{\tau} < \tilde{T}_{\tau} < T_{\tau+1} \right\}$, ***it is required*** to carry out a multifactor estimation of risk of any situation $S_k^{\tau} \in S_{\tau}$ and, based on the obtained results, to form and realize a decision on preventing and/or minimizing undesired consequences before the critical moment T_{cr} comes.

We point out a number of fundamentally important peculiarities of the formulated problem. In the given problem, as distinct from other typical optimization and recognition problems, it is taken into account that sets of risk factors and sets of situations are in principle unlimited. They change quantitatively and qualitatively in the course of functioning of complex engineering objects, forming the corresponding sequences of enclosed sets in the form (7.7). An important peculiarity of the given problem is a qualitatively new type of restriction, namely, a threshold restriction of time for decision forming. This restriction differs fundamentally in its significance and properties from the restrictions proper for typical system analysis and optimization problems. It is absolutely a top priority and unavoidable, and its violation is categorically inadmissible in practice, because it results in irreversible catastrophic consequences. It should be noted that the problem is not completely formalized, since indicators of multifactor risk estimation and criteria of multipurpose risk minimization are not determinate. Such a formalization technique is used purposely for the following reasons. A choice of the mentioned indicators and criteria is a nonformalized problem. It depends on many objective and subjective factors, and depends above all on the conditions and peculiarities of ensuring the safety of a specific complex system and the decision maker's subjective estimation of the corresponding requirements and available capabilities of a complex system. Indeed, the problem is presented in a generalized statement that gives a decision maker some freedom in adapting it to practical needs in a specific subject domain by concrete definition of the aforementioned indicators and criteria.

Based on the decomposition principle, we shall represent the general problem of analysis of multifactor risk as a sequence of the following system-coordinated, informationally interconnected problems [134]:

1. System multifactor classification of revealed and predicted risk situations.
2. System multifactor recognition of revealed and predicted risk situations.
3. System multicriterion ranking of situations.
4. Multipurpose risk minimization of a predicted set of abnormal situations.
5. Rational multipurpose optimization of informedness level in recognition of abnormal situations in the process of a complex system's functioning.
6. Rational coordination of margin of permissible risk of a predicted set of abnormal situations.
7. Definition of a rational informedness level at the threshold time limitation in the process of a complex system's functioning.
8. System estimation of margin of permissible risk under the dynamics of abnormal mode.

Such an approach is determined by the complexity of the general problem of system analysis. Also, it is orientated toward computer realization of the main procedures of system analysis of risks, which, in its turn, determines the necessity, possibility, and expediency of the module principle of organizing problem solving.

Let us consider peculiarities of two important problems: the problem of classifying of a set of risk factors and the problem of recognizing risk situations. Let us show that these problems differ in a basic way from the problem of classification and recognition in the theory of pattern recognition. It is caused by a fundmental difference in their properties [40, 90, 209].

In fact, an analysis of different classes of typical problems of pattern recognition theory shows that they possess the following properties:

- A set M of the analyzed objects ω_j allows representation in the form of aggregation of a finite number of nonintersecting classes Ω_i.
- Each element $\omega_j \in M$ is characterized by one and the same set of attributes containing their finite number.
- Each class Ω_i is characterized by a finite number of the attributes values.
- A priori learning information J is assigned in the form of a definite description of attributes of a finite number of objects in each class.

From here follow important peculiarities of the traditional recognition problems:

(a) All properties of objects in such problems are determined by a finite number of the corresponding attributes or elements.

(b) All classes $\Omega_i \in M$, $i = \overline{1,m}$ and all objects under investigation are characterized by one and the same finite set of attributes, and the difference in classes and objects is determined only by the difference in numerical values of the attributes.
(c) The required condition of workability of traditional methods and algorithms of recognition is the presence of learning information J on the attributes of each class Ω_i. It should be noted here that for learning a rather large volume of information is required and the learning process must be long enough.
(d) Recognition and classification processes are realized on the basis of precise estimation and are characterized by the following conditions and restrictions: $\alpha_{ij} = 1 \Rightarrow \omega_j \in \Omega_i$; $\alpha_{ij} = 0 \Rightarrow \omega_j \notin \Omega_i$. All other variants are described as $a_{ij} = \Delta$, where Δ means "unknown".
(e) The algorithms used to solve recognition problems are based on informal models that are formed in the process of training on the basis of standard learning information. The question of choosing and justifying the completeness of learning information remains open. In practice, learning information is chosen subjectively, and the level of its completeness and credibility depends on the experience, knowledge, and intuition of a researcher. This approach does not exclude many cases where recognition of an object is impossible and a solution is coded by the sign Δ.

Problems of recognition of risk situations in the process of complex system functioning possess fundamentally different properties and peculiarities:

1. A set of attributes of risk situations is in principle unlimited. This peculiarity results directly from an unlimited set of risk factors whose actions cause abnormal situations to occur.
2. Different abnormal situations are characterized by various sets of peculiarities. But no pairs of situations are characterized by the same attributes. For example, for engineering systems such a state is conditioned by the difference in properties of material, defects, and aspects of real materials, aging, and destruction mechanisms and environmental influence.
3. It is in principle impossible to represent a set of abnormal situations in the form of a finite set of classes, where each class is characterized by a definite individual set of attributes or a definite individual set of numerical values of attributes. It is determined by the unlimited set of attributes of abnormal situations.
4. For problems of abnormal-situation recognition it is in principle impossible to create standard learning information in the form of a finite set of data on attributes of each class of the accepted classification, which is determined by the unboundedness and changeability in time of a set of abnormal-situation attributes.
5. There is a threshold time restriction on the recognition of abnormal situations in the process of diagnostics that is conditioned by high dynamics of

abnormal-situation development and the possibility of consecutive transition from a normal mode to an extreme situation, accident, or catastrophe in a short period of time. That is why the recognition should be carried out within a period of time that would allow one to carry out the procedures for forming and realizing a decision before a critical moment T_{cr} comes.

The analysis of the presented properties and peculiarities allows us to make the following important conclusions:

1. In problems of recognition of abnormal situations it is in principle impossible to fulfill restrictions and admissions accepted in typical problems of image recognition.
2. The traditional techniques of classification, based on a single finite set of attributes for all classes, are not acceptable for classification of abnormal situations, because a set of attributes of abnormal situations is unlimited and each abnormal situation has its own unique set of peculiarities.
3. Typical methods, models, and algorithms of the recognition theory cannot be used directly for recognition of abnormal situations, since the formation of standard learning information is in principle impossible.

Let us formalize the indicated problems with certain practically realizable assumptions. Thus, assume that for some complex system for every $T_\tau \in T^{\pm}$ there is a possibility of choosing a finite set of risk factors and characterizing each factor by a finite set of attributes, describing each attribute with a finite fuzzy information vector according to the condition of practically acceptable credibility and accuracy of solving the general problem of system analysis of risks.

Problem 1. A system multifactor classification of risk situations. Information vectors I_q in the form of (7.6) for all risk factors ρ_q of set M_0, defined by the form (7.1), are a priori ***known***. The values of risk degree η_q and risk level W_q at definite values of components I_{qj} of information vector I_q of each risk factor $\rho_q \in M_0$ are also known. In the process of complex system functioning fuzzy information vectors I_k^τ, their components in the form of (7.15), (7.16), and numerical values of risk degree η_{qk}^τ and risk level W_{qk}^τ for definite values of components of vector I_k^τ are determined for each set $M_k^\tau \in M_\tau$.

It is required: (a) on the set M_0 to build an a priori subset into a finite number κ_i of nonintersecting classes (subsets) Ω_i, where each class Ω_i is characterized by a definite set $M_i \in M_0$ of risk factors and definite half-intervals of admissible values η_i and W_i:

$$\eta_i^- < \eta_i \leq \eta_i^+ , \quad W_i^- < W_i \leq W_i^+ ; \tag{7.17}$$

(b) in the process of complex system functioning using information about the set M_τ of risk factors and based on an a priori classification, to form such a subset of a finite number κ_i^τ of nonintersecting classes (subsets) Ω_i^τ, where each class Ω_i^τ is characterized by a defined set $M_i^\tau \in M_\tau$ of risk factors and has defined half-intervals of admissible values η_i^τ and W_i^τ, which are determined by an a priori classification when $\kappa_i^\tau = \kappa_i$, and defined by the relationship

$$\eta_i^- \leq \eta_{i\tau}^- < \eta_i^\tau \leq \eta_{i\tau}^+ < \eta_i^+; \quad W_i^- \leq W_{i\tau}^- < W_i^\tau \leq W_{i\tau}^+ < W_i^+ .$$

when $\kappa_i^\tau > \kappa_i$.

Let us point out the principal differences between the proposed system of risk classification and typical classification systems. Standard systems of risk classification have few gradations with a verbal description of each that does not contain a quantitative risk estimation and does not provide each gradation with an interconnection between risk attributes and a concrete quantitative risk estimation. For example, economic risks are usually classified into four zones: risk-free zone, admissible-risk zone, critical-risk zone, catastrophic-risk zone. In the proposed classification each class is characterized by a priori determined half-intervals of values of risk degree η_i and risk level W_i, and an interconnection of risk attributes and concrete quantitative estimation of risk for each class is provided by information vectors I_k^τ and their components.

This problem statement differs from the statement of typical classification problems, for example, problems of pattern recognition theory. In typical classification problems a general structure of classification is proposed since the structure, number, and composition of attributes of each class are invariable. In the given statement of the classification problem, the number of classes in the general classification structure and composition and the number of attributes and boundaries of half-interval values η_i and W_i of each class vary in the process of complex system functioning. In particular, each class may be represented in the form of a definite structure of subclass. This make it possible to realize a multilevel, hierarchical classification structure with the required gradation discreteness at each level, and, as a result, it ensures the required precision and credibility of classification and recognition of risk situations. It allows one to adapt the classification to concrete conditions of complex system functioning and, eventually, considerably increase the efficiency of its control.

Problem 2. Multifactor recognition of risk situations. The set M_k^τ of risk factors for situation $S_k^\tau \in S_\tau$ in the form (7.8), which contains a subset $\hat{M}_k^\tau$ of risk factors $\tilde{\rho}_k^\tau$ that are not elements of set M_0 ***is known***. For each risk factor from the set M_k^τ a fuzzy information vector I_k^τ and its components in the form (7.15), (7.16) ***are known***.

It is required to determine the risk degree and risk level of situation S_k^τ and the class Ω_i^τ to which situation S_k^τ belongs.

Solving this problem allows one to define a risk class Ω_i^τ for a concrete situation $S_k^\tau \in S_\tau$ by the classification formed by creating a subset of set

M_k^τ into classes Ω_i^τ. In risk forecasting some set S_τ of risk situations is considered, and set M_k^τ of risk factors is defined for each of them. That is why a practically important problem arises of ranking the situations $S_k^\tau \in S_\tau$ by the degree and level of risk. This problem should be considered as the evolution of a recognition problem, as its generalization onto the set S_τ. We shall state this problem with the same conditions as in the risk recognition problem.

Problem 3. Ranking situations by degree and level of risk. Let the set M_k^τ of risk factors in the form (7.8) ***be known*** for each situation $S_k^\tau \in S_\tau$. All M_k^τ, $k = \overline{1, K_\tau}$ sets contain subsets $\hat{M}_k^\tau$, elements of which are not elements of the set M_0. Fuzzy information vectors I_k^τ, $k = \overline{1, K_\tau}$ and their components ***are known***. The dependences of risk degree $\eta_{q_k}^\tau$ and risk level $W_{q_k}^\tau$ on components of vector I_k^τ ***are known***.

It is required to rank situations $S_k^\tau \in S_\tau$ by classes Ω_i^τ by risk degree and risk level at the moment of observation.

A natural continuation and development of the problem of situation ranking by risk degree and risk level is the problem of multipurpose risk minimization. A practical importance of this problem is obvious, since the ultimate goal of the system analysis of multifactor risks is the minimization of the risk degree and risk level. At first sight, it may seem that the minimization of risks may be ensured in the course of ranking situations by risk degree and risk level. But a ranking procedure allows one to reveal only those situations that in some sense are practically equivalent or have some advantage over others. This procedure may be considered as finding a Pareto set. But the objects belonging to a Pareto set are not adequate in the entire set of their attributes. That is why by introducing in accordance with the Godel complementarity principle an additional criteria, it is possible to find the only element of the Pareto set that is the best by the introduced additional criteria. From the position of this approach let us state the problem of multipurpose risk minimization under the conditions accepted at the stage of formalization of the problem of ranking situations by risk degree and risk level.

Problem 4. Multipurpose rational risk minimization. Information vector I_k^τ and its components in the form (7.15), (7.16) ***are known*** for each situation $S_k^\tau \in S_\tau$. Numerical values of risk degree $\eta_{q_k}^\tau$ and risk level $W_{q_k}^\tau$ at certain numerical values of the components of information vector I_k^τ for all risk factors $\rho_{q_k}^\tau \in M_k^\tau$ ***are known***. Characteristics of variation of all vector I_k^τ components within a specified time period ***are known***.

It is required to define class Ω_i^τ, to which each situation S_k^τ belongs, reveal the functional dependences of risk degree $\eta_{q_k}^\tau$ and risk level $W_{q_k}^\tau$ of each risk factor $\rho_{q_k}^\tau \in M_k^\tau$ on vector I_k^τ components, choose and justify the system of criteria of risk minimization, and minimize the risk degree and risk level of $S_k^\tau \in S_\tau$ situations, the risk of which may exceed a permissible level.

This problem statement differs from the traditional multipurpose minimization problem, because the system of criteria is not a priori assigned, but is justified in the course of solving the problem. On the one hand, this gives some freedom not only in the physical interpretation of minimization criteria, but in choosing a method for solving the optimization problem. But, on the other hand, it complicates the problem, since choosing criteria is a nonformalizable operation when stating an optimization problem.

Note that this statement is a simplified version of the risk minimization problem in the sense that a solution is being found for the predefined informedness level of a decision maker. But qualitative and quantitative characteristics of a decision maker's informedness are known to change in the course of forming a solution.

That is why an actual, practical problem of defining rational qualitative and quantitative characteristics of a decision maker's informedness arises. Let us state some alternatives to the given problem.

Problem 5. Choosing a rational level of the decision maker's informedness in situation recognition. The set M_k^τ of risk factors $\rho_{q_k}^\tau$ in the form (7.8)–(7.14) for each situation $S_k^\tau \in S_\tau$ ***is known***. Each set M_k^τ contains a subset $\tilde{M}_k^\tau$ whose elements are not elements of a set M_0. Only some components of vectors I_k^τ for $\rho_{q_k}^\tau \in \tilde{M}_k^\tau$, $k = \overline{1, \tilde{K}_\tau}$ and values of a risk degree $\eta_{q_k}^\tau$ and risk level $W_{q_k}^\tau$ at certain values of components of information vector I_k^τ for those risk factors of each set M_k^τ that are not elements of a subset $\tilde{M}_k^\tau$ ***are known***. It is known that the credibility of recognition of $S_k^\tau \in S_\tau$ and the credibility of estimation of the risk degree and risk level are increasing functions of the integral indicator of informedness $\bar{U}_\tau$. Information uncertainty regarding risk factors $\rho_{q_k}^\tau \in \tilde{M}_k^\tau$ creates the condition $\bar{U}_\tau < U^-$, which excludes the possibility of recognition $S_k^\tau \in S_\tau$ and determination of risk degree and risk level of situation $S_k^\tau \in S_\tau$.

It is required to define a rational level $\bar{U}_\tau^*$ of integral informedness indicator $\bar{U}_\tau$ for a decision maker, at which recognition of each situation $S_k^\tau \in S_\tau$ and defining a degree $\eta_{q_k}^\tau$ and level $W_{q_k}^\tau$ of risk of situation $S_k^\tau \in S_\tau$ are ensured with a practically permissible level of credibility.

Problem 6. Rational coordination of margin of permissible risk of an abnormal mode of a complex system. The sequence of embedded sets (7.7) for some period of time $T_r^\pm = [T_r^-, T_r^+]$ of diagnostics of a complex system ***is known***. Set M_k^τ in the form (7.8) is known for each $S_k^\tau \in S_\tau$ situation at every moment $T_\tau \in T_r^\pm$, $\tau = \overline{1, n_\tau}$ of diagnostics of a complex system. All sets M_k^τ, $k = \overline{1, K_\tau}$ contain subsets $\tilde{M}_k^\tau$, whose elements are not elements of set M_0. Fuzzy information vectors I_k^τ and their components for all factors of sets M_k^τ, $k = \overline{1, K_\tau}$ are known. Numerical values of the risk degree $\eta_{q_k}^\tau$ and risk level $W_{q_k}^\tau$ for the specified values of vector I_k^τ are also known.

It is required to define and ensure the rational coordination of the margin of permissible risk of abnormal, critical, extreme, emergency, and catastrophic situations for the predicted period of a complex system operation $T_{r+1}^\pm = [T_{r+1}^-, T_{r+1}^+]$ at $T_{r+1}^- \geq T_r^\pm$, $T_{r+1}^+ > T_{r+1}^-$.

The above stated problems allow one to analyze risks in the process of variation of a decision maker's informedness level. This level is characterized by an integral informedness indicator, which is a function of partial indicators, namely, completeness, credibility, and timeliness of informedness. In their turn, partial indicators are functions of information vectors and time. However, in the given problem statements the dependence on time of a decision maker's informedness integral indicator is not taken into consideration. At the same time, this indicator has a complex time dependence, since indicators of completeness and credibility are increasing functions of time, and the timeliness indicator is a decreasing function of time with a threshold restriction. The threshold restriction is in principle unavoidable due to the presence of a critical moment T_{cr}, when an emergency or catastrophic situation occurs. From here it follows that a decision on the prevention of unwanted consequences should be formed, taken, and realized on time, before a critical moment arises. Otherwise, any, even the best, decision (e.g., a global extremum considerably exceeding most of local extremums) will become useless.

The indicated circumstances determine the topicality and practical importance of investigation of risk situations at a threshold time restriction on forming and realizing a decision. We shall consider some problems of this class.

Problem 7. Determining a rational informedness level at a threshold time restriction. A mathematical statement of this problem corresponds to problem 5 with the additional conditions that at a moment T_0 ***the threshold restriction*** $\bar{U}_\tau(T_0) < U^-$ ***is fulfilled*** and the integral informedness indicator $\bar{U}_\tau$ is a threshold time function defined by the condition

$$\bar{U} = \begin{cases} \bar{U}_\tau(T_\kappa), & \text{of} \quad T_\kappa \leq \hat{T}_\kappa^* < T_{cr}, \\ 0, & \text{of} \quad T_\kappa > \hat{T}_\kappa^* \geq T_{cr}. \end{cases}$$

Here, T_0 and T_κ are, respectively, the moments of beginning and termination of additional information searches while solving a problem, $\hat{T}_\kappa$ is the duration of an additional information search, and $\hat{T}_\kappa = T_\kappa - T_0$ and $\hat{T}_\kappa^*$, $\bar{U}_\kappa^*$ are rational values of $\hat{T}_\kappa$ and $\bar{U}_\tau(T_\kappa)$, respectively.

It is required to define a rational compromise between the informedness level $\bar{U}_\kappa^*$ and duration T_κ^* of additional information searches when the recognition of each situation $S_k^\tau \in S_\tau$ and the determination of the risk degree and risk level $S_k^\tau \in S_\tau$ are ensured with a permissible level of credibility before a critical moment T_{cr} arises.

Problem 8. System estimation of margin of permissible risk in abnormal mode dynamics. A sequence of embedded sets (7.7) for some period of time $T_r^\pm = [T_r^-, T_r^+]$ of diagnostics of a complex system ***is known***. Set M_k^τ in the form (7.8) is known for each situation $S_k^\tau \in S_\tau$ at every moment $T_\tau \in T_r^\pm$, $\tau = \overline{1, n_\tau}$ of diagnostics of a complex system. All sets M_k^τ, $k = \overline{1, K_\tau}$ contain $\tilde{M}_k^\tau$ subsets, whose elements are not elements of set M_0. Fuzzy information vectors I_k^{τ} and their components for those risk factors of each set M_k^{τ} that are not elements of subset $\tilde{M}_k^{\tau}$ ***are known***. Numerical values of the risk degree η_{qk}^{τ} and risk level W_{qk}^{τ} for the known values of components of vector I_k^{τ} for those risk factors of each set M_k^{τ} that are not elements of subset $\tilde{M}_k^{\tau}$ ***are known***.

Also it is known that recognition $S_k^{\tau} \in S_{\tau}$ and estimation of the risk degree and risk level depend on the value of the integral informedness indicator $\bar{U}_{\tau}$ and is possible under the condition $\bar{U}_{\tau} > U^{-}$, where U^{-} is a detection threshold. The uncertainty of information about risk factors $\rho_k^{\tau} \in \tilde{M}_k^{\tau}$ creates the condition $\bar{U}_{\tau} < U^{-}$ at $T_0 \in T_r^{\pm}$, $T_0 > T_r^{-}$, when the possibility of recognition $S_k^{\tau} \in S_{\tau}$ and determination of a risk degree and risk level $S_k^{\tau} \in S_{\tau}$ are excluded.

It is required on the basis of a priori information and data of the current technical diagnostics determine the fundamental possibility of a critical and/or extreme situation occurring before a critical moment T_{cr} and defined as follows:

- A rational level $\bar{U}_{\tau}^{*}$ of integral informedness indicator $\bar{U}_{\tau}$, at which the recognition for determining the risk degree η_{qk}^{τ} and risk level W_{qk}^{τ} of each situation $S_k^{\tau} \in S_{\tau}$ and determining then moment T_{cr} is ensured with a practically admissible level of credibility.
- The margin of permissible risk of abnormal, critical, extreme, emergency, and catastrophic situations before a critical moment T_{cr} arises, and for the predicted period of complex system operation $T_{r+1}^{\pm} = \left[T_{r+1}^{-}, T_{r+1}^{+}\right]$ at $T_{r+1}^{-} \geq T_r^{\pm}$.

In conclusion let us note that the suggested methodological apparatus of system analysis and multicriterion minimization of the risk degree and risk level provides the possibility of fundamentally increasing the safety, survivability, and functioning efficiency of complex engineering systems due to the timely detection of causes and factors of critical and catastrophic risks and operative and timely prevention of their becoming accidents and catastrophes.

7.2 Properties and Peculiarities of Complex Engineering Systems Operating Under Conditions of Multifactor Risks

Experience in recent years has shown that the risk of occurrence of human-caused and natural catastrophes with consequences on a regional, national, and global scale continues to rise [33, 86, 88], promoted by various subjective and objective conditions and factors [167]. Among them the determining ones are the principles, approaches, and methods of control for technically and ecologically hazardous objects and ways of ensuring their safety in the presence of multifactor risks [52, 79, 98, 178]. This concerns, above all, nuclear power plants, large industrial agglomerations, complex technological and technical systems of various purpose, where safety requirements are of primary importance. In particular, in transportation systems, where under conditions of unexpected occurrence of catastrophic risk factors an emergency stop is in principle impossible. All types of aviation technique belong to such systems. The necessity of developing a qualitatively new approach to solving the problems of modern complex engineering system safety follows from the above considerations.

A new approach is based on the suggested conceptual foundations of system analysis, multicriterion estimation, and forecast of risk situations. The main idea of the suggested concept consists in the replacement of the typical principle of detection of the operability state turning into the inoperability state based on detection of failures, malfunctioning, faults, and forecast of reliability of an object by a qualitatively new principle [134,135]. The essence of this principle is the timely detection and elimination of the causes of a possible changeover of an operability state to an inoperability state based on the system analysis of multifactor risks of abnormal situations, a credible estimation of margin of permissible risk for different modes of operation of a complex engineering object, and a forecast of the main operability indicators of an object during the assigned operating period.

To achieve the above set goal it is necessary to define the basic notions and main subjects of investigation.

Let us consider the definitions that characterize the properties and peculiarities of a complex system under conditions of normal and abnormal modes of functioning [134, 135].

Functioning mode R is such a sequence $E_1, E_2, \ldots, E_\kappa, \ldots$ of the states of a complex system in which each state E_κ is characterized by certain indicators $(Y_\kappa,\ X_\kappa,\ U_\kappa)$ of system processes and certain indicators Ξ_κ of the influence of the environment and risk factors:

$$
\begin{aligned}
&E_\kappa = \{(Y_\kappa \in Y) \wedge (X_\kappa \in X) \wedge (U_\kappa \in U) \wedge (\Xi_\kappa \in \Xi)\}\,,\\
&Y_\kappa = \hat{Y}\,[T_\kappa]\,,\quad X_\kappa = \hat{X}\,[T_\kappa]\,,\quad U_\kappa = \hat{U}\,[T_\kappa]\,,\quad \Xi_\kappa = \hat{\Xi}\,[T_\kappa]\,,\\
&T_\kappa = \{t_\kappa | t_\kappa > t_{\kappa-1}\}\,,\quad T_\kappa \in T^{\pm}\,,\\
&T_0 = \left\{t_0 | t_0 = t^-\right\}\,,\quad T^{\pm} = \left\{t | t^- \le t \le t^+\right\}\,,\\
&Y = \left(\dot{Y}_i | i = \overline{1,m}\right)\,,\quad X = \left(\dot{X}_j | j = \overline{1,n}\right)\,,\\
&U = \left(\dot{U}_q | q = \overline{1,Q}\right)\,,\quad \Xi = \left(\dot{\Xi}_p | p = \overline{1,P}\right)\,.
\end{aligned}
\tag{7.18}
$$

Here, Y is a set of external parameters $\dot{Y}_i$ (technical, economic, and other indicators of the quality of system operation), X is a set of internal parameters $\dot{X}_j$ (structural, technological, and other indicators), U is a set of control parameters $\dot{U}_q$, Ξ is a set of parameters $\dot{\Xi}_p$ of the influence of the environment and risk factors, $\hat{Y}[T_\kappa]$, $\hat{X}[T_\kappa]$, $\hat{U}[T_\kappa]$, and $\hat{\Xi}[T_\kappa]$ are sets of values of the corresponding parameters at the time moment T_κ, and $T^\pm$ is the assigned or predicted period of complex engineering system functioning.

Normal mode R_{sd} is a functioning mode of an object for which $\forall T_\kappa \in T^\pm_{sd}$ all indicators Y_κ, X_κ, U_κ, and Ξ_κ are in the a priori defined range $D^\pm_Y$, $D^\pm_X$, $D^\pm_U$, and $\bar{D}^\pm_\Xi$:

$$\begin{aligned}
&\forall T_\kappa \in T^\pm_{sd} \Rightarrow\\
&\quad R_{sd} = \left\{(Y_\kappa \in D^\pm_Y) \wedge (X_\kappa \in D^\pm_X) \wedge (U_\kappa \in D^\pm_U) \wedge \left(\Xi_\kappa \in \overline{D}^\pm_\Xi\right)\right\},\\
&\quad T^\pm_{sd} = \left\{t | t^-_{sd} \le t \le t^+_{sd}\right\}, \quad T^\pm_{sd} \in T^\pm,\\
&\quad D^\pm_Y = \left(D^\pm_i | i = \overline{1,m}\right), \quad D^\pm_i = \left(\dot{Y}_i \,|\, \dot{Y}^-_i < \dot{Y}_i < \dot{Y}^+_i,\, i \in \overline{1,m}\right),\\
&\quad D^\pm_X = \left(D^\pm_j | j = \overline{1,n}\right), \quad D^\pm_j = \left(\dot{X}_j \,|\, \dot{X}^-_j < \dot{X}_j < \dot{X}^+_j,\, j \in \overline{1,n}\right),\\
&\quad D^\pm_U = \left(D^\pm_q | q = \overline{1,Q}\right), \quad D^\pm_q = \left(\dot{U}_q \,|\, \dot{U}^-_q < \dot{U}_q < \dot{U}^+_q,\, q \in \overline{1,Q}\right),\\
&\quad \bar{D}^\pm_\Xi = \left(\bar{D}^\pm_p | p = \overline{1,P}\right), \quad \bar{D}^\pm_p = \left(\dot{\Xi}_p \,|\, \dot{\Xi}^-_p < M\dot{\Xi}_p < \dot{\Xi}^+_p,\, p \in \overline{1,P}\right).
\end{aligned} \tag{7.19}$$

Abnormal mode R_{os} is a functioning mode of an object for which for $\forall T_\kappa \in T^\pm_{os}$ some indicators or some combinations of indicators or all indicators Y_κ, X_κ, U_κ, and Ξ_κ are beyond the a priori defined range $D^\pm_Y$, $D^\pm_X$, $D^\pm_U$, and $\bar{D}^\pm_\Xi$. In the general case, state R_{os} is described by the following relationships:

$$\begin{aligned}
&\forall T_\kappa \in T^\pm_{os} \Rightarrow R_{os} = A_1 \vee A_2 \vee A_3 \vee A_4 \vee A_5 \vee \ldots \vee A_N,\\
&\quad T^\pm_{os} = \left\{t | t^-_{os} \le t \le t^+_{os}\right\}, \quad T^\pm_{os} \in T^\pm,\\
&\quad A_1 = \left\{(Y_\kappa \notin D^\pm_Y) \wedge \left[(X_\kappa \in D^\pm_X) \wedge (U_\kappa \in D^\pm_U) \wedge \left(\Xi_\kappa \in \overline{D}^\pm_\Xi\right)\right]\right\},\\
&\quad A_2 = \left\{(X_\kappa \notin D^\pm_X) \wedge \left[(Y_\kappa \in D^\pm_Y) \wedge (U_\kappa \in D^\pm_U) \wedge \left(\Xi_\kappa \in \overline{D}^\pm_\Xi\right)\right]\right\},\\
&\quad A_3 = \left\{(U_\kappa \notin D^\pm_U) \wedge \left[(Y_\kappa \in D^\pm_Y) \wedge (X_\kappa \in D^\pm_X) \wedge \left(\Xi_\kappa \in \overline{D}^\pm_\Xi\right)\right]\right\},\\
&\quad A_4 = \left\{\left(\Xi_\kappa \notin \overline{D}^\pm_\Xi\right) \wedge \left[(Y_\kappa \in D^\pm_Y) \wedge (X_\kappa \in D^\pm_X) \wedge (U_\kappa \in D^\pm_U)\right]\right\},\\
&\quad A_5 = \left\{\left[(Y_\kappa \notin D^\pm_Y) \wedge (X_\kappa \notin D^\pm_X)\right] \wedge \left[(U_\kappa \in D^\pm_U) \wedge \left(\Xi_\kappa \in \overline{D}^\pm_\Xi\right)\right]\right\},\\
&\quad \cdots\cdots\cdots\cdots\cdots\cdots\cdots\cdots\cdots\cdots\cdots\cdots\cdots\cdots\\
&\quad A_N = \left\{(X_\kappa \notin D^\pm_X) \wedge (Y_\kappa \notin D^\pm_Y) \wedge (U_\kappa \notin D^\pm_U) \wedge \left(\Xi_\kappa \notin \overline{D}^\pm_\Xi\right)\right\}.
\end{aligned} \tag{7.20}$$

Transition mode $\overset{\cap -}{R_{tr}}$ is an uncontrollable functioning mode that is caused by the influence of set $\hat{M}_{tr}$ of risk factors ρ_k, $k \in \tilde{K}^-$ and that during period $\overset{\cap\pm}{T}_{tr} \in T^{\pm}$ results in a ***transformation*** of normal mode R_{sd} into abnormal mode R_{os}. Mode $\overset{\cap-}{R}_{tr}$ is described by the following relationships:

$$\overset{\cap-}{R}_{tr}\colon R_{sd} \overset{\hat{M}_{tr}}{\to} R_{os}, \tag{7.21}$$

$$\forall T \in T_{tr}^{\pm} \Rightarrow R_{tr}^{-} = R_{tr}^{b} \vee \tilde{R}_{tr} \vee R_{tr}^{e},$$

$$T = \{t | t \in T^{\pm}\}, \quad \overset{\cap\pm}{T}_{tr} = \left\{t | \overset{\cap-}{t}_{tr} < t < \overset{\cap+}{t}_{tr}\right\},$$

$$\overset{\cap b}{R}_{tr} = \left\{\left(\exists! T = T_{tr}^{-} = \tilde{T}_{sd}^{+}\right) \wedge \left(\exists! E_T = \overset{\cup-}{E}_{tr} = \overset{\cap+}{E}_{sd}\right) \wedge \right.$$

$$\left. \wedge \left[\exists! \mu_{R_{tr1}}(E_T) = \mu_{R_{sd}}(E_T) = 1\right]\right\},$$

$$E_T = E[T], \quad E_{tr}^{-} = E\left[\overset{\cap-}{T}_{tr}\right], \quad \overset{\cap+}{E}_{sd} = E\left[\overset{\cap+}{T}_{sd}\right],$$

$$\overset{\cap-}{T}_{tr} = \left\{t | t = \overset{\cap-}{t}_{tr}\right\}, \quad \overset{\cap+}{T}_{sd} = \left\{t | t = \overset{\cap+}{t}_{sd} \le t_{sd}^{+} \in T^{\pm}\right\},$$

$$\tilde{R}_{tr} = \left\{\left(\forall T \in \overset{\cap\pm}{T}_{tr} \Rightarrow \left(\tilde{E}_{tr}[T] = [E_T, \mu_{R_{sd}}(E_T)] \vee [E_T, \mu_{R_{os}}(E_T)]\right) \wedge \right.\right.$$

$$\left.\left. \wedge \left[\mu_{R_{sd}}(E_T) + \mu_{R_{os}}(E_T) = 1\right]\right)\right\},$$

$$R_{tr}^{e} = \left\{\left(\exists! T = \overset{\cap+}{T}_{tr}\right) \wedge \left(\exists! E_T = \overset{\cap+}{E}_{tr} = \overset{\cap+}{E}_{os}\right) \wedge \right.$$

$$\left. \wedge \left[\exists! \mu_{R_{tr}}(E_T) = \mu_{R_{os}}(E_T) = 1\right]\right\},$$

$$\overset{\cap-}{E}_{os} = E\left[\overset{\cap+}{T}_{tr}\right], \quad \overset{\cap+}{T}_{tr} = \left\{t | t = \overset{\cap+}{t}_{tr}\right\}.$$

Transition mode $\overset{\cup+}{R}_{tr}$ is a controllable functioning mode conditioned by the **control action** U_{tr} of the safety control system that during period $\overset{\cup\pm}{T}_{tr}$ results in a transformation of abnormal mode R_{os} into normal mode R_{sd}. Mode $\overset{\cup\pm}{R}_{tr}$ is described by the following relationships:

$$\overset{\cup+}{R}_{tr}\colon R_{os} \overset{U_{tr}}{\to} R_{sd}, \tag{7.22}$$

$$\forall T \in \overset{\cup\pm}{T}_{tr} \Rightarrow \overset{\cup+}{R}_{tr} = \overset{\cup b}{R}_{tr} \vee \tilde{R}_{tr} \vee \overset{\cup e}{R}_{tr},$$

$$T = \{t | t \in T^{\pm}\}, \quad \overset{\cup\pm}{T}_{tr} = \left\{t | \overset{\cup-}{t}_{tr} < t < \overset{\cup+}{t}_{tr}\right\},$$

$$\overset{\cup b}{R}_{tr} = \left\{ \left(\exists! T = \overset{\cup -}{T}_{tr} = \overset{\cap +}{T}_{tr} \right) \wedge \left(\exists! E_T = \overset{\cup -}{E}_{tr} = \overset{\cap +}{E}_{tr} \right) \wedge \right.$$

$$\left. \wedge \left[\exists! \mu_{R_{tr}} (E_T) = \mu_{R_{os}} (E_T) = 1 \right] \right\}, \tag{7.23}$$

$$E_T = E\left[T\right], \quad \overset{\cup -}{E}_{tr} = E\left[\overset{\cup -}{T}_{tr}\right] = \overset{\cup +}{E}_{os}\left[\overset{\cup +}{T}_{tr}\right],$$

$$\overset{\cup -}{T}_{tr} = \left\{ t | t = \overset{\cup -}{t}_{tr} = \overset{\cap +}{t}_{tr} \right\}, \quad \overset{\cup +}{T}_{sd} = \left\{ t | t = \overset{\cup +}{t}_{sd} \leq t^{+}_{sd} \right\},$$

$$\tilde{R}_{tr} = \left\{ \left(\forall T \in \overset{\cup \pm}{T}_{tr} \Rightarrow \left(\tilde{E}_{tr}\left[T\right] = \left[E_T, \mu_{R_{sd}} (E_T)\right] \vee \left[E_T, \mu_{R_{os}} (E_T)\right] \right) \wedge \right. \right.$$

$$\left. \left. \wedge \left[\mu_{R_{sd}} (E_T) + \mu_{R_{os}} (E_T) = 1 \right] \right) \right\},$$

$$R^{e}_{tr} = \left\{ \left(\exists! T = \overset{\cup +}{T}_{tr} = \overset{\cup -}{T}_{sd} \right) \wedge \left(\exists! E_T = \overset{\cup -}{E}_{sd} \right) \wedge \right.$$

$$\left. \wedge \left[\exists! \mu_{R_{tr}} (E_T) = \mu_{R_{sd}} (E_T) = 1 \right] \right\},$$

$$\overset{\cup -}{E}_{sd} = E\left[\overset{\cup +}{T}_{tr}\right] = E\left[T^{-}_{sd}\right], \quad \overset{\cup +}{T}_{tr} = \left\{ t | t = \overset{\cup +}{t}_{tr} = \overset{\cup -}{t}_{sd} \right\}.$$

Abnormal situation S_{os} is an abnormal functioning mode at moment T_{os} for which some indicators of the system quality or the environmental indicators are beyond the normal mode intervals within those limits where there is no danger of an emergency or catastrophe.

Critical situation S_{cs} is an abnormal functioning mode at moment T_{os} for which the quality indicators of a system or the environmental indicators are beyond the normal mode intervals within those limits where there appears a real danger of an emergency or catastrophe.

Extreme situation S_{es} is an abnormal functioning mode at the moment T_{os} for which the quality indicators of a system or the environmental indicators are beyond the normal mode intervals within those limits where an emergency or catastrophe are practically inevitable.

Emergency situation S_{as} is such abnormal functioning mode, during which an engineering system turns from an operative state into such inoperative, emergency state, when repair is necessary to return it to the initial state.

Accident: outcome of an emergency situation.

Catastrophic situation S_{ds}: abnormal functioning mode during which an engineering system turns from an operative state to an inoperative, catastrophic state when the return to an operative state is in principle impossible.

Catastrophe: outcome of a catastrophic situation.

Safety control: special form of complex system control realized in dynamics of complex system functioning in the form of a set of decisions and actions for the purpose of ensuring its survivability and prevention of critical and extreme situations, accidents, and catastrophes.

Survivability: property of a complex system that represents its ability to maintain the normal functioning mode and exclude the possibility of an accident or catastrophe in predictable and unpredictable conditions of the influence of destabilizing, nondestructive risk factors.

Serviceability: property of a complex system that represents its ability to perform its functions in the normal mode with the assigned indicators of quality and efficiency.

Risk degree η_i: probability of occurrence of undesirable consequences of the influence of ***any*** risk factors at any point in time $T_i \in T^{\pm}$ in the process of complex system functioning.

Risk level W_i: the size of damage caused by the influence of ***any*** risk factors at any point in time $T_i \in T^{\pm}$ in the process of complex system functioning.

Margin of permissible risk T_o: duration of complex system functioning period in a certain mode when the risk degree and risk level will not exceed the a priori assigned permissible values under the possible influence of risk factors.

Margin of permissible risk of abnormal situations T_{os}: duration of complex system functioning period in the normal mode when the risk degree and risk level of an abnormal situation will not exceed the a priori assigned values.

Margin of permissible risk of a critical situation T_{cs}: duration of complex system functioning period in the normal mode when the risk degree and risk level of a critical situation will not exceed the a priori assigned values.

Margin of permissible risk of an extreme situation T_{es}: duration of complex system functioning period in the normal mode when the risk degree and risk level of an extreme situation will not exceed the a priori assigned values.

Margin of permissible risk of an emergency situation T_{as}: duration of complex system functioning period in the normal mode when the risk degree and risk level of an emergency situation will not exceed the a priori assigned values.

Margin of permissible risk of a catastrophic situation T_{ds}: duration of complex system functioning period in the normal mode when the risk degree and risk level of a catastrophic situation will not exceed the a priori assigned values.

Margin of permissible risk of abnormal mode T_{ar}: duration of complex system functioning period in the normal mode when the risk degree and risk level of a critical, extreme, emergency, and catastrophic situation will not exceed the a priori assigned values.

Note that the main indicators of risk are the risk degree, risk level, and margin of permissible risk of a abnormal mode. The margin of permissible risk of the abnormal mode determines the duration of the complex engineering system functioning period when the risk degree and/or risk level of any abnormal situation should not exceed the permissible values specified by the approved specifications. And it is a resource whose value is determined not by simply summing up the duration of functioning during the specified term, but by the working capacity, survivability, and safety during the specified term taking into account the real operating conditions. Therefore, the real influence of various factors, including the influence of functioning and risk factors, are considered.

7.3 Analysis of Multifactor Risks of Accidents and Catastrophes

Let us describe the main concepts of accidents and catastrophes and the conditions of guaranteed safety that follow from these concepts.

Assertion 1. An accident or catastrophe is the final result of a consecutive transition of the normal mode functioning of complex system into an emergency or catastrophic situation. The transition is carried out by realization of the following interconnected abnormal mode stages. **Stage 1** represents a transitive mode from a normal state into an abnormal situation. **Stage 2** is caused by a transitive mode from an abnormal situation into a critical situation. **Stage 3** represents a transitive mode from a critical situation into an extreme situation. **Stage 4** describes a transitive mode from an extreme situation into an emergency or catastrophic situation. The duration of each stage depends on the magnitude of a margin of permissible risk of the corresponding an abnormal-mode situation. Stages may be reversible or irreversible, evolutional or threshold. Evolutional processes may be slow or fast, accelerating or decelerating.

Assertion 2. The risk of occurrence of an accident or catastrophe rises when an abnormal mode is passing from stage 1 to stage 4. The rate of risk increase at each stage is directly proportional to the rate of a decrease of the margin of permissible risk in the mode of the corresponding stage. The margin of permissible risk of occurrence of an accident or catastrophe reaches its maximum value in the normal mode and decreases in the course of passing through the stages of the abnormal mode and becomes equal to zero at stage 4.

Assertion 3. The degree of risk of an accident or catastrophe is an increasing function of time in the process of a consecutive transition of the normal mode of complex system functioning into an abnormal, critical, and extreme situation and reach their value limits in an emergency or catastrophic situation. The properties of functions at each stage are determined by properties of the transition mode of the corresponding stage. The maximum rate of the degree of risk increase corresponds to threshold processes of the stages. The minimum rate corresponds to reversible, slowing down processes of the stages.

The degree of risk is minimum at stage 1, and it reaches its maximum value at stage 4.

Assertion 4. The level of risk of an accident and catastrophe within the limits of a margin of the permissible risk may remain constant during all stages, and it may rise during the transition from stage 1 to stage 4.

Assertion 5. The necessary condition for guaranteeing the safety of a complex system is the fulfillment the following functions and requirements by the system of safety control:

- Identification, recognition, and prediction of risks of abnormal, critical, extreme, emergency, and catastrophic situations with a required level of credibility in real time.
- Prediction and realization of the required margin of permissible risk of the abnormal mode with the assigned permissible values of a degree and level of risk of abnormal, critical, extreme, emergency, and catastrophic situations.
- Justification and realization of those required indicators of the transition mode $\overset{\cup+}{R}_{tr}$ at which the transformation of the abnormal mode into the normal mode is reached within the limits of the margin of permissible risk of abnormal, critical, and extreme situations.

From the above presented statements we can draw a conclusion about the practical necessity of carrying out a system analysis of the basic features and properties of an abnormal mode and its components as the information basis for the formation and justification of the means and methods of ensuring the guaranteed safety of complex systems. To provide such information support a quantitative measure of the information is required that takes into account the specificity of an abnormal mode and its components.

At the same time it has been established [135] that a typical Shannon measure of the quantity of information [171] that is used in the theory of communication [20, 21, 48, 62, 170], in the theory of complex systems and processes [66, 95, 96, 194], in the information theory of control [73, 74, 146], in the theory of pattern recognition [40, 208]) does not correspond to conditions, properties, and features of the abnormal mode. The point is that this measure was obtained by Shannon on the assumption that the analyzed random events constituted a complete group of events. Under this condition, the realization of one event from the group makes the realization of other events from this group in principle impossible. Consequently, if one of the events really occurs, the probability of other events becomes equal to zero. In particular, this is characteristic of a consecutive transmission of information in letters through a communication system [21, 170, 171].

However, fundamentally different conditions and factors occur in abnormal situations. They are characterized by the following axiom [135].

The axiom of risk situations. A set of risk situations is in principle not a complete group of random events. Therefore, the possibility of simultaneous occurrence of several events from the given set is not excluded, and, consequently, it is in principle possible that the probability of several events is equal to unit. The occurrence of one or several events from the given set of risk situations does not exclude the possibility of occurrence of other events from this set. Therefore, the uncertainty of occurrence of any event from a set of risk situations does not become equal to zero upon realization of one or several events of the set.

Here and later on the concept of "risk situation" is treated in a broad sense and comprises different classes of abnormal, critical, extreme, emergency, and catastrophic situations. The axiom statements follow from the analysis of numerous accidents [102] caused by simultaneous or consecutive occurrence of several malfunctions or failures. An example is the simultaneous failure of all engines of an An-124 "Ruslan" plane that resulted in its crash on 6 December 1997 near the town of Irkutsk–2 [102].

Let us indicate one important feature of the Shannon measure: the quantity of information is defined as the difference between the object uncertainty before the information is obtained and the uncertainty after the information is obtained. As a result, there is an important condition that determines the ability to measure the quantity of information by a unified numerical value, irrespective of the information variety, which is of great importance for communication systems and a number of other applications. However, such a measure is unacceptable for the analysis of risk situations, as it has an essential disadvantage: it does not allow one to estimate simultaneously the quantity and the quality of information. The quality of information is considered to be characteristic of the information usefulness for solving concrete system problems of risk-situation analysis in conditions of threshold time restrictions. The presence of such a threshold is caused by the possibility that a risk situation may turn into a catastrophe or accident within the extremely limited time. It is obvious that the decision on preventing such a transition is to be formed, taken, and fulfilled before an accident happens; otherwise, it becomes useless. Thus the necessity of efficient and timely estimation of such information quality indicators as completeness, credibility, and timeliness follows from the above.

Taking into account the axiom of risk situations and the listed indicators of information quality, fundamentally new requirements for the quantitative measure of information are established. They can be described briefly as ensuring system coordination of information quality and quantity in conditions of threshold time restrictions. Therefore, it is expedient to estimate the **quantity** of information by the volume of data useful for forming and justifying a decision, and the **quality** by the meaningfulness of data as to their credibility and completeness of representation of risk situations and the timeliness of delivery of practically useful information for a decision maker. Note that in modern conditions of a wide application of information technologies and computer engineering it is expedient to accept **a byte** as a unit of information quantity in any field of practical activity.

On the basis of the formulated statements (1–5) and the introduced notions of quantity and quality of information we consider the peculiarities of processes of formation of an abnormal mode and briefly introduce the processes of emergency and catastrophe development, analyze the role and place of information support in the formation of decisions for recognizing and assessing situations.

First, let us consider the interconnections and interdependencies of normal and abnormal modes and the transformation of abnormal situations into catastrophes and emergencies. Taking into account that a complex system will largely function in normal mode, we will consider this as the initial state. In addition, the components of an abnormal mode are displayed in the block diagram in Fig. 7.1 in the order of their possible transformation into a catastrophe or emergency. In the analysis we will take into account information uncertainty of abnormal situations. Therefore, the sequence of transition of abnormal situations to an abnormal mode (blocks 1–5.3) is supplemented by an analysis of how abnormal situations transition to a normal mode (block 4). This procedure is intended for forming and realizing decisions in three possible variants of a state of complex systems:

Variant 1. The transition to a normal mode of functioning is possible. It is ensured by the procedures of analysis of the transition conditions (block 5.1), choosing a transition strategy (block 9.1), and realizing a transition strategy (block 10.1).

Variant 2. The transition to a normal mode of functioning is impossible. The conclusion follows that in this situation the abnormal mode of a complex situation (block 5.3) is in operation. An analysis of factors, conditions, and possibilities of a transition of an abnormal mode to a critical, extreme, emergency, or catastrophic situation, or to a failure and accident (blocks 6.1 and 6.2; 7.1 and 7.2; 8.1 and 8.2) is consistently carried out.

Variant 3 is the least defined and most risky. It occurs if at the analyzed moment the volume of fuzzy information about the condition of a complex system's operation is insufficient for taking a decision (question mark in diagram). In such a case the main objective of a decision maker is to prevent an emergency or catastrophe. This decision is formed on the basis of the available a priori and diagnostic information (block 5.2) and is directed at the development and realization of a rational strategy of actions (blocks 9.2 and 10.2).

In addition to the previous block diagram we shall present the main risk indicators and display one variant of consecutive transition of a normal mode into an accident or catastrophe (Fig. 7.2). We shall note that risk indicators for various components (blocks 2–8) of abnormal mode may differ essentially. Therefore, for every component, values of degree and level (blocks 2.1–8.1) and a margin of permissible risk (blocks 2.2–8.2) are determined. On the basis of these results an analysis of degree and level (block 9) of risk and a margin of permissible risk (block 10) is performed, and its results are the bases for forming and realizing decisions.

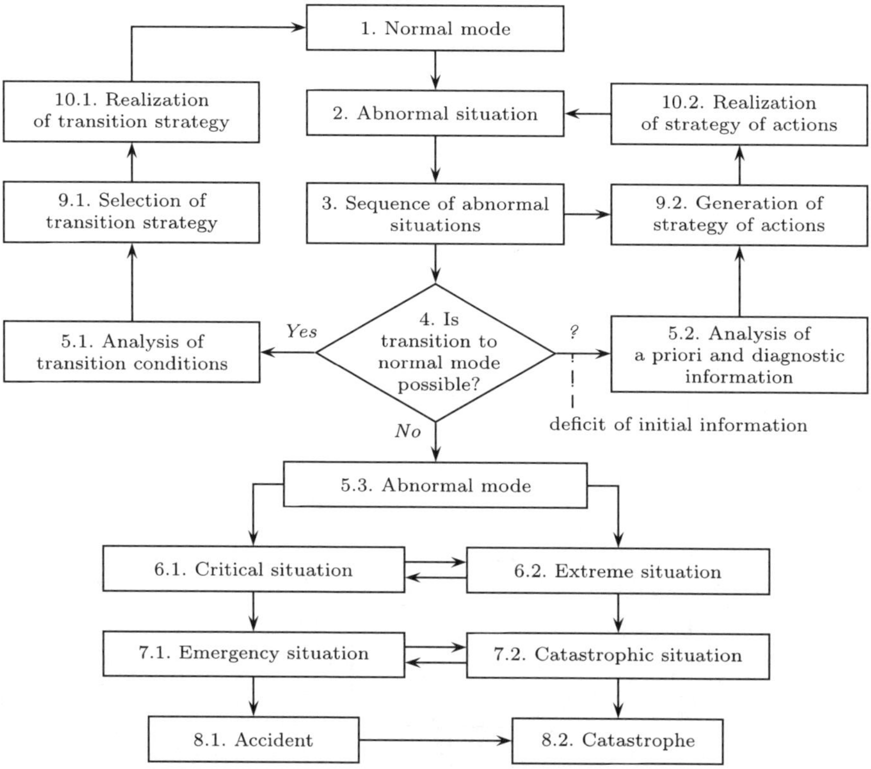

Fig. 7.1. Block diagram of interconnections and interdependencies of components of an abnormal mode

Note that in the considered block diagram values of risk indiactors for a concrete moment in time $T_k \in T^{\pm}$ are determined, but their changes in time are not taken into account. Such an approach is accepted on the condition that the state of the object is characterized by the consumption of a serviceability resource that makes a hundred thousand corresponding units, for example, kilometers for ground mass transport or hours of a run for small transport, sanitary, or agricultural aircraft.

However, for many objects it is impossible to restrict the investigation to a consideration of the serviceability resource consumption only; a more thorough analysis of the technical state and a more correct estimation of risk are required. Unique products of modern technology are nuclear and chemical reactors, thermonuclear plants, mining complexes, powerful air, ground, surface, underwater vehicles, etc.

For the analysis of such class of problems the algorithm presented in Fig. 7.3 is proposed. The peculiarity of the given algorithm is the parallel analysis of factors, properties, and features of processes of irregular situations (blocks 3.1, 3.3, 4.1–4.3) on the basis of the current information (block 1) and

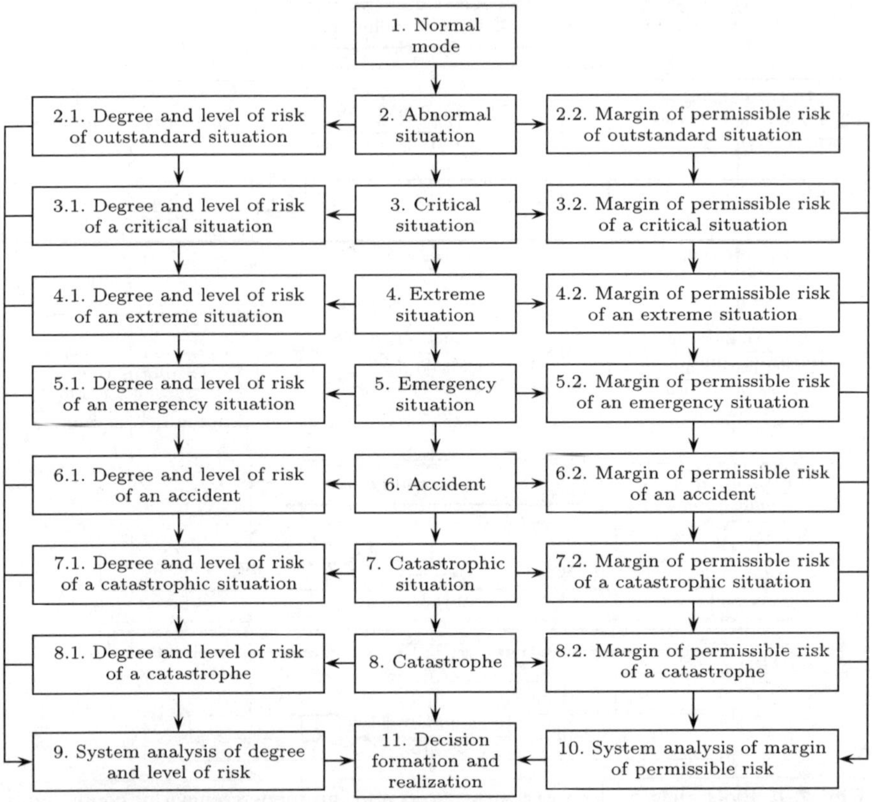

Fig. 7.2. Block diagram of interconnections between risk indicators and components of abnormal mode

the a priori information of the a priori knowledge base (block 3.2). Such an approach is realized for the purpose of reducing the time for recognizing abnormal situations (blocks 4.1–4.3) and for creating conditions for effective estimating and forecasting of risk situations, degree and level of risk, and margin of permissible risk (blocks 5.1–5.3, 6.1–6.3). Results of estimating and forecasting are used for forming and validating decisions (block 7). In real conditions of uncertainty some variants of decisions that are estimated and ranking by preference on separate risk indicators (blocks 8.1–8.3) are developed. Then decisions are estimated and ranged from a position of the system coordinated detection of the most dangerous risk factors and those risk indicators that change most quickly (block 9). On the basis of these decisions recommendations for the decision maker (block 10) are developed.

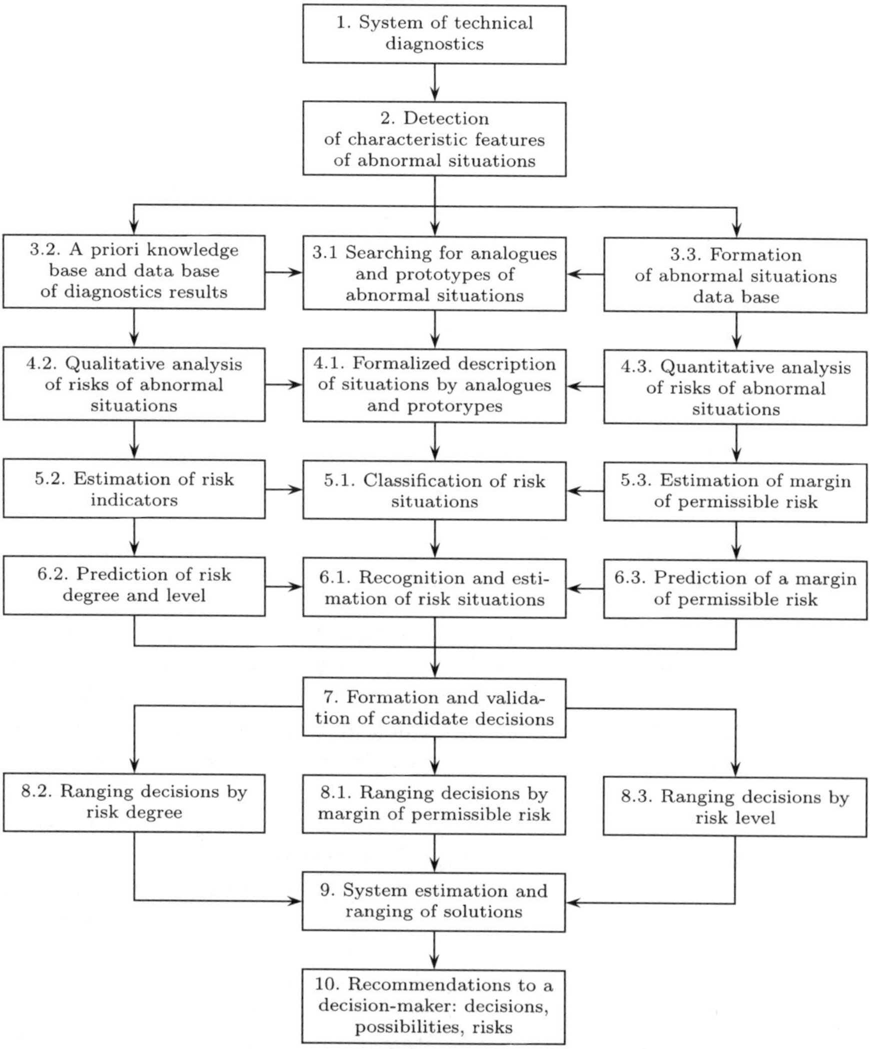

Fig. 7.3. Block diagram of risk analysis and estimating algorithm for abnormal modes of a complex system

7.4 Basic Principles and Peculiarities of Control of Complex-system Safety

The analysis of abnormal modes of various complex systems and structures and research on the causes and peculiarities of catastrophes and emergencies show that the known principles and methods of ensuring safety of complex systems do not completely correspond to modern requirements [33]; approaches and principles of estimation of survivability do not reflect the entire range of factors that influence the destruction of materials and structures [180]. Among the most important reasons for the present situation the following should be mentioned. The occurrence of situations of risk, emergencies, and catastrophes is not taken into account in traditional models of determinate and random processes. The reason is that the aging and destruction mechanisms of materials and structures are characterized by high complexity and nonstationarity [135]. The following groups of multidimensional risk factors are of special importance [180]:

- Properties of materials (tension, deformation, temperature, internal energy, magnetization, entropy, chemical potential, etc.).
- Defects and peculiar features of real materials (dispersion and heterogeneity of properties, instability of surface layer state, heterogeneity and instability of structure and phase composition, various technological defects, etc.).
- Processes of aging and destruction (cyclic, low-cyclic, thermal, radiating, corrosive, and others).
- Processes of environmental influence (force, temperature, chemical, radiating and others).

Only a small fraction of properties and features of one kind of destruction process is enumerated here. In real complex systems and objects (transport: planes, sea vessels, etc.; engineering structures: bridges, tanks, pipelines etc.) a great number of various processes of deformation, aging, and destruction take place, which interact with each other and the environment in a complicated way.

However, the theory of complex systems and the theory of control study only determinate and random processes of functioning of objects, assuming that probabilistic characteristics of random processes are given. An exception is the game approach [73,74]. The theory of destruction [180] studies separate processes of aging and destruction mechanisms without taking into consideration their system interconnections and connections with processes of functioning of a system. Synergy studies general properties of nonlinear processes in open systems and chaotic processes in open nonlinear systems [56, 75, 87]. Catastrophe theory studies threshold processes and other mechanisms for certain classes of models [2, 148, 179]. A number of disciplines and research directions similarly investigate other kinds of processes and phenomena. The indicated disciplines study separate processes and objects at various initial preconditions and assumptions that are not always coordinated and are not always fulfilled in practice. The main reasons for such a situation are:

- Absence of a united approach to analyzing various mechanisms of destruction.
- Insufficient accounting for real conditions and factors of abnormal situations.

- Absence of a united methodological apparatus of the system analysis of properties of a normal mode, risk factors, processes, and components of abnormal situations.

This is why the generalized representation of a functioning mode of a real complex object is needed that would allow one to reveal both the mechanisms of its functioning and serviceability in a normal mode and the mechanisms of occurrence and development of an abnormal mode from unified positions. It is expedient to express such generalized representation in the form of an empirical statement, called "a hypothesis about the mode of a complex system" [135].

Hypothesis about the mode of a complex system. The mode of functioning of any complex system at any moment in time is determined by an action of a set of its internal processes and a set of processes of its interaction with the environment. The mode properties depend on ordering, randomness, and chaotic state of internal and external processes. The mode contains the following components:

- A determinate component created by the action of typical determinate processes of a normal mode of functioning and typical processes of standard control of a system and described by the determinate regularities.
- A random component created by the action of various random factors of the external and internal environment of a system and ruled by certain probabilistic and statistical regularities.
- A chaotic component created by irregular, abnormal, unpredictable disturbing actions of various nonlinear evolutionary and threshold mechanisms and not ruled by known determinate or statistical regularities.

The indicated components are interconnected through a set processes and phenomena. This set includes various kinds and types of connections: direct and reverse, positive and negative, linear and nonlinear, determinate and random, reversible and irreversible, deliberate and parasitic. Joint actions of components at any moment in time determine the properties and features of the mode of functioning of a complex system.

The practical importance of the given hypothesis lies in its determining the necessity, possibility, and expediency of searching various experimental and empirical approaches and methods of understanding reality. Let us present the realization of the basic ideas of the hypothesis in the form of the following assertions.

Assertion 6 (on the computational irreducibility of complex systems). All real complex systems are computationally irreducible in the sense that their description in the form of any physical, mathematical, logic, simulation, algorithmic, and/or other models and formal descriptions constructed on the basis of any axiomatic theory is characterized by fundamentally irremovable incompleteness and by the unprovability of noncontradictoriness.

Assertion 7 (on the limitation of computational reducibility of complex systems). A real complex system may be **computationally reducible** in the specified limits in the sense that, based on physical, chemical, mechanical, computational, and/or other experiments and experimentally credible data and knowledge bases it is possible to build physical, mathematical, logical, simulation, algorithmic, and/or other models and formalized descriptions that reflect objective regularities of processes and system properties with fundamentally limited levels of completeness, credibility, and accuracy.

Assertions 6 and 7 allow one to formulate the following important assertions.

Assertion 8 (on the fundamental limitation of credibly describing complex systems). A real complex system in principle cannot be described with absolute completeness, absolute credibility, and absolute accuracy using methods and models of any axiomatic theory and/or using methods and models based on the use of knowledge and data bases obtained and verified experimentally.

Assertion 8 is the generalized conclusion that follows directly from assertions 6 and 7, from the hypothesis on the mode of a complex system, and from the axiom of risk situations. On its basis the following practically important assertions about properties of controllable and uncontrollable risk of a real complex system are formulated.

Assertion 9 (on the fundamental noneliminability of complex-system risk). The risk of an abnormal mode of any real complex system is **fundamentally unavoidable when the time of its functioning is unlimited**.

This assertion reflects a known fact: it is impossible to create a complex system capable of endless functioning in a normal mode. This assertion follows from assertions 6–8, namely, from the fundamental limitation of a level of completeness and credibility of description of a real complex system follows the fundamental impossibility of excluding risk situations for an unlimited time. But according to assertions 6–7 the possibility of limiting a risk within the boundaries of a finite period of a complex system while it is functioning is not excluded.

Assertion 10 (on the limitation of controlled risk of a complex system). For any real complex system it is possible to limit the risk of any abnormal mode by rational safety control and ensure the permissible value of degree and level of risk within the limits of a certain finite period of functioning of this complex system on the basis of timely detection and removal of causes of possible risk situations.

Assertion 11 (on the unboundedness of uncontrollable risk of complex systems). Without safety control of a real complex system it is fundametally impossible to limit the risk of any abnormal mode to a level of the permissible values of the degree and level of risk within the boundaries of the assigned finite period.

Assertions 10 and 11 are a concretization of assertion 9 and define the essence and differences of properties of controllable and uncontrollable risk of a real complex system. Assertion 10 follows from assertions 7–9, according to which it is possible to realize a system of safety control in the process of functioning of a real complex object capable of timely detection, recognition, and prediction of risk situations, as well as their minimization due to timely elimination of causes of occurrence of possible abnormal modes. Assertion 11 follows from assertion 10, in that, when there is no safety control, the degree and level of risk are uncontrollably changing and, as a consequence, within the limits of the finite period of functioning of a complex system the risk may exceed the assigned, permissible values.

Assertion 12 (on the level of a certain margin of permissible risk of a complex system). For any real complex system, by providing a rational safety control in the course of its functioning, it is possible to ensure a certain margin of permissible risk with the required permissible values of the degree and level of risk or the required margin of permissible risk with definite values of the degree and level of risk or with definite requirements for the safety control system to ensure the required margin of permissible risk with the required permissible values of the degree and level of risk.

Assertion 13 (on the level of uncontrollable risk margin of a complex system). Without safety control it is impossible to ensure the required margin of permissible risk in the process of functioning of any real complex system with the required permissible values of the degree and level of risk.

Assertions 12 and 13 directly follow from assertions 10 and 11, respectively. They define the essence and differences in possibilities for realization of the requirements for margin of permissible risk respectively in the presence and absence of safety control of a complex system.

On the basis of the formulated assertions the following principles of formation and realization of the guaranteed safety of complex systems are suggested.

Principle 1 (on the system coordination of safety-ensuring measures). The guaranteed safety of complex systems must be provided by coordinated engineering activities at all stages of a product's life cycle. And the measures of all stages must be system coordinated according to goals, tasks, terms, resources, and expected results from the position of guaranteeing the assigned indicators of survivability and the assigned indicators of serviceability during the given operation period.

Principle 2 (on system coordination of safety control). The guaranteed safety in the process of a complex system's operation must be provided by a safety control system. The indicators of quality and efficiency of this system must be system coordinated with the corresponding indicators of a control system of a complex system's operation for the purpose of simultaneously ensuring the assigned margin of permissible risk and assigned system serviceability during the predictable operation period with the required credibility.

Principle 3 (on guaranteeing timely realization of decisions in the process of safety control). The safety control system must guarantee timely formation and realization of decisions in the presence of an unremovable threshold time restrictions on a cycle of control in abnormal modes with the required credibility.

Principle 4 (on system-guaranteed diagnostics of risk situations). Indicators of quality and efficiency of a technical diagnostic system must be system coordinated with the corresponding indicators of the safety control system based on the condition of guaranteeing prevention of failures of a complex system within the limits of margin of permissible risk of various modes with the required credibility.

Principle 5 (on the timeliness and credibility of estimation and forecast of risk situations). A technical diagnostic system must ensure timely detection, recognition, and estimation of risk of an abnormal mode for the predicted period of a complex system operation with the required credibility based on the condition of guaranteeing timely elimination of risk causes before failures occur.

The proposed principles are the basis of the strategy of control of serviceability and safety of complex systems and, in particular, of human-caused and ecologically dangerous engineering systems and objects [110, 112, 134, 135]. This strategy allows for the creation of a qualitatively new level of safety and survivability of complex systems. For example, for air engineering in real conditions of flight there appears the possibility of timely detection of factors of considerable risk and their timely elimination during the preflight period, which excludes the chance of their transition into factors of a critical or catastrophic risk, or into a mode of failure. A significant increase in the level of safety and survivability for other types of complex systems is similarly provided.

Let us formulate now the main positions of a strategy for guaranteeing the safety of complex systems.

The general idea of the strategy is to ensure the safety of such systems by coordinated engineering activities (*concurrent engineering* in US terminology, *simultaneous engineering* in the terminology of EU) at all stages of a product's life cycle: from the stage of formation of an idea and concept of a product up to the stage of recycling of an object.

The main goal of the proposed strategy is to guarantee a rationally justified reserve of survivability of a complex system in real conditions of fundamentally irremovable information and time restrictions.

The main idea of the strategy is to ensure the timely and credible detection, recognition, and estimation of risk factors, forecasting their development during a definite period of operation in real conditions of a complex system's operation, and on this basis ensuring timely elimination of risk causes before the occurrence of failures and other undesirable consequences.

The main approaches and principles of the strategy for providing guaranteed safety of complex systems should be formed on the basis of the following principles [135]:

- System coordination according to the goals, tasks, resources, and expected results of measures aimed at ensuring the safety of a complex system.
- Mutual coordination of goals, tasks, resources, and expected results of control of serviceability and safety of a complex system.
- Timely detection, guaranteed recognition, and system diagnosis of factors and situations of risk.
- Efficient forecasting and credible estimation of abnormal and critical situations.
- Timely formation and efficient realization of decisions of safety control in the process of preventing abnormal and critical situations.

Therefore, the most important and obligatory requirement of the strategy is system coordination of decisions and actions at all stages of a product's life cycle according to its goals, tasks, terms, resources, and expected results. The coordination must be provided simultaneously from the position of guaranteeing both the required indicators of safety and survivability and the required indicators of serviceability during the given period of operation [98, 168].

The analysis of positive and negative experience of development, manufacture, and operation of complex engineering systems and actions of operators in various critical and extreme situations shows that it is possible to reveal some general principles of a rational strategy of a decision maker's actions. Briefly, the main idea is the formation and realization of a rational decision for practically acceptable time within the limits of irremovable time restrictions. In this connection it is expedient to form the action strategy of a decision maker on the basis of the following principles [134, 135]:

- Rationality and justification of decisions, timeliness, and efficiency of actions.
- Reasonable caution and rational risk during formation and realization of decisions.
- Rationality of levels of completeness, credibility, and timeliness of information support during the formation and justification of decisions.
- Rational compromise between the level of credibility and justification of a decision and the level of expenditures of time and other resources for its formation and justification.
- Rational combined use of intuition, experience, and knowledge of a person (expert, system analyst, manager, and/or decision maker) and computation and intellectual capabilities of computer systems of decision support in the process of forming, realizing, and controlling decisions.
- Rational combined use of capabilities of interactive and iterative modes during formation and justification of decisions.

The main goal of the above formulated principles is to ensure formation and realization of a rational compromise decision in the conditions of a priori unknown time limits, with a set of objective and subjective contradictions, on the basis of rational use of the present resources and possibilities. An important feature and fundamental difference between these principles and typical principles of the optimal control of complex engineering systems in the normal functional mode is accounting for qualitatively new restrictions, namely, fundamentally irremovable information and time restrictions. In the formation and realization of a decision, especially important is the presence of an irremovable threshold time restriction. If this restriction is broken, irreversible catastrophic consequences may occur, for example, explosion of an object, plane crash, or other calamities. The second important difference is a choice of the principle of decision rationality instead of the typical optimality principle. Such a choice is conditioned by the indicated restrictions and by a number of contradictions, in particular, contradictions between requirements of credibility and justification of a decision and requirements of timeliness of formation and realization of a decision. Here "rationality" (from Latin "rationalis") is understood as a definite, reasonably justified compromise in attaining contradictory goals, taking into account a set of contradictory restrictions, among which the threshold time restriction is of primary importance.

It is expedient to realize the strategy for ensuring the safety of complex systems on the basis of the above stated principles at all stages of a product's life cycle (from the stage of forming an idea and conception of a product up to the stage of utilizing an object) in the form of a sequence of the following main system-coordinated procedures [134, 135]:

- Rational choice of a system of criteria and indicators of estimation of conditions, dynamics, and consequences of multifactor risks.
- Development and realization of methodological tools of technology foresight and mathematical apparatus of multicriterion estimation of dynamics and consequences of multifactor risk influence.
- Development of a multilevel hierarchical system of information support of the system analysis of risks.
- Preliminary estimation of the degree and level of risk of the predicted set of abnormal situations.
- Preliminary estimation and optimization of margin of permissible risk of abnormal modes for the period of operation of a complex system based on the data of tests and physical and computational experiments.
- Preliminary estimation and optimization of margin of permissible risk of an abnormal mode for the prediction period of operation of a complex system on based on information of technical diagnostics and monitoring.
- Preliminary estimation and forecast of indicators of a complex system's serviceability for a definite operation period.

- Estimation and forecast of reserve of serviceability of a complex system in dynamics of a concrete abnormal mode at any period of operation.
- Preliminary estimation and forecast of a safety resource of a complex system for a definite period of operation.
- Estimation and forecast of a safety resource of a complex system in the dynamics of a concrete abnormal mode during operation.
- Formation and realization of decisions on minimizing the degree and level of risk for a definite period of operation of a complex system.
- Formation of decisions and realization of actions aimed at minimizing the degree and level of risk in the dynamics of an abnormal mode during operation of a complex system.
- Formation of decisions and realization of actions aimed at the guaranteed prevention of emergencies and catastrophes in the dynamics of an abnormal mode during operation of a complex system.

It is necessary to note some important features of the formulated procedures. The procedures are complicated enough, since their realization and practical application should be carried out in conditions of incompleteness and uncertainty of information about situations of risk. They cannot be realized on the basis of typical means of diagnostics and safety, as these means are focused solely on timely detection of failures. Practical realization of the procedures can be provided by the proposed concept of safety ensuring. It is focused on the timely detection and elimination of causes of possible malfunctions and failures.

Here the procedures are stated in the generalized form. A concrete definition of each procedure and their interconnections should be given for each subject domain in the form of the corresponding information, mathematical tools, and software. The necessity of such an approach to realizing the strategy is determined by the fundamental differences of dynamics of risk situations in different spheres of practical activities. For example, a critical situation in an industrial enterprise due to insufficient quantity of fuel and its consequences essentially differs from a critical situation and consequences caused by a lack of fuel in the dynamics of plane flight. From here follows an essential difference of priorities in estimating the causes and conditions of risk situation action and, as a consequence, the difference in formalization and algorithmization of similar procedures for different practical applications.

7.5 Example of Solving System Analysis Problems of Multifactor Risks

Let us consider the most actual and the most difficult question: how to get a credible solution to practically important problems of timely estimation and prevention of possible risk situations that really act in any field of human activity.

System analysis of characteristics and properties of problems. The main peculiarity of the general problem of system analysis of risks is that it is the basis of a sequence of problems of several types that are traditionally studied by various scientific disciplines. ***The first problem*** is multifactor classification of risks—a problem of cluster analysis. ***The second problem*** is the multifactor recognition of risks—a problem of recognition theory. ***The third problem*** is risk situations ranking—a problem of decision-making theory. ***The fourth problem*** is the problem of multipurpose minimization of risk. The integrity of a sequence of the given problems consists of the following. The first problem solution provides the basic data for the solution to the second problem, and solutions to the first and second problems are the basic data for the solution to the third problem. At the same time, the initial information and purpose of the first problem are determined by the goal and general purpose of the fourth problem. And the fourth problem is the basic unit of the system analysis of risks and determines the final result of analysis. At the same time, it is necessary to take into account that every problem has its own features that distinguish it from a standard statement of typical problems of the corresponding theories and research directions.

The most important differences are the following. Problems are not completely formalized in a standard statement because of incompleteness, uncertainty, and fuzziness of the initial information about risks; the time for solving a problem is fundamentally limited by an a priori unknown time threshold of transition of a risk situation to an emergency or a catastrophe. Thus the necessity arises of elaborating a methodological, mathematical, and programming apparatus of intellectual decision support. This apparatus should ensure efficient processing of a priori and dignostic information in the dynamics of risk situations; the possibility of forming a well-justified decision within a comparatively short time period; practical realization of a solution up to a critical moment T_{cr}, when a risk situation is transformed into an emergency or a catastrophe.

Examples. Let us now consider the practical application of some methods and ways of solving problems of system analysis of risks. Let us show the capability of the earlier presented principles and methods by examples of solving risk-classification problems, risk-recognition problems for complex engineering systems, and problems of ranking risk situations. These problems are formed in the process of testing and further operation of complex engineering systems and, therefore, are described by a unified mathematical statement.

Mathematical statement of problem. In the process of preliminary testing of a complex engineering system the risk situations were revealed and a standard set M_{st} of risk factors ρ_q, $q \in N_{st}$, $N_{st} = [1; n_{st}]$ was formed:

$$M_{st} = \left\{ \rho_q | q = \overline{1, n_{st}} \right\} . \tag{7.24}$$

Each factor ρ_q is characterized by a set L_q of attributes l_{qj} determined by the relationship

$$L_q = \left\{ l_{qj} | q \in N_{st};\ j = \overline{1, n_q} \right\} . \tag{7.25}$$

Each attribute l_{qj} is characterized by an informational vector

$$l_{qj} = \left\{ x_{qjp} | x_{qjp} \in H_{qjp};\ q \in N_{st};\ j \in N_q;\ p = \overline{1, n_{qj}} \right\} ;$$
$$N_q = [1, n_q] , \tag{7.26}$$
$$H_{qjp} = \left\{ x_{qjp} | x^-_{qjp} \le x_{qjp} \le x^+_{qjp} \right\} . \tag{7.27}$$

Each factor ρ_q is characterized by an informational vector

$$I_q = \left\{ I_{qj} | q \in N_{st};\ j = \overline{1, n_q} \right\} , \tag{7.28}$$

which, taking into account (7.26) and (7.27), is transformed into

$$I_q = \left\{ x_{qj} | x_{qj} = \left\langle x_{qjp} | p = \overline{1, n_{qj}} \right\rangle ;\ x_{qjp} \in H_{qjp};\ j = \overline{1, n_q} \right\} . \tag{7.29}$$

The set M_{st} is characterized by a standard informational vector

$$I_{st} = \left\{ I_q | q = \overline{1, n_{st}} \right\} , \tag{7.30}$$

which can be transformed into

$$I_{st} = \left\{ x_q | x_q = \left\langle x_{qj} | j = \overline{1, n_q} \right\rangle ;\ q = \overline{1, n_{st}} \right\} \tag{7.31}$$

on the basis of (7.29).

In the process of further testing and operation of a complex engineering system the numerical values $I^k_{qr}[t_\tau]$, $t_\tau \in T_k$ of informational vectors I^k_{qr} of risk factors, acting during the period T_k, are determined based on the data of technical diagnostics:

$$I^k_{qr} = \left\{ I^k_{qrj} = I^k_{qr}[t_\tau] \,|q \in N^k_{st};\ r = \overline{1, n_k};\ N^k_{st} \subset N_{st} \right\} , \tag{7.32}$$
$$M_k = \left\{ \rho_{qr} | q \in N^k_{st};\ r = \overline{1, n_k} \right\} , \tag{7.33}$$
$$T_k = \left\{ t_\tau | t^- \le t_\tau \le t^+;\ \tau = \overline{1, m_k} \right\} . \tag{7.34}$$

Here N^k_{st} describes a definite set of values q from N_{st} that characterizes a set M_k, n_k is the total number of factors in M_k, r is the sequence number of a factor ρ_{qr} in set M_k, and period T_k is a discrete set of moments t_τ of measurements.

In the process of diagnostics at each moment $t_\tau \in T_k$ all indicators for all predictable risk factors and risk factors revealed for the first time are measured.

Measuring results for each factor ρ_{qr} are formed as an array of quantitative values $I_{qr}^{k}[t_\tau]$, $t_\tau \in T_k$ of informational vectors I_{qr}^{k}:

$$I_{qr}^{k}[t_\tau] = \left\{ I_{qrj}^{k}[t_\tau] \,|\, q \in N_{st}^{k};\ r \in [1, n_k];\ j = \overline{1, n_q} \right\},$$
$$I_{qrj}^{k}[t_\tau] = \left\{ X_{qrj}^{k}[t_\tau] \,|\, X_{qrj}^{k} = \left\langle X_{qrjp}[t_\tau];\ j \in [1, n_q];\ p = \overline{1, n_{qj}} \right\rangle \right\}. \quad (7.35)$$

Here $X_{qrjp}^{k}[t_\tau]$ determines the value of the pth indicator of the jth attribute of the q_rth risk factor ρ_{qr} at the moment t_τ where q is a factor number in the set M_{st} and r is a sequence number in the set M_k. All values $X_{qrjp}^{k}[t_\tau]$ are limited and are formed on the basis of fuzzy variables

$$X_{qjp}^{k} = \left\{ \tilde{x}_{qjp}^{k} \,|\, \tilde{x}_{qjp}^{k} = \left\langle x_{qjp}^{k};\ \mu_{qjp}^{k} \right\rangle;\ x_{qjp}^{k} \in H_{qjp};\ \mu_{qjp} \in [0;1] \right\},$$
$$H_{qjp} = \left\{ x_{qjp} \,|\, \left\langle x_{qjp}^{-} \le x_{qjp} \le x_{qjp}^{+};\ q = \overline{1, n};\ j = \overline{1, n_q};\ p = \overline{1, n_{qj}} \right\rangle \right\}. \quad (7.36)$$

Based on the results of estimation $\hat{Q}_k = [1; Q_k]$ of risk factors ρ_q for each situation $S_k^\tau \in S_\tau$ of set S_τ, the fuzzy informational vector is formed at the moment t_τ

$$I_k^\tau = \left\{ X_k^\tau \,|\, X_k^\tau = \bigcup_{q=1}^{Q_k} X_q^k;\ X_q^k = \left\langle \tilde{x}_{qjp}^{k} \,|\, q \in N_{st}^{k};\ j \in [1; n_q];\ p \in [1; n_{qj}] \right\rangle; \right.$$
$$\left. \tilde{x}_{qjp}^{k} = \left\langle x_{qjp}^{k};\ \mu_{qjp}^{k} \right\rangle \right\}. \quad (7.37)$$

It is required:

1. To ensure the formation of a finite set Ω_0 of nonintersecting classes $\Omega_{i_1 i_2}$ on the basis of set M_k. Each of these classes is characterized by definite half-intervals of permissible values of the degree of risk η_{i_1} and level of risk W_{i_2}:

$$\eta_{i_1}^{-} < \eta_{i_1} \le \eta_{i_1}^{+};\quad W_{i_2}^{-} < W_{i_2} \le W_{i_2}^{+}. \quad (7.38)$$

2. To estimate, for each situation $S_k^\tau \in S_\tau$, the level and degree of risk and determine a class $\Omega_{i_1 i_2}$ to which the situation belongs on the basis of numerical values $I_{qr}^{k}[t_\tau]$, $t_\tau \in T_k$, informational vectors I_{qr}^{k} of the set S_τ of risk situations, and carry out ranking of all $S_k^\tau \in S_\tau$ situations by degree and level of risk.

Explanations and comments regarding the problem. 1. In formula (7.36) set H_{qjp} defines potentially possible values x_{qjp}. The indicator x_{qjp} numerically characterizes the pth manifestation of attribute l_{qj}. Each attribute l_{qj} can be described by n_{qj} indicators x_{qjp} and $p = \overline{1, n_{qj}}$. These peculiarities correspond to real conditions of abnormal situations. The simplest example—the leakage of gasoline from a parked car—is characterized by a distinctive smell, a wet spot at the point of failure, and the size of the spot. In relationship (7.24) set N_{st}^{k} describes a definite set of values q from

N_{st} that characterizes set M_k. For example, M_k is formed from factors that in M_{st} have the numbers 1, 3, 6, 8, 11; and in set M_k they have the sequence numbers 1, 2, 3, 4, 5, respectively; their quantity is $n_k = 5$. Thus we get $M_k = \{\rho_{1,1}, \rho_{3,2}, \rho_{6,3}, \rho_{8,4}, \rho_{11,5}\}$; $N_{st}^k = \{1, 3, 6, 8, 11\}$. In relationships (7.35), (7.36) the quantitive values of the pth indicator of the jth attribute of the q_rth risk factor ρ_{qr} at moment t_τ presented in the form $X_{qrjp}^k[t_\tau]$ can be determined by fuzzy values.

2. The degree of risk η_k and level of risk W_k of an abnormal situation S_k are conditioned by the influence of risk factors of set M_k. Values η_k and W_k are determined on the basis of two categories of information: (1) a priori known conditions and predictable influence of risk factors upon the complex engineering system under investigation and (2) the results of its technical diagnostics in the period of operation depending on the properties and characteristic features of the system and conditions of its operation. The degree of risk η_k and level of risk W_k may be predicted for a definite period of the system operation on the basis of results of technical diagnostics.

3. According to the principles of functioning of real complex systems of technical diagnostics it is taken into account that the results of technical diagnostics in the investigated period T_k are known for each moment $t_\tau \in T_k$.

4. Taking into account the experience of operation of real complex engineering systems, it is assumed that predictable conditions and the influence of risk factors are characterized by the following properties and features:

$D1_0$: The degree of risk of influence of each factor ρ_{qr} is minimal in the center of interval (7.38) and increases while approaching the boundary of an interval.

$D2_0$: Actions of risk factors are mutually independent of probability, but their dependence on the results of influence and their consequences is not excluded.

$D3_0$: Risk factor set M_k consists of two subsets, M_k^+ and M_k^-, characterized by the following properties:

$$M_k^+ \bigcup M_k^- = M_k; \quad M_k^+ \bigcap M_k^- = 0\,.$$

$D4_0$: Set M_k^+ consists of two dynamically synchronous risk factors whose risk maximum is reached on the boundaries of x_{qjp}^+.

$D5_0$: Set M_k^- consists of two dynamically synchronous risk factors whose risk maximum is reached on the boundaries of x_{qjp}^-.

$D6_0$: Situation S_k has a degree of risk $\eta_k = 1$ when all factors of set M_k^+ simultaneously reach the boundaries of x_{qjp}^+ or when all factors of set M_k^- simultaneously reach the boundaries of x_{qjp}^-.

$D7_0$: The probability of action of each risk factor ρ_{qr} increses with an increase in the quantity of attributes and with the values $X_{qrjp}[t_\tau]$ approaching the boundaries x_{qjp}^+ (if $\rho_{qr} \in M_k^+$) or the boundaries x_{qjp}^- (if $\rho_{qr} \in M_k^-$).

$D8_0$: A degree of risk determined by an attribute of risk factor ρ_{qr} according to condition $D2_0$ is characterized by the relationship

$$\eta_{qrj}\left[t_\tau\right] = 1 - \prod_{p=1}^{n_{qj}} \left(1 - \eta_{qrjp}\right) , \tag{7.39}$$

where

$$\eta_{qrjp} = \eta^{+}_{qrjp}\left[t_\tau\right] = \left(\frac{X_{qrjp}\left[t_\tau\right] - x^{0}_{qrjp}}{x^{+}_{qrjp} - x^{0}_{qrjp}}\right)^{v_{qrjp}} , \quad \forall \rho_{qr} \in M_k^{+} ; \tag{7.40}$$

$$\eta_{qrjp} = \eta^{-}_{qrjp}\left[t_\tau\right] = \left(\frac{x^{0}_{qrjp} - X_{qrjp}\left[t_\tau\right]}{x^{0}_{qrjp} - x^{-}_{qrjp}}\right)^{v_{qrjp}} , \quad \forall \rho_{qr} \in M_k^{-} ; \tag{7.41}$$

$$x^{0}_{qrjp} = \frac{x^{+}_{qrjp} + x^{-}_{qrjp}}{2} .$$

Indicators v_{qrjp} must correspond to certain limitations, for example, to conditions $0,5 \le v_{qrjp} \le 4$.

$D9_0$: The degree of risk determined by the influence of a factor ρ_{qr} is described by the formula

$$\eta_{qr}\left[t_\tau\right] = 1 - \prod_{j=1}^{n_q} \left(1 - \eta_{qrj}\left[t_\tau\right]\right) . \tag{7.42}$$

$D10_0$: The degree of risk determined by the influence of risk factors of set M_k^{+} is described by the relationship

$$\eta^{+}\left[t_\tau\right] = 1 - \prod_{k_1=1}^{k^{+}} \left(1 - \eta^{k_1}_{qr}\left[t_\tau\right]\right) . \tag{7.43}$$

Similarly for M_k^{-}:

$$\eta^{-}\left[t_\tau\right] = 1 - \prod_{k_2=1}^{k^{-}} \left(1 - \eta^{k_2}_{qr}\left[t_\tau\right]\right) . \tag{7.44}$$

Here k^{+} and k^{-} are the quantity of elements of sets M_k^{+} and M_k^{-}, respectively, at the moment of mesurement t_τ.

$D11_0$: Degree of risk η_k of situation S_k, which is conditioned by the influence of all risk factors of set M_k, in accordance with conditions $D2_0$ and $D3_0$, is determined by the relationship

$$\eta_k\left[t_\tau\right] = 1 - \left(1 - \eta^{+}\left[t_\tau\right]\right)\left(1 - \eta^{-}\left[t_\tau\right]\right) . \tag{7.45}$$

$D12_0$: An indicator of risk level W_k of the situation S_k conditioned by the influence of all risk factors of set M_k does not satisfy the condition $D2_0$

in the general case. The value W_k determines the total damage that may be caused by actions of risk factors simultaneously or consequently. Such damages can be summed up or mutually increase according to the principle of a chain reaction. The typical example is a chain reaction in a traffic accident. One driver's misactions result in damage to several automobiles and injury of those involved in the accident. Therefore, the following formulas are used for estimation of the level of risk:

$$W_{qrj}\left[t_\tau\right] = \sum_{p=1}^{n_{qj}} \left(\beta_{qrjp} W_{qrjp}\right) , \tag{7.46}$$

where

$$W_{qrjp} = W^+_{qrjp}\left[t_\tau\right] = \left(\frac{X_{qrjp}\left[t_\tau\right] - x^0_{qrjp}}{x^+_{qrjp} - x^0_{qrjp}}\right)^{\chi_{qrjp}} , \quad \forall \rho_{qr} \in M^+_k ; \tag{7.47}$$

$$W_{qrjp} = W^-_{qrjp}\left[t_\tau\right] = \left(\frac{x^0_{qrjp} - X_{qrjp}\left[t_\tau\right]}{x^0_{qrjp} - x^-_{qrjp}}\right)^{\chi_{qrjp}} , \quad \forall \rho_{qr} \in M^-_k ; \tag{7.48}$$

$$x^0_{qrjp} = \frac{x^+_{qrjp} + x^-_{qrjp}}{2} .$$

In (7.47) and (7.48) indicators χ_{qrjp} are determined on the basis of experimental and predictable values of damage.

$D13_0$: Values x^k_{qjp} and $\mu^k_{qjp}\left(x_{qjp}\right)$ depend on the level of informedness U and are determined by a priori known functional dependences $x^k_{qjp} = \Phi_1\left(U\right)$; $\mu^k_{qjp}\left(x_j\right) = \Phi_2\left(U\right)$. There are known functional dependences of the degree and level of risk $\eta = F_\eta\left(I^\tau_k\right)$, $W = F_W\left(I^\tau_k\right)$, where a fuzzy informational vector I^τ_k is determined as in (7.30).

$D14_0$: In the system of classification by degree and level of risk each class Ω_i must be characterized by a definite interval of variation in the degree and level of risk. General properties and descriptions of the classes are characterized by the following relationships:

$$\Omega_{i_1 i_2} = \left\{ \langle \eta_{i_1}, W_{i_2} \rangle \left| \eta_{i_1} \in \left[\eta^-_{i_1}, \eta^+_{i_1}\right), W_{i_2} \in \left[W^-_{i_2}, W^+_{i_2}\right) \right. \right\} ,$$
$$i_1 \in N_1; \quad i_2 \in N_2; \quad N_1 = \left[1, n_1\right]; \quad N_2 = \left[1, n_2\right] ,$$
$$\Omega_i \bigcap \Omega_{i+1} = 0, \qquad \bigcup_{i=[1;N_0]} \Omega_i = \Omega_0 ,$$
$$\eta^-_1 = 0; \quad \eta^+_{n_1} = 1; \quad w^-_1 = 0; \quad w^+_{n_2} = 1;$$
$$N_0 = N_1 \bigcup N_2; \quad N_0 = \left[i | i = \langle i_1; i_2 \rangle\right] , \quad n_0 = n_1 n_2 .$$

The quantity of classes and requirements for each interval of each class are determined by conditions of a particular problem.

Initial information.

1. Informational vectors of risk situations S_k have the following form:

$$S_1 : I_1 = \{\langle x_1; 0.2\rangle ;\ \langle x_2; 0.9\rangle ;\ \langle x_3; 0.4\rangle ;\ \langle x_4; 0.2\rangle ;\ \langle x_5; 0.4\rangle ;\ \langle x_6; 0.4\rangle\} ;$$
$$S_2 : I_2 = \{\langle x_1; 0.1\rangle ;\ \langle x_2; 0.2\rangle ;\ \langle x_3; 0.7\rangle ;\ \langle x_4; 0.3\rangle ;\ \langle x_5; 0.1\rangle ;\ \langle x_6; 0.25\rangle\} ;$$
$$S_3 : I_3 = \{\langle x_1; 0.3\rangle ;\ \langle x_2; 0.1\rangle ;\ \langle x_3; 0.43\rangle ;\ \langle x_4; 0.12\rangle ;\ \langle x_5; 0.6\rangle ;\ \langle x_6; 0.9\rangle\} .$$

2. The quantity of intervals for degree and level of risk are $n_1 = 5$ and $n_2 = 5$, respectively. Intervals for the degree and level of risk are determined by dividing the initial interval $[0; 1]$ into equal half-intervals.

3. Total quantity of classes: $n_0 = n_1 \times n_2 = 25$.

Problem solution.

1. Interval forming for the degree and level of risk is fulfilled through consecutive dividing of the initial interval $[0, 1]$ into five equal half-intervals. As a result, the following values of half-intervals are obtained:
For degree of risk:

$$I_\eta^1 = [0, 0.2); \quad I_\eta^2 = [0.2, 0.4); \quad I_\eta^3 = [0.4, 0.6);$$
$$I_\eta^4 = [0.6, 0.8); \quad I_\eta^5 = [0.8, 1.0) ;$$

For level of risk:

$$I_W^1 = [0, 0.2); \quad I_W^2 = [0.2, 0.4); \quad I_W^3 = [0.4, 0.6);$$
$$I_W^4 = [0.6, 0.8); \quad I_W^5 = [0.8, 1) .$$

2. Classes $\Omega_{i_1 i_2}$, $i_1 = \overline{1,5}$, and $i_2 = \overline{1,5}$ are formed by consecutive enumeration of all possible combinations of indicators i_1, i_2 in intervals $I_\eta^{i_1}$, $W_W^{i_2}$.

3. The correspondence of intervals of classes of degree and level of risk to the intervals of values x_j and a_j for η_α, W_α is established according to the assigned functional dependences.

4. The correspondence of intervals of classes of degree and level of risk to the assigned values of informational vectors is established.

5. To solve a problem of multifactor classification of risk situations, testing of the degree of fuzzy homogeneity of risk-factor description in each class is carried out. A problem solution is realized by the following algorithm.

5.1. Each class is divided into two nonintersecting sets K_1 and K_2. The degree of fuzzy enclosure K_1 into K_2 and K_2 into K_1 is determined in the form

$$v\left(K_1, K_2\right) \quad \text{and} \quad v\left(K_2, K_1\right) .$$

The degree of fuzzy enclosure of a class Ω_j into class Ω_i is determined in the form

$$v\left(\Omega_j, \Omega_i\right) = \bigcap_{x \in I_f} \max\left(1 - \mu_{\Omega_j}(x) ;\ \mu_{\Omega_i}(x)\right) .$$

5.2. Each informational vector is divided into two nonintersecting vectors I_{iq}^1 and I_{iq}^2. The degree of fuzzy nonuniformity is determined in an interval from θ^- to θ^+ for lower and upper boundaries, respectively. The lower boundary of fuzzy nonuniformity is $\theta^- = \left[v\left(I_{iq}^1, I_{iq}^2\right)\right] \bigcap \left[v\left(I_{iq}^2, I_{iq}^1\right)\right]$. The upper boundary of fuzzy nonuniformity is $\theta^+ = \left[v\left(I_{iq}^1, I_{iq}^2\right)\right] \bigcup \left[v\left(I_{iq}^2, I_{iq}^1\right)\right]$.

For most of the classes scattering between the lower and upper interval boundaries is considerable. For 12% of all classes the lower and upper interval boundaries coincide. Consequently, it is sufficient to analyze only the lower boundary. For 60% of events the lower interval boundary equals zero. Consequently, for these classes the informational set does not ensure the uniformity of description of characteristic features of risk factors. It indicates the necessity of introducing more rational classification, i.e., informational vectors are to be reformed in such a way that they are given uniformity in the description of risk-factor properties.

6. The degree of community of class elements is determined. The degree of community of the properties of indicators is determined by the formula

$$\mu_c\left(I_{iq}\right) = \min\left\{\mu_{I_{iq}}\left(x_1\right), \mu_{I_{iq}}\left(x_2\right), \ldots, \mu_{I_{iq}}\left(x_{q_0}\right)\right\}.$$

If the degree of community ≥ 0.5, then it is considered that there is a community of properties that are characteristic of all x from the considered set X. The required results are presented in Table 7.1.

As is seen from Table 7.1, in the considered case only 40% of classes do not have a community of properties.

7. Checking of the given requirements for unambiguous recognition of the given situation S_k in class Ω_{ij} is carried out. Then, a set of classes $D(\alpha)$ is considered, taking into account level $\alpha = 0.1$. The degree of fuzzy enclosure of situation S_k into each class Ω_{ij} is determined. The required results are presented in Table 7.2.

Table 7.1. Results of determining the degree of community of class elements

Class number by η	Class number by W 1	2	3	4	5
1	0.740	0.730	0.211	0.348	0.740
2	0.740	0.730	0.148	0.333	0.333
3	0.890	0.650	0.792	0.710	0.389
4	0.637	0.730	1.000	0.370	0.370
5	0.740	0.346	0.148	0.361	0.660

As follows from the results given in Table 7.2, situation S_1 is fuzzily enclosed in classes $I_{3.4}$ and $I_{4.4}$, situation S_2 is fuzzily enclosed in classes $I_{3.3}$ and $I_{5.3}$, and situation S_3 is fuzzily enclosed in class $I_{2.1}$.

The results of analysis of fuzzy enclosure of situations S_1, S_2, and S_3 in classes $\Omega_{11}, \ldots, \Omega_{55}$ allows for the following conclusions. The values of some indicators of enclosure of situations S_1, S_2, and S_3 into classes $\Omega_{11}, \ldots, \Omega_{55}$ exceed the threshold level of 0.5. Consequently, situations S_1, S_2, and S_3 are recognized unambiguously in the indicated classes and are risk situations.

Table 7.2. Results of determing the degree of fuzzy enclosure of situations S_1, S_2, and S_3 in classes Ω_{ij}

	Class number by η	Class number by W 1	2	3	4	5
$v(S_1, \Omega_{ij})$	1	0.100	0.265	0.100	0.348	0.100
	2	0.100	0.100	0.100	0.100	0.100
	3	0.100	0.265	0.100	0.600	0.100
	4	0.100	0.265	0.100	0.600	0.100
	5	0.100	0.100	0.100	0.100	0.100
$v(S_2, \Omega_{ij})$	1	0.300	0.300	0.300	0.300	0.300
	2	0.300	0.300	0.300	0.300	0.300
	3	0.300	0.385	0.700	0.300	0.300
	4	0.300	0.300	0.300	0.300	0.300
	5	0.300	0.385	0.700	0.300	0.300
$v(S_3, \Omega_{ij})$	1	0.100	0.100	0.100	0.100	0.100
	2	0.564	0.346	0.148	0.400	0.100
	3	0.100	0.100	0.100	0.100	0.100
	4	0.100	0.100	0.100	0.100	0.361
	5	0.346	0.148	0.361	0.100	0.100

Situation S_2 has the highest enclosure indicator value, 0.7. This situation is included in two classes, $I_{33} \subset \Omega_{33}$ and $I_{53} \subset \Omega_{53}$. Situation S_3 has the lowest enclosure indicator value, 0.564. This situation is included in class $I_{21} \subset \Omega_{21}$. Situation S_1 has an average enclosure indicator value of 0.6. This situation is included in classes $I_{34} \subset \Omega_{34}$ and $I_{44} \subset \Omega_{44}$.

8 System Control of Complex Objects

This chapter describes the system problems of coordinated control of serviceability and the safety of complex hierarchical systems (CHSs) in real functioning conditions characterized by uncertainty and risks. On both the structural-functional analysis level and the control level there is a need to find a rational solution under conditions of conceptual uncertainty regarding the object's shape, structure, and properties. In addition, it is necessary to take into account other factors [18, 83, 162, 175, 194, 198, 199, 201, 202, 207]. In particular, in practice there is no any a priori information that guarantees a reliable evaluation of constructional and technological abilities and predefined requirements to CHS main properties, indicators, and conditions of usage. The conditions of CHS usage are determined by various factors depending on the CHS's purpose. However, nowadays for most CHS types, the risks connected with accidents and catastrophes have the utmost importance [33, 180].

Specificity investigation and the development of the main principles for solving the system control problems under conditions of uncertainty and risks will be described in this chapter.

8.1 System Control Problem Analysis and Classification

In human practice a lot of control problems arise. In these problems the evaluation and optimization of the quality of an object's control is performed using a single criterion. Such problems include a family of the following simple problems: the maximal operating speed problem, the problem of the highest signal reproduction quality, the finite state optimization problem of the vertically launched rocket, the problem of fuel consumption minimization in order to reach the required trajectory by a rocket, the problem of the energy consumption minimization to perform a given task, etc. All these control problems can be, in general, stated in the following way.

Substantial problem definition. Acceptable state set and acceptable control set of an object are ***defined***. Initial conditions and boundary conditions are ***defined***. It is ***necessary*** to find an acceptable control by which the extreme value of the specified control quality indicator is achieved and the controlled object is transferred from an initial state to a finite one, remaining in the admissible state domain.

Let us note some features and properties of the considered problems. They differ not only in complexity but also in practical significance. The class of maximal operating speed determination problems appeared at the same time as the simplest mechanical regulators were introduced. For example, Watt's regulator, introduced in the 1840s. This circumstance made the name of a control discipline a regulation theory.

The period 1940–1970 witnessed a qualitative change in the problems. This was caused by the rapid technological expansion of aeronautical engineering and also by the creation and usage of space-system engineering.

The principal difference between these problems and traditional regulation problems is as follows. In traditional problem definition the quality of control of slowly changing processes was investigated. In this way, the quality of autopilot compensation of a random disturbing process in real-life flight conditions was investigated. In this problem the flight time is much longer than the time of transition of a random disturbing compensation. Yet, the flight time of a rocket with a working engine is an example of a transient process, because the continuous changes in rocket mass take place. The control problem is complicated by many other factors. In particular, the high price of fuel and high consumption rate of the useful load determines the necessity of minimizing fuel consumption while selecting the flight trajectory. All these factors resulted in the definition of qualitatively new problems that are fundamentally different from traditional regulation problems. Such differences are as follows: the presence of a set of conflicting objectives of control optimization; incompleteness, uncertainty, and fuzziness of initial information about controlled object service and usage in various conditions; and the necessity of control being optimally coordinated throughout the object's entire life cycle.

The necessity and features of applied system problems of control of a complex modern technique in different stages of its life cycle are determined by many factors. It is necessary to emphasize factors of uncertainty and risk in the CHS functioning process. These factors are the root cause of many accidents and catastrophes attended by the loss of lives and running into billions of dollars in damage [33, 180]. It follows that it is practically necessary to find new principles and approaches to the control of modern CHSs, which have a variety of purposes. Technologically and ecologically dangerous technical systems belong to such CHs. In finding a solution to this problem, in addition to risk factors it is necessary to take into account that functioning effectiveness and the possibility of an accident or a catastrophe directly depend on control structure properties of a given system. Thus, in working out a solution it should be taken into account that the controllability, adaptability, stability, coordination capability, survivability, and effectiveness of modern systems with their varying purposes are defined in many respects by the properties, possibilities, and effectiveness of a control.

Among other factors a continuous increase in requirements for technical systems is significant. This stimulates wide application of high science-intensive technologies at all stages of a system life cycle. It follows that it is practically necessary to perfect processes of forming, substantiating, and making rational decisions in strategic planning and control, in designing and manufacturing innovative and other promising technical products, and also in the allocation of resources and money for their development and manufacture. However, the versatile interconnection and multifarious interrelations among strategic decisions at different stages of life cycles of complex systems and other promising technical products have essentially complicated the problem of forming and making a decision at different levels of hierarchical control systems, particularly in evaluating practical necessity, technological possibility, and economical availability.

As a result there appeared multiple classes of nonformalizable or hardly formalizable control problems. For their solution, it is necessary to provide system unity of control principles, approaches, and criteria.

Thus the classical Winner's definition of control obtained two principally different interpretations. ***The first interpretation*** defines the control as a general category that includes all forms and kinds of changes in properties and behavior of controlled systems. In such an interpretation the control includes such processes as the transformation of the structure, directions, goals, and criteria of specified sorts of human activity. It also includes forming and making a decision about the evaluation, planning, and realization of promising new directions in human activity.

The second interpretation defines the control as specialized purposeful changes in operations and properties of systems within predetermined limits. The first interpretation is typical for control in industrial, economical, social, and other scopes of activity. The second one is typical for more specialized kinds of activity, for example, for medium and small-scale business of any kind, including firms that are developing tools of small-scale automation, automatic regulators, and dedicated controllers. The class of dedicated controllers includes airplane and car onboard computers and other similar specialized computers; each of them manages the specified technological process in the current situation. For example, a car computer can provide rational fuel consumption in different driving conditions. It is evident that the second interpretation does not fully suit general purposes, problems, and principles of a system control. Thus, later on it is expedient to consider problems of a system control according to the first interpretation, because it is more common and based on a broader understanding of the essence, principles, and problems of system control. In accordance with this approach the following formulation of the term "control" is offered:

A control is any purposeful action directed toward the achievement of specified useful changes or transformations in a controlled complex system.

Actions in different kinds of human activity may differ in many properties and indicators, in particular in principles and manner of influence on controlled object abilities. Because of the broad meaning of the term "control" in the first interpretation it is necessary to introduce the following definitions:

Behavior control (C1). This control provides the transition of an object from one state to another on limited set of possible states. C1 is realized during process of simple interactions between a controlled object and an environment, in a form of specified reaction of an object on an external action.

Property control (C2). This control provides changes in the forms and methods of forming a reaction to an external action in accordance with a current situation. C2 provides adaptation of an object to the current situation by means of changing a reaction to an external action in accordance with an external action peculiarity.

Structure control (C3). This control provides changes in the composition, structure, and interconnections of an object in accordance with an external action. C3 allows an object to adapt to a current situation by means of structural changes of an object in accordance with newly formed object goals.

Evolution control (C4). This control provides purposeful changes in the goals, properties, structure, and activity forms of an object in accordance with changes in conditions of interaction with the environment. C4 is the main kind of control during planning of the compound production complex development, complex space system development (i.e., "Mir" space station, GPS satellite navigation system, etc.), complex weapons, and defense technology (i.e., aircraft-carriers, submarines, etc.).

Purpose control (C5). This control provides changes in the objectives and problems of an object's operation in current conditions. C5 is the kind of control that is active in situations where previously predicted conditions of an object's operation are changing so fundamentally that no one of the aforementioned kinds of control allows the object to adapt to new conditions.

Note that the practical significance and complexity of control problems increase while switching from class C1 to class C5. This is conditioned by many reasons and factors, among which is, firstly, an increasing scale of human activity and of complexity. Such an increase is conditioned by the application of new technologies and essential enhancement of the possibilities and scope of use of various modern systems.

The main control purposes are enhancement of the effectiveness of a controlled object and ensuring of its adaptation to a higher level of uncertainty and complexity of external actions. Indeed, C1 class provides optimal control only within boundaries of a previously defined set of actions under a priori known external action restrictions. If the current conditions in which a real object is used differ essentially from accepted, possible conditions of external actions, then the control may be so ineffective that the object will be quite uncontrollable.

The consequences of such a state may be extremely serious. In particular, similar conditions accompanied the Chernobyl nuclear power plant disaster in Ukraine in 1986. At the same time the control of class C2 or higher allows one to react to unforeseen situations more adequately and promptly. This is realized by changing the properties (C2) or the structure (C3) of an object and during the process of fast evolving changes in a normal mode or unexpected external factors (wind velocity, temperature, etc.).

Other principles are realized in the implementation of the C4 or C5 class of control because their purposes and problems are qualitatively different from those of C1, C2, and C3. In addition, classes C4 and C5 are oriented toward a sufficiently long time period. In fact, fundamental changes in the direction of an object's evolution under C4 or changes in purpose under C5 require a large volume of previous examinations and researches, and transforming them into real products and technologies. Thus, the C4 and C5 class controls are implemented in the process of slowly changing strategic situations. The global experience with the development of various complex systems proves it. In particular, there is the experience with creating, using, and developing space systems of different purpose (GPS, satellite communications, aerography, earth remote sensing, etc.) The classical example of prolonged development and purposeful expansion of the abilities of a complex system is the Mir space station, which was used and developed in the USSR for more than 15 years.

It is significant to note one more necessary feature of the considered control classes: while switching from class C1 to class C5 the control problem complexity is fundamentally increasing. The main factor in complexity increase is the transition from very limited formalized control problem set of class C1 to a wide nomenclature of problems of class C5, by which an avalanchelike increase in the quantity of nonformalizable or hardly formalizable problems in the total volume of control problems occurs.

It is significant to take into account one more necessary factor in the development of system analysis methodology. Its essence is that at the same time as system analysis methodology development the qualitative change of the notion of "control" takes place. From a narrow interpretation of this notion, which follows from the goals and objectives of control problems of the C1 class, there occurs a gradual transition to a more broad, system understanding of control problems of the C5 class. However, these classes remain essentially within the bounds of a typical narrow understanding of a control and, therefore, do not include a number of practical necessary aspects of behavior of complex systems of varying purposes. In particular, there is the barest necessity of creation of a unified methodology of a system investigation and optimization of a control of modern complex systems in real conditions of uncertainty and risks. Therefore, assume it is expedient not to limit the classification by classes described above and to supplement it with a system control class.

System control (C6). This control provides purposeful changes in the functioning goals and/or other properties of an object under investigation for the sake of system-coordinated guarantee of a required level of operational capability, safety, and efficiency in real conditions of conceptual uncertainty and multifactor risks.

Taking into account that the substantial definition of the notion of "system control" is missing, we introduce the following definition.

The system control is an unlimited sequence of procedures of forming, substantiating, selecting, and realizing system interconnected and functionally interdependent decisions and actions that are coordinated under the goals, objectives, time constraints, resources, and expected results for the sake of ensuring the required level of operational capability, safety, and efficiency of the controlled object and/or achievement of specified changes in the controlled object in the presence of multifactor risks and uncertainty of external influences.

The notion of ***"unlimited sequence"*** means the sequence of procedures that are executed as needed in accordance with the current circumstances, conditions, and goals of the controlled object's operation. Thus the quantity of procedures is not limited a priori.

Now we consider the properties and features of complex object control problems in detail.

First, let us consider the C1 problems class by the example of a solution to the problem of dynamic object control. The main purpose of such an approach is to reveal the properties and features of the formalization and solution to control problems of this class. This will allow us not only to discover similarities and differences between approaches (accepted in control theory and in system analysis) to a control problem solution, but also to analyze the possibilities, merits, and demerits of typical system analysis apparatus. To this end it is necessary to analyze the specificity and conditions of problem formalization. Thus, let us consider the features of a formalized definition of an optimal control problem [99].

Formalized problem definition. Let the dynamic object movement be described by a system of differential equations represented in a vectorial form:

$$\frac{\mathrm{d}\bar{x}}{\mathrm{d}t} = \bar{f}\left(\bar{x}, \bar{u}, t\right) , \tag{8.1}$$

where $\bar{x}$, $\bar{u}$, and $\bar{f}$ are vectors with components x_i, u_i, and f_i, respectively; $\bar{x}(t)$ is an object-state vector (phase coordinate vector) that defines an object's state at moment t; $\bar{u}(t)$ is the control vector. Vectors $\bar{x}(t)$ and $\bar{u}(t)$ may change in some admissible domain:

$$\bar{x}\left(t\right) \in G_x , \tag{8.2}$$

$$\bar{u}\left(t\right) \in G_u , \tag{8.3}$$

or briefly

$$(\bar{x}, \bar{u}) \in G; \quad G = G_x \times G_u . \tag{8.4}$$

Condition (8.2) is used to name a state limitation (or phase limitation), and condition (8.3) to name a control limitation. In addition to limitations, the initial and finite states of a controlled object are also defined:

$$\bar{x}(t_0) \in E_0, \quad \bar{x}(T) \in E_T, \tag{8.5}$$

where t_0 and T are respectively the time of an initial state and the time of a finite state of a controlled object and E_0 and E_T are a set of admissible values of, respectively, an initial state and a finite state that are assumed to be defined.

As an example let us represent (8.1) in an explicit form that describes a spaceship moving while achieving an orbit:

$$\dot{x}_1 = x_3, \quad \dot{x}_2 = x_4, \quad \dot{x}_3 = \frac{1}{m}(\varphi_1 + u_1 \cos u_2),$$
$$\dot{x}_4 = \frac{1}{m}(\varphi_2 + u_1 \sin u_2), \quad \dot{m} = -F(u_1). \tag{8.6}$$

Here x_1 and x_2 are the current position coordinates of a spaceship, x_3 and x_4 are the current velocity values, m is the total carrier rocket and spaceship mass while orbitting. u_1 is the thrust value, u_2 is the angle between the thrust direction and the x_1-axis, $F(u_1)$ is the instant fuel consumption and thus characterizes the total mass change. Gravity influence, atmospheric resistance, and other hampering factors are taken into account in φ_1 and φ_2 variables, where φ_1 and φ_2 are the x_1-axis and x_2-axis projections of the aforementioned forces. Set G is a Cartesian product of sets G_x and G_u. Set G_x is a region of space around the Earth, within the limits of which the spaceship trajectory should pass.

In particular, trajectory limitations are evident when the pilot spacecraft starts. For example, the trajectory should not pass through the Earth's radiation belts and should not cross the ground surface. Set G_u is defined taking into account that the spacecraft flight control should be realized by changing the rocket thrust vector direction and value. The control vector is defined by thrust u_1 and angle u_2. Thus, G_u is a set of allowed values u_1 and u_2 defined by the construction, technological, and other constraints including an allowed human acceleration value.

The initial and the finite state should be defined in the form (8.5) in accordance with the spacecraft launch purpose. For the problem (8.6) solution it is necessary to define corresponding conditions for each equation. Therefore, there should be defined an initial spacecraft position, starting speed, and mass values. In this case it is possible to consider the initial state in the form:

$$x_1(t_0) = x_{01}, \quad x_3(t_0) = x_{03},$$
$$x_2(t_0) = x_{02}, \quad x_4(t_0) = x_{04}, \quad m(t_0) = m_0. \tag{8.7}$$

Here x_{01}, x_{02}, x_{03}, x_{04} and m_0 are specified fixed values and m_0 is the total starting mass of the spacecraft and carrier rocket.

The finite state after achieving an orbit is defined by orbit parameters. When a circular orbit is achieved with the specified radius R, the finite state of the spaceship will be defined with the following conditions:

$$\begin{aligned} x_1^2(T) + x_2^2(T) &= R^2, \\ x_3(T)\,x_1(T) + x_4(T)\,x_2(T) &= 0, \\ x_3^2(T) + x_4^2(T) &= V_R^2, \end{aligned} \tag{8.8}$$

where V_R is a specified velocity of the spacecraft in orbit. The first condition (8.8) means that the point with coordinates (x_1, x_2) at the moment $t = T$ is located on the circle of the specified radius R. The second condition means vectors $\bar{r}$ and $\bar{V}$, with components (x_1, x_2) and (x_3, x_4), respectively, are orthogonal, i.e., the spacecraft velocity vector $\bar{V}$ at the moment $t = T$ is directed at a tangent to a circle of the specified radius and is orthogonal to vector $\bar{r}$. The third condition shows us that the velocity of the spacecraft in orbit should be equal to a specified value. Thus, condition (8.8) guarantees that a free movement of the spacecraft in a circular orbit (with radius R and velocity V_R) will occur if the engine shuts down at the moment $t = T$.

Now, it is necessary to formalize a criterion of control efficiency. It is evident that the process by which a spacecraft achieves a specified orbit, i.e., the process of transition of a spacecraft from an initial state (8.7) to a finite state (8.8) may be realized in many different ways, where each of them is characterized by a specific control program. We can propose various criteria of program efficiency evaluation. As described earlier, one of the most necessary practical problems is the fuel-consumption-minimization problem. The mathematical representation of the given problem can be found on the basis of the last equation in (8.6). Since the value $F(u_1)$ defines the instant fuel consumption, then the total fuel consumption as the spacecraft approaches its target orbit will be defined by the integral

$$I(\bar{u}) = \int_{t_0}^{T} F(u_1)\,\mathrm{d}t\,. \tag{8.9}$$

It follows that the optimal program $\bar{u}^*$ of a spacecraft approaching it target orbit corresponds to the condition

$$I(\bar{u}^*) = \min_{U \in G_U} I(\bar{u}), \quad \bar{u}^* = \operatorname{Arg}\min_{u \in G} I(\bar{u})\,. \tag{8.10}$$

For short, condition (8.10) will be used in the form

$$I(\bar{u}) \to \min \tag{8.11}$$

or, if the criterion is maximized, in the form

$$I(\bar{u}) \to \max\,. \tag{8.12}$$

In the given problem control optimization is achieved via the selection of a control vector function. In a more common definition the optimization may get complicated. In particular, it may be provided also by selecting a control vector function and a state vector function, i.e., at the expense of the $\bar{u}(t)$ and the $\bar{x}(t)$ selection under conditions (8.11) or (8.12).

The given conditions of formalization of the optimal control problem allow one to represent the mathematical definition of the mentioned problem in the following, more common, form.

Mathematical optimal control problem definition. It is required to determine such vector functions $\bar{x}^*(t) \in R^n$ and $\bar{u}^*(t) \in R^m$ while $t \in [t_0, T]$ that provide the minimization of the functional

$$I = I(\bar{x}, \bar{u}) \tag{8.13}$$

in the form of the condition

$$I(\bar{x}, \bar{u}) \to \min \quad \text{or} \quad I(\bar{x}, \bar{u}) \to \max \tag{8.14}$$

under differential constraints

$$\dot{x} = \bar{f}(\bar{x}, \bar{u}, t) , \tag{8.15}$$

conditions

$$(\bar{x}, \bar{u}) \in G; \quad t \in [t_0, T] , \tag{8.16}$$

and edge conditions

$$(\bar{x}, t_0) \in E_0, \quad (\bar{x}, T) \in E_T , \tag{8.17}$$

where G is some specified region of space $R^n \times R^m$, and E_0 and E_T are regions defined in the space $R^n \times R^1$.

Conditions (8.13)–(8.17) define different types of optimal control problems. These problems can be divided into three groups in accordance with methods whose use allows one to set the specified problem properties and features.

For ***group 1*** the different forms of functional (8.13) presets are typical. It contains the following problem types:

1.1. Lagrange
1.2. Mayer
1.3. Boltz
1.4. Operating speed

Group 2 contains different forms of conditions (8.16). It contains the following problem types:

2.1. Control restrictions
2.2. State restrictions
2.3. Combined control and state restrictions
2.4. Integral restrictions (isoperimetric problem)

Group 3 contains different forms of the presets of edge conditions (8.17). It contains the following problem types:

3.1. Fixed trajectory ends
3.2. Free trajectory end
3.3. Traveling ends

The solution methods for the optimal control problems are described in the book by Moisejev [99].

8.2 System Control Problems of Complex Objects' Operational Capability and Safety

New approaches to the creation of modern engineering define qualitatively new requirements for technological and ecological safety. Such a demand is conditioned not only by the fact that the loss from partial or full destruction of a piece of machinery or a construction project may be ten times more than its production cost but also by the fact that catastrophes may have international or global scales of influence on inhabitants and the environment. Furthermore, several catastrophes with losses in the millions and billions of dollars have taken place in the last decades of the 20th century in almost all industrially developed countries. This proves that the present principles and tools of complex object safety control do not meet modern requirements [33, 180].

Thus, there is a practical necessity to qualitatively change the principles and the structure of operational-capability controls and the safety of modern engineering systems in real conditions of multifactor risk influence. First of all, the control of complex objects should be systemwide which means that there should be system coordination of operability control and safety control not merely by the corresponding goals, tasks, resources, and expected results but also, importantly, by the immediacy and effectiveness of interaction in real conditions of abnormal situations. Such coordination should provide immediate and effective interaction between the mentioned control systems. On the one hand, there should be provided effectiveness of the safety system for timely detection of abnormal situations, evaluation of risk degree and level, and the definition of an permissible risk margin during the process of forming recommendations about immediate actions given to the decision maker. On the other hand, the system of operational capability control after receiving a signal about abnormal situations should, in an effective and operative manner, make a complex object ready for an emergency transition to an offline state and should make it possible to effect this transition within the limits of permissible risk. The main objectives of the system control of complex objects follow this principle.

Let us show the mathematical formulation of this problem with a priori set variation intervals of main indicators of the system in the normal mode and predefined permissible boundes of the influence of external factors.

Mathematical formulation of complex object system control problem. It is known that system functioning is characterized by the following sequence of complex system states: $E_1, E_2, \ldots, E_k, \ldots$. Every state E_k is characterized by specified indicators of system function processes (Y_k, X_k, U_k) and specified indicators of external environmental influence and risk factors Ξ_k:

$$E_k = \{(Y_k \in Y) \wedge (X_k \in X) \wedge (U_k \in U) \wedge (\Xi_k \in \Xi)\} \,, \tag{8.18}$$

where the meaning of indicators at the moment $T_k \in T^{\pm}$ is defined by the following relations:

$$\begin{aligned}
&Y_k = \hat{Y}\,[T_k]\,, \quad X_k = \hat{X}\,[T_k]\,, \quad U_k = \hat{U}\,[T_k]\,, \quad \Xi_k = \hat{\Xi}\,[T_k]\,,\\
&T_k = \{t_k | t_k > t_{k-1}\}\,, \; T_k \in T^{\pm}, \; T_0 = \left\{t | t_0 = t^-\right\}, \; T^{\pm} = \left\{t | t^- \le t \le t^+\right\},\\
&Y = \left(Y_i | i = \overline{1, m}\right), \quad X = \left(X_j | j = \overline{1, n}\right),\\
&U = \left(U_q | q = \overline{1, Q}\right), \quad \Xi = \left(\Xi_p | p = \overline{1, P}\right).
\end{aligned} \tag{8.19}$$

Here Y is a set of external parameters Y_i that includes technical, economic, and other indicators of system-function quality; X is a set of internal parameters X_j that includes constructional, technological, and other indicators; U is a set of control parameters U_q; Ξ is a set of external environmental influence parameters and parameters of risk factor influence Ξ_p; $\hat{Y}\,[T_k]\,, \hat{X}\,[T_k]\,, \hat{U}\,[T_k]$, and $\hat{\Xi}\,[T_k]$ are sets of meanings of appropriate parameters at the moment T_k; and $T^{\pm}$ is a specified or predicted complex object functioning period.

It is required to find such meanings of indicators of the degree of risk η_i and level of risk W_i and also a margin of permissible risk T_{ar} that provide the possibility of transition from the mode $\overset{\cup +}{R}_{tr}$ during the time period $\overset{\cup \pm}{T}_{tr}$ in the normal mode before the critical moment T_{cr} of an abnormal mode becomes an emergency situation or a catastrophe. Here, the mode $\overset{\cup +}{R}_{tr}$ is a controlled functioning mode conditioned by the control influence U_{tr} of a safety control system. During the time period $\overset{\cup \pm}{T}_{tr}$ this mode leads to the reduction of the abnormal mode R_{os} to the normal mode R_{sd}. The mode $\overset{\cup +}{R}_{tr}$ is characterized by the following functional:

$$\overset{\cup +}{R}_{tr}\colon R_{os} \overset{U_{tr}}{\to} R_{sd}\,, \tag{8.20}$$

which defines the process of the reduction of the abnormal mode R_{os} to the normal mode R_{sd} under the influence of the control system U_{tr}.

The main system property is an operational capability characterized by given quality indicators defined by the set Y.

We will consider the ***notion "system safety"***. Here, this means an ability to timely prevent a consecutive transfer from a normal mode to an accident or a catastrophe on the basis of timely detection of essential risk factors and elimination of the possibility of their conversion into catastrophic risk factors. Safety is characterized by the following indicators: degree of risk η_i and risk level W_i; the reserve T_{ar} of permissible risk of an abnormal mode; the reserve T_{as} of permissible risk of an accident; and the reserve T_{ds} of permissible risk of a catastrophe.

The quantitative values of safety indicators are defined on the basis of the general problem of multifactor risk analysis, whose mathematical definition is described in this chapter.

Strategy for solving the problem of system control of complex objects. First, note the principal differences between the given problem and typical control problems. The main difference is that the initial information about a complex object contains only a small part of information about its state, properties, functioning processes, and operational capability characteristics. This information represents only the state and work characteristics of such objects in normal mode. Undoubtedly, this information is enough for decision making during the complex object control only on the condition that the normal mode continue for a long time. However, in real objects in view of existing technical diagnosis systems, oriented toward failure and malfunction detection, it is impossible to ensure that a malfunction or a failure will not appear within the next 5–10 min. It is a priori unknown how much time it will take to repair a malfunction. It may take from a few minutes up to several hours or even days and months. And, consequently, the possible damage is a priori unknown, and thus the safety control system is, essentially, a recorder of information about facts and damage.

A fundamentally different approach can be realized on the basis of the system control of complex objects. The essence of such control is systemically coordinated evaluation and adjustment of the operational capability and safety during the functioning process of an object.

The general strategy of such an approach is shown in Fig. 8.1. The "?" symbol means that there is incomplete, fuzzy information about the object functioning state at the moment $T_k \in T^{\pm}$. This information is not enough for decision making. This implies the significant property of such an approach. This property means that the situation analysis and decision making are provided not only in typical conditions of exact recognition of a normal or an abnormal system mode, but also in conditions where there is only fuzzy, incomplete information about a situation. It is significant that this approach, in conditions of fuzzy information about a situation, allows one, if necessary, to make a timely decision on emergency stop of the system operation. In the following control strategy in blocks 1–3 there are realized procedures of com-

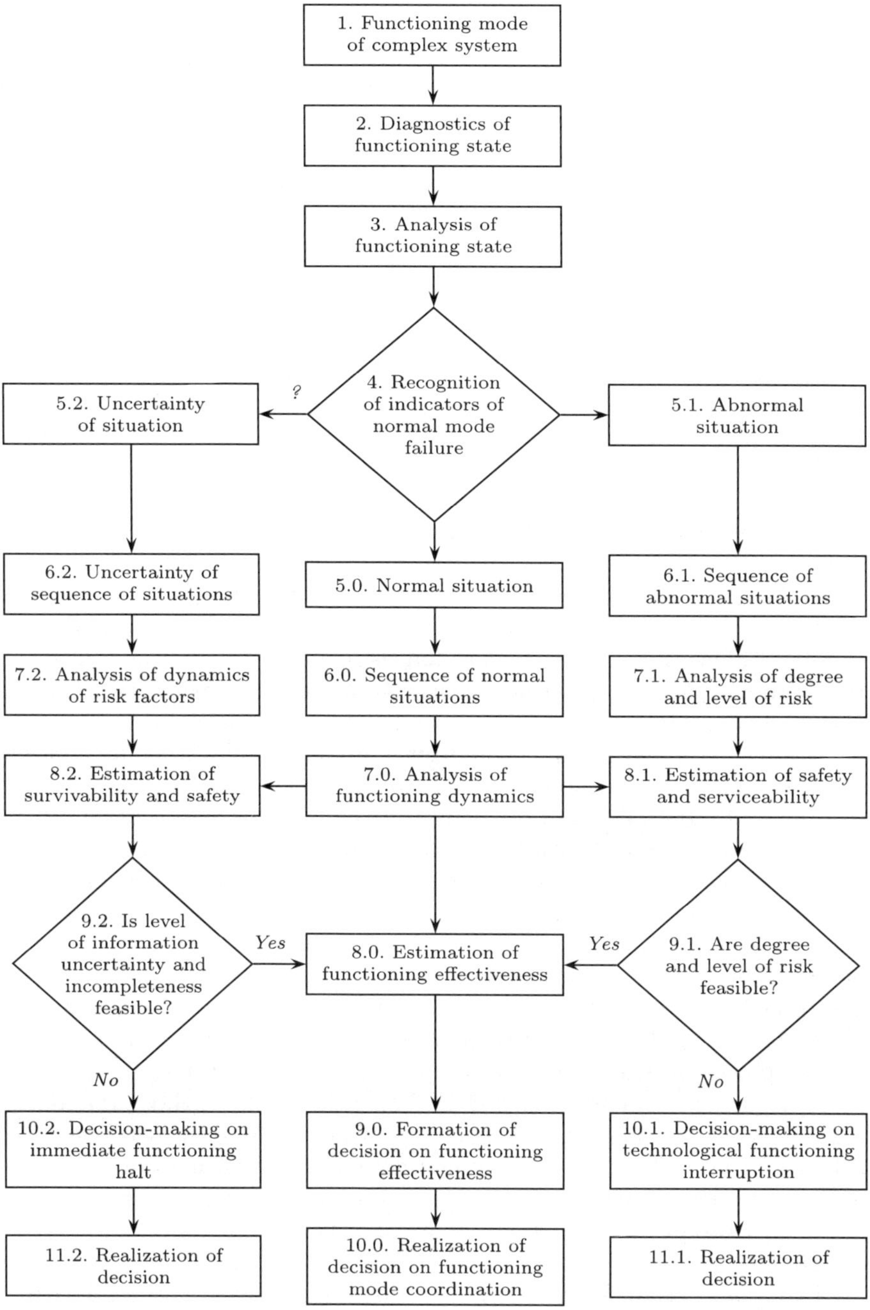

Fig. 8.1. System control strategy of a complex objects' capability and safety

plex object functioning diagnostics and analysis. In block 4, on the basis of the results of the execution of the procedures of blocks 2 and 3, the recognition of a normal functioning mode occurs. During this process, three possible variants of a complex object state are analyzed: the normal functioning mode remains (transition of a control to block 5.0); signs of a normal mode violation appears that make it possible to reveal that at time $T_k \in T^{\pm}$ the situation is abnormal (transition of a control to block 5.1); or at time $T_k \in T^{\pm}$ the situation becomes undefined (transition of a control to block 5.2).

In the first variant the system operation in a normal mode, and quality control is executed (blocks 6.0–10.0).

In the second variant on the basis of the sequence of abnormal situations the following actions are realized. The risk degree and level of an abnormal-situation sequence is analyzed, and the safety and operational capability of complex objects are evaluated (blocks 5.1–9.1), and a decision is made regarding the scheduled stop of the complex object functioning (transition of a control to blocks 10.1 and 11.1) or a decision is made regarding the continuation of the complex object functioning if the values of risk degree and level are acceptable (transition of a control to blocks 8.0–10.0).

In the third variant, an evaluation of survivability and safety of the system in conditions of uncertain information about abnormal situations is made. For this, the following sequence of operations is realized. An analysis is made of risk factors sequence of abnormal situations, on the basis of which the complex object survivability and safety are evaluated (blocks 5.2–9.2). If a certain uncertainty level and incompleteness level are acceptable, then the decision about the continuing functioning of an object is made (transition of a control to blocks 8.0–10.0). Otherwise, the decision on emergency stop of an object functioning is made (transition of a control to blocks 10.2 and 11.2).

In the following figures we will consider in detail the developed algorithm for system control of complex objects operational capability and safety in conditions of multifactor risks. Let us firstly consider the structure scheme of an algorithm of safety control in abnormal situations (Fig. 8.2), which details procedures of blocks 5.1–9.1 (Fig. 8.1). In this scheme blocks 2.0 and 3.0 realize diagnostic procedures and evaluation of abnormal situations during the process of transition of a complex system normal functioning mode into a sequence of abnormal situations. Using the results of these procedures a database and a scenario of a sequence of abnormal situations is formed (blocks 4.0 and 6.0). The obtained information is used for making a decision about the following actions (block 7). In this block the possibility of a transition of a complex object from abnormal situations to a normal mode is determined.

Three variants are realized: the transition is possible (variant 1); the transition is impossible (variant 2); there is not enough information about abnormal situations to make a decision (variant 3). If the transition to a normal mode is possible, then a procedure for evaluating an object's survivability

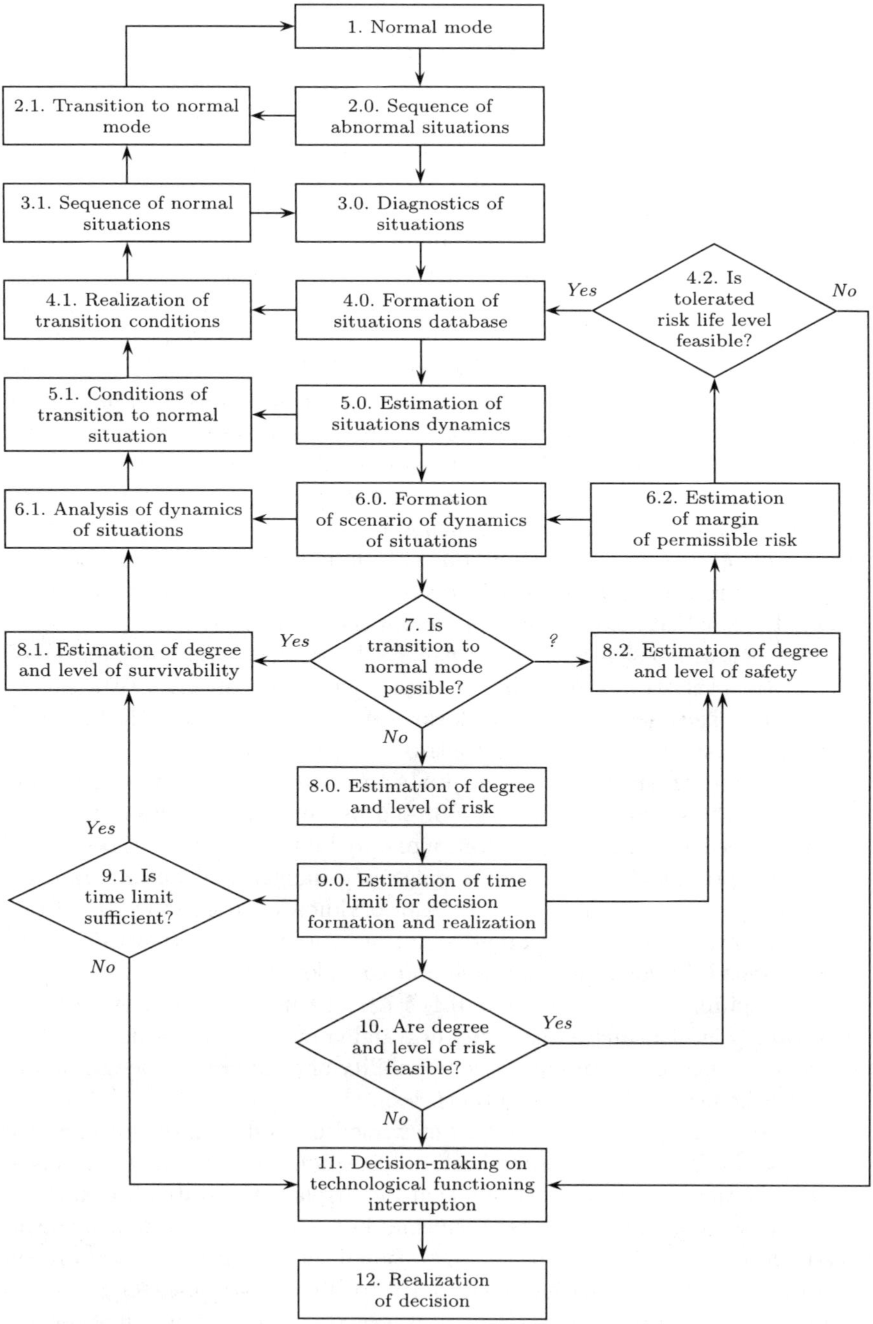

Fig. 8.2. Structure scheme of an algorithm of a complex object safety control in abnormal situations

degree and level is executed (block 8.1). If the transition to a normal mode is impossible, then a procedure for evaluating the risk degree and level of an abnormal situation sequence is executed (block 8.0). If there is not enough information about abnormal situations for making a decision about the possibility or impossibility of a transition to a normal mode, then the procedure for evaluating the safety degree and level of an object is executed (block 8.2).

The following actions of a control system in conditions of abnormal situations are oriented toward the elimination of the possibility of an accident or a catastrophe. According to variant 1, measures are taken for the transition of abnormal situations to a normal mode on the basis of step-by-step execution of procedures that are defined in blocks 9.1 to 2.1. According to variant 2 measures are taken to evaluate the risk degree and level and the time limit for forming and realizing a decision on scheduled stop of the object operation. This evaluation is made following the step-by-step execution of procedures described in blocks 9.0 and 10–12. According to variant 3 measures are taken to evaluate the safety degree and level, margin of permissible risk, time limit on forming and realizing a decision about scheduled stop of the object operation. This evaluation is made following the step-by-step execution of procedures described in blocks 9.0 and 8.2–4.2. All variants are oriented toward accident prevention before the moment T_{cr}.

Then let us consider the structure of a knowledge base and a database (Fig. 8.3). This structure is the detailed content of block 4.0 of a safety-control algorithm (Fig. 8.2) and blocks 7.0–9.0, 7.1–9.1, and 7.2–9.2 of an operational-capability and safety-control algorithm (Fig. 8.1).

Let us note the most important properties and features of this structure. It provides the forming, accumulation, and usage of three types of information in the system control of operational capability and safety. These three types are a priori knowledge, current diagnostic information, and information about complex-object control results. A priori knowledge is formed by system 1, which consists of a casual knowledge base, an empirical knowledge base, an experimental knowledge base, and an expert knowledge base. This knowledge is used in blocks 4.0, 5.0, 6.0, 6.1–3.1, and 8.0–8.2 of the safety-control algorithm (Fig. 8.2) and is especially demanded when an abnormal situation first emerges because it provides the possibility of comparing processes under way with known analogs and prototypes.

Real-time diagnostic information is formed by system 2 during the process of system object functioning both in a normal mode and in conditions of abnormal situations. Diagnostic information displays the state of technological processes, technological mechanisms and constructions, technical control means, processes of complex object operational capability, and safety control. Diagnostic results are used in blocks 4.0, 5.0, 6.0, 6.1–3.1, and 8.0–8.2 of the safety control algorithm (Fig. 8.2) and are especially demanded in abnormal situations, providing not only the possibility of analysis of processes that are under way, but also margins of permissible risk, and a time limit on forming

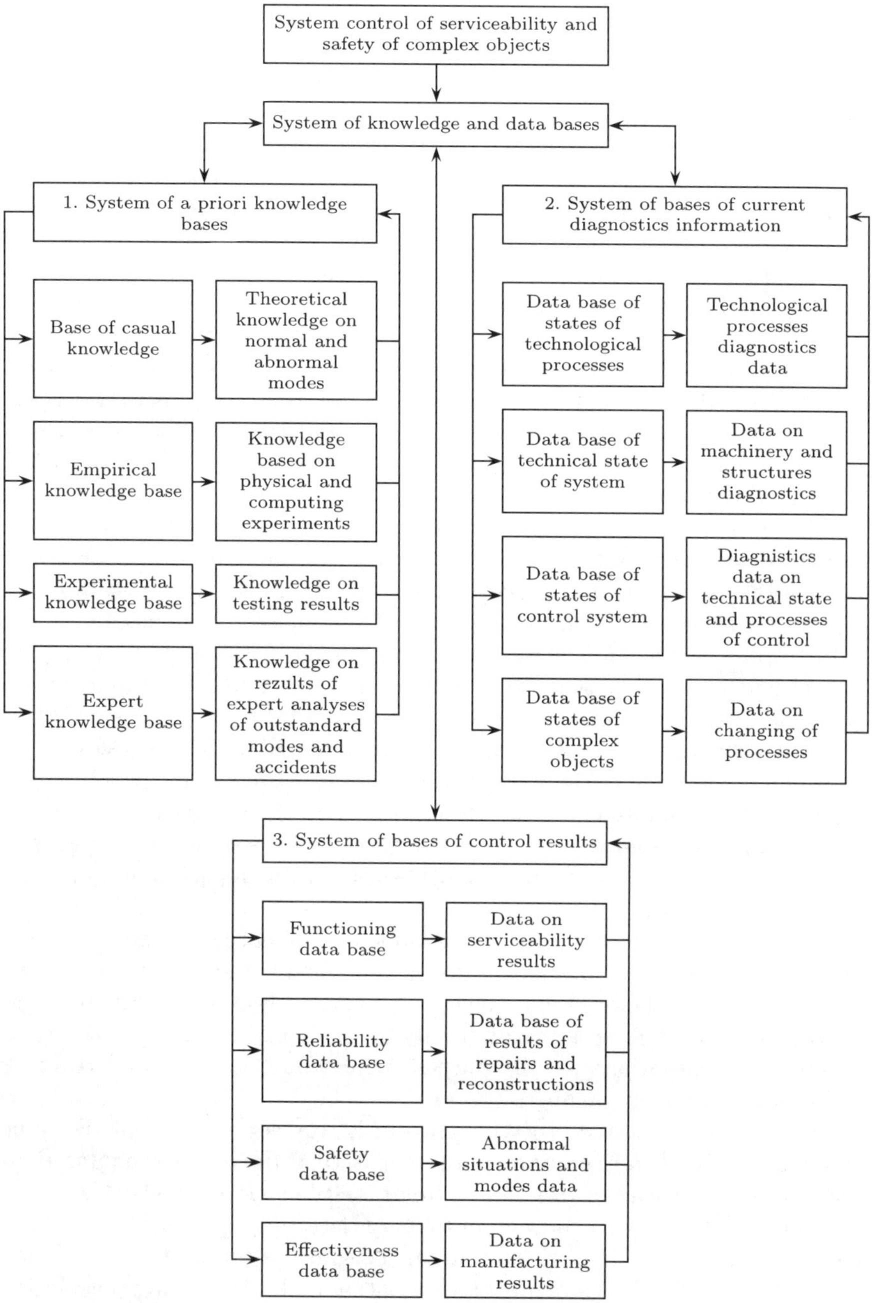

Fig. 8.3. Structure of a knowledge base and a database for the system control of the operational capability and safety of a complex object

and realizing a decision about schedulled stop of an object operation. A measurement discreteness may essentially vary during diagnosis depending on changing of abnormal situation. The purpose of such variation is, firstly, to reduce the time of schedulled stop of the object functioning, and, secondly, to create conditions to timely realize this procedure before the moment T_{cr}.

System 3 forms information about the results of object control. Information represents the results of an object's operational capability, of reconstructions and repairs, of actions in conditions of abnormal situations, and of manufacture. This information also represents the effectiveness of different functioning processes, the specificity of different planned preventive and emergency repair tasks, and also other features of a specific object.

Information obtained from systems 1–3 is used in a system-control algorithm of an object's operational capability and safety (Fig. 8.1) and is the basis for making a decision in the correction of object functioning processes in normal mode, and for making a decision of safety control for the purpose of accident and catastrophe prevention.

However, we will notice that the presence of the aforementioned information is a necessary, but insufficient, condition for accident and catastrophe prevention in abnormal situations. Practical experience demonstrates that abnormal situations during some period of time may become accidents and catastrophes if adequate measures are not taken. Thus, it is necessary to provide such control system characteristics that will guarantee a timely transition of an abnormal situation into a normal one before the moment T_{cr}, or a timely schedulled stop of an object functioning. For this it is necessary to provide on the basis of the aforementioned information a timely solution of a sequence of risk analysis tasks during control of the complex objects operational capability and safety. The list of these tasks is shown in a scheme of interrelations between risk analysis tasks during the complex object's control process (Fig. 8.4).

In this scheme the main attention is focused on specific safety tasks and objectives. The main objective is to prevent possible accidents and catastrophic situations before the moment T_{cr}. The main tasks are reliable evaluation and online forecasting of the dynamic of main safety indicators in different risk situations for the purpose of forming, making, and realizing a decision in a timely manner.

To achieve the stated objectives two task groups are solved at the same time: ***analysis of degree and level of risk of different abnormal situations and analysis of margin of permissible risk analysis***. The initial data are the results of a solution of tasks of detecting signs of abnormal situations and the results of risk situation recognition (blocks 1 and 2). Then there follow tasks of a risk category evaluation and a risk change evaluation (blocks 3.1 and 3.2), an evaluation of risk degree and level and a margin of permissible risk evaluation (blocks 5.1 and 5.2; blocks 8.1 and 8.2), ranking of problems by degree of importance and level of complexity (blocks 10.1

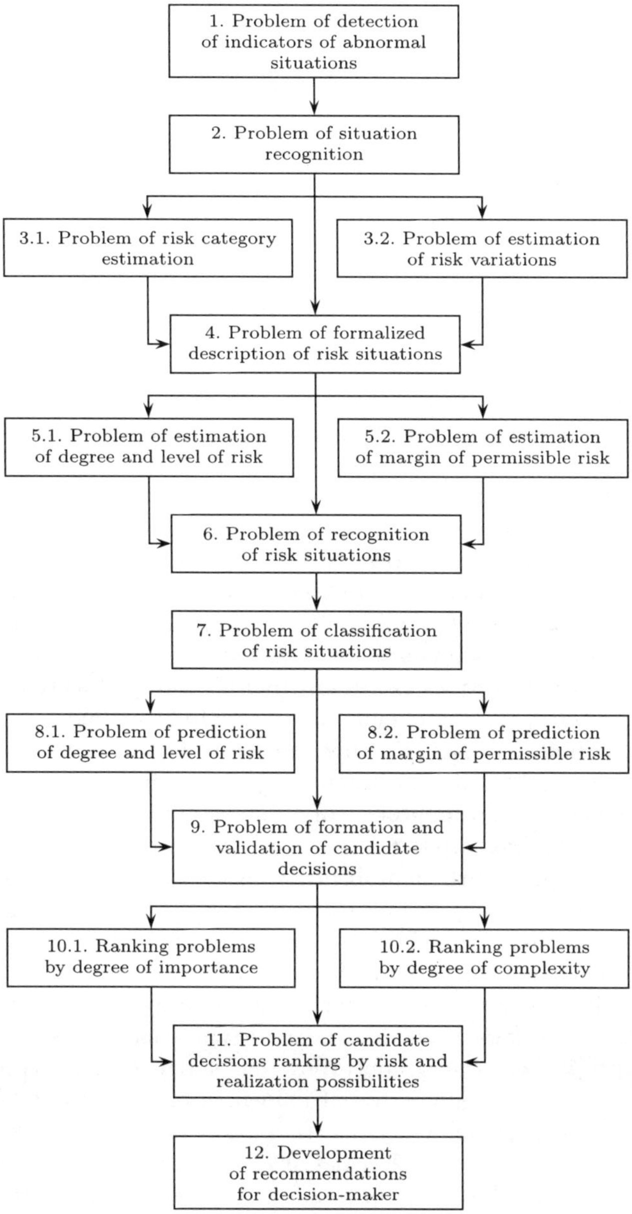

Fig. 8.4. Interrelations between risk analysis tasks during complex object's control process

and 10.2). In conclusion the following tasks are solved: the task of ranking of solution variants by risk and possibility of realization and the task of developing recommendations for the decision maker (blocks 11 and 12). The other tasks (blocks 4, 6, 7, and 9) are used by way of "supporting points", in which the results of the aforementioned task solutions are synthesized and generalized.

8.3 System Control of Complex-Object Structure and Properties

Let us proceed to examine the properties and features of other classes of control problems. First of all, it is necessary to discuss factors that determine the practical necessity of creating and using more complex classes of object control problems.

The problems considered earlier related to operational capability and safety control include many problems of complex object control in conditions of normal functioning mode. But at the same time the given class of control problems has limited capabilities. These limitations follow from the mathematical definition of an optimal control problem, where the number of practically significant factors are not taken into account, while these factors are related to real conditions of modern complex system functioning. In particular, real interactions between an object and its environment are not fully taken into account. For example, such important features as incompleteness, uncertainty, and inaccuracy of initial information, information lag, control lag, and the non-Markovian process of an object state change.

These disadvantages are taken into account in classes of control problems of higher levels, particularly in problems connected with the necessity of complex object properties and structure control. First of all, let us consider some techniques for eliminating incompleteness, uncertainty, and inaccuracy of initial information in control systems. In real conditions of designing the system of complex object control the insufficient level of decision maker awareness may be caused by different factors. The most typical situations are as follows.

Situation 1. During the control system design it may be found that properties and indicators of external influences on a controlled object are partially or completely unknown, and thus many indicators of a control system are virtually unknown. Such situations were typical in various spacecraft. Particularly, in Moon-buggy research information about an environment was virtually missing. There was only one property known: there is no atmosphere on the Moon. But there was no important information about the Moon's ground mechanical, physical, chemical, and other properties and indicators. In such conditions the demands made on a control system are principally different with the demands accepted in developing class C1 control systems.

In particular, the control system should meet the principally new requirement that is missing in class C1 control systems. The essence of this requirement is as follows: During a functioning process the control system should provide missing information about an environment and, using it, realize a decision about appropriate changes in a controlled object's parameters, properties, and structure. Therefore, it is required that the control system, during the functioning process, be able to perform control functions of class C1 (the behavior control), class C2 (the properties control), and class C3 (the structure control). During this the procedures for making changes to properties and structure may be single and multiple.

In the case of realization of a single procedure, the control system works in the following mode: missing information is provided, correction of properties and structure is carried out. Such a procedure is executed on conditions that environmental characteristics are almost constant while a controlled object is in use. This variant of a procedure is typical, particularly in cross-country vehicle usage in harsh climatic conditions (swamp, sand, firm soil, viscous soil, etc.). An example would be use of a cross-country vehicle in different seasons in transpolar terrain.

Situation 2. There is sufficient initial information about environmental properties and a controlled object for the creation of a control system. However, while a controlled object is in operation, environmental or object properties may change dramatically as a result of the influence of various factors. For example, in an aircraft during flight its mass changes because of fuel consumption, and thus the center of mass changes its position. At the same time as flying altitude changes, air density changes as does aerodynamics. The mentioned object property changes and nonstationary turbulent atmosphere processes may cause the control system, designed on the basis of initial information, will not provide required qualitative indicators of an object in operation during changes to controlled object properties and the environment. In such conditions there appears the necessity of providing continuous change in the properties and structure of the object that will be adequate to changes in external influences. Such a control system should ensure the presence of the required qualitative indicators of objects in operation at any moment of a current situation.

Situation 3. This situation is a generalization of two previous ones and is characterized by insufficient awareness about external conditions and factors at the stage of development of a controlled object, and also by the possibility of unpredictable, nonstationary changes in external conditions while an object is in operation.

The general property of the situations considered earlier is the necessity of adapting to new conditions while the situations are changing. Such accommodation is realized in different ways: by changing only the properties, by changing only the structure, or by changing both the structure and properties of a controlled object. The ability of a nonstationary system to adapt to environmental changes or object characteristic is called ***adaptation***. System that possess the ability to adapt to conditions change are called ***adaptive***.

Here, by nonstationary system we mean an entire object that consists of a controlled object and a control system that are structurally interconnected and functionally interacting for the sake of specified goals.

Thus, the above brief analysis shows that in real conditions of incompleteness, inaccuracy, and contradictoriness of initial information the effective functioning of modern complex systems is possible only if those systems are able to adapt to changes as they occur. Adaptation is possible only given a necessary level of informedness about external environmental properties and controlled object properties while in operation. It follows that adaptation and optimization problems are narrowly interconnected with problems of adaptive optimal information processing while a current situation is undergoing changes. Such problems are outside the scope of approaches of control theory class C1. The given above stimulated the evolution of different tools and methods of adaptive control and promoted the birth of methods of control classes C2 and C3.

At the same time, the continuous increase in the volume and level of demands on modern manufacturing makes for fundamentally new, essentially more complex theoretical and practical control problems. Only the first steps have been made toward the theoretical investigation of more actual control problems of classes C4 and C5. However, practical problems of class C4, particularly problems of complex multilayer, multipurpose system evolution, have been solved in many respects intuitively and empirically for the last few decades.

Class C5 control problems are the next step in the development of class C4 problems. This complication lies in the fact that, during the controlled object evolution its purpose is changed, and thus object goals are changed. In such conditions the problem becomes more difficult than developing the same object. In fact, the developer of a new object may choose the form, structure, functions, and elements of each hierarchical level and the object as a whole. In the case of purpose control, the situation in which the problem is solved is fundamentally different. The object already exists, already has established technologies, an established system of production supply and sale, and also an existing collective of factory and office workers with specific knowledge of practice in a specific environment and a specific vocational training. The main object property, its purpose, must be changed, but at the same time one must keep as much as possible of the existing manufacturing infrastructure and maintain its maximal effective usage in the new scope of practical activity. It is evident that providing an effective and timely solution to organizational, technological, economic, scientific and technical, social, and other problems that arise in such situations is possible only in the presence of their system coordination by objectives, tasks, terms, resources, and expected results and in the presence of multilayer control.

To understand all the complexity of a solution of system control problems of the C3, C4, and C5 classes we will briefly consider a class C3 problem.

This problem belongs to a group of simplest, but has necessary practical significance in real conditions of usage of technologically and ecologically dangerous objects. It is solved at the stage of creation and testing of control systems of complex hierarchical objects.

The formulation of a problem of a rational control of a hierarchical system in conditions of multifactor risks is based on the results of structural-functional analysis, considered in Chap. 6. We will cite a substantial formulation and mathematical definition of a considered problem.

Substantial problem formulation. For a complex hierarchical system there ***are defined*** structures of all levels and project decisions about functional elements (FEs) of all hierarchical levels in accordance with preliminary specifications. There are known requirements for main properties and indicators of control of system functioning processes in fuzzy and incompletely specified conditions.

It is required to find the form and structure of a control system, to develop project decisions about FEs of all hierarchical levels based on conditions of achievement of a required system control quality in normal, abnormal, and critical situations.

Mathematical problem definition. For a complex multilayer system there is a ***specified*** structural interconnection between FEs of different hierarchical levels in the following form:

$$f_q : V_Q \to \bigcup_{q=1}^{Q} V_q ; \quad f_{qp} : V_q \to \bigcup_{p=1}^{P_0} V_{qp} , \tag{8.21}$$

$$f_{qp}^{-1} : \bigcup_{p=1}^{P_0} V_{qp} \to V_q ; \quad f_q^{-1} : \bigcup_{q=1}^{Q} V_q \to V_Q , \tag{8.22}$$

where V_Q is a set of FEs of an object as a whole, V_q is a set of FEs of the qth hierarchical level, and V_{qp} is the pth FE of the qth hierarchical level. The functional interconnection of indicators of an entire object's functioning quality with FE parameters is defined in the form of a function with monotonous variables including:

$$\bar{K}_Q = \tilde{F}_Q \left(\bar{X}_Q , \tilde{F}_{Q-1} \left(\bar{X}_{Q-1} , \tilde{F}_{Q-2} \left(\ldots \tilde{F}_l \left(\bar{X}_l , \tilde{F}_{l-1} \left(\ldots \tilde{F}_1 \left(\bar{X}_1 \ldots \right) \right) \ldots \right) \right) \right) \right) , \tag{8.23}$$

where $\bar{K}_Q$ is a quality indicator vector, $\bar{X}_Q$ is a vector of FE parameters of a higher level of a hierarchical structure (level of an object as a whole), $\bar{X}_q$ is a vector of FE parameters of the qth hierarchical level, $q = 1, 2, \ldots, l, \ldots, Q$, $\bar{X}_1$ is a vector of FE parameters of a lower hierarchical level (for example, the module level), $\tilde{F}_Q, \ldots, \tilde{F}_l, \ldots, \tilde{F}_1$ are functions of interconnection between indicators of FE of different hierarchical levels. These functions are specified inaccurately, fuzzily, and incompletely.

It is required to build a sequence of transformations

$$\Phi_q : W_Q \to \bigcup_{q=1}^{Q_W} W_q ; \quad \Phi_{qp} : W_q \to \bigcup_{p=1}^{P_W} W_{qp} , \tag{8.24}$$

$$\Phi_{qp}^{-1} : \bigcup_{p=1}^{P_W} W_{qp} \to W_q ; \quad \Phi_q^{-1} : \bigcup_{q=1}^{Q_W} W_q \to W_Q , \tag{8.25}$$

that define the structure of a control system and parameters of its FEs from the condition of achievement of a necessary quality level of an object's functioning at any time $t \in [0; T]$ in predictable risk situations $S_k \in S_0, k = \overline{1, N}$.

In this formulation we will draw attention to the interconnection between reference (8.21) and (8.22), and also (8.24) and (8.25). Naturally, there arises the question as to why it is necessary to have two kinds of interconnection between FEs of a hierarchical structure. The first kind of interconnection is defined by references (8.21) and (8.24), and the second is defined by references (8.22) and (8.25). The practical necessity follows from real conditions and techniques of complex system development. The point is that in preliminary development specifications there are specified requirements for an object as a whole. The customer is not interested in what structure and what FEs will be used to create an object. He is interested in seeing that the object meets the specified requirements. At the same time, the designer should efficiently transform general requirements for an object into requirements for a specific FE of each hierarchical level of a designed system. For this purpose it is necessary to have a description of the interconnection between FEs in the form (8.21). Therefore, references (8.21) are the basis for a decomposition procedure realization.

The necessity of an interconnection between FEs in the form (8.22) follows from the technology of a real complex hierarchical system design. System projection begins from the development of the FEs of the lowest level (modules or components); then higher-level FEs are designed, and then the final object composition is designed. For example, for radio systems designed for different purposes the design procedure is performed in the following sequence: functional module (chip) $\mapsto$ functional block (high-frequency amplifier) $\mapsto$ functional unit (receiver) $\mapsto$ functional system (receiving chain as a system of N receivers, N receiving antennas, and other functional units) $\mapsto$ object as a whole (radio-relay station). Consequently, formula (8.22) is the basis for an aggregation procedure realization in a system analysis.

For the control system, relation (8.24) is similar to the formula (8.21) in purpose, and relation (8.25) to relation (8.22). Note also that formulas (8.21) and (8.22) define a structural interconnection between the FEs of an object, and formula (8.23) defines a functional interconnection between FEs through the quality indicators.

General solution strategy. The fundamental feature of the given problem is that it possess at the same time properties of a structural-functional analysis problem (control system structure selection) and a multifactor-risk system analysis problem (analysis and minimization of risk situations, ensuring the required system quality in specified risk situations). This feature should be taken into account while forming a solution strategy. First of all, we will consider that this problem in tasks related to the selection of a system structure and its FE parameter selection is similar to a structural-functional analysis problem, whose solution techniques were given above. Thus, we will consider the strategy of a control influence optimization in conditions of multifactor risks.

We will assume relation (8.23) as a basis. It defines the interconnection between object quality indicators $\bar{K}_Q$ and FE parameters $\bar{X}_q$, $q = 1, 2, \ldots, l, \ldots, Q$ of all hierarchical levels. The given relation must be transformed to take into account risk factors and control influence. For this we will represent the relation between the quality-indicator vector and the FE parameter vector, the risk-factor vector, and the control-influence vector in the generalized form:

$$\bar{K}_Q = \tilde{F}_Q \left(\bar{\chi}_Q, \tilde{F}_{Q-1} \left(\bar{\chi}_{Q-1}, \tilde{F}_{Q-2} \left(\ldots \tilde{F}_l \left(\bar{\chi}_{l-1} \left(\ldots \tilde{F}_1 \left(\bar{\chi}_1\right) \ldots\right) \ldots\right) \ldots\right)\right)\right), \tag{8.26}$$

where $\bar{\chi}_q$ is a tuple for the qth hierarchical level of a complex hierarchical system, which is defined by the relation:

$$\bar{\chi}_q = \left\langle \bar{X}_q, \overline{u}_q, \bar{\rho}_q \right\rangle . \tag{8.27}$$

Here $\bar{X}_q$ is a FE parameter vector of the qth hierarchical level, $\bar{u}_q$ is a control-influence vector, and $\bar{\rho}_q$ is a risk-factor vector.

We will take into account that all variables, defined by relation (8.27), are time functions and are represented at the time moment $t = t_r$ in the form

$$\begin{aligned} \bar{X}_q^r &= \bar{X}_q \left(t_r\right) , \\ \bar{u}_q^r &= \bar{u}_q \left(t_r\right), \quad \bar{\rho}_q^r = \bar{\rho}_q \left(t_r\right) . \end{aligned} \tag{8.28}$$

In the general case the complex system control problem, both as a problem of object control as a whole and as a problem of its qth hierarchical level control, lies in the simultaneous fulfillment of two requirements:

1. To provide a stable meaning of specified quality indicators $\bar{K}_{1q}$ at any fixed time moment $t_r \in [0, T]$.
2. To ensure time changes of certin quality indicators $\bar{K}_{2q}$ by set requirements.

The first requirement in the ideal case means $\bar{K}_{1q} = \text{const}\ \forall t_r \in [0, T]$. In real conditions one is allowed to make designated changes to indicators under the influence of disturbing risk factors on the qth level and a correcting influence in a specified interval. The given requirement is defined by the relation:

$$\bar{K}_{1q}\left(\bar{u}_q^r,\bar{\rho}_q^r,t_r\right)\in\left[\bar{K}_{1q}^+,\bar{K}_{1q}^-\right]\quad\forall t_r\in[0,T]\,,\quad q=\overline{1,Q}\,,\tag{8.29}$$

where $\bar{K}_{1q}^+$ and $\bar{K}_{1q}^-$ are predefined values. The second requirement in the ideal case means that every component of vector $\bar{K}_{2q}$ is a specified function of time. In real conditions under the influence of disturbing risk factors and a programmed control influence at any time moment $t_r\in[0,T]$ one must meet the following condition:

$$\bar{K}_{2q}\left(\bar{u}_q^r,\bar{\rho}_q^r,t_r\right)\in\left[\bar{K}_{2q}^+(t_r)\,,\bar{K}_{2q}^-(t_r)\right]\quad\forall t_r\in[0,T]\,,\quad q=\overline{1,Q}\,.\tag{8.30}$$

Here $\bar{K}_{2q}^+(t_r)$ and $\bar{K}_{2q}^-(t_r)$ are, respectively, specified upper and lower boundaries of an allowed variation range of a function $\bar{K}_{2q}(t_r)$ at the moment t_r.

To justify the strategy we will describe in detail the control procedure using a bilevel system as an example. In this case quality indicators $\bar{K}_Q$ are defined by the relation:

$$\bar{K}_Q=\tilde{F}_Q\left(\bar{\chi}_Q,\tilde{F}_1(\bar{\chi}_1)\right)\,.\tag{8.31}$$

Here on the basis of (8.27) for the lower level $q=1$ and upper level $q=Q$ we have:

$$\bar{\chi}_Q=\left\{\bar{X}_Q,\bar{u}_Q,\bar{\rho}_Q\right\};\quad\bar{\chi}_1=\left\{\bar{X}_1,\bar{u}_1,\bar{\rho}_1\right\}\,.\tag{8.32}$$

Now we specify control $\bar{u}_Q^*$ and $\bar{u}_1^*$ from condition (8.30), assuming quality indicators and parameter behavior in time is known, based on a structure-functional analysis problem solution, in the form:

$$\bar{K}_Q=\bar{K}_Q^*(t)\,,\quad\bar{K}_1=\bar{K}_1^*(t)\,,\tag{8.33}$$

$$\bar{X}_Q=\bar{X}_Q^*(t)\,,\quad\bar{X}_1=\bar{X}_1^*(t)\,.\tag{8.34}$$

As an initial solution we will assume a solution that is found without taking into account risk factors. Then, $\overline{\rho}_Q=0,\overline{\rho}_1=0$. In this case from (8.31) and taking into account (8.32) we get:

$$\bar{K}_Q=\tilde{F}_Q\left(\bar{X}_Q,\bar{u}_Q,\tilde{F}_1(\bar{X}_1,\bar{u}_1)\right)\,.\tag{8.35}$$

We will define a discrete series on the interval $[0,T]$:

$$t_i=t_0+i\Delta t;\quad\Delta t=\frac{T}{M_0};\quad i=\overline{0,M_0};\quad t_0=0;\quad t_{M_0}=T\,.\tag{8.36}$$

Then the problem of a structure control definition is reduced to an equality $\bar{K}_Q = K_Q^*(t)$ ensuring at any time moment t_i [from relations (8.35) and (8.36)]. As a result we get the following system of equations:

$$\tilde{F}_Q\left(\bar{X}_Q(t_i), \bar{u}_Q(t_i), \tilde{F}_1\left(\bar{X}_1(t_i), \bar{u}_1(t_i)\right)\right) = K_Q^*(t_i), \quad i = \overline{0, M_0}. \quad (8.37)$$

Here, components of control vectors $\bar{u}_Q(t_i)$ and $\bar{u}_1(t_i)$ are unknown, but other variables are known, because their corresponding functions are defined in a problem of structural-functional analysis.

The system (8.37) solution defines discrete values of desired control functions at points specified in (8.36). Based on a solution to system (8.37) it is possible to build continuous functions $\bar{u}_Q(t_i)$ and $\bar{u}_1(t_i)$ using approximation methods. In the system of equations (8.37) the number of components of vectors $\bar{X}_Q, \bar{X}_1$ and $\bar{u}_Q, \bar{u}_1$ is specified. So, the number of unknown variables N is equal to the total number of components of vectors $\bar{u}_Q$ and $\bar{u}_1$. The number of equations M is specified by a quantity of discrete values in (8.34) and is equal to $M = M_0 + 1$. Because the quantity of discrete values is selected during the problem solution, it is necessary to take into account the following features:

1. It is expedient to choose the number of equations $M \geq N$, i.e., equal or greater than a number of unknown variables.
2. With increasing M the approximation error for functions $\bar{u}_Q(t)$ and $\bar{u}_1(t)$, based on discrete values $\bar{u}_Q(t_i)$ and $\bar{u}_1(t_i)$, $i = \overline{0, M_0}$ will decrease. But system (8.37)'s computational complexity will increase.

The system (8.37) solution selection method depends on the problem dimension, the ratio between M and N, and features of functions in (8.37). Further, based on (8.26) it is necessary to take into account risk factors $\bar{\rho}_Q(t)$ and $\bar{\rho}_1(t)$ in function (8.35) and to reduce the task of control-function improvement to one of well-known forms of a problem of disclosing situation uncertainty or to a problem of minimization of degree and level of multiple factor risks.

Determination of the controlling functions in cases where the total number of levels $Q > 2$, may be reduced to a sequence of tasks for a bilevel system or directly reduced to a system like (8.37), but with the number of variables equal to the total number of components of all control vector functions $\bar{u}_q(t)$, $q = \overline{1, Q}$.

Thus the strategy of complex hierarchical system control for which the structural-functional analysis problem is solved may be reduced to the following sequence of tasks:

1. Solution of a system of nonlinear equations (like (8.37)).
2. Approximation of the desired functions based on their discrete values of previous problem solution results.

3. Taking into account risk factors based on methods of disclosing situational uncertainty or reduction to a multifactor risk analysis problem.
4. Development of design solutions of control FE.

In practice it is not always possible to obtain an acceptable solution to a system control problem as a result of a single iteration. Derived control functions may require FE structure corrections and function corrections for some hierarchical levels of an object or its control system. This will require the next iteration execution in an object's structural-functional analysis problem and, consequently, the solution correction in control problems of an object under investigation.

Solutions of a structural-functional analysis problem and a control problem allow one to proceed to the next problem, which lies in a technical-economical analysis of the functionality and operational capability of a complex object.

8.4 Technical-Economical Analysis of Complex Object System Control

The technical-economical analysis is the final problem in a sequence of system problems of researching the properties, structure, serviceability, and control of complex hierarchical systems under conditions of multifactor risks.

In previous sections we considered problems whose solution defined the shape and the structure of an object and its control system, as well as design solutions regarding FEs of all hierarchical levels. The considered problem should give an answer to the following question: how rationally the previous problems have been solved by comparing the gained effect and expenditures of resources for its achievement. The practical importance of the problem is obvious—only a comparison of the cost and efficiency of solutions makes it possible to estimate such major characteristics of the developed systems as the technical and economic level of complex object production or levels of product efficiency and competitiveness.

The problem statement of technical and economic analysis of system control in conditions of multifactor risks is determined by the general purpose of system research of the given class of complex objects in the process of their design, production, operation, and control in real conditions of multifactor risks.

Contensive statement of problem. Structure and design solutions regarding the FEs of all levels of a complex hierarchical system ***are assigned***, properties and quality indicators of an object ***are defined***, the structure and design solutions of a multilevel control system of an object are defined, and the design solutions and qualitative indicators of functional elements of a control system ***are known***.

It is required to determine the technical and economic efficiency of the system control of an object under real conditions of multifactor risks.

The mathematical statement of the problem follows from the general functional interconnection of indicators of an object quality as a whole and parameters of FEs of all hierarchical levels, risk factors and control actions of a control system, as well as interconnections of the cost of component parts, custom-designed modules, and structures of FEs of all hierarchical levels with technical indicators of the design solutions.

Mathematical statement of problem. For complex multilevel hierarchical systems the interconnections between the quality indicators and an object's FE parametrs and the technical means of a control system **are given** in the form:

$$\bar{K}_Q = \tilde{F}_Q\left(\bar{\chi}_Q, \tilde{F}_{Q-1}\left(\bar{\chi}_{Q-1}, \tilde{F}_{Q-2}\left(\dots \tilde{F}_l\left(\bar{\chi}_{l-1}\left(\dots \tilde{F}_1\left(\bar{\chi}_1\right)\dots\right)\dots\right)\dots\right)\right)\right), \tag{8.38}$$

where $\bar{\chi}_q$ is a tuple for the qth hierarchical level $(q = 1, 2, \dots, l, \dots, Q)$ of the complex system, which is defined by the relationship:

$$\bar{\chi}_q = \left\langle \bar{X}_q, \bar{u}_q, \bar{\rho}_q \right\rangle .$$

Here $\bar{X}_q$ is a vector of parameters of FEs of the qth hierarchical level, $\bar{u}_q$ is a vector of controls, and $\bar{\rho}_q$ is a vector of risk factors.

It is required to establish the indicators that will allow one to determine in a generalized form the effectiveness of the object's operation; to determine their dependence on FE parameters; to find the interconnection of the total cost of the designed product and the technical, structural, and technological parameters of FEs of each qth hierarchical level of an object itself and its control system; to determine the technical and economic efficiency of an object's operation as a whole in conditions of multifactor risks.

General strategy of problem solution. The general strategy of estimation of technical and economic efficiency of system control of complex objects in conditions of multifactor risks is based on the following basic regulations:

- Techniques and methods of value analysis.
- Techniques and methods of control of complex-object functionality.
- Techniques and methods of analysis and minimization of risks.
- Techniques and methods of system control of safety and serviceability.

We assume that there is no need to deal with the principles, techniques, and methods of value analysis since they are considered in detail in special subdisciplines of economics (microeconomics, economic productive efficiency, economic design calculations, etc.). Let us consider only the general idea of obtaining parameters and characteristics that are to be determined according to the mathematical statement of the problem. First of all, the interconnections between control vector $\bar{u}_q$ and the control system's FE parameters are defined in the form:

$$
\begin{aligned}
\bar{Y}_Q &= \bar{f}_{u_Q}\left(\bar{u}_Q, \bar{Y}_{Q-1}\right), \\
\bar{Y}_{Q-1} &= \bar{f}_{u_{Q-1}}\left(\bar{u}_{Q-1}, \bar{Y}_{Q-2}\right), \\
&\cdots\cdots\cdots\cdots \\
\bar{Y}_l &= \bar{f}_{u_l}\left(\bar{u}_l, \bar{Y}_{l-1}\right), \\
&\cdots\cdots\cdots\cdots \\
\bar{Y}_2 &= \bar{f}_{u_2}\left(\bar{u}_2, \bar{Y}_1\right), \\
\bar{Y}_1 &= \bar{f}_{u_1}\left(\bar{u}_1\right),
\end{aligned}
\tag{8.39}
$$

where $\bar{Y}_q (q = 1, 2, \ldots, Q)$ is a vector of parameters of FEs of the qth hierarchical level of a control system and $\bar{u}_q$ is a vector function of control influences at the qth hierarchical level.

Then, an interconnection is established between the cost of development, production, and operation of functional elements of the qth hierarchical level and the parameters of an object itself and its control system:

$$
C_c = \tilde{f}_c\left(\bar{X}_Q, \tilde{f}_{c_Q}\left(\bar{X}_{Q-1}, \tilde{f}_{c_{Q-1}}\left(\ldots \tilde{f}_{cl}\left(\bar{X}_l, \ldots\left(\tilde{f}_{c1}, \left(\bar{X}_1\right)\ldots\right)\ldots\right)\ldots\right)\right)\right), \tag{8.40}
$$

$$
C_u = \tilde{f}_u\left(\bar{Y}_Q, \tilde{f}_{u_Q}\left(\bar{Y}_{Q-1}, \tilde{f}_{u_{Q-1}}\left(\ldots \tilde{f}_{ul}\left(\bar{Y}_l, \ldots\left(\tilde{f}_{ul}, \left(\bar{Y}_1\right)\ldots\right)\ldots\right)\ldots\right)\right)\right). \tag{8.41}
$$

Here relationship (8.40) establishes the interconnection between the cost C_c and parameters of FEs of all hierarchical levels of a controlled object, and relationship (8.41) establishes a similar interconnection for a control system.

In the general case the total cost of an object C_Σ as a cost function of FEs of system C_c and control system C_U is defined as:

$$
C_\Sigma = f_\Sigma\left(C_c, C_U\right). \tag{8.42}
$$

The effectiveness of the object's operation as a generalized estimate of its serviceability results for a definite period is defined through some generalized indicators, which in quality control are traditionally referred to as an integral object quality indicator. It is expressed through quality indicators in the form:

$$
K_\Sigma = F_\Sigma\left(\bar{K}_Q\right). \tag{8.43}
$$

Then the technical and economic efficiency defined as a ratio of the generalized effectiveness of the object's operation to the total cost for the period of the life cycle of an object may be represented in the following form:

$$
E_\Sigma = \frac{K_\Sigma}{C_\Sigma}. \tag{8.44}
$$

It is necessary to pay attention to one practically important feature of the determination of total cost. The expenditures at the stage of development and production of an object begin to be covered only at the operation stage. Therefore, financial, material, and other resources spent at the stages of development and production of an object for some time equal to the duration of these stages are "frozen" and do not make any profit. This results in an increase in the total cost from the moment of the onset of operation. This circumstance is taken into account by a certain coefficient that depends on the duration of margin "freezing". For example, such accounting can be made on the basis of the following approximate relationship:

$$C_t = \frac{C_0}{(1-E)^t},$$

where C_0 is the cost value at the beginning of an object elaboration ($t = 0$), C_t is the cost value over t years, E is a normative coefficient, frequently assumed to be $E = 0,1,\ldots,0,15$.

The strategy of technical and economic efficiency estimation can be formulated briefly as follows:

1. The total cost of a product is determined taking into account the expenditures for the development, production, and operation of a technical system along with a control system and technical diagnostics system.
2. The total effectiveness of a complex object's operation is determined on the basis of an integral indicator, which is expressed through indicators of serviceability and quality of an object's operation, taking into account the losses on liquidation of abnormal situations, failures, and malfunctions.
3. Technical and economic efficiency is determined as a ratio of effectiveness to costs.
4. The degree of conformity of technical and economic efficiency to the requirements given in the requirements specification is verified.
5. The degree and level of risk are determined on the basis of the general principles of the system analysis of risk.
6. The degree to which the requirements are met with regard to technical and economic efficiency is estimated. If the requirements are not met, a new iteration of estimating and choosing functional elements and solving a structural-functional analysis begins.

When the required indicators of effectiveness and the technical-economic efficiency of a complex object's operation based on the data of its functioning during an a priori set period are obtained, one can assume that the general problem of a complex hierarchical system analysis is solved.

It should be noted that the above considered approach of the estimation of effectiveness and technical-economic efficiency of a complex object's operation is the simplest one and does not take into account a number of practically important factors. A more general and sufficiently detailed representation of the system estimation of the technical and economic efficiency of complex objects is given by the example of the serviceability of modern information-telecommunications systems [165, 166].

General strategy of technical-economical analysis. Technical-economical analysis is a major component of system analysis of functioning effectiveness of various complex objects.

It is obvious that the practical necessity of estimating the influence of various factors and realizing the basic principles of ensuring technical and economic efficiency requires a certain strategy of optimization of the properties and capabilities of complex objects. It is possible to give the two most common formulations of the optimization problem. The first formulation consists in the development of a strategy for choosing a rational compromise of contradictory requirements in the interests of attaining the highest level of technical and economic efficiency with the assigned restrictions on effectiveness indicators. The second formulation of the optimization problem requires provision of the maximal effectiveness with the assigned restrictions of technical and economic efficiency.

In spite of a certain distinction of the content and accepted goal functions, these problems have one practically important general property. Its essence lies in the fact that only the realization of the systemness of technical and economic efficiency can ensure a high credibility of the obtained estimation. At first sight, reasoning about the systemness of efficiency analysis seems to be absurd. It seems very simple—increase the numerator and reduce the denominator in formula (8.44) and the required level of credibility and estimation effectiveness will be guaranteed. Reducing the denominator means that it is necessary to reduce the expenses at each stage of a product's life cycle. However, experience has shown that so straightforward an approach may lead to absurd results. To understand this, one should pay attention to one feature of the accepted indicator of formula (8.44). The value E_{Σ} remains constant if the numerator and denominator in (8.44) are increased or reduced proportionally and simultaneously.

Also, it is necessary to pay attention to one more important feature of indicator (8.44)—the effectiveness value is a nonlinear and a priori unknown function of costs. And, first of all, as the experience shows, the maximum payback of costs is achieved for the earliest stages of a life cycle and, mainly, when the appearance, principles of operation, and other initial preconditions of a product design are chosen. From here it follows that it is necessary to provide a multipronged system analysis of accounting of factors that determine various aspects of the technical and economic efficiency of a product.

We list only those areas where it is required to realize the systemness of technical-economical analysis.

1. Systemness of costs.

1.1. Systemness of costs of an object's structure.
1.2. Systemness of costs of an object's life cycle.
1.3. Systemness of analysis of reasons and factors that increase costs.
1.4. Systemness of analysis of reasons and factors that decrease costs.
1.5. Systemness of estimation of risks during life cycles of complex objects.

2. Systemness of effectiveness analysis.

2.1. Systemness of effectiveness of decisions concerning the object's structure.

2.2. Systemness of effectiveness of decisions concerning the object's life cycle.
2.3. Systemness of conditions and factors of increases in efficiency.
2.4. Systemness of conditions and factors of decreases in efficiency.
2.5. Systemness of conditions and factors of reducing multifactor risks.

3. Systemness of analysis of application conditions.
3.1. Systemness of technical and economic efficiency.
3.2. Systemness of production and control efficiency.
3.3. Systemness of the analysis of possible market demand and supply.
3.4. Systemness of the analysis of competition.
3.5. Systemness of estimation of product development.

8.5 Example of Solving a Problem of System Control of Serviceability and Safety of a Complex Engineering Object

We will consider a problem of serviceability and safety control of a complex engineering object in the process of its transition from a normal to abnormal mode by the example of an aircraft engine [127].

Contensive statement of problem. We believe it expedient to consider a safety control problem as a system problem involving the detection of risk factors whose influence may result in abnormal situations. At the same time it is taken into consideration that safety and functioning control is implemented under conditions of incompleteness and uncertainty of dynamics of the transition of a normal to abnormal mode. The normal mode is not permanent and changes considerably at different stages of a system operation cycle. These conditions are met by the operation of an aircraft engine, the modes of which during takeoff, cruising, and landing differ fundamentally. The influences most critical and sensitive to risk factors are the takeoff and landing modes. These modes are in their physical essence transitional between two stationary modes: the first is an off mode, when airplane stands still; the second is a cruising flight mode, when the airplane flies at a specified altitude and the engine works in a permanent mode during the flight period.

The normal functioning mode consists of a number of basic stages. Crucial are the following stages: the basic stage is a stationary cruising flight mode, the transitional stages are non-stationary takeoff and landing modes. The engine parameters vary synchronously during a transition from one mode to another and in transitional modes. Among these parameters are traction, fuel mass, oil pressure, air supply, etc. These parameters vary in different modes as time functions depending on the purposeful actions of a control system and uncontrolled influence of external factors of climatic, geographical, geological, technological, and other processes [13, 14, 102, 191].

General properties and peculiarities of an object under investigation. The main properties, indicators, and the characteristics of a normal functioning mode of an object are known. The mode is determined by the following basic stages and properties:

- Stage 1 → $[t_0^-, t_1]$ is the period of engine start-up and warm-up.
- Stage 2 → $[t_1, t_2]$ is the preparation for takeoff, starting up and testing the engine transition from a normal power mode to an afterburning mode.
- Stage 3 → $[t_2, t_3]$ is takeoff execution in the engine afterburning mode, which is characterized by the engine power increasing by 1.2–1.3 times in comparison with the standard power.
- Stage 4 → $[t_3, t_4]$ is a transition process from the afterburning to cruising mode, characterized by a reduction in the engine power level to 0.75–0.85 of its standard value.
- The main stage 5 → $[t_4, t_5]$ is the cruising mode; the main functioning indicators are 0.75–0.85 of their standard values.
- Stages 6–8 are similar to takeoff stages by general indicators, so we represent them in a simplified version. Stage 6 is a transition from the cruising mode and the landing mode. Stage 7 is the landing mode, stage 8 is an engine-shutdown mode, which ends with the transition to ***the mode of waiting and preparation*** to the next start-up. Typical functioning modes are presented in Fig. 8.5.

Characteristics and peculiarities of modes are known and determined by the following properties and accepted tolerances.

1. Each stage is characterized by its specific duration, the initial and final values of each indicator y_i, determined respectively at the initial and final moment of the time period. Variations y_i during the period are determined by a corresponding model. The period duration and initial and final values are a priori assigned.

2. All indicators y_i are synchronous and cophased.

These properties consist in a simultaneous increase or decrease of indicators without a time delay under the influence of various factors.

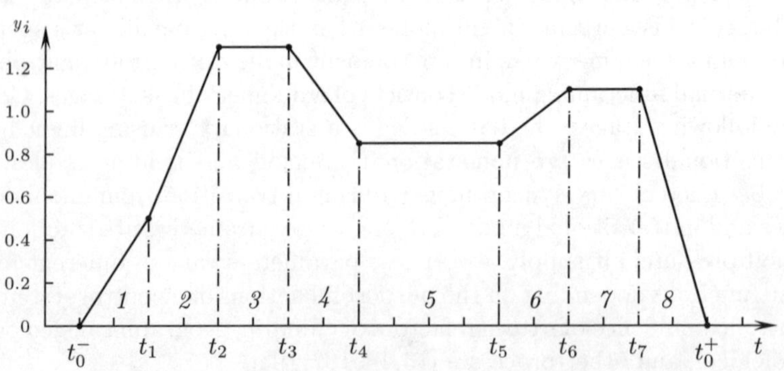

Fig. 8.5. Typical modes of an aircraft engine functioning

3. Control action $U = \left(U_j | j = \overline{1, m}\right)$ is inertialess, i.e., without a time delay between a control action and the object's reaction to it.

4. Risk factors $\bar{\rho} = \left(\rho_q | q = \overline{1, n_q}\right)$ change the value of influence on the engine in time, and with an increase in time of influence the degree and level of risk increase.

5. Control action can slow down risk-factor influence or stop their negative influence on the controlled object under the condition that the speed of control exceeds the growth rate of the level of risk-factor influence.

6. Termination of negative influence of risk factors can be ensured only under the condition that a decision is developed, made, and realized before the critical moment T_{cr} comes, when risk-factor influence results in irreversible consequences of an accident or a catastrophe.

The recognition of abnormal situations will be considered **on the basis of the given above properties and features** as a problem of recognition of risk factors whose influence may result in an abnormal situation. We shall take into account the main differences between normal and abnormal modes, which are enumerated in Chap. 7.

Mathematical statement of problem. It is known that an object's normal functioning mode is characterized by a set:

$$Y = \left\{ y_i | y_i^- \leq y_i \leq y_i^+, \ i \in N_0, \ N_0 = \overline{1; n_0} \right\} . \tag{8.45}$$

Values of the object's functioning indicators $y_i \in Y$ are limited by the control system action $\bar{U} = \left(U_j, j = \overline{1, J_0}\right)$ and random influences of risk factors $\bar{\rho} = \left(\rho_q | q = \overline{1, n_q}\right)$. The dependence of indicators on the control system actions and influences of random risk factors is defined by the relationship:

$$y_i = f_i \left(\bar{U}, \bar{\rho}\right) , \tag{8.46}$$

where $f_i(\bar{U}$ and $\bar{\rho})$ are unknown functions to be determined by discrete data of the diagnostics system.

The parameters y_i are measured in the known discrete time moments $t_k = t_0 + k\Delta t$ with known values U_j and unknown values ρ_q. The measurement results are known in the form of the finite sample of observations during the given period $[t_0, t_{k_0}]$ and the finite sample of observations during the period $[t_{k_1}, t_{k_0}]$, where $t_0 < t_{k_1} < t_{k_0}$. The samples are represented in the form of the following arrays:

$$B_0 = \left\{ \left\langle \bar{y}\left(t_k\right), \bar{U}\left(t_k\right) \right\rangle | t_k = t_0 + k\Delta t; \ k = \overline{0; k_0} \right\} , \tag{8.47}$$

$$B_{k_1} = \left\{ \left\langle \bar{y}\left(t_k\right), \bar{U}\left(t_k\right) \right\rangle | t_k = t_0 + \hat{k}\Delta t; \ \hat{k} = \bar{k}_1; k_0 \right\} , \tag{8.48}$$

where $\bar{y} = \left(y_i | i = \overline{1, n_0}\right)$.

There is a set $\hat{Y}$ of dynamically synchronous indicators:

$$\hat{Y} = \left\{ \hat{y}_{i_1} | \left\langle U_{j_1} \uparrow, \ j_1 \in \hat{M}_1 \right\rangle \Rightarrow \left\langle \hat{y}_{i_1} \uparrow, \ i_1 \in \hat{N}_1 \right\rangle \right\} ,$$
$$\hat{y} \subseteq Y, \ M_1 = \overline{1; J_0}; \ j \leq J_0; \ \hat{N}_1 \subseteq N_0; \ \hat{N}_1 = \overline{1; n_1}; \ n_1 \leq n_0 . \tag{8.49}$$

The values of all control vector components at the moment t_0 are known:

$$\bar{U}_0 = \{U_j^0 | U_j^0 = U_j(t_0)\}, \quad j = \overline{1, J_0}, \tag{8.50}$$

and the required values of normal mode indicators, without the influence of uncontrolled risk factors, are also known:

$$\bar{y}_0 = \left\{y_j^0 | y_j^0 = \hat{f}(\bar{U}_0);\ y_i^0 \in Y;\ i = \overline{1; n_0}\right\}. \tag{8.51}$$

For the specified period of operation $\overset{\cup}{T} = [t_0; t]$ permissible deviations of parameters y_i under the influence of uncontrolled factors are defined in the form of the following restrictions:

$$y_i^- \leq \overset{\cup-}{y_i} < \overset{\cup+}{y_i} \leq y_i^+.$$

If this condition is not fulfilled simultaneously for several synchronous indicators $\hat{y}_{i_1} \in \hat{Y}$ and $\hat{y}_{i_2} \in \hat{Y}$, the situation is abnormal. Formally, the condition for an abnormal situation will be described by the following relationship:

$$\left\{\left[\left(\hat{y}_{i_1} \in \hat{Y}\right)\right] \wedge \left[\left(\hat{y}_{i_1} < y_{i_1}^-\right) \vee \left(y_{i_1}^+ < \hat{y}_{i_1}\right)\right]\right\} \wedge$$
$$\wedge \left\{\left[\left(\hat{y}_{i_2} \in \hat{Y}\right)\right] \wedge \left[\left(\hat{y}_{i_2} < y_{i_2}^-\right) \vee \left(y_{i_2}^+ < \hat{y}_{i_2}\right)\right]\right\}. \tag{8.52}$$

It is required:

- To ensure the recognition of an abnormal situation on the basis of samples B_0 and B_{k_1}.
- To define the degree and level of risk at each moment t_k.
- To find the moment and rate of possible transition from a normal mode to an abnormal one.

Initial values of indicators

1. The quantity of indicators $\hat{y}_{i_1} \in \hat{Y}$; $\hat{y}_{i_2} \in \hat{Y}$; $\langle i_1; i_2 \rangle \in \hat{N}_1$; $N_1 = \{i | i = \overline{1, n_1}\}$ of an object is limited by $n_0 = 2$; $i_1 = 1$; $i_2 = 2$. The indicators are dynamically synchronous and time dependent $\hat{y}_{i_1} = y_1(t)$, $\hat{y}_{i_2} = y_2(t)$ and correspond to condition (8.49). The maximal changes take place during the transition modes of stages 1 and 8, where conditions (8.52) are most probable. The limits of variation for these stages are assigned:

$$y_1[t_0] = 0; \quad y_1[t_k] = 0.45; \quad y_2[t_0] = 0; \quad y_2[t_k] = 0.35.$$

Henceforth the partial values of indicators are normalized by their general values during the flight.

2. The mode variation is performed by a control system whose indicators $U_j \in \bar{U}$; $\bar{U} = \left(U_j,\ j = \overline{1, J_0}\right)$ are time functions $U_j = U_j(t)$. Conditions (8.50) and (8.51) are fulfilled at the moment t_0. It is assigned $J_0 = 3$, and

the functions $\langle U_1(t), U_2(t), U_3(t)\rangle \in \bar{U}$ are restricted and represented in the form:

$$\begin{aligned}
&0 \le U_1(t) \le 1; \quad && U_1(t) = d_{10} + d_{11}t + d_{12}t^2;\\
&0 \le U_2(t) \le 0.7; \quad && U_2(t) = d_{20} + d_{21}t;\\
&0 \le U_3(t) \le 0.5; \quad && U_3(t) = d_{30} + d_{31}t + d_{32}t^2 .
\end{aligned} \tag{8.53}$$

3. In the system at the definite discrete time moments $t_k = t_0 + k\Delta t$ parameters $y_i = y_i[t_k]$ are measured with known values $U_j = U_j[t_k]$ and unknown values ρ_q. Measurement results are represented in the forms (8.47) and (8.48).

It is known that the measuring interval constitutes in minutes t the period $t \in [0; 0,3]$. Within this interval its discrete analog is formed, which fulfills the following conditions: $k = \overline{1;11}$; for $k = 1$, $t = t_0 = 0$; for $k = k_0 = 11$, $t = t_{k_0} = 0.3$. In the process of technical diagnostics a sequence of discrete intervals during the flight is formed, which includes intervals of all stages from 1 to 8.

4. On the basis of discrete intervals the values for functions $\langle U_1(t), U_2(t), U_3(t)\rangle \in \bar{U}$ are formed, which are defined by relationship (8.53). The values of functions are defined by the conditions

$$\begin{aligned}
&U_1(t_0) = 0; \quad U_1(0.3) = 1; \quad U_1(0.15) = 0.7;\\
&U_3(t_0) = 0; \quad U_3(0.3) = 0.5; \quad U_3(0.2) = 0.4;\\
&U_2(t_0) = 0; \quad U_2(0.3) = 0.7 .
\end{aligned} \tag{8.54}$$

5. The following properties and features are considered in the analysis of risk-factor influence on an object:

- Risk factors $\bar{\rho} = \left(\rho_q | q = \overline{1, n_q}\right)$ are independent and vary according to random rules whose distribution is a priori unknown.
- Risk factors can have an impact simultaneously on one, a few, or all indicators y_i.
- If under the influence of risk factors the values of at least two indicators y_i change simultaneously, synchronously, and cophased, and it is detected by the diagnostics system, this means that an object is passing from a normal to abnormal mode.

6. On the basis of the above indicated conditions, properties, and features, the influence of risk factors on the object under investigation is taken into account in the following way. Under the influence of risk factors the values of indicator y_i will be defined by the value $\tilde{y}_i$. At the moment t_k indicator $\tilde{y}_i[t_k]$ is defined by the equation:

$$\tilde{y}_i[t_k] = \frac{1}{J_0}\sum_{j=1}^{J_0} \tilde{b}_{ij} \sum_{r=0}^{R_j} a_{jr} T_r^*(U_j); \; \tilde{b}_{ij} = b_{ij} \cdot F_i(\rho_{qk}) . \tag{8.55}$$

Here, value $\rho_{qk} = \rho_q [t_k]$ defines the value of the qth risk factor at the moment t_k. Indicator b_{ij} characterizes the level of influence of the kth risk factor on the indicator y_i of an object at the moment t_k, implemented by acting on indicator U_j of the object control. The extreme values b_{ij} are also assigned: for $i = 1$; $j = \overline{1;3} \Rightarrow b_{11} = 0.2$; $b_{12} = 0.3$; $b_{13} = 0.5$; for $i = 2$; $j = \overline{1;3} \Rightarrow b_{21} = 0.1$; $b_{22} = 0.25$; $b_{23} = 0.4$. Function $F_i (\rho_{q_k})$ characterizes the level of influence of factor ρ_{q_k} on the ith indicator y_i. It is to meet the condition that in the absence of risk factors the equality $\tilde{y}_i = y_i$ is to be fulfilled (i.e., with $\rho_{q_k} = 0$). Several risk factors $\rho_{n_q} = \left\langle \rho_{qk} | \rho_{qk} = \rho_q [t_k];\ q = \overline{1; n_q} \right\rangle$ may simultaneously and independently affect the object. Function $F_i \left(\rho_{n_q}\right)$ is defined by the relationship:

$$F_i \left(\rho_{n_q}\right) = \prod_{q=1}^{n_q} (1 - c_{iq} \rho_{qk}) \,,$$

$$\rho_{n_q} = \left\{\rho_{qk} | \rho_{qk} = \rho_q (t_k);\ q = \overline{1; n_q}\right\} \,. \tag{8.56}$$

Here, the following value is assigned: $0 < c_{iq} \leq 1$; $i = \overline{1; 2}$; $q = \overline{1; n_q}$.

Problem solution.

Let us consider the algorithm of the problem solution, where, using the initial data (8.47) and (8.48) on the basis of (8.53)–(8.56), it is required to execute the following procedures:

1. Normalize each stage duration to the corresponding part of the total duration of engine operation $\left[t_0^-, t_0^+\right]$, typical operation modes of which are represented in Fig. 8.5.

Normalized duration τ_1 and τ_2 of stages 1 and 2 are defined by the relationships:

$$\tau_1 = 1 - \frac{t - t_0^-}{t_1 - t_0^-}; \quad t \in \left[t_0^-, t_1\right], \quad \tau_2 = 1 - \frac{t - t_1}{t_2 - t_1}; \quad t \in [t_1, t_2] \,.$$

Discrete values τ_1 and τ_2 are calculated in accordance with p. 3 of the initial data. Normalization is performed for all other stages in the same way. For each stage the value τ should vary within the interval [0;1]. Such an approach allows one to raise the accuracy of models due to an increase in the number of points for each interval while compiling the corresponding system of equations.

2. Perform function $y_i (t_k)$ ***indicators and control actions*** $U_j (t_k)$ ***normalization*** on the basis of the relationships:

$$\hat{y}_i (t_k) = \hat{y}_i = \frac{y_i (t_k)}{y_i^+}; \quad y_i^+ = \max_{t_k \in \left[t_0^-, t_0^+\right]} y_i (t_k) \,;$$

$$\hat{U}_j (t_k) = \hat{U}_j = \frac{U_j (t_k)}{U_j^+}; \quad U_j^+ = \max t_k \in \left[t_0^-, t_0^+\right] U_j \,.$$

Normalization should be performed corresponding to conditions (8.54).

3. *Form a system of models for control action indicators* $U_j(t_k)$. A system of models is formed in accordance with (8.53) and (8.54) and consists in determining the corresponding function coefficients $d_{01}, \ldots, d_{32}$. Determination of coefficients for the function $U_1(t) = d_{10} + d_{11}t + d_{12}t^2$ according to (8.53) and (8.54) is reduced to solving the following system of equations:

$$d_{10} + d_{11}t_0 + d_{12}t_0^2 = 0\,,$$
$$d_{10} + d_{11}0.15 + d_{12}(0.15)^2 = 0.7\,,$$
$$d_{10} + d_{11}0.3 + d_{12}(0.3)^2 = 1.0\,.$$

Note: In this system of equations there is a practically important peculiarity. The moment $t_0 = t_0^- = 0$ corresponds to an offline state of the engine. Thus, in the first equation, all items should be zero. Then, the last two equations have three unknown variables d_{10}, d_{11}, and d_{12}, which results in an unlimited set of solutions and, thus, is practically unacceptable. The solution to this paradox is possible after a fundamental change in the initial reference point. That is, the reference point should be a value of control action in an established mode, which is typical for the cruising stage. Then, value d_{jo} should be considered as an indicator of steady operation of the control system, d_{j_1} as the speed of control mode switching, and d_{j_2} as the acceleration of the control mode switching. With this approach the value d_{jo} is not only a priori determined, but can be specified directly during flight. Values of two other indicators are defined directly from the last two equations. As a result, we obtain a final solution in the form of the functions

$$U_2(t) = d_{20} + d_{21}t + d_{22}t^2\,,$$
$$U_3(t) = d_{30} + d_{31}t + d_{32}t^2\,,$$

for which the conditions $d_{20} = d_{30} = \hat{d}_{05}$ are fulfilled and $\hat{d}_{05}$ is an indicator of steady operation of the control system at stage 5 in the cruising mode.

4. *Form a system of models for indicators of an object functioning in a normal mode*. The basic models of an object's operation will be formed in a normal mode, taking into account the equality $\tilde{y}_i = y_i$ for all periods. In such conditions the equality $F_i\left(\rho_{n_q}\right) = 1$ is satisfied, since for a normal mode by definition $\rho_{q_k} = 0$; $q = \overline{1, n_q}$ and, therefore, according to (8.56), we get $\rho_{n_q} = 0$. Then, the models of object's operation in compliance with restrictions and recommendations of p. 4 of the initial data are defined by the relationship

$$y_i = \frac{1}{J_0} \sum_{j=1}^{J_0} b_{ij} \sum_{r=0}^{R_j} a_{jr} T_r^* (U_j)\,.$$

Systems of equations for defining unknown variables a_{ir} and b_{ij} should be formed on the basis of the discrete sample (8.47)

$$B_0 = \left\{ \left\langle \bar{y}(t_k),\ \bar{U}(t_k) \right\rangle | t_k = t_0 + k\Delta t;\ k = \overline{0, k_0} \right\}$$

of observations during the period $[t_0, t_{k_0}]$. The sample is defined by a sequence of discrete intervals in compliance with p. 3 of the initial data.

5. Form a system of models for indicators of an object functioning in an abnormal mode. Models of an object functioning in an abnormal mode should be formed for all periods on the basis of relationships (8.55) and (8.56):

$$\tilde{y}_i [t_k] = \frac{1}{m} \sum_{j=1}^{m} \tilde{b}_{ij} \sum_{r=0}^{R_j} a_{jr} T_r^* \cdot (U_j); \quad \tilde{b}_{ij} = b_{ij} \cdot F_i \left(\rho_{n_q}\right)$$

using the sample data (8.48)

$$B_{k1} = \left\{ \left\langle \bar{y} \left(t_k\right), \bar{U} \left(t_k\right) \right\rangle | t_k = t_0 + \hat{k} \Delta t; \hat{k} = k_1, k_0 \right\}$$

for the period $[t_{k_1}, t_{k_0}]$, in compliance with restrictions and recommendations of pp. 4–6 of the initial data. Systems of equations for finding the unknown variables a_{ir} and b_{ij} and the function $F_i \left(\rho_{n_q}\right)$ should be formed on the basis of a discrete sample defined by a sequence of discrete intervals in compliance with the requirements of p. 3.

6. Develop a system of versions of a demo system of control and estimation of abnormal mode risks. On the basis of the models of analysis of normal and abnormal process dynamics proposed above, a demo system of detection and estimation of abnormal situation risks was developed. The models allow one to define the values of the degree and level of risk for each moment t_k.

The above indicated problems for a normal mode of an object's operation are solved by using the simplex method for solving incompatible systems of linear equations. The initial data and results of mode correction with determined values of maximum deviation are given in Fig. 8.6.

The moment of possible transition of the normal mode of an object's operation into abnormal mode is shown in Fig. 8.7.

In conclusion we note that the presented theoretical studies and examples of problem solutions show that it is possibile to define, with sufficient practical accuracy, the moment of possible transition of a normal mode of a complex engineering system into an abnormal mode and, thus, ensure the safety of its operation.

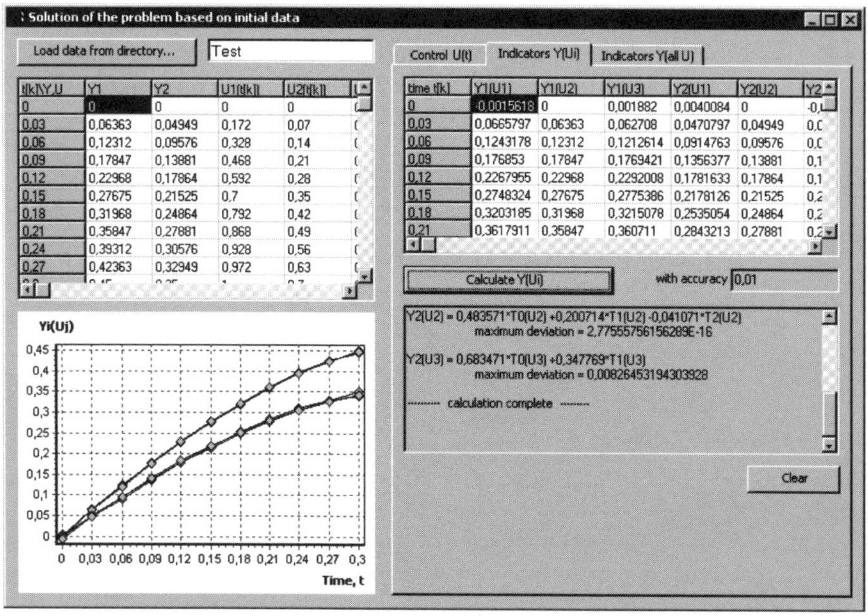

Fig. 8.6. Initial data and results of mode correction

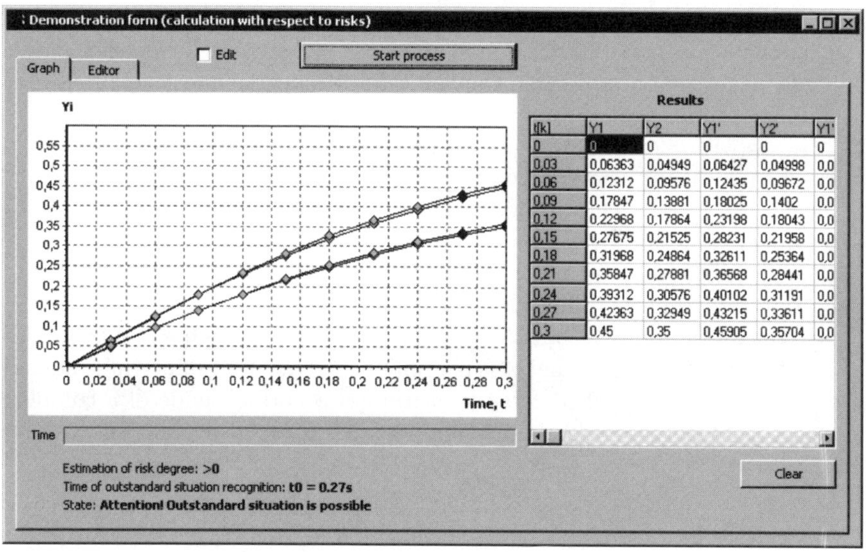

Fig. 8.7. Recognition of abnormal situation

9 Foresight System Methodology

Foresight, and above all technology foresight, has become the most important methodology of innovation and socioeconomic development of modern society, not only at the national or regional level, but also at the level of separate industry branches or large organizations and companies. This methodology to a large extent has realized the ideas of Glushkov [37] related to the construction of information networks and technologies for different areas of human activity, especially for the knowledge-based economy.

Foresight methodology, as it is applied in many countries, is based, above all, on the use of intuition, experience, knowledge, and expert skills in many subject domains when solving the problems of strategic planning and decision making. Many of the practical methods in use today are great as far as they go, but due to the novelty of foresight methodology they are all characterized by the absence of such methods as cybernetics, mathematics, and modern information technologies.

In this chapter we systematize and unify the most efficient practical methods in the foresight sphere and, above all, technology foresight. They are supplemented by formal procedures and methods of expert estimation of various objects. On this basis, a powerful toolkit such as scenario analysis methodology and a man–machine information platform for solving a given class of problems is created.

9.1 Foresight Importance and Objectives

Throughout human history, many thinkers and prophets have tried to "look" into the future to gain insight into what lies ahead. To achieve this insight, they have used various methods: from chiromancy and sorcery to astrological and science-fiction forecasts. Such forecasts were particularly subjective and could not withstand any criticism from the natural sciences.

Perhaps there is no better example of foreseeing the future than Leonardo da Vinci. More than anyone of his time, he understood the connection between science and art. Leonardo's fruitful mind gave birth to fantastic inventions of submarines, airplanes, bicycles, cutting machines, and even a tank. Not only did his ideas become possible later on, but they were similar to the real machines that appeared centuries later.

Challenges and threats appearing in modern society make it necessary to objectively foresee the future at least approximately. Foresight is required to form a rational and fault-free development strategy in any organized society (nation, country, organization, or company) in the world of intense competition.

Since World War II new technologies have been a powerful force in the economic development of all advanced countries of the world. In the 1980s, economists of the so-called new growth theory school showed that science and technology became the most important factors of social progress.

By the end of the 20th century, rapid technological advances and the burgeoning development of the global markets for products created by new technologies brought to life new risks and instabilities for large companies and countries because such changes had not been anticipated on time. For example, in the 1980–1990s, most office equipment and, in particular, traditional printing-press manufacturers suffered great losses due to not foreseeing the "avalanchelike" development of the personal computer market. We also know of some great failures of competing countries in space-exploration programs and in the development of modern systems of armament and collective security.

Thus, in the modern world, progress-minded technologies are the main moving force behind economic competitiveness. Foresight of economic competitiveness at early stages essentially influences on the development of future society. Especially now, at the beginning of the new millennium, foresight is very important. Facing the dwindling reserves of fossil fuels and other natural resources, the acceleration of economic globalization processes with many negative consequences (loss of national identity by less developed countries, collapse of national economies, extreme widening of the gap between the rich minority and the poor majority, and the aggravation of public and interethnic relations), humankind awaits a new paradigm of development with different social, economic, scientific, and technological methods at its basis.

Initial attempts to obtain objective knowledge about the future were mainly connected with the development of new method of foresight and application of traditional mathematical and statistical methods. Therefore, a group of powerful methods was created, such as the temporal series, one- and multidimensional regression analysis, simulation modeling, econometric model methods, and others.

All of these methods belong to the so-called ***methods of quantitative forecasting***. They are used for approximate "determination" of the future behavior of a certain parameter or a set of interconnected parameters in a given time interval.

These methods are computationally intensive. However, their practical application is essentially limited to the processing of quantitative retrospective data in processes that change monotonously. In other words, forecasting methods describe the future, which actually is the continuation or extrapolation of the past. This fact essentially limits the possibilities of such methods, first, because we live in a world where events happen that are qualitatively different from past events. First of all, many of the ***fractured and discontinuous changes*** belong to these methods through their connection with the ***monotony breaking*** of the processes and having the characteristics of nonlinear phenomena. Examples of such phenomena at the end of the 20th century include the disintegration of the Soviet Union, resulting in changes in the worldwide geopolitical balance of power and other global changes. No methods of quantitative forecast can predict such phenomena. Popular in the 1960s, forecasting has lost its status as the universal methodology. The reason for this was that forecasting methods could not predict the oil crisis in the early 1970s with its destructive consequences and the explosive development of information technologies in the 1990s, the financial crisis in southeast Asia and Russia in 1998, and many other global changes.

Therefore, today's conditions urgently require solving a new problem—***representing the future, which cannot be interpreted as a simple continuation of the past, as this future will have essentially different forms and structures compared to what was known in the past***.

This problem has received the name ***foresight*** [101]. This terminology was used as early as the end of the 1950s by ***Gaston Berger*** in the well-known magazine ***"Two Worlds"***, but the recognition of foresight as an independent science and technology methodology occurred only in the early 1990s.

Note that today there are no universal and perfect approaches to solving this problem. There are only attempts to build possible development scenarios of different events in the future. But the principal difference from the old practice of solving the same problems is that presently we use methods of a ***qualitative character***, but not quantitative. Some of them were known long ago and were developed and used for solving special, target-oriented tasks. For example, in the 1960s the American company *Rand Corporation* created a method allowing one to facilitate the so-called ***visualization*** of scenarios of development and application of new equipment in all technological aspects. This method was named ***Delphi*** (after the Greek oracle Delphos).

Today we know other methods of qualitative analysis that may be used at certain stages of foresight of future events. But none of them solves the problem entirely.

We think that ***foresight*** is *the process of decision making for complex systems with the human element concerning their possible future behavior.* This process applies individual methods in a certain sequence establishing definite interrelations between them. The process is formed with using the more universal methodology known as the ***scenario analysis*** [197].

Note that in international practice the concept of foresight is mainly applied to technology, which led to the wide usage of the term ***technology foresight***. The fact that technological changes in society influence development in other areas is undeniable. Thus, the invention of the steam engine, the discovery of the electric current, and vast changes in the field of information and telecommunications technologies had a profound effect on social development. For example, modern information and telecommunications technologies not only assist radical structural transformations of international trade but also radically change all economic functions of society.

Classical economic theory says that land, capital, and labor are the traditional factors of manufacturing. Under modern industrial conditions, the importance of ***new knowledge*** increases, considerably changing the essence of the production factor. In particular, the development of mathematical methodologies and software is the ***backbone*** in the area of information technologies. Therefore, knowledge becomes the major production factor when new information technologies are being created. At the same time, the development of information technologies has actually resulted in the appearance of a new field in economics—***economy of knowledge***, or the so-called ***weightless economy***.

Taking into account modern tendencies in production-factor transformations for any large company, branch of industry, or any country, it is not only desirable but necessary to develop technology foresight as the fundamental tool of policy and strategy formation, considering the significant changes, new challenges, and big risks that the future holds for humankind.

Moreover, not only countries and nations but also organizations, institutions, and companies, irrespective of their political or economic models and ownership types, require new knowledge to come up with an at least approximate prediction of possible future scenarios, especially in the most important directions of social, scientific, and technological development. This is due to two basic reasons:

1. Not to lag behind science and technological achievements that will determine the social development of the future society.
2. To have the opportunity to timely understand the advantages of those achievements that will allow countries, organizations, or companies to achieve the maximum benefits in competitive struggle by the modern world's rules.

According to data provided by the United Nations Industrial Development Organization (UNIDO), the main coordinator of these projects, today over 40 countries in the world, from the group of developed and those starting on the road of intensive development, have launched national programs on technology foresight. So-called ***regional programs on technology foresight*** deserve special mention. They are aimed at solving global problems, which are very real for some regions of the world, together with the groups of countries connected to these regions.

During the last 10 to 15 years technology foresight has become the obligatory tool for developed countries and for the most reputable international organizations: the European Union, UNIDO, and others. With its help they are solving the problems of short- and long-term planning and strategic decision making concerning the industrial and economic development of countries, certain regions of the world, and large transnational companies and enterprises.

9.2 Scenario Analysis as a Basic Foresight Methodology

Formally by the term foresight we shall understand a complex system with a human factor that may be presented by a company, an enterprise, an industry, or a country as a whole, uniting certain parts of society (humans or any social groups) with technological, ecological, economic, and other components that are characteristics of such systems.

The decision making regarding the future behavior of complex systems with a human factor is connected to a number of features schematically presented in Fig. 9.1.

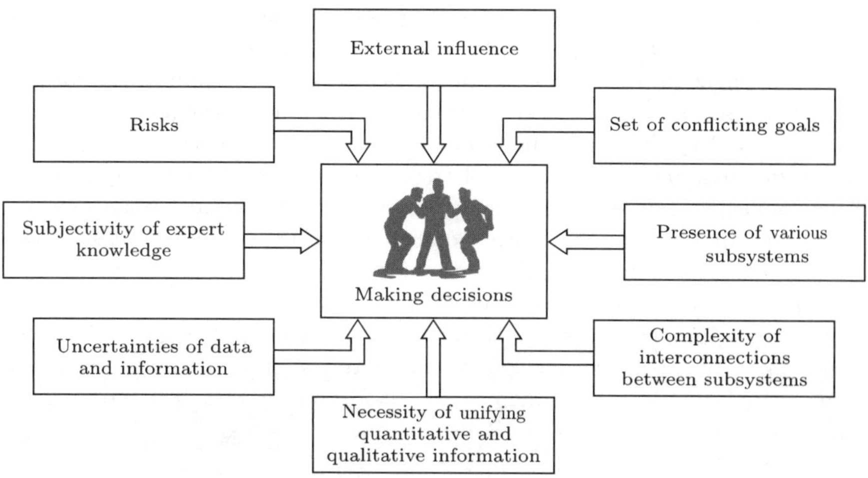

Fig. 9.1. Complex system with human factor

First of all, these systems may be subjected to various external influences and restrictions: legislative, political, economic, etc. Systems with a human factor consist of various subsystems with complex interconnections between them, of both quantitative and qualitative character. They function according to various purposes that, in most cases, clash among themselves.

For complex systems with a human factor considerable uncertainties of data and information are characteristic, and risks of various sorts are inherent in their behavior. Expert judgments concerning their qualitative character always have a subjective character. Thus, taking into account all peculiarities of such systems, regarding their behavior in the future quite definite decisions should be taken in the form of scenarios and strategies of their development.

At this time, there are no universal approaches and methods of foreseeing the future behavior of complex systems with a human factor and events connected to this behavior. There are only attempts at building possible scenarios of their realization. Qualitative and quantitative methods are used to solve such problems in a complicated human-machine procedure. At the same time, the advantages and disadvantages of each method are not taken into account, as well as the peculiarities of the system under investigation, the topology of connections between its inner elements, types of both quantitative and qualitative information that circulates inside the system, contradictions between goals and criteria on which the problem is being solved, the levels of information uncertainty, and so on. Problems of the same type should be solved using the system analysis methodology, which allows one to take into account all required properties and characteristics of the objects under investigation.

It is important to note that the influence of the human factor on foresight results determines the major level of subjectivity of the given procedure. It is connected with a combination of objective knowledge and subjective human attitudes to the object under investigation. Every expert taking part in the foresight process expresses his or her personal opinion in the form of subjective estimation; however, at the same time, he or she should be maximally guided by objective knowledge. At the same time, such a glance into the future requires hypotheses and assumptions. The convergence of objective knowledge and creative assumptions in an interactive human-machine procedure allows one to increase the accuracy of the development scenarios for processes, phenomena, and events under investigation.

Such scenarios can be constructed with the help of the universal set of means and approaches called ***scenario analysis methodology*** [197], which is a complex of mathematical, software, logical, and organizational means and tools for determining the sequence of method application, their interconnections, and the formation of the foresight process in general.

The conventional sequence of the main stages used in scenario analysis methodology realization is presented in Fig. 9.2.

In the first stage one should study the problem and the object of foresight using qualitative and quantitative analysis. After that qualitative and quantitative information is reduced to a unique platform. Then the consequences of the methods will be defined and the interconnection between them will be established. This makes it possible to form a complete foresight process and to develop a group of scenarios of the future behavior of the foresight object (complex system with a human factor).

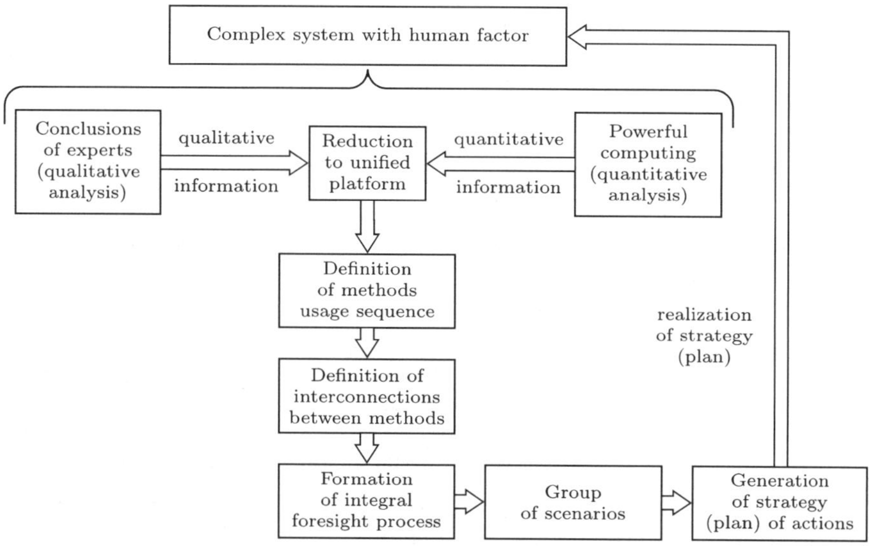

Fig. 9.2. Scenario analysis scheme

Analyzing characteristics and peculiarities of each of the developed scenarios the group of strategic decision makers chooses scenarios that interest them, develops a plan of action according to the foresight object, and ensures implementation of this plan.

In this methodology eight qualitative and quantitative analysis methods are selected and adapted for solving foresight problems. Their comparative characteristics are presented in Table 9.1. These methods are used at four stages of the foresight.

Stage 1: preliminary study of problem
Stage 2: qualitative analysis of problem
Stage 3: scenario construction
Stage 4: analysis and selection of scenarios

Let us examine these stages in more detail.

At the first stage for the preliminary study of the problem it is necessary to analyze its characteristics in more detail, define research directions (focuses or platforms), and form the most important criteria and goals for the given problem. The methods used here in their essence and organizational forms are very simple. But their correct application is of considerable importance, as the loss of important information at this stage will result in significant errors and unjustified expenses during fulfilment of the complex of works on technology foresight.

Let us consider two basic methods that are characteristic of the preliminary stage of the problem study.

Table 9.1. Methods of qualitative and quantitative analysis used in the foresight procedure

No	Method	Conditions and peculiarities of application	Requirement for software facilities
1	Scanning method	Availability of expert groups in different fields of knowledge	Network software and hardware for online operation
2	Brainstorming method	Availability of some expert groups in narrow fields of knowledge; application at first stage of scanning method	Network software and hardware for online operation
3	Delphi method	Availability of expert group in certain fields of knowledge; ease of collection, preparation, and processing of information	Network software and hardware for online operation
4	Cross impact method	Used at first stage of Delphi method; expert estimation of simple and conditional probabilities of events and scenarios; with the help of mathematical programming the computational values of possible event and scenario probabilities are searched	Powerful software facilities for modeling of complex systems
5	Saaty method (Analytic Hierarchy Process)	Should be based on the Delphi method application if the possible scenarios cannot be described verbally; highly qualified specialists in the field of network theory are required; with the help of mathematical programming the computational values of the probability of possible scenarios are searched	Powerful software for information processing in hierarchy network application
6	Morphological analysis method	Specialists in the theory of sets; calculation of possible scenarios, searching for crossings of morphological spaces of characteristic parameters for investigated systems	Powerful software application
7	Scenario writing method	Requires system analysts of high qualification; based upon qualitative analysis methods, introduction of empirical assumptions and system construction of integral scenario	Powerful software application for imitation of scenario modeling
8	Method of Bayes models	Requires analysts in the field of probability theory and random values; to be used in addition to other qualitative analysis methods for checking the degree of confidence of the constructed possible scenarios	Powerful software application

1. Scanning method. The scanning method is used at the first "contemplation" of a given problem in a wide range (focus) of ideas and approaches that may be applied to solve it. As a rule, this method is used for the preliminary study of new problems, regarding which there is no experience with practical solutions.

The procedure, which forms the basis of the scanning method, is as follows:

1. Creation of a group of experts—specialists in a wide area to which the investigated problem may be referred.
2. Each expert should "generate" an idea regarding the solution method of the considered problem or, at least, to characterize the possible approaches to its solution. The experts' opinions are made out as conceptual summaries. At this stage, the ideas formulated by each expert are not discussed. Sometimes it is preferable if the statement is made and the experts' judgements gathered anonymously.
3. Decision makers (some of whom may be in the group of experts) consider all experts' summaries. The purpose of this consideration is ***clustering*** (distribution by groups) of all ideas and opinions, "generated" by the experts.
4. Decision makers from a set of clusters select so-called ***constructive clusters***, which are studied and used at the following stages of foresight.

2. Brainstorm method. This method is intended for deep and intensive study of a problem in narrow directions, ranges, or focuses of ideas and approaches. Narrow directions are defined at first, for example, by clustering on the basis of use of the scanning method.

The brainstorm method procedure is reduced to the following steps:

1. Generate a problem in the given narrow focus (narrow statement of a problem or narrow platform).
2. Create a group of experts—specialists in a narrow field of knowledge according to the formulated problem.
3. In conditions of limited time and an assigned list of criteria the experts should "generate" a set of ideas and approaches to the problem solution for a certain range of possible solutions and relate them to time schedule. A discussion of the formulated opinions, with regard to their advantages and disadvantages, is not to be carried out. Sometimes experts' work is organized anonymously to ensure the independence of opinions of each expert.
4. The "generated" opinions are divided into two temporal groups: those that have urgency in the long term (for example, for a period of no less than 5 years) and those that are urgent in the immediate future, and, consequently, are not used for foresight.
5. Select and document those ideas and approaches to the problem solution that are to be used at subsequent stages of foresight. Such a selection can be carried out by another group of people responsible for decision

making. This group also assigns the list of criteria on the basis of which the experts should generate their opinions.

Let us give some examples of the basic criteria that are characteristic for the creation of new types of competitive technologies using the brainstorming method:

- Existing and potential markets for creating production and technologies used in competition.
- Direct influence on foreign trade.
- Social acceptability of created technologies and major incentives.
- Interests in support of competitive production.
- Vulnerability and risks of industrial production dependence.
- Contribution to national needs in the fields of defense, power generation, ecology, public health, and culture.
- Interconnections with national industry.
- The possibility of introducing technologies into the national industry and their general recognition.
- Stability in conditions of global influence of competition, etc.

Thus, at the stage of preliminary study of the problem it is expedient to use consistently both presented methods. During the first examination of the problem we use the scanning method for the purpose of formulating and clustering all constructive ideas and approaches to its solution in a wide range (focus) of research. After that, for each method, from the predefined directions of research, which correspond to their cluster, the problem is studied taking into account the assigned group of criteria (in narrow focus of research) using the brainstorming method. The selected and documented ideas and approaches to the formulated problem solution are used further for preparation of the solutions at subsequent stages of foresight.

At ***the second stage*** (the stage of qualitative analysis of the problem) we use the other group of methods. Let us present the most frequently used methods.

3. Delphi method. In the literature it is known as the ***expert estimations method***. For more than its 40 years of existence it has developed considerably and acquired various interpretations and wide practical application. But, despite its long life and numerous updates, its main idea remains unchanged. It consists in the necessity of obtaining a conclusion from the group of experts about the future behavior of one or several interconnected characteristics of the considered system. The obtained results are used to construct possible scenarios of behavior of the system under investigation. To this end at the first stage the so-called questionnaires are developed. They are applied in order to obtain optimum estimations of values of the studied characteristics offered by the experts.

The practical application of the Delphi method is reduced to the performance of the following steps:

1. Choose a group of experts according to the character and subject of the considered problem.
2. Formulate the goals that are expected to be achieved as a result of the problem solution.
3. Develop a questionnaire for the formed group of experts.
4. Ask experts' opinion according to the developed form.
5. Process the statistical data of the questioning in order to synthesize new results.
6. Analyze the obtained results of each expert and give to the expert an opportunity to take into account the answers and conclusions of the whole group.
7. If some experts correct their answers after stage 6, then data processing of the questioning is repeated in conformity with step 5.
8. Steps 5 to 7 are carried out until the experts stop correcting their answers. The obtained result is considered to be the ***consensus***. Sometimes, after the repeated performance of steps 5 to 7, the consistency is not achieved. It shows the impossibility of solving the formulated problem or that the selection of experts was not successful. In such cases it is necessary to return to step 1 and to repeat steps 1 to 8.
9. The consensus solution is also analyzed by the experts for its interpretation and to work out the development of scenarios of the system under investigation.

4. Cross impact method at the first stage is also based on the use of expert estimations concerning events that could describe the future for a certain time interval. For example, trying to investigate the future development of the telecommunications industry, first, it is necessary to define those important events that would reflect the scenarios of development of this industry in the fullest way.

As soon as the major events Π_i, $i = 1, \ldots, N$, capable of adequately characterizing the scenarios of the future in a concrete sector of an industry are determined, the probabilities of realization of each of these events are estimated. They are defined as simple or initial probabilities $P_0(\Pi_1), \ldots, P_0(\Pi_N)$.

The following step is the construction of a so-called matrix of cross impact. It has dimensions $N \times N$, where N is the number of chosen events. This matrix determines the interference of events.

Let the foresight procedure concern the development of some conditional branch of industry, for which four important future events were defined.

Let the first event (Π_1) influence the third event (Π_3) based on the opinions of experts with whom consultations were conducted. The second event (Π_2) influences the fourth (Π_4), which, in turn, influences the second one (Π_2). After that the probabilities of definite events are estimated: $P\Pi(\Pi_i)$, $i = 1, \ldots, N$.

For this purpose the methods of complex system modeling, in particular simulation modeling, stochastic modeling, and others, are used. After the probability estimations $P\Pi(\Pi_i)$, $i = 1, \ldots, N$ are found, one more procedure of modeling will be carried out with an application of the same methods of obtaining estimations for realizing each possible scenario $p_c(C_j)$, $j = 1, \ldots, 2n$, the quantity of which is determined by the number $2n$.

Let us note that the given method at the first stage is actually based on the application of the Delphi method, the results of which are used for calculation of probability estimations of possible future development scenarios according to the described procedure. A set of the most probable scenarios also serves as the basis for the foresight of the future.

5. Analytic hierarchy process. This method was developed by the American mathematician Thomas L. Saaty and is based on his known results in the field of ***"Nonstructured Decision Making"*** [152–155]. In spite of the fact that the Saaty method at early stages of application (the early 1990s) was considered only as an auxiliary tool for decision making, later on it was used for solving problems of "visualization of the future", which makes it rather interesting for technology foresight.

Unlike the other methods used in this area of knowledge, the idea of the Saaty method consists in an obligatory condition of ***"focusing"*** or ***"convergence"*** to something unified in relation to the conclusions of experts and actions of numerous executors of complex prospective process. This means that in this case the method is based on the "casual" dependencies of processes, laid down in the foundation of development of future scenarios.

The application of the method is based on the use of so-called ***hierarchical networks*** in the construction of a model intended for calculating the probabilities of each possible scenario in the future. And at the first stage a group of experts managing the research must define, at least verbally, the likely future scenarios. The probability of occurrence or the degree of uncertainty of each possible scenario is determined by application of algorithms of this method and a model of hierarchical networks. The model of hierarchical network used for this type of problems is shown in Fig. 9.3.

6. The method of morphological analysis was offered in 1969 by the Swiss mathematician and astronomer Fritz Zwicky for the study of new geometrical forms of created technological systems. The method is based on the application of the system approach, and in this respect requires identification of so-called ***describing parameters*** Π_i, $i = 1, \ldots, K$ for systems under investigation.

Thus, first of all, it is necessary to define a group of characteristic parameters that to the greatest extent influence the attainment of new qualities of a created system.

After selection of the describing parameters, it is necessary to define sets of values of each of them or so-called ***possible morphological spaces***.

The common possible morphological space for a created system is defined as the crossing of possible morphological spaces for all describing parameters $\mathrm{M}\{\text{system}\} = \mathrm{M}\{\Pi_1\} \cap \mathrm{M}\{\Pi_2\} \cap \cdots \cap \mathrm{M}\{\Pi_K\}$.

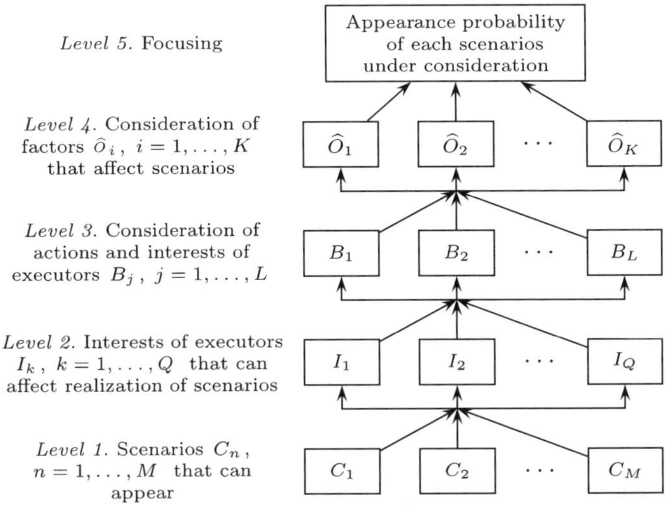

Fig. 9.3. Saaty hierarchical network structure

From the possible morphological space of a created system the so-called morphological space of real achievement of a system is singled out in which only those elements will remain that can truly be achieved. Thus, the technology foresight problem with an application of the method of morphological analysis is solved when the elements of possible morphological spaces for various describing parameters are compatible, i.e., the crossing of these spaces is not an empty space $\text{M}\{\text{system}\} \neq \varnothing$.

7. Scenario-writing method. At ***the third stage***, an empirical nine-step procedure of scenario construction presented in Fig. 9.4 can be used. In the literature this procedure is known as ***"The Method of Scenario Writing"***. It was developed by the known expert in the technology foresight from Manchester, UK, David Loveridge [84].

And, at the ***fourth stage***, the last one, the scenarios are presented to a group of people who are to take strategic decisions (Fig. 9.5), and a comprehensive analysis of these scenarios is carried out in accordance with the following procedure:

- Determine the reality and feasibility level for each scenario.
- Estimate event probability based on scenarios.
- Estimate risks connected with each of the scenarios.
- Construct simulation models.
- Select most acceptable scenarios based on above indicated criteria.

The realness and realizability level of each scenario (confidence level) is determined on the basis of searching for "contrary instance" or "antievent" for each event of the scenario. These "contrary instance" or "antievent" excludes the possibility of realization of the examined event. If such antievents

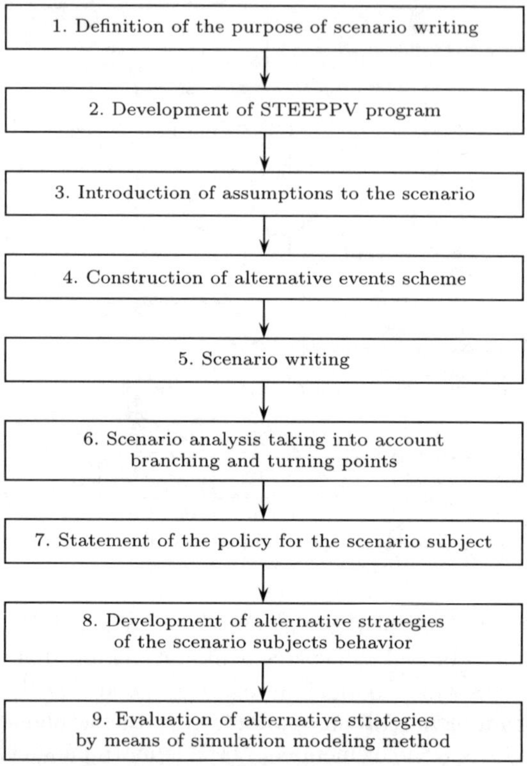

Fig. 9.4. Scenario-writing procedure

are found, the confidence level for the investigated scenario decreases. The procedure is based on the application of methods of ***combinatorial mathematics*** of variants of enumeration of combinations.

8. Bayes model method. An estimation of each scenario probability in the scenario analysis methodology is done using the Bayes method that determines the conditional probability of one or another event:

$$P^c\left(\frac{C_j}{\Pi_1,\ldots,\Pi_N}\right) = \frac{P^0(C_j)\,P^c\left(\frac{\Pi_1}{C_j}\right)P^c\left(\frac{\Pi_2}{C_j,\Pi_1}\right)\cdots P^c\left(\frac{\Pi_N}{C_j,\Pi_1,\ldots,\Pi_{N-1}}\right)}{\sum_{i=1}^{2n} P^0(C_i)\,P^c\left(\frac{\Pi_1}{C_i}\right)P^c\left(\frac{\Pi_2}{C_i,\Pi_1}\right)\cdots P^c\left(\frac{\Pi_N}{C_i,\Pi_1,\ldots,\Pi_{N-1}}\right)}, \quad (9.1)$$

where $P^c(\cdot)$ is the conditional probabilities of scenario N realization and $P^0(\cdot)$ is the a priori probability of each scenario.

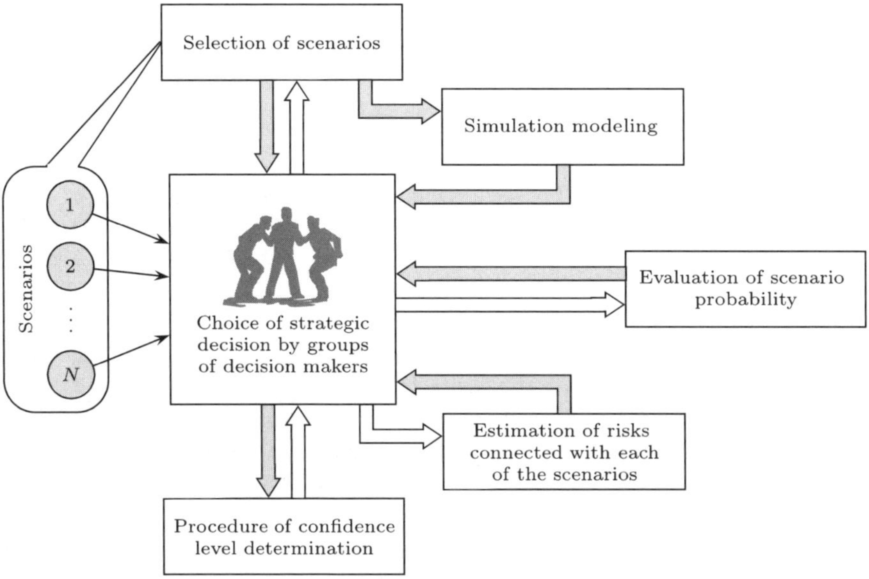

Fig. 9.5. Scheme of scenario submission to the group of people making strategic decisions

The application of the method is reduced to the following steps.

1. For a given problem of technology foresight the possible scenarios of the future for a certain time interval C_j, $j = 1, \ldots, 2n$ are formulated. This formulation or description is done either by application of other methods of technology foresight or verbally (description of what may take place). Such a description should correspond to two necessary conditions:
   a) The considered scenarios should mutually exclude each other, which means that the occurrence of one of them necessarily predetermines the impossibility of occurrence of others.
   b) All probable scenarios of the future should consider the whole spectrum of what may take place in the future more completely, i.e., to be maximally complete and exhaustive.
2. We shall carry out an estimation of a priori or initial probabilities of occurrence of each of these scenarios $P^0(C_j)$, $j = 1, \ldots, 2n$ using all available information about the possible scenarios of the future. It can be done, for example, by application of the Delphi method.
3. Define the list of the most important events Π_i, $i = 1, \ldots, N$ that may take place when the possible future scenarios are formulated.
4. Estimate repeatedly the probability of each of the formulated scenarios taking into account the most important events, determined in item 3, and using the widely known Bayes formulas. The obtained probabilities are called corrected and conditional by their character.

5. For visualization of the tendencies connected to each of the possible future scenarios, it is expedient to represent the obtained results graphically. On the basis of analysis of these results with involvement of experts for making final conclusions about which of the scenarios under investigation are most realistic.

Risk estimation with regard to each scenario is performed on the basis of using methods of multifactor risk analysis, presented in the previous chapters, for various combinations of factors and events forming the scenario.

9. Simulation modeling. This is done using a wide arsenal of methods and algorithms of modeling in order to conduct general research on each of the scenarios for various combinations of the external factors and probabilities of realization of events forming the scenario.

The choice of the most acceptable scenario is made by a group of people making strategic decisions taking into account the above-mentioned criteria and characteristics, and also other considerations pertaining only to the competence and responsibility of these people.

9.3 General Procedure of Expert Estimation in Technology Foresight Problems

Let us formalize the expert estimation problems and develop the general procedure to solve them [122, 138].

Contensive statement of problem. Let a complex system under investigation consist of the finite set of subsystems that differ in their properties, characteristics, and goals and in the character of their behavior.

It is required to develop a group of development scenarios of a complex system under investigation on the basis of the rational organization of each subsystem behavior and coordination of their interconnections.

To develop such scenarios and to achieve the set goal, it is necessary to solve a number of problems, among which system problems of multicriterion estimation of each subsystem's state, characteristics, and other functional elements of the active function cycle are very important. The peculiarity of the considered problems of estimation is the existence of latent properties whose indicators cannot be measured, calculated, or determined directly by other tools. Moreover, estimation in foresight may concern latent properties of the system under investigation characterized by information uncertainty and determined by future uncertain situations. The estimation of latent properties may be performed by a person on the basis of knowledge, experience, intuition, and foresight. However, under uncertain conditions, a person is not capable of making an exact estimation, but he or she can produce an estimation only with a certain degree of assumption and confidence in that estimation. That is why one must use formal expert estimation procedures that would strengthen a person's capability of solving such problems.

Table 9.2. Gradation s of scale for estimation of jth parameter of nth object

Level s number	Qualitative characteristics of sth level	Quantitative characteristics of sth level	Expert estimation of possibility of realizing sth level
1	Extremely low	$[0, 0.1]$	*Enter*
2	Very low	$[0.1, 0.25]$	*Enter*
3	Low	$[0.25, 0.4]$	*Enter*
4	Medium	$[0.4, 0.6]$	*Enter*
5	High	$[0.6, 0.75]$	*Enter*
6	Very high	$[0.75, 0.9]$	*Enter*
7	Extremely high	$[0.9, 1]$	*Enter*

Mathematical statement of problem. Let the system under investigation consist of the finite set of objects $O_0 = \{ O_n \mid n = \overline{1,N}\}$. Each object $O_n \in O_0$ is characterized by a finite set Q_0 of indicators Q_j, $Q_0 = \{ Q_j \mid j = \overline{1,J}\}$. The group E_0 of experts E_k, $E_0 = \{ E_k \mid k = \overline{1,K}\}$ is formed. Each expert $E_k \in E_0$ carries out estimation in the online mode on the basis of the information provided in the form of a scale of qualitative and quantitative estimation of indicators (Table 9.2). An expert performs estimation independently with no information on other experts' estimations. The estimation of the jth indicator of the nth object by the kth expert is defined for each level $s \in S_0$, $S_0 = [1, S]$ as a fuzzy variable $\widetilde{Q}^k_{njs} = \langle Q^k_{njs}, \mu^k_{njs}\rangle$. The estimation of the jth indicator for every $s = \overline{1,S}$ is formed by a computer system of experts' opinion survey in the online mode in the form

$$\widetilde{Q}^k_{nj} = \{\langle Q^k_{njs}, \mu^k_{njs}\rangle \mid n \in N_0,\, j \in J_0,\, s = \overline{1,S}\}, \quad N_0 = [1, N],\; J_0 = [1, J], \tag{9.2}$$

where μ^k_{njs} is defined by the estimation results of the kth expert on the possibility of realizing estimation Q^k_{njs} for $s = \overline{1,S}$ as given in Table 9.3.

Estimation μ^k_{njs}, for each value $s \in S_0$, $S_0 = [1, S]$, is carried out independently. Each estimation is chosen from the interval $[0, 1]$. The resulting estimation of every jth indicator is determined by normalization of all estimates for $s = \overline{1,S}$.

The estimation by the kth expert of the nth object as a whole is formed by the expert questioning system as an array

$$\widetilde{Q}^k_{n} = \left\{\widetilde{Q}^k_{nj} \mid n \in N_0,\, j = \overline{1,J},\, k \in K_0\right\}, \quad K_0 = [1, K]. \tag{9.3}$$

The expert results for the nth object by all experts are formed by the expert questioning system as an array

$$\widetilde{Q}_{n} = \left\{\widetilde{Q}^k_{n} \mid n \in N_0,\, k = \overline{1,K}\right\}. \tag{9.4}$$

It is required to determine the estimations for all objects O_n, $n = \overline{1,N}$ of set O_0 by all experts $k = \overline{1,K}$.

Table 9.3. Results of kth expert's estimation of jth indicator of nth object

s level number	Qualitative characteristics of sth level	Quantitative characteristics of sth level	Expert estimation of possibility of realizing sth level
1	Extremely low	$[0 \div 0.1]$	0.05
2	Very low	$[0.1 \div 0.25]$	0.15
3	Low	$[0.25 \div 0.4]$	0.35
4	Medium	$[0.4 \div 0.6]$	0.9
5	High	$[0.6 \div 0.75]$	0.45
6	Very high	$[0.75 \div 0.9]$	0.05
7	Extremely high	$[0.9 \div 1]$	0.01

Note that it is desirable to form the quantity of gradations in the expert estimation scale (Tables 9.2 and 9.3) according to a Miller's number equal to 7 ± 2.

Problem-solving method. In the process of real expert estimation in foresight problems, the rational choice of metrics and criteria, on the basis of which indicators of processes or objects under investigation are compared, is important. We take into account that, as a rule, the indicators of object quality take positive values. Many practical applications need direct comparison of indicators related to object quality, for example, when choosing technological equipment, the productivity, processing accuracy, and other indicators directly compared. Similar comparisons are made during the expert evaluation of complex projects. In such conditions, for comparisons of different indicators it is expedient to use the Chebyshev criterion and metrics. Then, the difference measure of any pair of expert estimations of the jth indicator of the nth object will be determined by the following relationship:

$$\widetilde{\rho}_{njs} = \left\{ |\widetilde{Q}^{p}_{njs} - \widetilde{Q}^{q}_{njs}| : p, q \in K_0,\, p \neq q,\, s \in S_0,\, n \in N_0 \right\} .$$

We take into account that, according to the initial information (Table 9.2), the quantitative characteristics of the sth level of the jth indicator of quality, for example competitiveness, for the nth product type depends only on $j \in J_0$, $s \in S_0$. Therefore, for every $n \in N_0$, $p, q \in K_0$, the condition $Q^{p}_{njs} = Q^{q}_{njs}$ is fulfilled, where p and q determine the numbers assigned to experts in group K_0. Hence, the difference of any pair of expert estimations (9.2) is determined only by the difference in experts' estimations and is characterized by the relationship

$$\widetilde{\rho}_{njs} = |\mu^{p}_{njs} - \mu^{q}_{njs}| \, .$$

Based on the above explanations and the introduced metrics, the solution to the problem can be reduced to a sequence of the following procedures.

1. Form the general estimation of the jth indicator of the nth object based on the estimation results by all $k = \overline{1,K}$ experts.
2. Find limiting interval boundaries for the jth indicator of the nth object using the following relationships:

$$\widetilde{\rho}_{nj}^{+} = \max_{p,q \in K_0} \left\{ \max_{s=\overline{1,S}} |\mu_{njs}^{p} - \mu_{njs}^{q}| : n \in N_0 \right\},$$

$$\widetilde{\rho}_{nj}^{-} = \min_{p,q \in K_0} \left\{ \min_{s=\overline{1,S}} |\mu_{njs}^{p} - \mu_{njs}^{q}| : n \in N_0 \right\}.$$

3. Find the center of an interval for the jth indicator of the nth object as half of the sum of interval boundaries

$$\hat{\widetilde{\rho}}_{nj} = \frac{\widetilde{\rho}_{nj}^{+} + \widetilde{\rho}_{nj}^{-}}{2}.$$

4. Find the limiting boundaries relative to the interval center for the jth indicator of the nth object type with the help of the following relationships:

$$\rho_{nj}^{+} = \max_{p \in K_0} \left\{ \max_{s=\overline{1,S}} |\hat{\widetilde{\rho}}_{nj} - \mu_{njs}^{p}| : n \in N_0 \right\},$$

$$\rho_{nj}^{-} = \min_{p \in K_0} \left\{ \min_{s=\overline{1,S}} |\hat{\widetilde{\rho}}_{nj} - \mu_{njs}^{p}| : n \in N_0 \right\}.$$

5. Execute procedures from 1 to 4 for the nth object for all other indicators $j \in J_0$.
6. Form the general estimation $\widetilde{Q}_n$ using relationship (9.4) for the nth object using its estimation results (9.3) by all $k = \overline{1,K}$ experts.
7. Execute procedures 1 to 6 for all the remaining objects $n \in N_0$.

Note that in the absence of information exchange between experts, it is most likely that the limiting interval boundaries for each indicator of each product type will turn out excessively wide. Therefore, the experts' estimations can become distributed almost uniformly over the whole interval or concentrate in just three areas: at the interval center and near its two boundaries (bottom and top), which reflects the opinions of cautious pragmatists, optimists, and pessimists, respectively. For real decision making, such results are usually unsuitable. Therefore, the correction of results is required based on coordination and clarification of experts' opinions. Two basic types of correction are possible: direct reduction of the interval's size by concentration or uniform distribution of estimates to the interval's center or by concentrating nonuniformly distributed estimations to the center of three areas, which are determined by the answers of the cautious pragmatists, optimists, and pessimists, respectively.

For this, based on the grouped sample theory, we shall define the mean of distribution and variance of the expert questioning array for the jth indicator of the nth object. Let us take into account a number of peculiarities of the initial data formation. Thus, due to independent estimations of the jth indicator by experts, the obtained results for each sth level form the grouped data, which are defined by the relationship

$$\widetilde{Q}^k_{njs} = \left\{ \langle Q^k_{njs}, \mu^k_{njs} \rangle \mid n \in N_0,\, j \in J_0,\, s \in S_0,\, k = \overline{1, K} \right\} . \qquad (9.5)$$

To calculate a parameter of the jth indicators, it is convenient to take the interval center $\hat{\widetilde{\rho}}_{nj}$ as the reference. Then, the average of distribution and variance are defined as follows:

$$\mathsf{M}\widetilde{Q}_{nj} = \hat{\widetilde{\xi}}_{nj} = \hat{\widetilde{\rho}}_{nj} + \frac{1}{K} \sum_{s=1}^{S} k_s \left| \hat{\widetilde{\rho}}_{nj} - \mu^{k_s}_{njs} \right| ,$$

$$\mathsf{D}\widetilde{Q}_{nj} = \sigma^2_{nj} = \frac{1}{K} \sum_{s=1}^{S} \left| k_s \cdot \left(\mu^{k_s}_{njs} \right)^2 - \left(\hat{\widetilde{\xi}}_{nj} \right)^2 \right| ,$$

where k_s is the number of experts whose estimation by the sth scale gradation falls within the limits of the middle interval with the center $\hat{\widetilde{\rho}}_{nj}$ and boundaries ρ^+_{nj}, ρ^-_{nj}.

Based on the given relationships, using the Delphi method [14] or other method of qualitative analysis, it is possible to perform the correction of the expert analysis results.

9.4 Scenario Analysis Information Platform

When solving foresight problems, in reality, one should use different methods of qualitative analysis in unified human-machine procedures. It should be done taking into account the advantages and disadvantages of each method, peculiarities of the system under investigation taking into account a connections between the inner elements, the character of information circulating in the system (quantitative or qualitative), the contradictoriness of criteria by which the problem is solved, and other aspects. It is possible to solve such problems only by using the ***system approach***, where the totality of properties and characteristics of the system under investigation, methods, and peculiarities of procedures used for their creation are taken into account.

For example, the most efficient system technique among the considered methods of qualitative analysis is the application of two groups of methods at two stages of the foresight process. At the first stage, it is expedient to use synchronously the Delphi method by combining Saaty methods, cross impact, or morphology analysis to find possible future scenarios, depending on the peculiarities of the problem. At the second stage, one should use the Bayes model method to find the most realistic scenario.

Based on a comparison of the characteristics of the qualitative analysis methods and the application, advantages, and disadvantages of each of them, foresight-problem researchers must choose the optimal combination of methods and find the correct order of their usage, taking into account all requirements of the systems under investigation and peculiarities of the problems being solved.

Thus, all methods considered in Sect. 9.2 by definition have a qualitative character, irrespective of the model's complexity and peculiarities of computational procedures. This is connected with the fact that the initial data for them are conclusions of experts and analysts involved in solving concrete technology foresight problems. Such conclusions are always made on the basis of ***knowledge, experience, intuition,*** and ***common sense*** of professionals of a given field. That is why all results obtained using the considered methods are only ***approximations*** of what will actually happen. As a result, the combined powerful human-machine procedures may be considered an efficient and powerful toolkit for ***constructing scenarios***, probably closer to the present events and future scenarios that we are trying to predict. Thus none of the results obtained in such a way may be considered as established fact in the future. At the same time, the construction of these scenarios may be based on system application of four groups of methods at different stages of the foresight process. The typical foresight cycle is shown in Fig. 9.6.

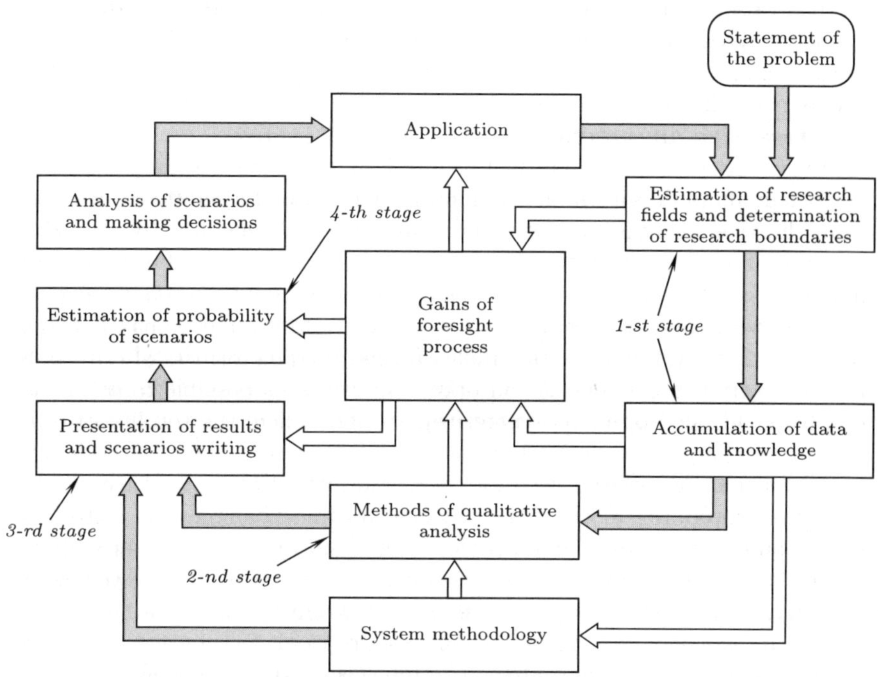

Fig. 9.6. Typical foresight cycle

When solving foresight problems it is important to find experts among the most skilled specialists in the specified subject domain, and also to use advanced mathematical support and powerful information technologies. This should guarantee the highest accuracy and adequacy of the possible future scenarios [196, 197, 206].

It should be remembered that the human factor influence on foresight results brings considerable subjectivity to this procedure. Procedure becomes both intuitive and rational. The main point lies in combining in foresight both objective and subjective attitudes to the object under investigation. This subjective component of foresight arises mostly in connection with assumptions that one should make in the process of scenario construction, on the basis of definite knowledge or doing it intuitively.

Methods and means used in foresight should reflect the connections between objects and subjective factors, proceeding from the fact that foresight may be done by experts or by groups of people by consultations with the interested audience or even with certain groups of the country's population.

The indicated characteristics of foresight determine its duality. In fact, the intuitive component of foresight does not depend on this process itself. However, it is possible to integrate different foresight processes into established programs. Let us consider the so-called ***foresight triangle*** [84] (Fig. 9.7) by an illustration of the ratio between ***objective knowledge*** and ***subjective creative work***, which comprise two vertices of a triangle. The third vertex of the triangle is connected with the interpretation of the results of interaction between knowledge and creative work in the process of policy formation by ***interactive alignment***.

Every expert taking part in the foresight activities may express his or her personal opinion in the form of subjective reasoning; however, as is seen from the triangle, the foresight procedure should be maximally guided by objective knowledge. At the same time, a glance into the future requires making assumptions. As a rule, the compromise between knowledge and creative work may result in a great "correction" of views, which should be taken into account in the policy-formation process. It will influence the interactive alignment or "bridging" between radical views and existing, acknowledged opinions regarding a strategy of any given studied process.

In the triangle we can see the disposition of different methods used in foresight procedures. Each of them is placed at a definite distance from the corresponding vertices. For example, science fiction, which is closely connected with a utopian vision of the future, is placed closer to the vertex "Creative work", while at the same time scenario testing with the help of objective tests and their presentation to a group of decision makers is near the vertex "Objective knowledge". The interior of the triangle shows us that a combination of methods may be used in the process of formation of experts' opinion and the presentation of results. These results must attain interactive alignment with the requirements of the interested audience, responsible party,

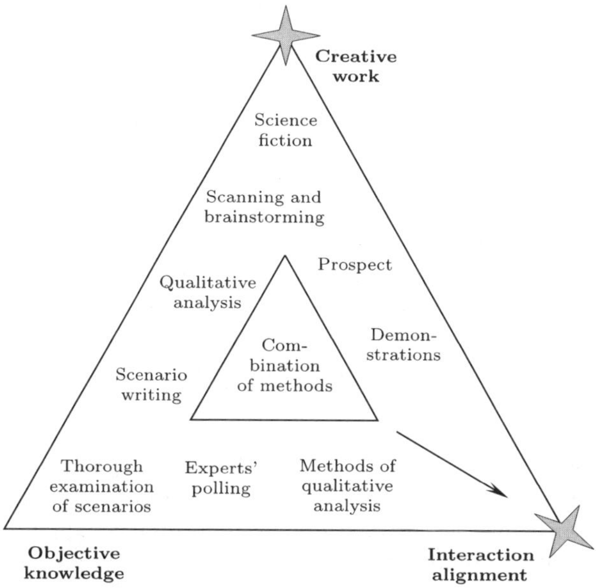

Fig. 9.7. Typical foresight triangle

or the society as a whole in order to make the foresight program gain confidence. Under such conditions it may be used in forming the corresponding policy.

Using both objective knowledge and creative assumptions in interactive human-machine procedures allows one to increase the credibility of scenarios of studied processes, phenomena, and events. The above-mentioned process is realized on the basis of creating and using a universal set of means and approaches, called the ***Scenario Analysis Information Platform*** (***SAIP***) [193, 203, 205, 206]. This platform is a complex of mathematical, software, logical, organizational, and technological means and tools for foresight process realization on the basis of a person's interaction with specially created software. It should be noted that in solving practical problems of foresight on the basis of SAIP, as a rule, the estimates are obtained in the online mode via the Internet.

The software used in this platform is the network-based information system of decision making for future scenario construction, including a special mathematical apparatus and a convenient and flexible user interface. With the help of the Internet, this interface allows one to involve experts from any part of the world, quickly obtain and process their decisions, and organize communication, interaction, and information exchange among them. All that, together with the employed analysis and expert information processing methods, allows us to obtain the proved high-quality solutions for problems related to each specific expert estimation. Based upon the chosen methods of system development, the main components of SAIP were implemented in the Java language. Today, Java technologies are used for the development of corporate systems to support decision making, Internet-based systems, and system working in a heterogeneous environment. This approach has a number of considerable advantages for the development of distributed systems of a client-server type. First of all, this is belongs the complete object-oriented component model that does not depend on the computer architecture and installed operating system, which allows any Internet user to use this system without thinking about hardware compatibility. In the process of development, special attention should be paid to further application of the system on the Internet with the possibility of involving experts from all over the world. Any Internet-oriented information system must take into account the peculiarities of working with users who are far apart from each other. In particular, while conducting a survey among experts who live in different parts of the world, one should take into account the difference in time zones (the system should be accessible in the online mode 24 hours a day), differences in the languages being used, and also differences in the level of the experts' computer literacy. Therefore, internationalization and multilingual support functionality is provided in SAIP. The Web-based interface will allow remote users to use the software that is most familiar to them.

The described information platform of system analysis is implemented as a distributed cross-platform system built on ***client-server*** architecture. The functional system architecture is shown in Fig. 9.8. The component approach allows us to present the system in the form of the following hierarchy levels.

The system technological control level carries out a dialog with the user. It contains the system control and expert interaction modules via Internet in the online mode. The plotting module is intended for visual presentation of expert examination results or any other analytical information. In addition, the experts and remote users make use of the operating system and Internet browser familiar to them. The multilingual support module will allow one to dynamically form a Web interface for remote users in their preferred language.

The system organization level implements the main algorithms of the system based on the series of functionally independent data processing methods, namely:

- It contains the mechanisms for system administration, project and expert estimation control tools, problem domain determination, and technology foresight method implementation tools.
- It is responsible for user management, the levels of access to the system for users, and project management.
- It includes all the necessary procedures for expert examinations: definition and formalization of the problem in the subject domain, invitation of experts and determination of their competence level, setting and realiza-

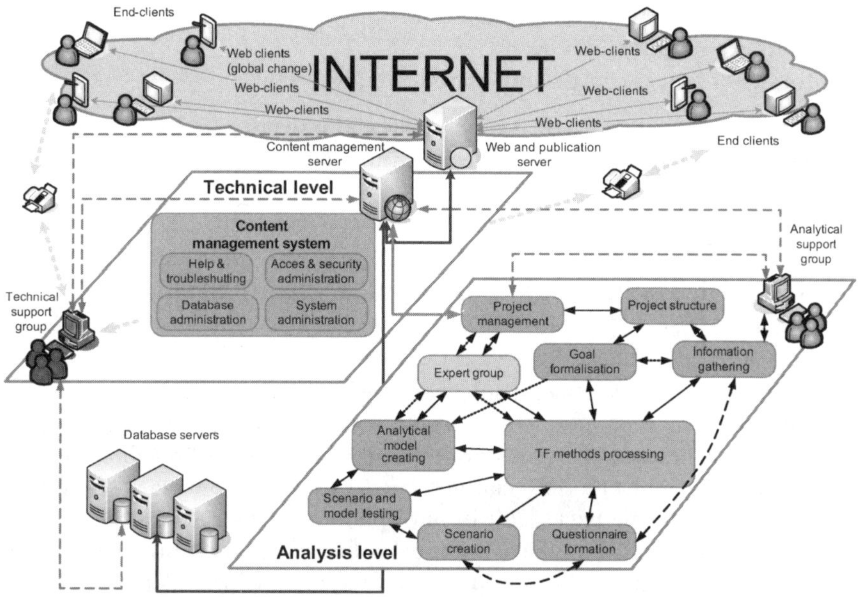

Fig. 9.8. Scenario Analysis Information Platform architecture

tion of expert examination rounds, surveying the experts and processing their answers.

- It carries out the formalization of the problem on the earlier developed algorithms, a definition of the problem expert estimations, determination of the questionnaire structure, scale construction, questionnaire generation in accordance with the assigned structure, and assigning the objects of expert examination.
- It contains a set of mathematical methods of information processing based upon fuzzy logic.

The data storage level uses one of the 12 most popular relational database management systems and is based on the objective-relational approach to data transformation, which allows for taking into account the advantages of both object-oriented and relational approaches. The main objectives of the data storage subsystem are as follows:

- Storage of information about all system users and about experts who are involved in expert examination in the online mode.
- Storage of data for description of the objects under study.
- Storage of hierarchy structure and questionnaire content.
- Storage of information connected with the realization of expert examinations: about all stages and expert examination results.
- Storage of experts' answers at all stages of expert examination and results of their processing.

- Storage of the proximity measure of answers and the chosen scales of expert examination.
- Storage of housekeeping and auxiliary information.

Thus, the described SAIP is a powerful and universal toolkit for foresight procedure fulfillment in different decision-making problems and for strategic planning.

9.5 Role of Technology Foresight System Strategy in Innovation

The modern stage of the world's development is characterized by high globalization rates of economic, social, ecological, and other processes. Global processes have created the new development effect, which French economist Godet has fully and precisely characterized by a short phrase: "The future ceased to look like the past" [39]. In our opinion, the essence of this effect is based on the fact that the typical approaches and methods of prognosis aimed at investigation of gradual evolutionary development cannot detect and foresee changes in processes that are quick in time and discontinuous. At the same time, such processes characterize the modern global dynamics of the world system. Such dynamic characteristics are conditioned by the continuous influence of a semistructured, multilevel, hierarchical, practically unlimited set of continuously changing positive and negative interconnections, interdependencies, and interactions of various processes, factors, and situations. Such dynamics results started to reveal themselves as "unforeseen and unpleasant consequences" [32].

Under the existing conditions of world dynamics, *innovation activity* becomes the determining tendency of economic and social progress. In more than 40 countries, the transition has occurred from the spontaneous development of selected industries to the planned formation and implementation of national programs of innovation development in industrial and social areas. However, such a transition not only results in positive qualitative changes in a country, but it also creates completely new organizational, technological, scientific and technical, and other problems, the solution to which is a necessary condition for realizing the programs and projects of innovation development [114,121,165,193].

The practical necessity has emerged in new systemically coordinated principles, approaches, and methods of unveiling possibilities and evaluating the prospects and trends in civilization and international economic development, as well as the directions and possibilities of innovative development of a given country. In some countries and international organizations, the new apparatus of qualitative and quantitative foresight of different dynamic processes has started to intensively develop; this is generally referred to as scientific-technological foresight or technology foresight. The practical necessity of developing technology foresight methodological apparatus in innovation development in Ukraine is demonstrated in [195, 196, 206].

The experience of developed countries shows that the transition to innovation development opens new possibilities in solving problems in the manufacturing and social fields of a given country under conditions of high globalization rates in world processes. The strategy of the country's innovation development has the main role in realizing these possibilities. Obviously, this strategy should determine the main development directions of strategically important industries, the prospects of competitiveness products, the sensible use of natural, human, industrial, and scientific resources, and the technological potential of the country, and the strategy should consider the needs of the world market of high technologies and hi-tech products.

While developing a strategy, one should consider intercommunications, interdependencies, and interactions between different processes, factors, and conditions of domestic and world markets of high technology that substantially affect the results of innovation activities. This will result in the practical necessity of solving certain systemic problems that allow one to determine their positive and negative influences and choose rational directions of innovation activities. The complexity of dynamics and the uncertainty of properties in given processes and factors require prior investigations using models that define the existing interconnections between the processes and practically required interconnections between the innovation activity subjects. A structural model (Fig. 9.9) defines the structure and goals of system relations between the main organizational innovation activity subjects (industry, science, education) themselves and between them and the market of intellectual products [121].

It should be specially noted that the innovation activity strategy of a country is totally different from the development program of that country under planned economy conditions. This is due not only to the underlying differences between the market economy and the planned economy, but also to the peculiarities of innovation activities under modern conditions of high dynamism and globalization of international processes. The main difference consists in the conceptual uncertainty; it is conceptual in the sense that, unlike information uncertainty, such uncertainty is understood as the unified system of uncertainty, ambiguity, and contradictoriness of interconnected and interdependent elements of a set of uncertainties of various types [120]. This set contains the uncertainty of development goals and perspectives of innovation product competitiveness; the dynamics of the demand and distribution markets of competitive products and active competitors' opposition; the situation risk uncertainty in the dynamics of developing, producing, distributing, and using the innovation products.

Under these conditions, the development and realization of the innovation activity strategy on the national, industry, or business scale should be based on new principles that systemically take into account the actions of

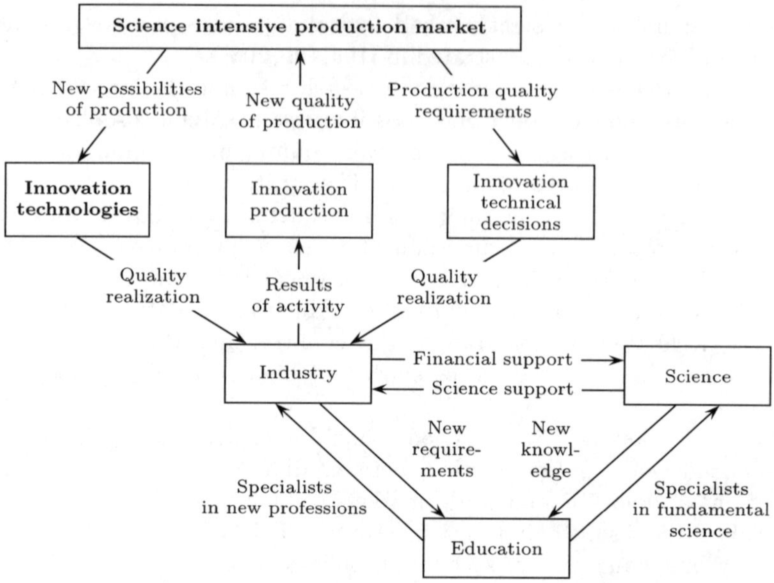

Fig. 9.9. System subject correlation structure of innovation activity

conceptual uncertainty, high dynamism of the modern market, and time and risk factors. The following list of principles is suggested [140]:

- Rationality of selecting solutions and actions in the dynamics of forming and implementing a strategy.
- Reasonable care and rational risk when forming and realizing strategic decisions.
- Rational simultaneous use of intuition, experience, and knowledge of humans (experts, system analysts, managers, and/or decision makers) and the computational and intellectual abilities of computers.
- Rational simultaneous use of abilities of interactive and iterative procedures when producing and justifying solutions.
- Rational compromise between the level of correctness and validity of an operative decision and the amount of time and other resources spent to form and validate this decision.

To realize these principles, we need to base the strategy development of a nation's innovation activities on the technology foresight strategy results, because the standard foresight approaches and results do not guarantee the practically accepted completeness and reliability of the results.

According to the main goals and tasks of innovation activities, technology foresight processes should be based on the strategy and methods of expert procedures [107]. However, expert estimation and realization in technology foresight differ completely from the typical expert procedures by a multitude of attributes and properties, mainly due to the requirements set by the goals, tasks, and peculiarities of innovation activities. At the same time, they have common requirements due to the factors related to every type of innovation activity.

Technology foresight expert-procedure requirements. Technology foresight expert procedures in innovation activities should meet the following requirements [204]:

1. Provide the sufficient compliance of research goals with real conditions of innovation activities.
2. Reach the practically acceptable reliability of the expert estimation of the innovation activity objects under investigation under real conditions of conceptual uncertainty.
3. Give each expert the opportunity to realize his or her valuable individual moral and psychological qualities, knowledge, intuition, and foresight as well as the possibility to prove and substantiate his or her opinions.
4. While forming the database, do not ignore any individual expert estimation based on the average behavior statistics for grouped data, substantial divergence from other estimations, or presence of only one estimation.
5. Reveal the full range of opinions (from pessimistic to optimistic) under every quantitative index, qualitative, and latent characteristic of the object under investigation.
6. Under conditions of group coordination of the final results, provide conditions for realizing the combined abilities of members of the expert group that will allow them to find a rational compromise in the process of forming and agreeing on estimations while rationally considering every expert's individual opinion and competence level.

Development of principles of technology foresight strategy. The formulated requirements raise a number of completely new problems related to the strategy of technology foresight; among these problems, first of all, we should underline those related to the development of new expert estimation principles in technology foresight. A technology foresight strategy should not only satisfy the new requirements, but also correspond to new conditions in innovation activities and correct itself in synchrony with the changing market for the corresponding product type. Adjustments of one's strategy should be adapted to dynamic market changes while preserving the strategic goals in the national manufacturing. In turn, strategic goals should be coordinated with the long-term development trends of the world market and the development dynamics of the corresponding national industry.

The necessity of developing new principles of expert procedures is dictated not only by the listed earlier factors and peculiarities of innovation activities, but also by the essential differences between technology foresight conditions and typical conditions of expert estimations. Typical expert procedures are oriented toward intuitive execution of the logical analysis of objects or processes of reality on the basis of comparing, ranking, systematizing, grouping, measuring, classifying, and other procedures [107]. In particular, the classification may be performed as the ranking of objects by increasing (or decreasing) values of a given parameter the quantitative (or qualitative) value of which is known for every object, and may be measured or calculated. Other expert procedures are implemented in the same way. Essentially, the ability to obtain initial information with the required completeness level means that an expert (each expert individually or the expert group as a whole) knows a priori that the given expert evaluation procedure can certainly be performed. Moreover, when information is provided with the use of expert evaluation results, it is possible to make a valid statement about the feasibility of implementing the given technologies, products, or other products, processes, and actions under investigation. Such conditions are satisfied in practice for a wide range of problems. From the theoretical viewpoint, this approach corresponds to *the potential realization principle*, which is accepted in intuitive logic [23]. The main point is that an expert knows a priori, proves, or postulates that the procedure is potentially realizable, and that is why he can ignore limitations in space and time margins. Furthermore, in the case of implementing the potential realization principle, the conditions are satisfied for *the invariance principle* of intuitive logic.

The main point of the invariance principle is the following: if the validity of any statement, opinion, or conclusion is determined or proved, it will holds in the future too [23]. Note that the conditions and principles mentioned earlier allow us not only to simplify the solutions to many important theoretical and practical problems, but it also allows us to create various automatic systems for classifying, comparing, measuring, and rejecting various products, making it possible to exclude human participation in expert procedures.

However, fundamentally different conditions are typical for *technology foresight expert procedures.* The high competition dynamism of innovation products in the world market place has created fundamentally different conditions of innovation activities, which are characterized not only by the conceptual uncertainty of market dynamics, but also by the multifactorial risk of overdue realization and quick obsolescence of an innovation product proposed in any project, and also due to the absence of technological possibilities for product realization realization. In particular, information incompleteness and uncertainty of many properties and peculiarities of attitudes toward the innovation product in the market are typical for an innovation project, for example, the information on how potential buyers and competitors feel about this product. Thus, the technical foresight expert procedures cannot be implemented for a typical expert estimation, which requires complete and valid information about the object under investigation. Therefore, technology foresight expert evaluation procedures should not only conform to the fundamentally new requirements, but it should also contain the fundamentally new characteristics and principles of organization and realization of expert procedures.

Expert procedures in technology foresight should be organized and realized on the basis of technologies that allow the insufficiency and uncertainty of the initial information about the investigated innovation object to be supplemented and compensated with knowledge, experience, intuition, and human foresight. In this case, the expert estimation results become dependent on many new risk factors that, in the group strategy of a typical expert estimation, are practically excluded. We need to emphasize that, among these factors, the expert estimation results depend on the peculiarities, knowledge, experience, and ability of a customer to determine the expert estimation goals and problems, a manager to choose the expert estimation strategy and form experts' groups, and, what is most important, the results depend on the abilities, knowledge, experience, intuition, and foresight of each expert. Note that, under conditions of uncertainty, the dependence arises on external factors and time, on the amount and level of research in scientific areas that have a direct or indirect effect on the innovation object under investigation. The essence of such dependence is the fundamental possibility to change estimations of an innovation object during a comparatively short period of time. If something is not known today, that does not mean that it will be impossible to realize an innovation object in the future. It may become known, possible to realize, and very useful and needed tomorrow. The most impressive example is a personal computer. But it is not excluded that tomorrow someone proves that a given innovative technological solution or a product is unquestionably impossible to realize.

Therefore, conditions and factors that exclude a priori postulation of potential realization of an innovation project or product take place in innovation activities. And, as a result, the possibility is excluded of realizing the invariance principle in time for the correctness of an expert statement, opinion, or conclusion. It follows that other principles that take into account its conceptual uncertainty must work for innovation activities; therefore, they differ completely in their properties from the principles of intuitive logic that are the potential realization principle and invariance principle [23].

Based on the expert procedure analysis of innovation activities, the following principles can be proposed. Instead of the potential realization principle, *the possible realization principle* is suggested [204].

Possible realization principle. For certain innovation objects (scientific ideas or technical solutions, or projects of industrial products, or manufacturing technologies), the initial expert estimation results cannot guarantee their practical feasibility or prove the impossibility of being realized.

This principle postulates that for the listed innovation objects, based on the expert estimation results of the presented information, a reliable estimation cannot be obtained a priori that would allow one, for the object under investigation, being grounded and valid, to exclude the possibility of being unrealizable. The estimation for an innovation project retains the uncertainty of the conclusion about the possibility of realization, until, theoretically or experimentally, the possibility of technical or technological realization is proved for the product.

Instead of the truth invariance principle, which postulates the invariability of the theoretical or technical statement, judgement, conclusion, or opinion about the object for a comparatively long time, a completely different principle is needed. Such a necessity is brought about by the previous principle and the innovation activity practice, as expert estimations under conditions of conceptual uncertainty cannot stay the same for a long time. In scientific research and experimental design processes, not only new knowledge about a product under development is accumulated, but the conception about the product's characteristics, use, and application areas may change; new inventions, technical solutions, and other know-how may emerge. Thus, the new principle must reflect the probabilistic characteristics of invariance in time of the initial estimation results of an innovation object, and that is why we shall call it the *probabilistic invariance principle* [204].

Probabilistic invariance principle. Initial expert estimation results of certain innovative ideas or technical solutions, industrial products, or manufacturing technologies are probabilistic and do not guarantee that they will be saved in time.

This principle postulates that initial expert estimation results obtained under conditions of conceptual uncertainty as positive or negative findings, proposals, or recommendations are not invariant and may substantially change, be confirmed, or be disproved as time passes. Therefore, we do not exclude in a certain time frame both safekeeping of truth expert statements, opinions, or findings, and the possibility of disproving them.

Based on principles described earlier, we shall formulate some postulates on the properties of technological foresight processes and results of innovation activities. Let us present them as the following assertions [204].

Assertion 1. During technology foresight or expert research of a new product of an innovation project that does not have analogies and prototypes of experimentally confirmed inventions or already existing objects, it is essentially impossible to generate an accurate estimation of the project's practical feasibility. Under such conditions, the estimation can define only the probability of potential opportunity of the project's practicability. The proof of potential practicability or impracticability of an innovation product of the project as a whole or in part should be based on theories confirmed a priori in practice or based on the first realized experiments whose reliability can be verified many times.

Assertion 2. The incompleteness and uncertainty of the initial information about an innovation object of research under conditions of conceptual uncertainty can be supplemented and/or compensated only partially by the knowledge, experience, intuition, and foresight of a person. A more complete disclosure of the conceptual uncertainty is possible on the basis of the first computational and technical experiments that are based on the situations of the theories, confirmed by practice a priori, whose reliability can be proven many times.

The assertions not only follow from the properties of conceptual uncertainty and the basic limitations of human knowledge, but are also proved by Godel's incompleteness theorem [182].

Assertion 3. The incompleteness and uncertainty of the initial information about a researched object under conditions of conceptual uncertainty can change in time when researching the object and accumulating knowledge in the appropriate subject domain.

The assertion follows from a person's ability to enlarge his or her knowledge over time and is proved by Godel's theorem on consistency.

Assertion 4. The incompleteness and uncertainty of the initial information about a researched object under conditions of conceptual uncertainty allow one to generate expert opinion on the practicability of the researched unrealized innovative technical solution, technology, or product only with a certain subjective degree of reliability and confidence of the expert.

Assertion 5. Incompleteness and uncertainty of the initial information about the innovation object of research do not allow guaranteeing, in time, the invariance of the results of their experts' estimation. However, the invariance of the validity of the expert judgments, opinions, or conclusions, and opportunity to refute them, are not excluded regardless of how much time has passed.

Assertion 6. For innovation ideas, technical solutions, or industrial technologies that do not have analogies and prototypes, an opportunity to be implemented or the occasion to be shown essentially unfeasible to implement cannot be excluded. The proof of potential practicability or impracticability of such innovation objects should be based on theories, a priori confirmed by practice, or on the first realized experiments whose reliability can be proven multiple times.

Technology foresight expert estimation procedure requirements. Assertions given earlier prove that it is possible for innovative objects that do not have analogies and prototypes of objects that are actually produced to be producible or nonproducible. Therefore, there is a series of certain practically important requirements for technology foresight expert estimation procedures.

First, while performing these procedures, one must not only estimate the properties, advantages, and disadvantages of innovative objects, but also estimate the probabilities of them being realized in practice. Then, it is advisable, taking into consideration the subjectivity of expert estimations, to introduce an additional index that should determine the expert's confidence in the estimation. The introduction of such indices opens the possibility to substantially improve the reliability of the total object estimation, because the group of experts is formed by taking into account the known a priori level of each expert's competence and the possibility of taking this level into account while processing the expert estimation results.

Second, while doing the expert estimation of innovation objects in the technology foresight process, it is advisable to introduce latent indices of project quality estimation and innovation objects proposed in this process. In particular, such indices characterize the *practical necessity, technology possibility, and economy expediency of an investigated object.* In order to obtain the coordinated qualitative and quantitative estimations of these indices, it is necessary to provide each expert with a simple understanding of the content and the sense of estimated characteristics of expert estimation objects. To accomplish this, it is necessary to provide an unambiguous interpretation for these concepts. The following definitions of latent indices of innovation products are proposed.

Practical necessity: the presence of a relatively high market need in innovation products that are proposed in the investigated project or that have a certain demand and sale in domestic and foreign markets.

Technology possibility: the presence or possibility to develop materials and components, equipment, and technologies for serial production of innovation products.

Economy expediency: the presence of real conditions and the proved prospects of the demand and selling market to obtain an acceptable level of technical and economic effectiveness of innovation products.

The main feature of the latent indices of innovation projects and products is their mutually exclusive dependence. It expresses itself in such a mutual systemic coordination of conditions for realizing characteristics that an absence of any of them excludes the necessity of realizing other characteristics. For example, an absence of market demand for innovation production excludes the necessity and expediency of its industrial production. The next main feature is the discrepancy of requirements for the internal and external characteristics of innovation products. This follows from the contrast of technical and economic interests of producers, which are expressed by internal characteristics (prime cost, technical and economic effectiveness, convenience, etc.), and the social and market interests of its potential customers, which are presented by external characteristics (price, aesthetics, quality, usage convenience etc.). Other features are the impossibility of direct qualitative estimation of combined characteristics, as they depend not only on qualitative and quantitative indices of innovation products, but also on sets of different factors. We cannot always control and forecast the structure and level of their influence. Among uncontrolled factors the most important are the various risk factors in the dynamics of the sequence of stages in the product's life cycle. We need to emphasize the importance of risks in the production cycle and the risks of market demand for new products. Their impact may substantially lessen the technical and economic effectiveness of production or even make the production unprofitable.

The formalization and realization of technology foresight expert procedures in innovative activities will be examined using an example of the foresight problem of availability of hi-tech products, suggested in innovation projects [122].

Mathematical statement. A finite set O_o of innovative projects O_n, $O_n \in O_o$, $O_o = \{O_n | n = \overline{1, N}\}$, $N_o = [1, N]$ of hi-tech products is presented for expert estimation. Each project is based on new ideas, which are realized as new technical solutions. Innovative solutions for each project differ and do not have analogies and prototypes. Products belong to the same branch of practical activity but differ in purpose, technical solution, prospective technological and economic effect, and costs for the life cycle. Each product $O_n \in O_o$ is characterized by finite set $Q_{on} = \{Q_{nj} | j = \overline{1, J_n}\}$ of indices Q_{nj} of quality.

It is **required** to quantitatively and qualitatively estimate the practical necessity, technological possibility, and economic expediency for each hi-tech product proposed in these projects.

Principal differences of the problem from typical expert problems. The formulated problem principally differs from a typical expert problem in many peculiarities and characteristics, which is conditioned by the specificity of given innovation activities. First, the problem is formulated only in the verbal description without formalizing the indicators on which basis the products in innovation projects can be estimated and compared. In particular, there are no characteristics defining the practical necessity, technical possibility, and economic expediency of product realization. The direct comparison of products by their quality characteristics is excluded, because they differ fundamentally in their purposes. Moreover, it is noted that innovative solutions for each project do not have analogs and prototypes; therefore, even demonstrative comparison of projected products with actually existing objects based on quality is excluded. Second, characteristics of such objects are to be qualitatively and quantitatively justified. They are defined by a variety of objective and subjective factors. Moreover, potential customers of innovation products accept these factors in different ways; in particular, they will be indispensable to some customers and unnecessary to others.

The list of such fundamental differences in the compared problems may be substantially widened, but the indicated comparisons are enough to draw a conclusion about the necessity of a fundamentally new approach to formalization and realization of expert procedures in an innovation activity process. First of all, we must develop mathematical models that will make it possible to conduct quantitative comparisons of basic object characteristics, presented for expert estimation. Such a comparison will be made on the basis of latent indicators, which define the most important product characteristics under conditions of global competitiveness and the essential variability of world market. Among such characteristics of innovation production, it is necessary, above all, to pay attention to the competitiveness and prospects of demand.

9.6 An Example of the Foresight Problem Solution for Multicriterion Estimation of Innovation Objects

Let us consider a practical ***example*** where the technology foresight methodology is used to solve problems of multicriterion estimation of innovation object characteristics.

Purpose and peculiarities of the example. The basic principles of expert estimation and coordination of judgements given in Sect. 9.3 demonstrate the main ideas of expert technology in foresight problems.

The purpose of this example is to disclose the possibilities of the expert estimation procedure while solving real practical problems in conditions of market competition.

The initial information for expert estimation is provided by the developer of the project in the form of an explanatory note, technical proposal, conceptual design, or other documentation. On this basis, the customer prepares additional documentation for experts, where the expert estimation goals and objectives, basic information on objects, indicators, and criteria of their estimation are stated.

Statement of multicriterion estimation problem. ***Let*** the proposals on the development of some objects $O_n \in O_0$, $n = \overline{1,N}$ be given to the customer. Assume $N = 4$. Each object O_n is estimated by a set Q_0 of indicators Q_j, the compositions and quantity of which depend on the purpose and properties of the objects under investigation. In the considered problem, the quantity of indicators $Q_0 = \left\{Q_j \mid j = \overline{1,J},\ J = 4\right\}$ is limited.

It is **required** to estimate the properties of the objects on the basis of the following latent indicators: K_1 is the indicator of the market competitiveness of a product, K_2 is the indicator of the availability of demand for a product on the market, K_3 is the indicator of technological complexity of the product's manufacturing process, K_4 is the indicators of economic efficiency of the product's sales cycle. Furthermore, all indicators have a double index $K_{11}, K_{12}, \ldots, K_{43}, K_{44}$, where the first index defines the order number of an object under investigation and the second defines the order number of an indicator.

The group E_0 of experts E_k is formed according to the purpose of the researched objects, goals, and objectives of expert estimation. To perform the expert estimation, the quantity and composition of experts are defined by the chief of expert estimation and agreed upon with the customer. In this case, it is accepted that $E_0 = \left\{E_k \mid k = \overline{1,K},\ K = 16\right\}$.

Formation of the resulting estimation of each parameter K_{11}, K_{12}, ..., K_{43}, K_{44} of each object under investigation is carried out by normalization of results for each level on the scale $s = \overline{1,S}$.

The main purpose of this example is to develop a rational classification of expert estimates. Taking into account a wide range of experts' opinions on each indicator, for provide a compact distribution of a whole spectrum of estimations it is expedient to limit the quantity of classes by the minimal value of the Miller interval $s \Rightarrow [5,9]$. Under such conditions the following classes are formed [204]: class Ω_1 of strongly pessimistic estimates; class Ω_2 of rationally pessimistic estimates; class Ω_3 of rationally pragmatic estimates; class Ω_4 of rationally optimistic estimates; class Ω_5 of highly optimistic estimates. Let us consider the mathematical description of the classes, using class Ω_1 as an example. The model of this class is presented as a sequence of relationships.

$$
\begin{aligned}
\Omega_1 &= \left\{\Omega_1^{n_1} \mid n_1 = \overline{1, N_1}\right\} , \\
\Omega_1^{n_1} &= \left\{\tilde{Q}_{n_1}^{k_1} \mid n_1 \in N_{01},\ k_1 = \overline{1, K_1}\right\} , \\
\tilde{Q}_{n_1}^{k_1} &= \left\{\tilde{Q}_{n_1 j_1}^{k_1} \mid n_1 \in N_{01},\ k_1 \in K_{01},\ j_1 = \overline{1, J_1}\right\} , \\
\tilde{Q}_{n_1 j_1}^{k_1} &= \left\{\langle\tilde{Q}_{n_1 j_1 s_1}^{k_1}\rangle \mid n_1 \in N_{01},\ k_1 \in K_{01},\ j_1 \in J_{01},\ s_1 = \overline{1, S_1}\right\} , \\
\tilde{Q}_{n_1 j_1 s_1}^{k_1} &= \left\{\langle Q_{n_1 j_1 s_1}, \mu_{n_1 j_1 s_1}^{k_1}\rangle \mid n_1 \in N_{01},\ k_1 \in K_{01},\ j_1 \in J_{01},\ s_1 \in S_{01}\right\} ,
\end{aligned}
$$

where the restrictions on the main indicators $Q_{n_1 j_1 s_1}$ and $\mu_{n_1 j_1 s_1}^{k_1}$ are defined by the relationships:

$$
\begin{aligned}
&Q_{s_1}^{-} < Q_{n_1 j_1 s_1} \le Q_{s_1}^{+}, \quad n_1 \in N_{01}, \quad j_1 \in J_{01}, \quad s_1 \in S_{01} , \\
&\mu_{s_1}^{-} \le \mu_{n_1 j_1 s_1}^{k_1} \le \mu_{s_1}^{+}, \quad n_1 \in N_{01}, \quad k_1 \in K_{01}, \quad j_1 \in J_{01}, \quad s_1 \in S_{01} . \qquad (9.6)
\end{aligned}
$$

In this model sets K_{01}, N_{01}, J_{01}, and S_{01} are respectively the subsets of sets K_0, N_0, J_0, and S_0 and consist of the variables k_1, n_1, j_1, and s_1, whose values satisfy conditions (9.6). The mathematical description of the other classes is fulfilled in a similar way.

The interval boundaries for indicators $Q_{n_1 j_1 s_1}$ and $\mu_{n_1 j_1 s_1}^{k_1}$ are formed taking into account the peculiarities of the multilevel scale. For classes $\Omega_1, \ldots, \Omega_5$, taking into account their interconnections, which are characterized by the following conditions:

$$\Omega_p \cap \Omega_{p+1} = \varnothing, \quad p = \overline{1,4}, \qquad (9.7a)$$

$$\Omega_p \cap \Omega_{p+1} \ne \varnothing, \quad p = \overline{1,4}, \qquad (9.7b)$$

the boundary-limit variants for indicators $Q_1, \ldots, Q_5$ are presented in Table 9.4.

The boundary values of indicators $\mu_{n_1 j_1 s_1}^{k_1}$ for all classes $\Omega_1, \ldots, \Omega_5$ are defined by the boundary values of interval $[0,1]$. It is conditioned by a wide range of results of expert estimation of the possibility of realizing the investigated indicators at each level of scale $s \in S_0$, $S_0 = [1, S]$, given in Table 9.3. It follows that fuzzy estimation of innovation objects includes two procedures: ***first***, estimation of each innovation object by the assigned indicators conducted by all experts; ***second***, formation of final estimates on each innovation object by the totality of expert estimates.

Table 9.4. Boundaries of estimation classes

Class	(a) $\Omega_p \cap \Omega_{p+1} = \varnothing$, $p = \overline{1,4}$		Class	(b) $\Omega_p \cap \Omega_{p+1} \neq \varnothing$, $p = \overline{1,4}$	
Ω_1	$Q_1^- = 0.001$	$Q_1^+ = 0.199$	Ω_1	$Q_1^- = 0.0$	$Q_1^+ = 0.25$
Ω_2	$Q_2^- = 0.2$	$Q_2^+ = 0.399$	Ω_2	$Q_2^- = 0.2$	$Q_2^+ = 0.45$
Ω_3	$Q_3^- = 0.4$	$Q_3^+ = 0.599$	Ω_3	$Q_3^- = 0.4$	$Q_3^+ = 0.65$
Ω_4	$Q_4^- = 0.6$	$Q_4^+ = 0.799$	Ω_4	$Q_4^- = 0.6$	$Q_4^+ = 0.85$
Ω_5	$Q_5^- = 0.8$	$Q_5^+ = 0.999$	Ω_5	$Q_5^- = 0.8$	$Q_5^+ = 1.0$

Estimation is carried out by each expert independently for each jth indicator of the nth project on the basis of the questionnaire as presented in Table 9.5. An expert must give a quantitative estimation within the limits of an interval $[0, 1]$ in each window replacing the word "enter".

On the basis of the received information in the first round of expert estimation, all 16 experts have given their opinions in the questionnaire (Table 9.5).

Due to the large quantity of received results, in Tables 9.6–9.9 only the results of estimation by all experts of one object O_1 are presented: for indicator K_{11} of market competitiveness, indicator K_{12} of product demand, indicator K_{13} of technological complexity, and indicator K_{14} of the economic efficiency of the product's sales cycle.

In some cases, it is expedient to create a more complex questionnaire structure; in particular, together with the estimation of the possibility of realizing each level of the expert evaluation scale (Table 9.3) we additionally introduce the indicator of the experts's confidence level in his estimation accepted on each level of the scale. An example of such a questionnaire, together with the answers of the 11th expert regarding estimation of only indicator K_{12} of market demand for an innovation product O_1 is given in Table 9.10.

Such construction of the scale is aimed at achievement of the main goals of expert estimation: to systemize estimation procedures of different kinds of innovation production and to provide a practically acceptable reliability level of estimations in conditions of conceptual uncertainty. The presented scale structure (Tables 9.3 and 9.10) for the description of quantitative and qualitative characteristics of an object is unified and equally used for estimation of quantitative indicators and qualitative and latent properties. From here follows the possibility to normally form a wide range of opinions on each characteristic of an innovation object, goods, or other production under investigation in the form of qualitative or quantitative information for each gradation level. The introduction of indicator ν_{njs}^k of confidence level of the kth expert in his accepted estimation of the jth indicator of the nth project on the s scale level allows us to take into account individual possibilities, competence, objectivity, and other personal expert properties. According to the expert estimation results such a scale allows one to determine the further decisions and actions of a decision maker, in particular, to

Table 9.5. Example of a questionnaire given to experts for innovation object estimation

Object	Level number s	Qualitative characteristics of sth level	Quantitative characteristics of sth level	K_{11}	K_{12}	K_{13}	K_{14}
O_1	1	Extremely low	$[0.00, 0.10]$	*Enter*	*Enter*	*Enter*	*Enter*
	2	Very low	$[0.10, 0.25]$	*Enter*	*Enter*	*Enter*	*Enter*
	3	Low	$[0.25, 0.40]$	*Enter*	*Enter*	*Enter*	*Enter*
	4	Middle	$[0.40, 0.60]$	*Enter*	*Enter*	*Enter*	*Enter*
	5	High	$[0.60, 0.75]$	*Enter*	*Enter*	*Enter*	*Enter*
	6	Very high	$[0.75, 0.90]$	*Enter*	*Enter*	*Enter*	*Enter*
	7	Extremely high	$[0.90, 1.00]$	*Enter*	*Enter*	*Enter*	*Enter*
				K_{21}	K_{22}	K_{23}	K_{24}
O_2	1	Extremely low	$[0.00, 0.10]$	*Enter*	*Enter*	*Enter*	*Enter*
	2	Very low	$[0.10, 0.25]$	*Enter*	*Enter*	*Enter*	*Enter*
	3	Low	$[0.25, 0.40]$	*Enter*	*Enter*	*Enter*	*Enter*
	4	Middle	$[0.40, 0.60]$	*Enter*	*Enter*	*Enter*	*Enter*
	5	High	$[0.60, 0.75]$	*Enter*	*Enter*	*Enter*	*Enter*
	6	Very high	$[0.75, 0.90]$	*Enter*	*Enter*	*Enter*	*Enter*
	7	Extremely high	$[0.90, 1.00]$	*Enter*	*Enter*	*Enter*	*Enter*
				K_{31}	K_{32}	K_{33}	K_{34}
O_3	1	Extremely low	$[0.00, 0.10]$	*Enter*	*Enter*	*Enter*	*Enter*
	2	Very low	$[0.10, 0.25]$	*Enter*	*Enter*	*Enter*	*Enter*
	3	Low	$[0.25, 0.40]$	*Enter*	*Enter*	*Enter*	*Enter*
	4	Middle	$[0.40, 0.60]$	*Enter*	*Enter*	*Enter*	*Enter*
	5	High	$[0.60, 0.75]$	*Enter*	*Enter*	*Enter*	*Enter*
	6	Very high	$[0.75, 0.90]$	*Enter*	*Enter*	*Enter*	*Enter*
	7	Extremely high	$[0.90, 1.00]$	*Enter*	*Enter*	*Enter*	*Enter*
				K_{41}	K_{42}	K_{43}	K_{44}
O_4	1	Extremely low	$[0.00, 0.10]$	*Enter*	*Enter*	*Enter*	*Enter*
	2	Very low	$[0.10, 0.25]$	*Enter*	*Enter*	*Enter*	*Enter*
	3	Low	$[0.25, 0.40]$	*Enter*	*Enter*	*Enter*	*Enter*
	4	Middle	$[0.40, 0.60]$	*Enter*	*Enter*	*Enter*	*Enter*
	5	High	$[0.60, 0.75]$	*Enter*	*Enter*	*Enter*	*Enter*
	6	Very high	$[0.75, 0.90]$	*Enter*	*Enter*	*Enter*	*Enter*
	7	Extremely high	$[0.90, 1.00]$	*Enter*	*Enter*	*Enter*	*Enter*

Table 9.6. Expert estimations of index K_{11} of market competitiveness of innovative object O_1

Expert number k	Level number s 1	2	3	4	5	6	7
1	0.05	0.1	0.25	0.45	0.35	0.15	0.05
2	0.15	0.35	0.65	0.45	0.35	0.25	0.1
3	0.15	0.2	0.45	0.5	0.35	0.25	0.1
4	0.15	0.25	0.45	0.55	0.5	0.35	0.25
5	0.1	0.45	0.55	0.5	0.25	0.15	0.05
6	0.2	0.35	0.65	0.55	0.45	0.35	0.25
7	0.2	0.25	0.4	0.55	0.5	0.35	0.15
8	0.15	0.35	0.45	0.6	0.75	0.25	0.15
9	0.2	0.4	0.65	0.6	0.45	0.25	0.1
10	0.15	0.45	0.75	0.65	0.35	0.25	0.1
11	0.2	0.25	0.55	0.7	0.35	0.15	0.1
12	0.65	0.85	0.75	0.7	0.25	0.15	0.1
13	0.35	0.55	0.85	0.75	0.6	0.35	0.2
14	0.4	0.5	0.8	0.75	0.55	0.25	0.15
15	0.4	0.55	0.85	0.8	0.55	0.25	0.1
16	0.25	0.65	0.95	0.85	0.65	0.45	0.25

Table 9.7. Expert estimations of index K_{12} of availability of demand for innovative object O_1

Expert number k	Level number s 1	2	3	4	5	6	7
1	0.1	0.25	0.55	0.45	0.25	0.1	0.05
2	0.05	0.15	0.55	0.65	0.25	0.15	0.1
3	0.1	0.15	0.5	0.55	0.25	0.15	0.05
4	0.15	0.2	0.45	0.55	0.35	0.1	0.05
5	0.1	0.45	0.55	0.5	0.45	0.25	0.1
6	0.15	0.25	0.55	0.45	0.4	0.15	0.05
7	0.05	0.35	0.65	0.55	0.5	0.25	0.1
8	0.15	0.3	0.45	0.55	0.45	0.25	0.15
9	0.25	0.35	0.45	0.65	0.35	0.15	0.1
10	0.15	0.35	0.55	0.45	0.35	0.25	0.1
11	0.05	0.15	0.35	0.45	0.25	0.15	0.1
12	0.25	0.3	0.55	0.5	0.35	0.25	0.15
13	0.35	0.45	0.75	0.65	0.55	0.25	0.1
14	0.3	0.35	0.65	0.6	0.4	0.15	0.05
15	0.4	0.55	0.65	0.8	0.65	0.5	0.1
16	0.25	0.35	0.55	0.95	0.75	0.35	0.15

Table 9.8. Expert estimations of index K_{13} of technological complexity of manufacturing innovative object O_1

Expert number k	Level number s 1	2	3	4	5	6	7
1	0.1	0.15	0.2	0.35	0.45	0.35	0.05
2	0.05	0.1	0.15	0.45	0.55	0.15	0.1
3	0.1	0.2	0.25	0.35	0.45	0.2	0.1
4	0.15	0.2	0.25	0.55	0.6	0.35	0.1
5	0.1	0.25	0.3	0.5	0.55	0.25	0.1
6	0.1	0.15	0.35	0.55	0.5	0.45	0.25
7	0.15	0.25	0.35	0.65	0.55	0.35	0.1
8	0.15	0.2	0.35	0.55	0.65	0.25	0.15
9	0.25	0.3	0.35	0.65	0.55	0.25	0.15
10	0.15	0.2	0.25	0.65	0.45	0.25	0.15
11	0.2	0.25	0.3	0.55	0.6	0.15	0.1
12	0.25	0.35	0.45	0.5	0.55	0.35	0.25
13	0.25	0.35	0.4	0.75	0.6	0.35	0.2
14	0.15	0.2	0.35	0.45	0.55	0.25	0.15
15	0.2	0.4	0.5	0.7	0.6	0.3	0.2
16	0.05	0.2	0.45	0.65	0.75	0.35	0.25

Table 9.9. Expert estimations of index K_{14} of economic efficiency of innovative object O_1 sales

Expert number k	Level number s 1	2	3	4	5	6	7
1	0.1	0.15	0.35	0.45	0.25	0.15	0.1
2	0.15	0.2	0.65	0.45	0.35	0.15	0.1
3	0.2	0.25	0.45	0.55	0.35	0.25	0.15
4	0.25	0.35	0.45	0.35	0.25	0.15	0.05
5	0.1	0.15	0.35	0.55	0.35	0.25	0.15
6	0.3	0.35	0.65	0.55	0.45	0.35	0.3
7	0.2	0.25	0.45	0.5	0.45	0.3	0.2
8	0.25	0.3	0.35	0.45	0.25	0.1	0.1
9	0.25	0.35	0.45	0.55	0.35	0.25	0.15
10	0.15	0.45	0.75	0.65	0.45	0.35	0.3
11	0.15	0.35	0.65	0.75	0.45	0.35	0.25
12	0.35	0.45	0.55	0.65	0.35	0.25	0.15
13	0.25	0.45	0.65	0.75	0.45	0.35	0.1
14	0.15	0.25	0.55	0.75	0.35	0.15	0.05
15	0.25	0.35	0.55	0.65	0.35	0.25	0.1
16	0.45	0.6	0.75	0.65	0.55	0.45	0.35

find the expediency of the collective discussion of expert estimation results. The main purposes of collective discussion may be the system analysis of causes and factors, which lead to an essential difference in estimations of an object by one or each indicator; analysis of the possibility and expediency of finding a rational compromise in estimating an object with considerable differences of estimations of an individual object and experts' confidence level; preference ranking of investigated objects in conditions of great differences

Table 9.10. Estimation of index K_{12} of market demand for innovation product O_1 by 11th expert

Level number s	Qualitative characteristics of sth level	Quantitative characteristics of sth level	Realization possibility estimation of μ^k_{njs} by kth expert	Estimation confidence level ν^k_{njs} of kth expert
1	Extremely low	$[0, 0.1]$	0.05	0.99
2	Very low	$[0.1, 0.25]$	0.15	0.85
3	Low	$[0.25, 0.4]$	0.35	0.65
4	Medium	$[0.4, 0.6]$	0.45	0.97
5	High	$[0.6, 0.75]$	0.25	0.75
6	Very high	$[0.75, 0.9]$	0.15	0.85
7	Extremely high	$[0.9, 1]$	0.1	0.95

in experts' individual estimations regarding each or some indicators on the basis of a collective system analysis of causes and factors that are taken into account by each expert in estimation; other complex situations of expert estimation with simultaneous estimation by the assigned set of indicators and preference ranking of some objects.

The computer realization of innovation-object estimation procedures is based on the following mathematical support. Estimation by the kth expert of the jth indicator of the nth object is formed on the basis of the following relationships:

$$\tilde{Q}^k_{nj} = \left\{ \langle \tilde{Q}^k_{njs} \rangle \mid n \in N_0,\ j \in J_{0n},\ k \in K_0,\ s = \overline{1, S} \right\},$$
$$\tilde{Q}^k_{njs} = \left\{ \langle Q_{njs}, \mu^k_{njs} \rangle \mid n \in N_0,\ j \in J_{0n},\ k \in K_0,\ s \in S_0 \right\}, \qquad (9.8)$$

where μ^k_{njs} defines the result of the jth indicator of the kth expert estimation of the nth object at the scale level $s \in S_0$. Integer sets $K_0 = \{ k \mid k = \overline{1, K} \}$, $N_0 = \{ n \mid n = \overline{1, N} \}$, and $J_{0n} = \{ j \mid j = \overline{1, J_n} \}$ characterize respectively the number of experts, the number of given objects, and the number of quality indicators of the nth object. An estimate of the kth expert of the nth object in total is formed on the basis of (9.8) and is presented in the form of an array:

$$\tilde{Q}^k_n = \left\{ \langle \tilde{Q}^k_{nj}, \nu^k_{nj} \rangle \mid n \in N_0,\ j = \overline{1, J_n},\ k \in K_0 \right\},$$
$$\nu^k_{nj} = \left\{ \nu^k_{njs} \mid n \in N_0,\ j \in J_{0n},\ k \in K_0,\ s = \overline{1, S} \right\}, \qquad (9.9)$$

where ν^k_{njs} defines the degree of confidence of the kth expert in an accepted estimate of the jth indicator of the nth object at the $s \in S_0$ level. The expert estimation results of the nth object by all experts are formed on the basis of (9.9), and the obtained results are compiled in an array:

$$\tilde{Q}_n = \left\{ \tilde{Q}^k_n \mid n \in N_0,\ k = \overline{1, K} \right\}. \qquad (9.10)$$

The sequential accumulation of expert estimation results in the form of (9.10) creates the basis for forming estimation results of all objects as the final array:

$$\tilde{Q}_{\Sigma} = \left\{ \tilde{Q}_n \mid n = \overline{1, N} \right\} . \tag{9.11}$$

The procedure for forming estimation results of all objects as the final array is the last stage of expert estimation. The array only creates the informational basis for the next decision-making step. The reason is that, if there is no information exchange between experts, the most probable outcome will be a casual distribution of estimations of each indicator almost throughout the whole interval $[0, 1]$ of each level of estimates. It follows that there is the practical necessity to develop the mathematical apparatus for estimation, coordination, and specification of the expert estimation results.

Taking into account that arrays (9.10) and (9.11) are divided a priori into seven levels, each of which is characterized by a definite interval. It is expedient to use the principles and approaches of the randomly grouped sample [23] for forming the mathematical apparatus of system analysis. We shall form the basic indicators of the sample consecutively starting with the estimation result of the kth expert of the jth indicator of the nth object. The average of distribution for the Q^k_{nj} indicator will be approximated by the sample average $\bar{Q}^k_{nj}$, which is defined by the relationship:

$$\bar{Q}^k_{nj} = \frac{1}{S} \sum_{s=1}^{S} \nu^k_{njs} \bar{Q}_{njs} . \tag{9.12}$$

Here the indicators ν^k_{njs} are random and define the confidence level of the kth expert in the jth indicator estimate for $j \in J_0$ at a priori known values $s = \overline{1, S}$ and $k = \overline{1, K}$. The value $\bar{Q}^k_{nj}$ is a consistent unbiased estimate for the general average, which defines the average of distribution as a functional describing the characteristics of a random sequence $\nu^k_{nj1} \bar{Q}^k_{nj1}, \nu^k_{nj2} \bar{Q}^k_{nj2}, \ldots, \nu^k_{njS} \bar{Q}^k_{njS}$. The indicator $\bar{Q}_{njs}$ defines the average interval value of quantitative characteristics of the sth scale level for the investigated jth indicator of an object. Indicators $\bar{Q}_{njs}$ for all $j \in J_0$ are determinate values and are defined by a priori known intervals for different values of $s = \overline{1, S}$, and they do not depend on expert judgments. In particular, for the indicator K_{12} of market demand availability of an object we have (Table 9.7)

$$\bar{Q}_{12s} = \langle 0.05, 0.175, 0.325, 0.5, 0.675, 0.825, 0.95 \rangle, \quad s = \overline{1, S}, \quad S = 7 . \tag{9.13}$$

Here the value 0.05 corresponds to $s = 1$ and defines the average value of the scale interval $[0, 0.1]$; 0.95 corresponds to $s = 7$ and the interval $[0.9, 1]$. The other values characterize the intermediate intervals. Such a choice of interval sequence corresponds to the structure and physical content of the scale in full. It is accepted that in the scale an interval of each level of indicator is coordinated with the practical value of the indicator estimation. The least average value of an interval corresponds to the worst estimate of an indicator, and the largest average value of the interval characterizes the best estimate. The value $\bar{Q}_{12s} = 0.5$ in (9.13) corresponds to the average value of the interval $[0.4, 0.6]$ at $s = 4$ for the indicator K_{12} for the object O_1.

From relatioship (9.12) at the initial conditions (9.13) and from an assumption about the absolute expert's confidence in estimate $\nu^k_{njs} = 1$ we get $\bar{Q}^k_{nj} = 0.5$, which corresponds to the structure and physical content of the scale in full. The introduction of the ν^k_{njs} indicator, which characterizes the confidence level of the kth expert, allows one to improve values $\bar{Q}^k_{nj}$ for $j \in J_0$ at different values $s = \overline{1, S}$, $S = 7$ and $k = \overline{1, K}$, $K = 16$.

We consider an example of estimating $\bar{Q}^k_{nj}$, taking into account the confidence level of the kth expert according to the data presented in Table 9.10.

$$\nu^k_{njs} = \langle 0.99, 0.85, 0.65, 0.97, 0.75, 0.85, 0.95 \rangle, \quad s = \overline{1, S}, \quad S = 7 . \tag{9.14}$$

On the basis of (9.12)–(9.14) we obtain the value of sample average:

$$\bar{Q}^k_{nj} = \langle 0.0495, 0.1488, 0.2113, 0.485, 0.5063, 0.7013, 0.9025 \rangle .$$

The next stage is the formation of the average of distribution for indices μ^k_{njs}. Taking into account the confidence level of the kth expert the average of distribution will be approximated by the sample average in the form of the relationship:

$$\bar{\mu}^k_{nj} = \frac{1}{S} \sum_{s=1}^{S} \nu^k_{njs} \mu^k_{njs} . \tag{9.15}$$

Taking into account the estimation and confidence level of the kth expert, which are presented in Table 9.10 for indicator μ^k_{njs}, we get the following average of distribution value:

$$\bar{\mu}^k_{nj} = \langle 0.1763, 0.2630, 0.3555, 0.5638, 0.3070, 0.1833, 0.0857 \rangle .$$

Relationships (9.12), (9.15), and (9.8) give us the possibility to define the average of distribution for latent fuzzy indicators of innovation objects. As an example, we shall define the average of distribution for the market demand indicator K_{n2} on the basis of its approximation by the sample average as the fuzzy value $\widetilde{K}^k_{n2}$. Estimations for latent indicators of innovation objects are formed on the basis of (9.8) and (9.9) by definition. That is why in order to estimate the indicator K_{n2} of the nth object under investigation by the kth expert the following correspondences are fulfilled:

$$K_{n2} \Rightarrow K^k_{n2} \Rightarrow \langle \widetilde{Q}^k_{nj} \mid j = 2 \rangle \Rightarrow \widetilde{Q}^k_{n2}, \quad \bar{\widetilde{Q}}^k_{n2} \Rightarrow \bar{\widetilde{K}}^k_{n2} . \tag{9.16}$$

From here according to (9.8) and (9.9) we have:

$$\widetilde{Q}^k_{n2} = \langle Q^k_{njs}, \mu^k_{njs} \rangle, \quad j = 2,\ s = \overline{1, S}, \quad S = 7\,. \tag{9.17}$$

Then the average of distribution of the kth expert estimation of the K_{n2} indicator according to (9.16) and (9.17) may be presented as:

$$\mathsf{M}\widetilde{Q}^k_{n2} = \bar{\widetilde{Q}}^k_{n2} = \langle \bar{Q}^k_{n2}, \bar{\mu}^k_{n2} \rangle\,.$$

In the general case the average of distribution of the kth expert estimation result of the random jth indicator of the nth object is characterized by the relationship:

$$\mathsf{M}\widetilde{Q}^k_{nj} = \bar{\widetilde{Q}}^k_{nj} = \langle \bar{Q}^k_{nj}, \bar{\mu}^k_{nj} \rangle\,, \tag{9.18}$$

where $\bar{Q}^k_{nj}$ and $\bar{\mu}^k_{nj}$ are defined by (9.12) and (9.15).

On the basis of (9.18) it is enough just to create a model that allows one to define the average of distribution of the kth expert estimation result of the nth object as a whole on assigned indicators:

$$\mathsf{M}\widetilde{Q}^k_{n} = \bar{\widetilde{Q}}^k_{n} = \frac{1}{J_n} \sum_{j=1}^{J_n} \beta^k_{nj} \bar{\widetilde{Q}}^k_{nj}\,. \tag{9.19}$$

Here β^k_{nj} is the value of the importance parameter of the jth indicator for the nth object that the kth expert defines on the interval $[0, 1]$. In practice the average of distributions as presented in (9.19) is used for preference ranking of some objects. The other variants of importance indicator estimation are possible. In particular, the importance indicator values may be assigned a priori or be corrected in the process of coordinating expert opinions. It is possible to develop another model for evaluation of an object as whole by many indicators. In it the indices are ranked by confidence degree and level. In this case the model is presented as a ranked consistency of indicators.

Model (9.19) is the basis for forming the final expert estimation results of the object under investigation, which should reflect the opinion of all experts. The average of distribution $\mathsf{M}\widetilde{Q}_n$ of the final estimate of the nth object by all experts within the whole group of the assigned indicators is formed on the basis of model (9.19) and is defined by the relationship:

$$\mathsf{M}\widetilde{Q}_n = \bar{\widetilde{Q}}_n = \frac{1}{K} \sum_{k=1}^{K} \chi^k_n \bar{\widetilde{Q}}^k_n\,, \tag{9.20}$$

where χ^k_n is the indicator of the kth expert competence in a practical area determinig the vital cycle of the nth object.

Model (9.20) is the basis for expert estimation of forming the final results of all presented objects. The results are to reflect the advantages and disadvantages of each object and to provide their ranking by an organized sequence of the assigned indicators. In the process of forming and coordinating the expert estimation final results changes and additions are possible in the composition of indicators and criteria of estimation. The expert estimation final result must be a sequence of objects under investigation that reflects their order by preference on the basis of the initial indicators or with the addition of indicators and criteria that estimate the degree of importance, level of importance, and other latent object parameters.

For example, the initial objects are ranked by preference in the form of a sequence:

$$O_{n_0}^{l_1} \succ \cdots \succ O_n^l \succ \cdots \succ O_{n_k}^{l_n} .$$

Here l characterizes the order number of the object in the sequence $l \in L_0$, $L_0 = [l_1, l_n]$, obtained by the expert estimation results; and $n \in N_0$, $N_0 = [1, N]$ defines the initial number of the object that it obtained during expert estimation preparation.

In evaluation of expert estimation quality a very important role belongs to characteristics of position and dispersion of expert estimation sample. Among position characteristics special attention should be paid to the center of distribution, which is described by average sample and approximates the average of distribution of the grouped sample.

The average of distribution is not only the precision center of experts' estimations by $s = \overline{1, S}$ levels for each jth indicator of each nth object, but it also defines the average final estimation in the form (9.19) for the object as a whole. Such estimation is sufficient for preliminary ranking of objects under investigation by preference. However, the average estimation for the object levels the values of some of its indicators. If there are more than two of them, then it may appear that, in ranking, those objects are most important that exceed the others by only one or two indicators but that, at the same time, surrender to the other indicators.

That is why it is necessary to take into account the other sample characteristics, in particular, the characteristics of sample scattering. Among the scattering characteristics it is expedient to single out variance, average absolute deviation, and sample range.

We shall define variance sequentially, starting from the estimation results of the kth expert of the jth indicator of the nth object. For indicator Q_{nj}^k the variance is defined by the relationship:

$$\mathsf{D}Q_{nj}^k = \frac{1}{S} \sum_{s=1}^{S} \nu_{njs}^k \left(\bar{Q}_{njs} - \mathsf{M}\bar{Q}_{nj}^k \right)^2 , \tag{9.21}$$

where $\mathsf{M}\bar{Q}_{nj}^k$ is the average value of interval qualitative characteristics of the sth scale level for the jth indicator. The average of distribution of indicator Q_{nj}^k is defined by relationship (9.12).

Variance of estimates μ_{nj}^k is defined in the same way. Taking into account (9.15) we get the relationship

$$\mathsf{D}\mu_{nj}^k = \frac{1}{S} \sum_{s=1}^{S} \nu_{nj}^k \left(\mu_{njs}^k - \mathsf{M}\mu_{nj}^k \right)^2 . \tag{9.22}$$

The estimation result variance of the kth expert of the nth object as a whole by all indicators is characterized by the relationship:

$$\mathsf{D}\bar{\tilde{Q}}_n^k = \frac{1}{J_n} \sum_{j=1}^{J_n} \beta_{nj}^k \left(\bar{\tilde{Q}}_{nj}^k - \mathsf{M}\tilde{Q}_n^k \right)^2 , \tag{9.23}$$

where $\bar{\tilde{Q}}_{nj}^k$ is defined by relationship (9.18).

The final variance of the nth object estimation by all experts is formulated on the basis of relationship (9.23) as:

$$\mathsf{D}\bar{\tilde{Q}}_n = \frac{1}{K} \sum_{k=1}^{K} \chi_n^k \left(\bar{\tilde{Q}}_n^k - \mathsf{M}\tilde{Q}_n \right)^2 . \tag{9.24}$$

Relationships (9.21)–(9.24) characterize the sample scattering of distribution of expert estimation results. Note that for average samples and variance there are simplified calculation methods that are convenient for defining the center of sample distribution. Sometimes at the beginning of counting out it is better to take the middle of an interval of each jth indicator. In particular, for indicator K_2 at the beginning of counting out one may take the middle of an interval for the $s = 4$ level. Then the point of counting out will be defined by the value 0.5. At the same time, for program realization of the examined calculations, the calculations mentioned above are more convenient.

Among other characteristics of scattering note the practical importance of the sample estimation range. In general, a random sample range characterizes the difference between its maximal and minimal values. In the expert estimation the sample value defines the difference level of experts' opinions while estimating the assigned indicators of objects under investigation. We shall quantitatively define the difference value of experts' opinions while estimating the jth indicator of the nth object for each level of scale $s \in S_0$, $S_0 = \overline{1, S}$ with the help of the relationship:

$$\widehat{Q}_{njs} = Q_{njs}^+ - Q_{njs}^-, \quad Q_{njs}^+ = \max Q_{njs}, \quad Q_{njs}^- = \min Q_{njs} . \tag{9.25}$$

Similarly, the difference between experts' opinions in estimating the jth indicator of the nth object and the object on all assigned indices shall be defined in the following way:

$$\begin{aligned} \widehat{Q}_{nj} &= Q_{nj}^+ - Q_{nj}^-, \quad Q_{nj}^+ = \max Q_{nj}, \quad Q_{nj}^- = \min Q_{nj} , \\ \widehat{Q}_n &= Q_n^+ - Q_n^-, \quad Q_n^+ = \max Q_n, \quad Q_n^- = \min Q_n . \end{aligned} \tag{9.26}$$

On the basis of the obtained normalized results of the estimation for each class of $\Omega_1, \ldots, \Omega_5$, using the Delphi method or any other method of qualitative analysis, we may correct the expert estimation results. The solution to this problem can be reduced to the following procedures.

1. Determine expert estimations $\widetilde{Q}^{k}_{nj}$ of the jth indicator of the nth product type that differ most from the fuzzy average of distribution $\mathsf{M}\widetilde{Q}_{nj}$ or increase the sample distribution in regard to the interval center $\mathsf{R}\widetilde{Q}_{nj}$. We shall determine such estimations on the basis of the following formulas:

$$\overset{\frown}{\widetilde{Q}}{}^{k_s}_{njs} = \max_{k \in K_0} |Q^{k_s}_{njs} - \mathsf{M}\widetilde{Q}_{nj}|, \quad n \in N_0,\ j \in J_0,\ s = \overline{1, S}\,, \tag{9.27}$$

$$\overset{\smile}{\widetilde{Q}}{}^{k_s}_{njs} = \max_{k \in K_0} |Q^{k_s}_{njs} - \mathsf{R}\widetilde{Q}_{nj}|, \quad n \in N_0,\ j \in J_0,\ s = \overline{1, S}\,. \tag{9.28}$$

2. Determine the relationships of values $Q^{k_s}_{njs}$ and $\mathsf{M}\widetilde{Q}_{nj}$ and of $Q^{k_s}_{njs}$ and $\mathsf{R}\widetilde{Q}_{nj}$. As is known, for real problems, tolerances are established that correspond to certain physical conditions of manufacture, operation, application, storage, or other restrictions for the given product. In expert estimation, tolerances are mathematically determined by the value of the highest possible deviation of the obtained estimates from the average of distribution $\mathsf{M}\widetilde{Q}_{nj}$ or from the center $\mathsf{R}\widetilde{Q}_{nj}$ of the allowable interval of indicator values. Therefore, the purpose of the procedure is to determine whether or not there is an agreement between the obtained expert estimates and the assigned allowable deviations. This may be determined by the analysis of relationships among the above-mentioned pairs of values. The analysis allows us reveal the following practically important outcomes of expert estimation results.

   a) The conditions for the agreement of values $\overset{\frown}{\widetilde{Q}}{}^{k_s}_{njs}$ and $\overset{\smile}{\widetilde{Q}}{}^{k_s}_{njs}$ are fulfilled:

$$0 < |\overset{\frown}{\widetilde{Q}}{}^{k_s}_{njs}| \le \overset{\frown}{\beta}_{nj} \mathsf{M}\widetilde{Q}_{nj}, \quad 0 < |\overset{\smile}{\widetilde{Q}}{}^{k_s}_{njs}| \le \overset{\smile}{\beta}_{nj} \mathsf{R}\widetilde{Q}_{nj}\,. \tag{9.29}$$

   b) The condition is fulfilled for $\overset{\frown}{\widetilde{Q}}{}^{k_s}_{njs}$, but not for $\overset{\smile}{\widetilde{Q}}{}^{k_s}_{njs}$:

$$0 < |\overset{\frown}{\widetilde{Q}}{}^{k_s}_{njs}| \le \overset{\frown}{\beta}_{nj} \mathsf{M}\widetilde{Q}_{nj}, \quad |\overset{\smile}{\widetilde{Q}}{}^{k_s}_{njs}| > \overset{\smile}{\beta}_{nj} \mathsf{R}\widetilde{Q}_{nj}\,. \tag{9.30}$$

   c) The condition is fulfilled for $\overset{\smile}{\widetilde{Q}}{}^{k_s}_{njs}$, but not for $\overset{\frown}{\widetilde{Q}}{}^{k_s}_{njs}$:

$$0 < |\overset{\smile}{\widetilde{Q}}{}^{k_s}_{njs}| \le \overset{\smile}{\beta}_{nj} \mathsf{R}\widetilde{Q}_{nj}, \quad |\overset{\frown}{\widetilde{Q}}{}^{k_s}_{njs}| > \overset{\frown}{\beta}_{nj} \mathsf{M}\widetilde{Q}_{nj}\,. \tag{9.31}$$

   In relationships (9.29)–(9.31), it is taken into account that the values $\overset{\frown}{\beta}_{nj}$ and $\overset{\smile}{\beta}_{nj}$ can be set by technical conditions for allowable deviations in product indicators or be defined by the experts in the process of coordinating opinions on concrete expert estimation objects. The allowable deviations with the symmetric tolerances are limited to:

$$0 < |\overset{\frown}{\widetilde{Q}}{}^{k_s}_{njs}| \le 0.5 \mathsf{M}\widetilde{Q}_{nj}, \quad 0 < |\overset{\smile}{\widetilde{Q}}{}^{k_s}_{njs}| \le 0.5 \mathsf{R}\widetilde{Q}_{nj}\,,$$

which is why values $\overset{\frown}{\beta}_{nj}$ and $\overset{\smile}{\beta}_{nj}$ should be limited by the following intervals:

$$0 < \overset{\frown}{\beta}_{nj} \leq 0.5\,, \quad 0 < \overset{\smile}{\beta}_{nj} \leq 0.5\,.$$

3. Execute procedures 1 and 2 for all $n = \overline{1, N}$ product types and for all $j = \overline{1, J}$ indicators of quality of each $n \in N_0$ product type. The necessity of the given procedure directly follows from the statement of the expert estimation problem.
4. Define conditions of termination for the expert estimation. The formally necessary and sufficient condition for the termination of the expert estimation is the realization of all $n = \overline{1, N}$ and $j = \overline{1, J}$ restrictions, which are characterized by relationships (9.28). However, it may happen that the fulfillment of such restrictions is not enough for the rational solution of real problems. In particular, the sample size will be insufficient for a practically acceptable level of credibilility of estimation for one or several product types. One of the possible reasons may be uneven distribution of expert estimation results within the limits of the assigned interval. A similar result is obtained while culling that part of a sample where the values of components $\overset{\frown}{\widetilde{Q}}{}^{k_s}_{njs}$, $\overset{\smile}{\widetilde{Q}}{}^{k_s}_{njs}$ lead to condition (9.29) or (9.30). There are also other issues that may result in a sample volume that does not allow one to obtain the required confidence interval for estimation of objects under study. Therefore, the following procedure may be necessary.
5. Conduct a collective discussion of the expert estimation results by the group of experts and make the decision on whether to terminate the expert estimation or to correct the sample size.
The acceptance of the decision based on the outcome of the discussion completes the first round of the expert estimation results coordination. If the decision confirms the necessity to increase the sample size, the following procedure is performed.
6. Correct the sample size. The correction can be carried out in the following ways:
   - By excluding a set of data $N_0^- \in N_0$, $J_0^- \in J_0$, defined by conditions (9.29) or (9.30), and their subsequent replacement by new estimation results, determined by the limited group of experts.
   - By performing an additional expert estimation for the purpose of increasing the sample size in order to increase the accuracy of parameters $\mathsf{R}\widetilde{Q}_{nj}$, $\mathsf{M}\widetilde{Q}_{nj}$, σ^2_{nj} for all $n = \overline{1, N}$ and $j = \overline{1, J}$.
   - By performing an additional expert estimation for the purpose of increasing the accuracy of parameters only for those $n \in N_0$ and $j \in J_0$ that determine values $\overset{\frown}{\widetilde{Q}}{}^{k_s}_{njs}$, $\overset{\smile}{\widetilde{Q}}{}^{k_s}_{njs}$.

Furthermore, a rational sample size may be defined with the help of the following procedure.

7. Provide a rational sample size allowing one to obtain a practically acceptable confidence interval for all $j = \overline{1, J}$ of each $n = \overline{1, N}$. Based on the corrected results, execute procedures 1–6 and make, at this stage, a final decision about the results of expert estimation on each object under investigation.

On the basis of the developed above apparatus, expert estimation procedure, and algorithm of its realization, the mathematical processing of data provided by 16 experts is presented in Tables 9.6–9.9. In particular, the estimations of parameters $K_{11}, K_{12}, \ldots, K_{43}, K_{44}$ were investigated for objects O_1, O_2, O_3, O_4, presented by 16 experts.

On the basis of the data given in Tables 9.6–9.9, estimations are studied on indices $K_{11}, K_{12}, K_{13}, K_{14}$ of object O_1.

The results of the study for index K_{11} of the market competitiveness of object O_1 are presented below.

CLUSTER N1: Formed by estimations of the 1st expert;
Cluster median: 0.05, 0.1, 0.25, 0.45, 0.35, 0.15, 0.05;
Confidence interval includes estimations of this expert;
Coefficient of agreement in this cluster is $\gamma = 1$.

CLUSTER N2: Formed by estimations of the 2nd, 3rd, 4th, 5th, 6th, 7th, 9th, 10th, and 11th experts;
Cluster median: 0.2, 0.4, 0.65, 0.6, 0.45, 0.25, 0.1;
Confidence interval includes estimations of the 2nd, 7th, 9th, 10th, and 11th experts;
Coefficient of agreement in this cluster is $\gamma = 0.55556$.

CLUSTER N3: Formed by estimations of the 8th expert;
Cluster median: 0.15, 0.35, 0.45, 0.6, 0.75, 0.25, 0.15;
Confidence interval includes estimations of this expert;
Coefficient of agreement in this cluster is $\gamma = 1$.

CLUSTER N4: Formed by estimations of the 12th expert;
Cluster median: 0.65, 0.85, 0.75, 0.7, 0.25, 0.15, 0.1;
Confidence interval includes estimations of this expert;
Coefficient of agreement in this cluster is $\gamma = 1$.

CLUSTER N5: Formed by estimations of the 13th, 14th, 15th, and 16th experts;
Cluster median: 0.35, 0.55, 0.85, 0.75, 0.6, 0.35, 0.2;
Confidence interval includes estimations of the 13th, 14th, and 15th experts;
Coefficient of agreement in this cluster is $\gamma = 0.75$.

For index K_{11} the following conclusion can be made:

THE MAJORITY OF THE EXPERTS (81%) EXPECT A LOW LEVEL OF MARKET COMPETITIVENESS. IN ADDITION, THE OPINIONS OF THREE EXPERTS DIFFER

FUNDAMENTALLY FROM ALL OTHERS; THEY ESTIMATE THIS PARAMETER AS VERY LOW, AVERAGE, AND HIGH, RESPECTIVELY.

According to the data given in Table 9.7, the estimations presented by 16 experts on index K_{12} of market demand for object O_1 are investigated.

CLUSTER N1: Formed by estimations of the 1st, 2nd, 3rd, 4th, and 11th experts;
Cluster median: 0.1, 0.15, 0.5, 0.55, 0.25, 0.15, 0.05;
Confidence interval includes estimations of the 2nd, 3rd, and 4th experts;
Coefficient of agreement in this cluster is $\gamma = 0.6$.

CLUSTER N2: Formed by estimations of the 5th, 6th, 7th, 8th, 9th, 10th, 11th, 12th, and 14th experts;
Cluster median: 0.15, 0.35, 0.55, 0.45, 0.35, 0.25, 0.1;
Confidence interval includes estimations of the 5th, 6th, 8th, 10th, and 12th experts;
Coefficient of agreement in this cluster is $\gamma = 0.625$.

CLUSTER N3: Formed by estimations of the 13th expert;
Cluster median: 0.35, 0.45, 0.75, 0.65, 0.55, 0.25, 0.1;
Confidence interval includes estimations of this expert;
Coefficient of agreement in this cluster is $\gamma = 1$.

CLUSTER N4: Formed by estimations of the 15th expert;
Cluster median: 0.4, 0.55, 0.65, 0.8, 0.65, 0.5, 0.1;
Confidence interval includes estimations of this expert;
Coefficient of agreement in this cluster is $\gamma = 1$.

CLUSTER N5: Formed by estimations of the 16th expert;
Cluster median: 0.25, 0.35, 0.55, 0.95, 0.75, 0.35, 0.15;
Confidence interval includes estimations of this expert;
Coefficient of agreement in this cluster is $\gamma = 1$.

For index K_{12} the following conclusion can be made:

HALF OF THE EXPERTS (50%) ESTIMATE MARKET DEMAND AS LOW AND 31% AS AVERAGE.

According to the data given in Table 9.8, the estimations presented by 16 experts on indicator K_{13} of the technological complexity of object O_1 manufacture are investigated.

CLUSTER N1: Formed by estimations of the 1st, 2nd, and 3rd experts;
Cluster median: 0.1, 0.2, 0.25, 0.35, 0.45, 0.2, 0.1;
Confidence interval includes estimations of the 1st and 3rd experts;
Coefficient of agreement in this cluster is $\gamma = 0.66667$.

```
CLUSTER N2: is formed by estimations of the 4th, 5th, 6th, 8th, 9th,
    10th, 11th, and 14th experts;
Cluster median: 0.1, 0.25, 0.3, 0.5, 0.55,; 0.25, 0.1;
Confidence interval includes estimations of the 4th, 5th, 8th, and
    11th experts;
Coefficient of agreement in this cluster is γ = 0.625 .

CLUSTER N3: Formed by estimations of the 7th expert;
Cluster median: 0.1, 0.15, 0.35, 0.55, 0.5, 0.45, 0.25;
Confidence interval includes estimations of this expert;
Coefficient of agreement in this cluster is γ = 1 ;

CLUSTER N4: Formed by estimations of the 12th, 13th, and 15th experts;
Cluster median: 0.2, 0.4, 0.5, 0.7, 0.6, 0.3, 0.2;
Confidence interval includes estimations of the 13th and 15th experts;
Coefficient of agreement in this cluster is γ = 0.66667 .

CLUSTER N5: Formed by estimations of the 16th expert;
Cluster median: 0.05, 0.2, 0.45, 0.65, 0.75, 0.35, 0.25;
Confidence interval includes estimations of this expert;
Coefficient of agreement in this cluster is γ = 1 .
```

For index K_{13} the following conclusion can be made:

```
HALF OF THE EXPERTS (50%) FORESIGHT A HIGH LEVEL OF COMPLEXITY OF THE
TECHNOLOGY PARAMETER WITH RESPECT TO PRODUCT MANUFACTURE.
```

According to the data given in Table 9.9, the estimations presented by 16 experts on indicator K_{14} of economic efficiency of object O_1's sales are studied.

```
CLUSTER N1: Formed by estimations of the 1st, 3rd, 4th, 5th, 6th, 8th,
    9th experts;
Cluster median: 0.2, 0.25, 0.45, 0.55, 0.35, 0.25, 0.15;
Confidence interval includes estimations of the 3rd, 5th, 6th, and 9th
    experts;
Coefficient of agreement in this cluster is γ = 0.57143 .

CLUSTER N2: Formed by estimations of the 2nd expert;
Cluster median: 0.15, 0.2, 0.65, 0.45, 0.35, 0.15, 0.1;
Confidence interval includes estimations of this expert;
Coefficient of agreement in this cluster is γ = 1 .

CLUSTER N3: Formed by estimations of the 7th, 10th, 11th, 12th, 13th,
    and 15th experts;
Cluster median: 0.25, 0.45, 0.65, 0.75, 0.45, 0.35, 0.1;
Confidence interval includes estimations of the 11th, 12th, 13th, and
    15th experts;
Coefficient of agreement in this cluster is γ = 0.66667 .
```

CLUSTER N4: Formed by estimations of the 14th expert;
Cluster median: 0.15, 0.25, 0.55, 0.75, 0.35, 0.15, 0.05;
Confidence interval includes estimations of this expert;
Coefficient of agreement in this cluster is $\gamma = 1$.

CLUSTER N5: Formed by estimations of the 16th expert;
Cluster median: 0.45, 0.6, 0.75, 0.65, 0.55, 0.45, 0.35;
Confidence interval includes estimations of this expert;
Coefficient of agreement in this cluster is $\gamma = 1$.

For index K_{14} the following conclusion can be made:

THE OVERWHELMING MAJORITY OF EXPERTS (81%) ESTIMATE THE ECONOMIC EFFICIENCY OF THE PRODUCT AS AVERAGE.

In Fig. 9.10 the estimation results on indices K_{11}, K_{12}, K_{13}, and K_{14} of object O_1 are presented. Here the values of a cluster median on the vertical axis are postponed, and on the horizontal axis values of gradation s are postponed.

On the basis of the estimation results, in order to find the common opinion of all 16 experts the average of distribution, variance, and coefficient of coordination are calculated. The values of these characteristics for the

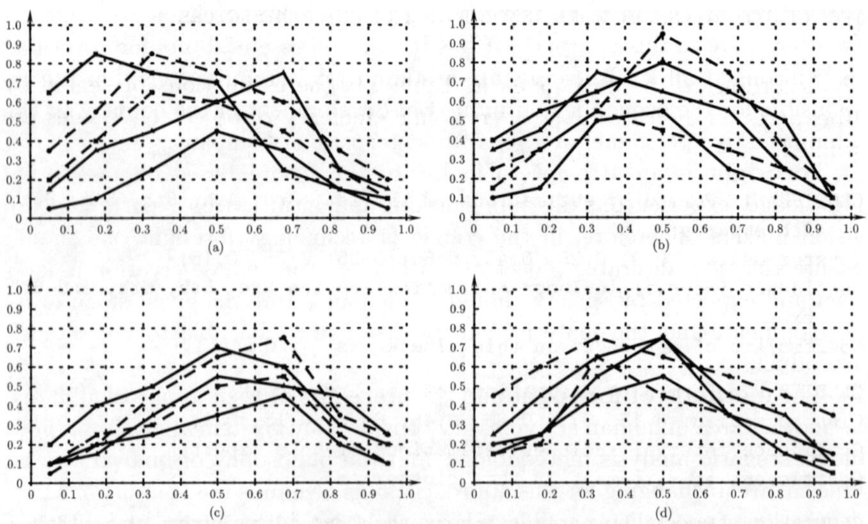

Fig. 9.10. Estimation results on parameters of object O_1: **a** Parameter K_{11} of market competitiveness. **b** Parameter K_{12} of market product demand. **c** Parameter K_{13} of technological complexity of object manufacture. **d** Parameter K_{14} of economic efficiency of object sales. The following classes are presented: 1st class (*circles, solid line*), 2nd class (*squares, dashed line*), 3rd class (*triangles, solid line*), 4th class (*squares, solid line*), 5th class (*circles, dashed line*)

$K_{12}, K_{13}, \ldots, K_{43}, K_{44}$ parameters of O_1, O_2, O_3, O_4 objects are presented in Table 9.11.

The results of the first round of the expert estimation procedure of object O_1 including parameters K_{11}, K_{12}, K_{13}, and K_{14} with recommendations for further actions are presented in Table 9.12. The expert estimation results on three other objects, O_2, O_3, O_4, can be presented similarly.

The final estimation results on all innovation objects O_1, O_2, O_3, O_4 are presented in Table 9.13.

Thus, after conducting the first round of expert estimation, based on judgments of the majority of the experts, objects 3 and 4 have the best prospects from the point of view of development.

By the decision of the expert working group, the decision on conducting further rounds of expert estimations under a similar scheme can be accepted for the purpose of specifying and coordinating of expert estimations to achieve a consensus.

Conclusion. **Foresight** is a complicated process that is characterized by a considerable scope of research of an interdisciplinary nature, arbitrary selection of criteria and objectives, and a significant amount of so-called empiricism and indistinct foresight process that causes certain risks of "not feeling" or "not seeing" an important situation in a scenario of future events or critical technology of the future. However, as has been shown by the experience of all developed countries of the world that are actively fighting for natural resources and markets for selling products and technologies, there is no alternative to this approach. It has become an essential tool for governing institutions at all levels, from state and industrial organizations that are responsible for the economic and industrial development of the country to the administrative personnel of a separate enterprise or company.

Foresight is an attempt to look into the future based on the actual knowledge, experience, skills, application, and intuition of experts and decision makers. Therefore, in the course of technology foresight one should ensure that any desirable object of material culture whose creation is foreseen and expected by society should be demanded at the given stage of its development, and the material culture itself must be able to create such a product.

The **methods of foresight** and the procedure of their application in one or another area of human activity vary considerably. System analysis methodology, scenario analysis methodology, and the philosophy of innovation and renewal are unchanging for this approach. This demands the training of a new generation of specialists capable of large-scale, interdisciplinary, and systemic thinking and able to effectively apply technology foresight with the purpose of innovation development of a country, region, or enterprise.

Some countries, which have only just begun to use the foresight methodology, including Ukraine, may wonder if, due to the methodological complexity and considerable expenses required for its practical imple-

Table 9.11. Values of average of distribution, variance, and coefficient of coordination for $K_{11}, K_{12}, \ldots, K_{43}, K_{44}$ parameters of O_1, O_2, O_3, O_4 objects

Object	Parameter	Class	Average of distribution	Variance	Coefficient of coordination
O_1	Market competitiveness of product K_{11}	1	0.49286	0	1
		2	0.45612	0.00423	0.77778
		3	0.32857	0.00475	0.6
		4	0.38571	0	1
		5	—	—	—
	Demand for product K_{12}	1	—	—	—
		2	0.34571	0.00489	0.625
		3	0.37429	0.00558	0.625
		4	—	—	—
		5	—	—	—
	Technological complexity of product K_{13} manufacture	1	—	—	—
		2	—	—	—
		3	0.37286	0.00437	0.83333
		4	0.30952	0.00453	0.9
		5	—	—	—
	Economic efficiency of product K_{14} selling	1	—	—	—
		2	0.46905	0.00276	0.6
		3	0.35238	0.00380	0.72727
		4	—	—	—
		5	—	—	—
O_2	Market competitiveness of product K_{21}	1	—	—	—
		2	0.40714	0.02058	0.75
		3	0.35306	0.00729	0.88889
		4	0.29643	0.00912	0.66667
		5	—	—	—
	Market demand for product K_{22}	1	—	—	—
		2	0.31429	0.00904	1
		3	0.34	0.01300	0.83333
		4	0.35612	0.01000	0.875
		5	—	—	—
	Technological complexity of product K_{23} manufacture	1	—	—	—
		2	—	—	—
		3	0.42619	0.00552	0.6
		4	0.44821	0.00449	0.72727
		5	—	—	—
	Economic efficiency of product K_{24} selling	1	—	—	—
		2	0.39796	0.01451	0.875
		3	0.35536	0.00745	0.8
		4	0.42143	0.00470	0.66667
		5	—	—	—

Table 9.11. Continued

Object	Parameter	Class	Average of distribution	Variance	Coefficient of coordination
O_3	Market competitiveness of product K_{31}	1	—	—	—
		2	0.47143	0.00195	0.33333
		3	0.37403	0.00258	0.84615
		4	—	—	—
		5	—	—	—
	Demand for product K_{32}	1	—	—	—
		2	—	—	—
		3	0.38095	0.00613	0.6
		4	0.40714	0.00379	0.72727
		5	—	—	—
	Technological complexity of product K_{33} manufacture	1	—	—	—
		2	—	—	—
		3	0.36071	0.00604	0.66667
		4	0.46161	0.00361	0.72727
		5	0.38214	0.00516	1
	Economic efficiency of product K_{34} sales	1	—	—	—
		2	—	—	—
		3	0.39	0.00773	0.83333
		4	0.41327	0.00380	0.6
		5	—	—	—
O_4	Market competitiveness of product K_{41}	1	—	—	—
		2	—	—	—
		3	0.45238	0.00283	0.6
		4	0.42143	0.00249	0.6
		5	0.42262	0.00288	1
	Demand for product K_{42}	1	—	—	—
		2	0.36071	0.02412	0.66667
		3	0.4375	0.00295	0.8
		4	0.41905	0.00375	0.75
		5	0.49524	0.00247	0.75
	Technological complexity of product K_{43} manufacture	1	—	—	—
		2	0.38571	0.00441	0.5
		3	0.34762	0.00418	0.625
		4	0.42619	0.00347	0.75
		5	0.38929	0.00298	1
	Economic efficiency of product K_{44} sales	1	—	—	—
		2	0.39643	0.00571	1
		3	0.46875	0.00329	0.72727
		4	0.41071	0.00335	1
		5	—	—	—

Table 9.12. Results of first round of expert estimation procedure of parameters K_{11}, K_{12}, K_{13}, and K_{14} of object O_1 and recommendations for further actions

Parameter of market competitiveness of product K_{11}

Class:	1	2	3	4	5
Level:	*Middle*	*Low*	*High*	*Very low*	*Low*
Conditions on expert estimation completion are fulfilled:	+		+	+	

Variants of further actions:
- Reject estimations not obtained in a confidence interval.
- Carry out reassessment by those experts whose results were not obtained in a confidence interval.
- Increase the amount of experts estimating this parameter.
- Finish the estimation procedure.

Parameter of market demand K_{12}

Class:	1	2	3	4	5
Level:	*Middle*	*Low*	*Low*	*Middle*	*Middle*
Conditions on expert estimation completion are fulfilled:			+	+	+

Variants of further actions:
- Reject estimations not obtained in a confidence interval.
- Carry out reassessment by those experts whose results were not obtained in a confidence interval.
- Increase the amount of experts estimating this parameter.
- Complete the estimation procedure.

Parameter of technological complexity of product K_{13} manufacture

Class:	1	2	3	4	5
Level:	*High*	*High*	*Middle*	*Middle*	*High*
Conditions on expert estimation completion are fulfilled:			+		+

Variants of further actions:
- Reject estimations not obtained in a confidence interval.
- Carry out reassessment by those experts whose results were not obtained in a confidence interval.
- Increase the amount of experts estimating this parameter.
- Finish the estimation procedure.

Table 9.12. Continued

Parameter of market demand K_{14}					
Class:	1	2	3	4	5
Level:	*Middle*	*Low*	*Middle*	*Middle*	*Low*
Conditions on expert estimation completion are fulfilled:		+		+	+

Variants of further actions:

- Reject estimations not obtained in a confidence interval.
- Carry out reassessment by those experts whose results were not obtained in a confidence interval.
- Increase the amount of experts estimating this parameter.
- Complete the estimation procedure.

Table 9.13. Expert estimation results of objects O_1, O_2, O_3, O_4

Index	Objects 1	2	3	4
Competitiveness	Low	Middle	High	Very high
Market demand	Low	High	High	Middle
Technological complexity of manufacture	High	High	High	Middle
Economic efficiency	Low	Low	High	Middle

mentation, it would be much easier to purchase the results of its application for similar areas from other countries and use the results. This is impossible. Groups of participants of innovation development and the connections between them in each country have their own peculiarities. Therefore, it is practically impossible to find two identical systems in the world. Therefore, the results of technology foresight and conclusions drawn for different systems would fundamentally differ. Therefore, each country, above all Ukraine, is destined to independently solve the problem of innovation development.

Finally, the generated technologies that will be created with the help of such an approach will have a revolutionary influence on industry, the economy, culture, and the environment over the next 10 years. These technologies depend on scientific and technological progress, which in turn also depends on them.

If these technologies are foreseen at the earliest stages, government along with other interested parties will be able to direct their resources at strategic research and the corresponding infrastructure that will assist in a rapid development of innovation. This is the main foresight task.

References

1. S. A. Ajvazyan, V. M. Buhshtaber, I. S. Enyukov, L. D. Meshalkin. *Applied Statistics. Classification and Dimension Reduction.* Mir, Moscow, 1983. In Russian.
2. V. I. Arnold. *The Theory of Catastrophes.* Nauka, Moscow, 1990. In Russian.
3. W. R. Ashby. *Introduction into cybernetics.* IL, Moscow, 1960. In Russian.
4. W. R. Ashby. Some peculiarities of complex systems. *Cybernetic Medicine*, 9(2):1–7, 1973.
5. L. von Bertalanffy. *General Systems Theory.* Mir, Moscow, 2nd edn, 1960. In Russian.
6. L. von Bertalanffy. General system theory – review of problems and results. *Systems Research. YB 1969*, pp 30–54. Institute of a history of natural sciences and technics, Moscow,1969. In Russian.
7. L. von Bertalanffy. The history and status of general systems theory. *Systems Research. YB 1973*, pp 20–37. Institute of a history of natural sciences and technics, Moscow,1973. In Russian.
8. I. V. Blauberg, E. M. Mirsky, V. N. Sadovsky. System approach and system analysis. *Systems Research. YB 1982*, pp 47–64. Nauka, Moscow, 1982. In Russian.
9. I. V. Blauberg, V. N. Sadovsky, E. G. Yudin. Philosophical principle of systemity and system approach. *Philosophical Questions*, (8):39–52, 1978. In Russian.
10. I. V. Blauberg, E. G. Yudin. *The Formation and Essence of the Systems Approach.* Nauka, Moscow, 1973. In Russian.
11. A. A. Bogdanov. *General Organizational Science or Tectology*, vol 1–3. Nauka, Moscow, 1913–1929. In Russian.
12. K. L. Boulding. General systems theory – the skeleton of science. *Management Science*, 2:197–208, 1956.
13. P. Bourgeois. Technology foresight for strategic decision-making. In: *Proceedings of the UNIDO Technology Foresight Conference for Central and Eastern Europe and the Newly Independent States*, p 24, Vienna, 4–5 April 2001.
14. W. L. Brockhaus, J. F. Mickelsen. An analysis of prior Delphi applications and some observations on its future applicability. *Technology Forecasting and Social Change*, 1:103–110, 1977.
15. N. P. Buslenko. *Modelling of Complex Systems.* Nauka, Moscow, 1968. In Russian.
16. P. B. Checkland. *Systems Thinking, Systems Practice.* Wiley, Chichester, 1986.
17. Y. V. Cornacchio. System complexity—a bibliography. *International Journal of General Systems*, 3(4):267–271, 1977.

18. V. S. Dejneka, I. V. Sergienko. *Optimal Control of Heterogeneous Distributed Systems.* Naukova Dumka, Kiev, 2003. In Russian.
19. A. I. Diveev, N. A. Severtsev. Transformation method in a scheme of the sequential analysis and variants sieving. In *Proceedings of International Conference on The Problems of Optimization of Computations XXVII*, pp 94–97, Kiev, 6–8 October 1997. In Russian.
20. R. L. Dobrushin. The general statement of the basic Shannon theorem in the information theory. *Mathematical Sciences Success*, 11(1(67)):17–75, 1956. In Russian.
21. R. L. Dobrushin. The general statement of the basic Shannon theorem in information theory. *Mathematical Sciences Success*, 14(6):3–104, 1959. In Russian.
22. S. I. Doroshenko. Science metric array indices of the soviet literature on system research. *System Research, YB 1978*, pp 127–135. Nauka, Moscow, 1978. In Russian.
23. A. G. Dragalin. *Mathematical Intuitionism. An Introduction to the Proof Theory.* Nauka, Moscow, 1979. In Russian.
24. S. V. Emelyanov, E. L. Nappelbaum. Basis principles of system analysis. In: *The Problems of a Scientific Approach to Socialistic Industry Management*, pp 92–99. Economics, Moscow, 1974. In Russian.
25. S. V. Emelyanov, E. L. Nappelbaum. Complex systems research methods. 1. The logic of rational choice. *Science and Technology Totals. Technical Cybernetics*, (8):5–101, 1977. In Russian.
26. S. V. Emelyanov, E. L. Nappelbaum. Complex systems research methods. 2. Choice under conditions of uncertainty. *Science and Technology Totals. Technical Cybernetics*, (9):169–242, 1977. In Russian.
27. A. C. Enthoven. System analysis – grand rules for constructive debate. *Air Force Magazine*, pp 33–40, January 1968.
28. Y. M. Ermoliev, T. Y. Ermolieva, G. J. MacDonald, V. I. Norkin, A. Amendola. A system approach to management of catastrophic risks. *European Journal of Operational Research*, 122:452–460, 2000.
29. Y. M. Ermoliev, V. S. Mihalevich. On research in a risk area. Academy of Sciences of Ukraine. V. M. Glushkov Institute of Cybernetics, Kiev, 1991. In Russian.
30. K. D. Forbus. Qualitative process theory. *European Journal of Operational Research*, 122 :85–168, 2000.
31. T. Forester. Megatrends or megamistakes? What ever happened to the information society? *Information Society*, 8(3):133–146, 1992.
32. J. Forrester. *The world's dynamics.* Nauka, Moscow, 1978. In Russian.
33. K. V. Frolov. *Catastrophe Mechanics.* International Institute for Complex Technical Systems Safety, Moscow, 1995. In Russian.
34. B. M. Gerasimov, Y. Ya. Samohvalov, A. I. Bobunov. Rational choice of the technical implementation of a complex system. *Control Systems and Machines*, 5:3–6, 1999. In Russian.
35. Ja. Gharajedaghi. *Systems Thinking: Managing Chaos and Complexity. A Platform for Designing Business Architecture.* Butterworth-Heinemann, Boston, 1999.

36. V. M. Glushkov. On system optimization. *Cybernetics*, 5:89–90, 1980. In Russian.
37. V. M. Glushkov. *The Basics of a Paperless Information Science.* Nauka, Moscow, 1982. In Russian.
38. V. M. Glushkov, V. S. Mihalevich, V. L. Volkovich, G. A. Dolenko. System optimization in multicriteria problems of linear programming at interval preference specifying. *Cybernetics*, 3:1–8, 15, 1983. In Russian.
39. M. Godet. Reducing the blunders in forecasting. *Futures*, 15(3):181–192, 1983.
40. A. L. Gorelik, N. B. Gurevich, V. A. Skripnik. *The Current State of the Recognition Problem.* Radio and Communication, Moscow, 1985. In Russian.
41. Y. M. Gorsky. *System Information Analysis of Control Processes.* Nauka, Novosibirsk, 1988. In Russian.
42. Ya. M. Grigorenko, N. D. Pankratova. *Numerical Methods in Applied Mathematics Problems.* Lybid, Kiev, 1995. Ukrainian.
43. Ya. M. Grigorenko, A. T. Vasilenko, N. D. Pankratova. *Problems of the Elasticity Theory of a Nonhomogeneous Body.* Naukova Dumka, Kiev, 1991. In Russian.
44. D. M. Gvishiani. International institute for applied system analysis: goals, main perspectives. *System Research. Methodological Problems. YB 1987*, pp 7–25, 1987. In Russian.
45. D. M. Gvishiani. Aurelio pechchei and the formation of global problems. *System Research. Methodological Problems (1992–1994)*, pp 7–28, 1996. In Russian.
46. A. Ya. Hinchin. 1. The concept of entropy in probability theory. *Uspehi matematicheskih nauk*, 8(3):3–20, 1953. In Russian.
47. A. Ya. Hinchin. 2. On basic theorems of information theory. *Uspehi matematicheskih nauk*, 11(1):17–75, 1956. In Russian.
48. A. Ya. Hinchin. On basic theorems of the information theory. *Uspehi matematicheskih nauk*, 14(6):3–104, 1959. In Russian.
49. A. Holl. *Methodology experience for systems engineering.* Soviet Radio, Moscow, 1975. In Russian.
50. Y. V. Kalashnikov. Representing and using the knowledge for computerized testing of complex dynamic objects functioning. *Foreign Radioelectronics*, 7:3–23, 1992. In Russian.
51. A. Karaoglou, G. Desmet, G. Kelly, H. Menze. *The Radiological Consequences of the Chernobyl Accident.* IIPSA, Luxembourg, 1996.
52. V. V. Kelle. Thinking over the system methodology: P. Cheklend's version. *System Research. Methodological Problems. YB 1995–1996*, pp 376–389. Editorial URSS, Moscow, 1996. In Russian.
53. G. J. Klir. Science on systems: new science measuring. *System Research: Methodological Problems. YB 1983*, pp 61–85. Nauka, Moscow, 1983. In Russian.
54. G. J. Klir. *Systemology. Automation of System Problems Solution.* Radio and Communication, Moscow, 1990. In Russian.
55. V. V. Klyuev, ed. *Technical Means for Diagnostics.* Mashinistroenie, Moscow, 1989. In Russian.
56. E. N. Knyazeva, S. P. Kurdyumov. Synergy as a new world vision: a dialogue with I. Prigozhin. *Philosophical Questions*, 12:3–20, 1992. In Russian.

57. A. N. Kolmogorov. On representing continuous functions of multiple variables as a superposition of continuous functions of one variable and summation. *Doklady AN SSSR*, 114(5):953–956, 1957. In Russian.
58. A. N. Kolmogorov. The theory of information transmission. In *Session of Academy of Sciences of the USSR on Scientific Problems of Manufaturing Automation 15–20 October 1956*, pp 66–69. Publishing House of AS of the USSR, Moscow, 1957. In Russian.
59. A. N. Kolmogorov. A new metric invariant of transient dynamic systems and automorphisms in Lebesgue spaces. In *Proceedings of Academy of Sciences*, vol 119, pp 861–864. 1958. In Russian.
60. A. N. Kolmogorov. On entropy to unit time as metric invariant of automorphisms. In: *Proceedings of Academy of Sciences*, vol 124, pp 754–755. 1959. In Russian.
61. A. N. Kolmogorov. Three approaches to defining the concept of "information quantity". *Information Transmission Problems*, 1(1):3–11, 1965. In Russian.
62. A. N. Kolmogorov. *The Theory of Information and the Theory of Algorithms.* Nauka, Moscow, 1987. In Russian.
63. A. N. Kolmogorov, A. G. Dragalin. *Mathematical Logic. Additional Chapters.* Moscow State University Press, Moscow, 1984. In Russian.
64. N. P. Kornejchuk. On research methods for extreme problems from the theory of the best approximation. *Mathematical Sciences Success*, 29(3):9–41, 1974. In Russian.
65. T. Kotarbinski. *Elemety Teorji Poznania, Logiki Formalnoj i Metodologji Nauk.* Wyd-wo Zakladu Narodowego imienia Ossolinshich, Lwiw, Ukraine, 1929.
66. V. N. Koval. *Applied System Analysis of Multivariate Processes.* Naukova Dumka, Kiev, 2002. In Russian.
67. I. N. Kovalenko, A. G. Nakonechny. *Approximate Calculation and Optimization of Reliability.* Naukova Dumka, Kiev, 1989. In Russian.
68. K. B. Krohn, Y. L. Rhodes. Complexity of finite semigroups. *Annals of Mathematics*, 88:128–160, 1968.
69. A. I. Kuhtenko. *Cybernetics and Fundamental Sciences.* Naukova Dumka, Kiev, 1987. In Russian.
70. B. Kuipers. Commonsense reasoning about causality intelligence. *Annals of Mathematics*, 88:169–203, 1968.
71. B. Kuipers. Qualitative simulation. *Annals of Mathematics*, 88:289–338, 1968.
72. S. Kulbak. *The Theory of Information and Statistics.* Nauka, Moscow, 1954. In Russian.
73. V. M. Kuntsevich. Adaptive control of dynamic objects. The game approach. In *Cybernetics Methods of Planning, Projecting and Control*, pp 122Ч130. Academy of Sciences of Ukraine. V. M. Glushkov Institute of Cybernetics, Kiev,1982. In Russian.
74. V. M. Kuntsevich, M. M. Lychak. *Synthesis of Optimal and Adaptive Systems of Control: The Game Approach.* Naukova Dumka, Kiev, 1985. In Russian.
75. S. P. Kurdyumov, G. G. Malinetsky, A. B. Potapov. *Synergy—New Directions.* Znanie, Moscow, 1989. In Russian.

76. B. Kurilin, N. Pankratova, O. Pugach. Method of purposeful choice of a rational structure for a complex construction. In: *XXXVIII Sympozjon "Modelowanie w mechanice"*, vol 9, pp 141–147, Gliwice, Poland, 1999.
77. B. I. Kurilin. To solving the Chebishev's problem of approximation for the incompatible system of nonlinear equations. *Journal of Calculus Mathematics and Mathematical Physics*, 10(1):3–14, 1970. In Russian.
78. B. I. Kurilin. Intelligent system of support of decision in dynamics of critical situations. In *Proceedings of the second world congress of structural and multidisciplinary optimization*, vol 1, pp 85–90, Zacopane, Poland, 26–30 May 1997. Institute of Fundamental Technological Research, Warsaw, Poland,1997.
79. I. I. Kuzmin, S. V. Romanov. Risk and safety from the viewpoint of system dynamics. *Radiation Safety and Atomic Power-Station Protection*, pp 82Ч105. Energoatomizdat, Moscow, 1991. In Russian.
80. E. Kveid. *Complex Systems Analysis.* Mir, Moscow, 1969. In Russian.
81. K. Lantsosh. *Practical Methods of Applied Analysis.* Fizmatgiz, Moscow, 1961. In Russian.
82. O. I. Larichev, E. M. Moshkovich. *Qualitative Methods for Decision Making. Verbal Solution Analysis.* Nauka and Fizmatlit, Moscow, 1996. In Russian.
83. J. L. Lione. *Optimal Control of Systems Described by Equations with Partial Derivatives.* Mir, Moscow, 1972. In Russian.
84. D. Loveridge. Technology forecasting and foresight: pedantry or disciplined vision. *Ideas in Progress*, (2): 3-9, 1997.
85. I. M. Gelfand, A. Y. Kolmogorov, A. M. Yaglom. To the general definition of data quantity. *DAN USSR*, 111(4):745–748, 1956. In Russian.
86. N. A. Mahutov, V. P. Petrov, N. I. Tartashev. Science and technical problems of man-caused safety. *Problems of Mechanical Engineering and Machine Reliability*, 5:164–170, 1998. In Russian.
87. B. B. Mandelbrot. *Fractals, from Chance and Dimension.* Freeman, San Francisco, 1977.
88. G. I. Marchuk, K. Ya. Kondratiev. *The Priorities of Global Ecology.* Nauka, Moscow, 1992. In Russian.
89. M. Mazur. *Qualitative Theory of Information.* Mir, Moscow, 1974. In Russian.
90. V. D. Mazurov. *Mathematical Methods of Pattern Recognition in Solving the Problems of Planning and Control.* Uralskoe izdatelstvo, Sverdlovsk, 1977. In Russian.
91. D. L. Meadows. *Limits to Growth.* Moscow State University Press, Moscow, 1991. In Russian.
92. D. L. Meadows. Beyond the limits to growth. In: *Global Environment and Local Action, The Club of Rome Conference* in Fukuoka, Japan, pp 35–46, 1992.
93. D. L. Meadows. Beyond the limits to growth. *System Research. Methodological Problems 1992–1994*, pp 29–40, 1996. In Russian.
94. A. N. Melihov, L. S. Bershtein, S. Ya. Korovin. *Situational Advising Systems with Fuzzy Logic.* Nauka, Moscow, 1990. In Russian.
95. M. Mesarovich, D. Macko, I. Takahara. *Hierarchical Multilevel Systems Theory.* Mir, Moscow, 1973. In Russian.
96. M. Mesarovich, Ya. Takahara. *The General Theory of Systems: Mathematical Foundations.* Mir, Moscow, 1978. In Russian.
97. V. S. Mihalevic, V. L. Volkovich. *Numerical Methods for Researching and Designing Systems.* Nauka, Moscow, 1982. In Russian.

98. V. S. Mihalevic, B. M. Yanenko, K. L. Atoev. An information system for estimating the risks of ecological disasters. *Functioning Condition Modelling of an Organism and its Control.* V. M. Glushkov Institute of Cybernetics, Kiev, pp 52–74, 1993. In Russian.
99. N. N. Moiseev. *Mathematical Problems of System Analysis.* Nauka, Moscow, 1981. In Russian.
100. N. N. Moiseev. *Giving Up Simplicity.* Agraf, Moscow, 1998. In Russian.
101. J. E. A. Morales. The most commonly applied methodologies in technology foresight. In *Proceedings of the UNIDO Technology Foresight Conference for Central and Eastern Europe and the Newly Independent States*, pp 170–178, Vienna, 4–5 April 2001.
102. I. A. Muromov. *100 Great Catastrophes.* Veche, Moscow, 2003. In Russian.
103. E. Nagel, D. Newman. *Godel's Theorem.* Znanie, Moscow, 1970. In Russian.
104. E. L. Nappelbaum. System analysis as a program for scientific research: Structure and key concepts. In: *System Research. Methodological Problems. YB 1979*, pp 55–77, 1980. In Russian.
105. S. P. Nikanorov. System analysis: the stage of decision making methodology development in the USA. In: *System Analysis for Solving the Business and Industrial Problems*, pp 7–43. Mir, Moscow, 1969. In Russian.
106. A. V. Palagin. The concept of system integration in innovation strategies. *Control Systems and Machines*, 2:6–9, 2001. In Russian.
107. L. A. Pankova, A. M. Petrovsky, M. V. Shnejderman. *Organizing the Expertise and Analysing the Experts' Information.* Nauka, Moscow, 1984. In Russian.
108. N. Pankratova. Deformation of the anisotropic non-homogeneous cylinder. *Journal of Theoretical and Applied Mechanics*, 34(4):733–748, 1996.
109. N. Pankratova. Elastic equilibrium of anisotropic non-homogeneous hollow bodies. *Zhurnal Polimery i Kompozyty Konstrukcyjne*, pp 82–88, 1996.
110. N. D. Pankratova. System minimization of multifactor risks in dynamics of critical situations. In *Proceedings of the 2nd World Congress on Structural and Multidisciplinary Optimization*, vol 2, pp 619–624, Zacopane, Poland, 26–30 May 1997. Institute of Fundamental Technological Research, Warsaw, Poland, 1997.
111. N. D. Pankratova. Thermo-stressed state of anisotropic non-homogeneous constructive elements. In Y. W. Kwon (ed) *Recent Advances in Solids/Structures and Application of Metallic Materials*, vol 369, American Society of Mechanical Engineers, New York, pp 203–210, 1997.
112. N. D. Pankratova. A system analysis of multifactor risks in conditions of uncertainty. In *Proceedings of the 15th IFIP World Computer Congress*, Vienna/Austria and Budapest/Hungary, 31 August–4 September 1998. CD-ROM, file: /doc/000/000/665.htm.
113. N. D. Pankratova. The thermo-stressed state of composite constructive elements. *Zhurnal Polimery i Kompozyty Konstrukcyjne*, pp 61–68, 1998.
114. N. D. Pankratova. General tendencies and system problems of information technologies development. *Problems of Control and Informatics*, 1:58–68, 1999. In Russian.
115. N. D. Pankratova. Forming the target functions in the system problem of conceptual uncertainty. *Proceedings of the National Academy of Sciences of Ukraine*, 9:67–73, 2000. In Russian.

116. N. D. Pankratova. System designing of complex constructions. In *IV Scientific-technical Conference "Polimery i Komposyty Konstrukcyjne"*, pp 31–42, Ustron, Poland, 2000.
117. N. D. Pankratova. Pareto's set formation in system problem of conceptual uncertainty. In *Proceedings of the National Academy of Sciences of Ukraine*, 12:65–70, 2001. In Russian.
118. N. D. Pankratova. System optimization of structural elements of modern technical equipment. *Cybernetics and System Analysis*, 3:119–131, 2001. In Russian.
119. N. D. Pankratova. Formation and development of system analysis as an applied science discipline. *System Research and Information Technologies*, 1:65–94, 2002. In Russian.
120. N. D. Pankratova. Rational compromise in system problems of conceptual uncertainty. *Cybernetics and System Analysis*, 4:162–180, 2002. In Russian.
121. N. D. Pankratova. Market formation problems of science intensive production. In *Approval of Innovation Model of Economic Development of Ukraine*, pp 337–346, Kiev, 2003. NTUU "KPI". In Ukrainian.
122. N. D. Pankratova. The mathematical software of the technology foresight problems with reference to industry. *System Research and Information Technologies*, 1:26–33, 2003. In Russian.
123. N. Pankratova, A. Vasilenko. Stress-deformed state of laminated anisotropic environment with cavity or inclusion. In *Proceedings of the 4th International Symposium on Mine Planning and Equipment Selection*, pp 1023–1027, Calgary, Canada, 1995.
124. N. Pankratova, L. Gracheva. Deformation of structural elements from anisotropic composite materials under thermal loading. In *Proceedings of the 2nd International Symposium on Thermal Stresses and Related Topics, Thermal Stresses '97*, pp 1012–1018, Rochester, NY, 1997.
125. N. D. Pankratova, I. A. Ivanchenko. To estimating the integral level of informedness when solving system analysis problems of abnormal and critical situations. In *Proceedings of the Second National Scientific and Practical Conference on System Analysis and Information Technologies*, pp 42–46, NTUU "KPI", 2000. In Russian.
126. N. D. Pankratova, I. A. Ivanchenko. Determining the quantitative and qualitative indicators of informedness in problems of system forecasting the dynamics of risk situations. In *Proceedings of the 3rd National Scientific and Practical Conference on System Analysis and Information Technologies*, pp 109–112. NTUU "KPI", 2001. In Russian.
127. N. D. Pankratova, I. I. Katkov. Toward the question of technical diagnostics of an airplane engine. In *Proceedings of the 2nd National Scientific and Practical Conference on System Analysis and Information Technologies*, pp 49–53. NTUU "KPI", 2000. In Russian.
128. N. D. Pankratova, I. Y. Katkov, E. L. Oparina, A. L. Chervinsky-Ivashura. Robust information technologies of detection of legitimacies on multiparameter statistical and empirical data. *Reporter of National Technical University "KPIHPI"*, 22:3–14, 2001.
129. N. D. Pankratova, M. S. Krivosheeva. Choice and system substantiation of qualitative indices of a complex system. In *Proceedings of the 3rd National Scientific and Practical Conference on System Analysis and Information Technologies*, pp 112–114. NTUU "KPI", 2001. In Russian.

130. N. D. Pankratova, B. I. Kurilin. System modeling and optimization of nonhomogeneous anisotropic constructive elements. In *35th Symposium on Modeling in Mechanics*, pp 268–272, Gliwice, Poland, 1996.
131. N. Pankratova, B. Kurilin. System modeling of abnormal situations at designing composite constructions. In *Proceedings of the 36th International Symposium on Modeling in Mechanics*, pp 275–284, Gliwice, Poland, 1997.
132. N. D. Pankratova, B. I. Kurilin. The rational choice of a hierarchical structure for a complex system. In *Proceedings of the National Academy of Sciences of Ukraine*, 9:54–59, 1999. In Russian.
133. N. D. Pankratova, B. I. Kurilin. System modeling of abnormal situations at designing of complex constructions. In *Proceedings of the National Academy of Sciences of Ukraine*, 3:56–60, 1999.
134. N. Pankratova, B. Kurilin. Conceptual foundations of the system analysis of risks in dynamics of control of complex system safety. Part 2. The general problem of the system analysis of risks and the strategy of its solving. *Journal of Automation and Information Sciences*, 33(4):1–14, 2001.
135. N. Pankratova, B. Kurilin. Conceptual foundations of the system analysis of risks in dynamics of control of complex system safety. Part 1. Basic statements and substantiation of approach. *Journal of Automation and Information Sciences*, 33(2):15–31, 2001.
136. N. D. Pankratova, B. I. Kurilin. Mathematical basis of information analysis of system problems. In *Proceedings of the 5th International Conference on Intellectual Analysis of Information*, pp 224–233, Kiev, 17–20 May 2005. In Russian.
137. N. D. Pankratova, E. L. Oparina. Recognition and minimization of risks in dynamics of operation for complex ecological systems. In *Miedzynarodova Konferencja "V Jubileuszowa Szkola Geomechaniki" Materialy Konferencyine, Czesc II*, pp 77–86, Gliwice-Ustron, 2001.
138. N. D. Pankratova, E. L. Oparina. Information maintenance of the technology foresight problems for the branch in dynamic development. In *Miedzynarodowe Sympozjum "Geotechnika – Geotechniks 2003"*, pp 25–31, Ustron, Poland, 2003.
139. N. D. Pankratova, E. L. Oparina. The multifactor dependencies recovery under the conditions of conceptual uncertainties. *System Research and Information Technologies*, 3:103–114, 2004. In Russian.
140. N. D. Pankratova, E. L. Oparina. Problems and principle of system strategy in innovation activity. In *11th Miedzynarodowe Sympozjum "Geotechnika – Geotechniks 2004"*, pp 89–96, Ustron, Poland, 2004.
141. N. D. Pankratova, O. V. Pugach. The rational choice of structure of multilevel hierarchical technical systems. In *Proceedings of the 3rd National Scientific and Practical Conference on System Analysis and Information Technologies*, pp 117–119. NTUU "KPI", 2001. In Russian.
142. N. D. Pankratova, E. L. Oparina, A. Shapoval. Recovery of functional dependencies at uncertainty disclosing in automated testing and control systems. In *International Conference on Control, Automation-2000*, pp 194–199, Lvov, Ukraine, 2000. In Russian.
143. Ya. G. Panovko, I. I. Gubanov. *Stability and Vibration of Elastic Systems: Modern Concepts, Paradoxes, and Mistakes.* Nauka, Moscow, 4th edn, 1987. In Russian.

144. A. Pechei. *Human Qualities.* Progress, Moscow, 2nd edn, 1985. In Russian.
145. E. Pestel. *Beyond The Growth Limits.* Progress, Moscow, 1988. In Russian.
146. B. N. Petrov, V. V. Petrov, G. M. Ulanov, et al. Introduction to the Theory of Information Control. In *Totals of Science and Technology "Technical Cybernetics"*, pp 221–352. VINITI, Moscow, 1970. In Russian.
147. V. V. Podinovsky, V. D. Nogin. *Pareto - Optimal Solutions of Multicriterion Problems.* Nauka, Moscow, 1982. In Russian.
148. T. Poston, I. Stewart. *Catastrophes Theory and Its Applications.* Pifmann, London, 1978.
149. A. Rapoport. Various approaches to the general system theory. *System Research. YB 1969*, pp 55–79. Nauka, Moscow, 1969. In Russian.
150. E. Ya. Remez. *The Basics of Numerical Methods of Chebyshev's Approximation.* Naukova Dumka, Kiev, 1969. In Russian.
151. T.L. Saaty. *Decision Making. Analytic Hierarchy Process.* Radio and Communication, Moscow, 1993. In Russian.
152. T.L. Saaty. How to make and justify a decision: the analytic hierarchy process (AHP). Part 1. Examples and applications. *System Research and Information Technologies*, 1:95–107, 2002.
153. T.L. Saaty. The analytic network process, examples, Part 2.3. *System Research and Information Technologies*, 4:7–23, 2003.
154. T.L. Saaty. Theory of the analytic hierarchy and analytic network processes—examples, Part 2.2. *System Research and Information Technologies*, 2:7–34, 2003.
155. T.L. Saaty. Theory of the analytic hierarchy process, Part 2.1. *System Research and Information Technologies*, 1:48–72, 2003.
156. T.L. Saaty, K. Kerns. *Analytical Planning. System Organization.* Radio and Communication, Moscow, 1991. In Russian.
157. V. N. Sadovsky. Paradoxes of system thinking. *System Research, YB 1972*, pp 133–146. Nauka, Moscow, 1972. In Russian.
158. V. N. Sadovsky. Selected essential problems in developing the general systems theory. *System Research, YB 1971*, pp 35–54. Nauka, Moscow, 1972. In Russian.
159. V. N. Sadovsky. *Foundations of the General Theory of Systems.* Nauka, Moscow, 1974. In Russian.
160. V. N. Sadovsky. System approach and general systems theory: The status, main problems, and development perspectives. *System Research. Methodological Problems. YB 1979*, pp 29–54. Nauka, Moscow, 1980. In Russian.
161. V. N. Sadovsky. The paradigm shift in system thinking. *System Research. Methodological Problems. YB 1992–1994*, pp 64–78. Nauka, Moscow, 1980. In Russian.
162. J. Seige, J. L. Mele. *Identification of Control Systems.* Nauka, Moscow, 1974. In Russian.
163. J. Seige, J. L. Mele. *Estimation Theory and its Application in Communication and Management.* Svyaz, Moscow, 1976. In Russian.
164. I. V. Sergienko. On general directions of informatics development. *Cybernetics and System Analysis*, 6:3–93, 1997. In Russian.
165. I. V. Sergienko. *Informatics in Ukraine: Formation, Development, Problems.* Naukova Dumka, Kiev, 1999. In Ukrainian.
166. I. V. Sergienko. *Informatics and Computer Technologies.* Naukova Dumka, Kiev, 2004. In Ukrainian.

167. I. V. Sergienko, V. P. Shilo. *Discrete Optimization Problems. Solution Methods. Research.* Naukova Dumka, Kiev, 2003. In Russian.
168. I. V. Sergienko, V. M. Yanenko, K. L. Atoev. The general concept for controlling the risks of ecological, technological, and social catastrophes. *Cybernetics and System Analysis*, 2:65–86, 1997. In Russian.
169. N. A. Severtsev, A. I. Diveev. Optimal choice of technical product variant. *Problems of Mechanical Engineering and Machine Safety*, 5:3–8, 1995. In Russian.
170. C. Shannon. *Mathematical Communication Theory.* Izd-vo inostr. lit., Moscow, 1963. In Russian.
171. C. Shannon. *Works on Information Theory and Cybernetics.* Foreign Language Publisher, Moscow, 1963. In Russian.
172. C. E. Shannon. A mathematical theory of communication. Pt I, II. *Bell Systems Technology Journal*, 27(3):379–423, 1948.
173. T. Yu. Sidorina. Growth limits. 20 years later. *System Research. Methodological Problems. 1992–1994*, pp 41–46. Editorial URSU, Moscow, 1996. In Russian.
174. H. A. Simon. *Administrative Behavior.* Free Press, New York, 1957.
175. V. V. Skopetsky, O. A. Marchenko, N. A. Lezhnina. Objects' system analysis influenced by interactive processes. *Cybernetics and System Analysis*, 6:54–66, 2001. In Russian.
176. V. I. Skurihin. On conception formulation. The "Four I's" conception. *Upravlyayushchie sistemy i mashiny*, 2:7–12, 1989. In Russian.
177. R. L. Stratonovich. On values of information. *Proceedings of the Academy of Sciences of the USSR. Technical cybernetics*, 5:3–12, 1965. In Russian.
178. A. Tetelman, P. Bezuner. An application of risk analysis to the investigation of a brittle fracture and fatigue of steel structures. *Mechanics of constructions destruction*, pp 7–30. Mir, Moscow, 1980. In Russian.
179. J. Tompson. *Instability and Catastrophes in Science and Technology.* Mir, Moscow, 1985. In Russian.
180. V. T. Troshchenko. *Resistance of Materials to Straining and Destruction. Part 1.* Naukova Dumka, Kiev, 1993. In Russian.
181. A. I. Uemov. *The Logic and Methodology of System Research.* Vyshcha Shkola, Kiev, 1977. In Russian.
182. V. A. Uspensky. Godel's theorem in elementary statement. *Uspehi Matematicheskih Nauk*, (1(82)):3–47, 1974. In Russian.
183. V. N. Vapnik. *Renewal of Dependences on Basis of Empirical Data.* Nauka, Moscow, 1979. In Russian.
184. V. I. Vasiliev, I. V. Surovtsev. Inductive methods for patterns of relationship recognition based on the reduction theory. *Control Systems and Machines*, 5:3–13, 1998. In Russian.
185. V. I. Vernadsky. *Biosphere Studies and its Gradual Transition to Noosphere.* Nauka, Moscow, 3rd edn, 1978. In Russian.
186. S. Vertiev, A. Egorov, A. Zolotov, et al. A family of automation testing systems for flying machines. *The World of Computer Automation*, 1:68–71, 1995. In Russian.
187. V. L. Volkovich, A. F. Voloshin. On one scheme of sequential analysis method and variants' sieving. *Cybernetics*, 4:98–105, 1978. In Russian.
188. W. Weaver. Science and complexity. *American Scientist*, 36, pp.536–544, 1968.

189. N. Wiener. *Cybernetics*. MIT Press, Cambridge, MA, 1948.
190. A. S. Yakovlev. *The Purpose of Life (Notes from an Aircraft Designer)*. Politizdat, Moscow, 1972. In Russian.
191. B. G. Yudin. Some peculiarities of system research development. *System Research. Methodological Problems. YB 1981*, pp 7–22. Nauka, Moscow, 1981. In Russian.
192. M. Zgurovsky. Technology foresight in Ukraine. In *Proceedings of the UNIDO Technology Foresight Conference for Central and Eastern Europe and Newly Independent States*, pp 140–151, Vienna, 4–5 April 2001.
193. M. Zgurovsky. The scenario analysis platform as a methodological base of the national foresight program of Ukraine. *System Research and Information Technologies*, 1:7–25, 2003.
194. M. Z. Zgurovsky. *Integrated Systems of Optimal Control and Design*. Vishcha Shkola, Kiev, 1990. In Russian.
195. M. Z. Zgurovsky. State and perspectives of development of system analysis methodology in Ukraine. *Cybernetics and System Analysis*, 1:101–109, 2000. Ukrainian.
196. M. Z. Zgurovsky. *Foresight System Methodology*. Politechnika, Kiev, 2001. Ukrainian.
197. M. Z. Zgurovsky. The scenario analysis as the system methodology of foresight. *System Research and Information Technologies*, 1:7–38, 2002. In Russian.
198. M. Z. Zgurovsky, P. I. Bidyuk. *Analysis and Control of Large Space Structures*. Naukova Dumka, Kiev, 1997. In Russian.
199. M. Z. Zgurovsky, V. A. Denisenko. *Discretely Continuous System with Controlled Structure. Theory, Modelling, Application*. Naukova Dumka, Kiev, 1998. In Russian.
200. M. Z. Zgurovsky, A. V. Dobronogov. *System Analysis Methodology Application to the Problems of Provision of Pensions in Ukrainian Transition Economy*. Naukova Dumka, Kiev, 1998. In Russian.
201. M. Z. Zgurovsky, V. S. Melnik. *Nonlinear Analysis and Control of Infinite Systems*. Naukova Dumka, Kiev, 1999. In Russian.
202. M. Z. Zgurovsky, A. N. Novikov. *Analysis and Control by One-sided Physical Processes*. Naukova Dumka, Kiev, 1996. In Russian.
203. M. Z. Zgurovsky, N. D. Pankratova. The information platform of scenario analysis for the technology foresight problems. *Cybernetics and System Analysis*, 4:112–125, 2003. In Russian.
204. M. Z. Zgurovsky, N. D. Pankratova. System strategy of the technology foresight in innovation activity. *System Research and Information Technologies*, 3:7–24, 2003. In Ukrainian.
205. M. Z. Zgurovsky, N. D. Pankratova. *System Analysis. Problems, Methodology, Application*. Naukova Dumka, Kiev, 2005. In Russian.
206. M. Z. Zgurovsky, N. D. Pankratova. *Technology Foresight*. ICP Vidavnitstvo Politechnika, Kiev, 2005. In Russian.
207. M. Z. Zgurovsky, V. S. Melnik. *Nonlinear Analysis and Control of Physical Processes and Fields*. Springer, 2004.
208. Y. I. Zhuravlev. On algorithmic approach to solving the recognition and classification problems. In *Problems of Cybernetics*, pp 5–68. Nauka, Moscow, 1971. In Russian.

209. Y. I. Zhuravlev, I. B. Gurevich. Methods and means of transforming and processing the information in pattern recognition and image analysis problems. In *Parallel Information Processing*, vol 5, pp 218–318. Naukova Dumka, Kiev, 1990. In Russian.
210. S. I. Zuhovitsky, A. I. Avdeeva. *Linear and Convex Programming.* Nauka, Moscow, 1967. In Russian.

Index

Printing: Krips bv, Meppel
Binding: Stürtz, Würzburg

The advantage of theranostics — the dual functionality of diagnostics and therapy within one carrier — is mainly that two goals can be achieved with a single administration, potentially saving time and effort from multiple administrations. However, there have been complications with multiple ligands or moieties being conjugated to the same carrier. Nanoparticles and other carriers are limited in size and often do not allow multiple ligand conjugation; steric hindrance being one factor and chemical interference the other. Furthermore, the synthetic difficulty increases with a corresponding decrease in efficiency upon addition of more groups, which can also require a significant expenditure of time and labor. To remedy this, theranostic systems could take advantage of materials that inherently have multifunctional properties, such as gold or iron which are biocompatible and have both imaging capabilities (Raman scattering) and therapeutic capabilities (thermal ablation). In short, the field of smart theranostics has shown immense promise in preclinical studies so far. Every indication is that it will continue to evolve with innovations for eventual translation.

Acknowledgement

This work is supported by NIH (HL109442, EB015300, UH2TR000505, and AI096305), NSF EEC-0425626, AOSpine Foundation, DoD W81XWH-12-1-0261.

References

1. Tannock, I. F. and Rotin, D. (1989). Acid pH in tumors and its potential for therapeutic exploitation. *Cancer Research*, **49**, 4373–4384.
2. Xiao, H. H. Y., Matson, V. Z., Javadi, A., Xu, W. and Yang, Y. (2012). Gold nanorods conjugated with doxorubicin and cRGD for combined anticancer drug delivery and PET imaging. *Theranostics,* **2**, 757–768.
3. Wang, C., Ravi, S., Martinez, G. V., Chinnasamy, V., Raulji, P., Howell, M., Davis, Y., Mallela, J., Seehra, M. S. and Mohapatra, S. (2012). Dual-purpose magnetic micelles for MRI and gene delivery. *Journal of Control. Release,* **163**, 82–92.
4. Jing, L., Liang, X., Li, X., Yang, Y. and Dai, Z. (2013). Covalent attachment of Mn-porphyrin onto doxorubicin-loaded poly(lactic acid) nanoparticles for potential magnetic resonance imaging and pH-sensitive drug delivery. *Acta Biomater.,* **9**, 9434–9441.
5. Guo, J., Hong, H., Chen, G., Shi, S., Zheng, Q., Zhang, Y., Theuer, C. P., Barnhart, T. E., Cai, W. and Gong, S. (2013). Image-guided and tumor-targeted drug delivery with radiolabeled unimolecular micelles. *Biomaterials,* **34**, 8323–8332.
6. Bae, Y., Nishiyama, N., Fukushima, S., Koyama, H., Yasuhiro, M. and Kataoka, K. (2004). Preparation and biological characterization of polymeric micelle drug carriers with intracellular pH-triggered drug release property: Tumor permeability, controlled subcellular

drug distribution, and enhanced *in vivo* antitumor efficacy. *Bioconjugate Chem.,* **16**, 122–130.

7. Huang, H., Li, Y., Li, C., Wang, Y., Sun, Y. and Wang, J. (2013). A novel anti-VEGF targeting and MRI-visible smart drug delivery system for specific diagnosis and therapy of liver cancer. *Macromol. Biosci.,* **13**, 1358–1368.
8. Lim, E. K., Sajomsang, W., Choi, Y., Jang, E., Lee, H., Kang, B., Kim, E., Haam, S., Suh, J. S., Chung, S. and Huh, Y. M. (2013). Chitosan-based intelligent theragnosis nanocomposites enable pH-sensitive drug release with MR-guided imaging for cancer therapy. *Nanoscale Res. Lett.,* **8**, 467.
9. Koo, H., Huh, M. S., Sun, I. C., Yuk, S. H., Choi, K., Kim, K. and Kwon, I.C. (2011). *In vivo* targeted delivery of nanoparticles for theranosis. *Acc. Chem. Res.,* **44**, 1018–1028.
10. Jung, S., Nam, J., Hwang, S., Park, J., Hur, J., Im, K., Park, N. and Kim, S. (2013). Theragnostic pH-sensitive gold nanoparticles for the selective surface enhanced raman scattering and photothermal cancer therapy. *Anal. Chem.,* **85**, 7674–7681.
11. Kai, Y., Liangzhu, F., Xiaoze, S. and Zhuang, L. (2013). Nano-graphene in biomedicine: Theranostic applications. *ChemInform.,* **44,** 530–547.
12. Xia, Y., Li, W., Cobley, C. M., Chen, J., Xia, X., Zhang, Q., Yang, M., Cho, E. C. and Brown, P. K. (2011). Gold nanocages: From synthesis to theranostic applications. *Acc. Chem. Res.,* **44**, 914–924.
13. Zhang, Z., Wang, L., Wang, J., Jiang, X., Li, X., Hu, Z., Ji, Y., Wu, X. and Chen, C. (2012). Mesoporous silica-coated gold nanorods as a light-mediated multifunctional theranostic platform for cancer treatment. *Adv. Mater.,* **24**, 1418–1423.
14. Lukianova-Hleb, E. Y., Hanna, E. Y., Hafner, J. H. and Lapotko, D. O. (2010). Tunable plasmonic nanobubbles for cell theranostics. *Nanotechnology,* **21**, 085102.
15. Melancon, M. P., Zhou, M. and Li, C. (2011). Cancer theranostics with near-infrared light-activatable multimodal nanoparticles. *Acc. Chem. Res.,* **44**, 947–956.
16. Lukianova-Hleb, E. Y., Santiago, C., Wagner, D. S., Hafner, J. H. and Lapotko, D. O. (2010). Generation and detection of plasmonic nanobubbles in zebrafish. *Nanotechnology,* **21**, 225102.
17. Lee, J. E., Lee, N., Kim, T., Kim, J. and Hyeon, T. (2011). Multifunctional mesoporous silica nanocomposite nanoparticles for theranostic applications. *Acc. Chem. Res.,* **44**, 893–902.
18. Ma, X., Tao, H., Yang, K., Feng, L., Cheng, L., Shi, X., Li, Y., Guo, L. and Liu, Z. (2012). A functionalized graphene oxide-iron oxide nanocomposite for magnetically targeted drug delivery, photothermal therapy, and magnetic resonance imaging. *Nano Research,* **5**, 199–212.
19. Thevenot, J., Oliveira, H., Sandre, O. and Lecommandoux, S. (2013). Magnetic responsive polymer composite materials. *Chem. Soc. Rev.,* **42**, 7099–7116.
20. Yoo, D., Lee, J. H., Shin, T. H. and Cheon, J. (2011). Theranostic magnetic nanoparticles. *Acc. Chem. Res.,* **44**, 863–874.
21. Ho, D., Sun, X. and Sun, S. (2011). Monodisperse magnetic nanoparticles for theranostic applications. *Acc. Chem. Res.,* **44**, 875–882.
22. Rasmussen, J. W., Martinez, E., Louka, P. and Wingett, D. G. (2010). Zinc oxide nanoparticles for selective destruction of tumor cells and potential for drug delivery applications. *Expert Opin. Drug Deliv.,* 7, 1063–1077.

23. Oliveira, H., Pérez-Andrés, E., Thevenot, J., Sandre, O., Berra, E. and Lecommandoux, S. Magnetic field triggered drug release from polymersomes for cancer therapeutics. *J. Control. Release,* **169**, 165–170.
24. Shubayev, V. I., Pisanic, T. R., 2nd and Jin, S. (2009). Magnetic nanoparticles for theragnostics. *Adv. Drug Deliv. Rev.,* **61**, 467–477.
25. Cole, A. J., Yang, V. C. and David, A. E. (2011). Cancer theranostics: The rise of targeted magnetic nanoparticles. *Trends Biotechnol.,* **29**, 323–332.
26. Wadajkar, A. S., Menon, J. U., Kadapure, T., Tran, R. T., Yang, J. and Nguyen, K. T. (2013). Design and application of magnetic-based theranostic nanoparticle systems. *Recent Pat. Biomed. Eng.,* **6**, 47–57.
27. Sanson, C., Diou, O., Thévenot, J., Ibarboure, E., Soum, A., Brûlet, A., Miraux, S., Thiaudière, E., Tan, S., Brisson, A., Dupuis, V., Sandre, O. and Lecommandoux, S. (2011). Doxorubicin loaded magnetic polymersomes: Theranostic nanocarriers for MR imaging and magneto-chemotherapy. *ACS Nano.,* **5**, 1122–1140.
28. He, H., David, A., Chertok, B., Cole, A., Lee, K., Zhang, J., Wang, J., Huang, Y. and Yang, V. C. (2013). Magnetic nanoparticles for tumor imaging and therapy: A so-called theranostic system. *Pharmaceutical Research,* **30**, 2445–2458.
29. Hayashi, K., Nakamura, M., Sakamoto, W., Yogo, T., Miki, H., Ozaki, S., Abe, M., Matsumoto, T. and Ishimura, K. (2013). Superparamagnetic nanoparticle clusters for cancer theranostics combining magnetic resonance imaging and hyperthermia treatment. *Theranostics,* **3**, 366–376.